STOCHASTIC DYNAMICS OF STRUCTURES

STOCHASTIC DYNAMICS OF STRUCTURES

Jie Li and Jianbing Chen
Tongji University, China

John Wiley & Sons (Asia) Pte Ltd

Visit our Home Page on www.wiley.com

Other Wiley Editorial Offices

John Wiley & Sons, Ltd, The Atrium, Southern Gate, Chichester, West Sussex, PO19 8SQ, UK

John Wiley & Sons Inc., 111 River Street, Hoboken, NJ 07030, USA

Jossey-Bass, 989 Market Street, San Francisco, CA 94103-1741, USA

Wiley-VCH Verlag GmbH, Boschstrasse 12, D-69469 Weinheim, Germany

John Wiley & Sons Australia Ltd, 42 McDougall Street, Milton, Queensland 4064, Australia

John Wiley & Sons Canada Ltd, 5353 Dundas Street West, Suite 400, Toronto, ONT, M9B 6H8, Canada

Wiley also publishes its books in a variety of electronic formats. Some content that appears in print may not be available in electronic books.

Library of Congress Cataloging-in-Publication Data

Li, Jie, 1957 Oct.-
Stochastic dynamics of structures/Jie Li, Jianbing Chen.
p. cm.
Includes bibliographical references and index.
ISBN 978-0-470-82424-5 (cloth)
1. Structural dynamics–Mathematics. 2. Stochastic processes. I. Chen, Jianbing. II. Title.

TA654.L487 2009
624.1'71–dc22 2008045603

ISBN 978-0-470-82424-5 (HB)

Typeset in 10/12pt Times by Thomson Digital, Noida, India.

To Min Xie, My wife

Jie Li

To My Parents

Jianbing Chen

Contents

Foreword xiii

Preface xv

1 Introduction 1

1.1 Motivations and Historical Clues 1
1.2 Contents of the Book 5

2 Stochastic Processes and Random Fields 7

2.1 Random Variables 7
2.1.1 Introduction 7
2.1.2 Operations with Random Variables 8
2.1.3 Random Vectors 12
2.1.4 Decomposition of Correlation Matrix 16
2.2 Stochastic Processes 17
2.2.1 Specification of Stochastic Processes 17
2.2.2 Moment Functions of a Stochastic Process 19
2.2.3 Spectral Description of a Stochastic Process 23
2.2.4 Some Operation Rules about Expectation, Correlation and Spectrum 25
2.2.5 Karhunen–Loève Decomposition 27
2.3 Random Fields 29
2.3.1 Basic Concepts 29
2.3.2 Correlation Structures of Random Fields 31
2.3.3 Discretization of Random Fields 32
2.3.4 Decomposition of Random Fields 34
2.4 Orthogonal Decomposition of Random Functions 36
2.4.1 Metric Spaces and Normed Linear Spaces 36
2.4.2 Hilbert Spaces and General Orthogonal Decomposition 37
2.4.3 Orthogonal Decomposition of Random Functions 40

3 Stochastic Models of Dynamic Excitations 43

3.1 General Expression of Stochastic Excitations 43
3.1.1 Dynamic Excitations and Modeling 43
3.1.2 Models of Stationary and Nonstationary Processes 44
3.1.3 Random Fourier Spectrum Model 46

3.2 Seismic Ground Motions 48
3.2.1 One-Dimensional Model 48
3.2.2 Random Field Model 50
3.2.3 Physical Stochastic Model 53
3.3 Fluctuating Wind Speed in the Boundary Layer 56
3.3.1 Structural Wind Pressure and Wind Speed 56
3.3.2 Power Spectral Density of Fluctuating Wind Speed 58
3.3.3 Random Fourier Spectrum of Fluctuating Wind Speed 60
3.3.4 Random Fourier Correlation Spectrum 63
3.4 Wind Wave and Ocean Wave Spectrum 65
3.4.1 Wind Waves and Wave Forces 65
3.4.2 Power Spectral Density of Wind Waves 68
3.4.3 Direction Spectrum 70
3.5 Orthogonal Decomposition of Random Excitations 72
3.5.1 Orthogonal Decomposition of a Stochastic Process 72
3.5.2 Hartley Orthogonal Basis Function 74
3.5.3 Orthogonal Expansion of Seismic Ground Motions 75
3.5.4 Orthogonal Expansion of Fluctuating Wind Speed Process 77

4 Stochastic Structural Analysis 79
4.1 Introductory Remarks 79
4.2 Fundamentals of Deterministic Structural Analysis 80
4.2.1 The Basic Idea of Finite-Element Analysis 80
4.2.2 Element Stiffness Matrix 81
4.2.3 Transformation of Coordinates 84
4.2.4 Static Equations 87
4.2.5 Dynamic Equations 88
4.3 Random Simulation Method 90
4.3.1 Monte Carlo Method 90
4.3.2 Sampling of Random Variables with Uniform Distribution 91
4.3.3 Sampling of Random Variables with General Probability Distribution 93
4.3.4 Random Simulation Method 95
4.3.5 Accuracy of Random Simulation Method 97
4.4 Perturbation Approach 99
4.4.1 Deterministic Perturbation 99
4.4.2 Random Perturbation 101
4.4.3 Random Matrices 102
4.4.4 Linear Expression of Random Matrices 103
4.4.5 Dynamic Response Analysis 107
4.4.6 Secular Terms Problem 110
4.5 Orthogonal Expansion Theory 113
4.5.1 Orthogonal Decomposition and Sequential Orthogonal Decomposition 113
4.5.2 Order-Expanded System Method 116

4.5.3 Proof of the Order-Expanded System Method 120
4.5.4 Dynamic Analysis 124
4.5.5 Recursive Condensation Algorithm 129

5 Random Vibration Analysis 133
5.1 Introduction 133
5.2 Moment Functions of the Responses 133
5.2.1 Response of a Single-Degree-of-Freedom System in the Time Domain 134
5.2.2 Response of MDOF Systems in the Time Domain 140
5.3 Power Spectral Density Analysis 146
5.3.1 Frequency Response Function and Power Spectral Density 146
5.3.2 Evolutionary Spectral Analysis 156
5.4 Pseudo-Excitation Method 160
5.4.1 Pseudo-Excitation Method for Stationary Stochastic Response Analysis 161
5.4.2 Pseudo-Excitation Method for Evolutionary Stochastic Response Analysis 163
5.4.3 Notes on Sections 5.2–5.4 164
5.5 Statistical Linearization 164
5.5.1 Statistical Linearization Approximation 164
5.5.2 Random Vibrations of Hysteretic Structures 167
5.5.3 Notes on Arguments and Some Special Issues 171
5.6 Fokker–Planck–Kolmogorov Equation 174
5.6.1 Stochastic Differential Equation 174
5.6.2 Fokker–Planck–Kolmogorov Equation 181
5.6.3 Solution to the Fokker–Planck–Kolmogorov Equation 184

6 Probability Density Evolution Analysis: Theory 191
6.1 Introduction 191
6.2 The Principle of Preservation of Probability 192
6.2.1 Functions of Random Variables and their Probability Density Function Revisited 192
6.2.2 The Principle of Preservation of Probability 195
6.3 Markovian Systems and State Space Description: Liouville and Fokker–Planck–Kolmogorov Equations 198
6.3.1 The Liouville Equation 198
6.3.2 Fokker–Planck–Kolmogorov Equation Revisited 204
6.4 Dostupov–Pugachev Equation 209
6.4.1 From Equation of Motion to Random State Equation 209
6.4.2 The Dostupov–Pugachev Equation 210
6.5 The Generalized Density Evolution Equation 213
6.5.1 Derivation of the Generalized Density Evolution Equation 213
6.5.2 Linear Systems: Uncoupling of the Dostupov–Pugachev Equation 217
6.5.3 Initial and Boundary Conditions 219
6.5.4 Physical Sense of the Generalized Density Evolution Equation 219

6.6 Solution of the Generalized Density Evolution Equation 221
6.6.1 Analytical Solution 221
6.6.2 Numerical Solving Flow of the Generalized Density Evolution Equation 226

7 Probability Density Evolution Analysis: Numerical Methods 231
7.1 Numerical Solution of First-Order Partial Differential Equation 231
7.1.1 The Finite-Difference Method 231
7.1.2 Dissipation, Dispersion and Total Variation Diminishing Schemes 240
7.2 Representative Point Sets and Assigned Probabilities 251
7.2.1 Sphere Packings, Covering and Partition of Space 251
7.2.2 Representative Point Sets and Assigned Probabilities 256
7.2.3 First- and Second-Order Discrepancies of Point Sets 260
7.2.4 Two-Step Procedure of Constructing Representative Points 261
7.3 Strategy for Generating Basic Point Sets 262
7.3.1 From Sphere Packings: Tangent Sphere Method 262
7.3.2 From Thinnest Covering: Lattices Approach 266
7.3.3 Number Theoretical Method 270
7.4 Density-Related Transformation 272
7.4.1 Affine Transformation 272
7.4.2 Density-Related Transformation 273
7.4.3 Radial Decay Distribution: Spherical Sieving and Expansion–Contraction Transformation 274
7.5 Stochastic Response Analysis of Nonlinear MDOF Structures 278
7.5.1 Responses of Nonlinear Stochastic Structures 278
7.5.2 Stochastic Seismic Response of Nonlinear Structures 282

8 Dynamic Reliability of Structures 285
8.1 Fundamentals of Structural Reliability Analysis 285
8.1.1 Structural Reliability 285
8.1.2 Dynamic Reliability Analysis of Structures 286
8.1.3 Global Reliability of Structures 287
8.2 Dynamic Reliability Analysis: First-Passage Probability Based on Excursion Assumption 288
8.2.1 Excursion Rates 288
8.2.2 Excursion Assumption and First-Passage Probability 290
8.2.3 First-Passage Probability Considering Random Thresholds 292
8.2.4 Pseudo-Static Analysis Method 293
8.3 Dynamic Reliability Analysis: Generalized Density Evolution Equation-Based Approach 295
8.3.1 Absorbing Boundary Condition Method 295
8.3.2 Extreme-Value Distribution of the Stochastic Dynamical Response 296
8.3.3 Extreme-Value Distribution-based Dynamical Reliability Evaluation of Stochastic Systems 299

8.4 Structural System Reliability 300
8.4.1 Equivalent Extreme-Value Event 300
8.4.2 Inherent Correlation Property of Equivalent Extreme-Value Event 304
8.4.3 Differences between the Equivalent Extreme-Value Event and the Weakest Link Assumption 305
8.4.4 Evaluation of Structural System Reliability 308

9 Optimal Control of Stochastic Systems 313

9.1 Introduction 313
9.2 Optimal Control of Deterministic Systems 315
9.2.1 Optimal Control of Structural Systems 315
9.2.2 Linear Quadratic Control 318
9.2.3 The Minimum Principle and Hamilton–Jacobi-Bellman Equation 320
9.3 Stochastic Optimal Control 325
9.3.1 Stochastic Optimal Control of Nonlinear Systems: Classical Theory 325
9.3.2 Linear Quadratic Gaussian Control 328
9.3.3 Probability Density Evolution Analysis of Stochastic Optimal Control Systems 330
9.4 Reliability-Based Control of Structural Systems 338
9.4.1 Reliability of Controlled Structural Systems 338
9.4.2 Determination of Control Criterion 340

Appendix A: Dirac Delta Function 343

A.1 Definition 343
A.2 Integration and Differentiation 344
A.3 Common Physical Backgrounds 346
A.3.1 Probability Distribution of Discrete Random Variables 346
A.3.2 Concentrated and Distributed Loads 347
A.3.3 Unit Impulse Function 347
A.3.4 Unit Harmonic Function 348

Appendix B: Orthogonal Polynomials 349

B.1 Basic Concepts 349
B.2 Common Orthogonal Polynomials 351
B.2.1 Hermite Polynomials $H_{e_n}(x)$ 351
B.2.2 Legendre Polynomials $P_n(x)$ 352
B.2.3 Gegenbauer Polynomials $C_n^{(\alpha)}(x)$ 354

Appendix C: Relationship between Power Spectral Density and Random Fourier Spectrum 355

C.1 Spectra via Sample Fourier Transform 355
C.2 Spectra via One-sided Finite Fourier Transform 357

Appendix D: Orthonormal Base Vectors **361**

Appendix E: Probability in a Hyperball **377**
E.1 The Case s is Even 378
E.2 The Case s is Odd 378
E.3 Monotonic Features of $F(r, s)$ 380
E.3.1 Monotonic Feature of $F(r, s)$ with Respect to the Radius r 380
E.3.2 Monotonic Feature of $F(r, s)$ with Respect to the Dimensions 381

Appendix F: Spectral Moments **383**

Appendix G: Generator Vectors in the Number Theoretical Method **385**

References and Bibliography **391**

Index **405**

Foreword

It is a great pleasure to introduce *Stochastic Dynamics of Structures* by Jie Li and Jianbing Chen. The book begins with a brief history of the early discovery and developments of the field, starting with Einstein's introduction of the Brownian motion, followed by the classical developments, including the mathematical formulations of Fokker, Planck, and Kolmogorov. It is a timely and much needed exposition of the existing state of knowledge of stochastic dynamics and its potential applications in structural dynamics and the reliability of dynamical systems.

The topical coverage of stochastic dynamics starts properly with an introduction of the fundamentals of random variables, random vectors, and stochastic processes including random fields, which are the essentials necessary for the study of random vibration and stochastic structural analysis, and culminates with the presentation of the probability density evolution theory and its corollary the equivalent extreme value distribution; the latter is especially significant for evaluating the dynamic reliability of structures and other engineering systems.

This book is a valuable contribution to the continuing development of the field of stochastic structural dynamics, including the recent discoveries and developments by the authors of the probability density evolution method (PDEM) and its applications in the assessment of the dynamic reliability and control of complex structures through the equivalent extreme-value distribution. The traditional analytical approach to such a dynamic reliability problem is to formulate it as a "barrier-crossing problem" that leads to the solution of the Fokker-Planck equation; the limitations of this approach are well known, even for single-degree-of-freedom systems. The authors thoroughly discuss this classical approach and show its limitations, following with the PDEM, including the numerical solution of complex multi-degree-of-freedom systems. These are preceded with new insights, derivations, and interpretations of the classical formulations and solutions—such as the Liouville equation, the Kolmogorov equation, and the Itô stochastic equations—are provided through the concept of the preservation of probability.

Besides elucidating the principles of stochastic dynamics from an engineer's viewpoint, the most significant contribution of this book is its lucid presentation of the PDEM and its applications for the assessment of the dynamic reliability and control of structures under earthquake excitations and wind and wave forces. In this regard, the PDEM should serve to spur further developments of stochastic structural dynamics; with the PDEM, solutions to the dynamic reliability of multi-degree-of-freedom systems can be evaluated numerically, including non-linear systems. Innovative numerical schemes are proposed; besides finite difference schemes, spherical packing schemes are also suggested for solutions of highly complex problems.

In other words, this book includes a novel approach to the field of stochastic dynamics with special emphasis on the applications to the dynamic response and reliability of structures. It should serve well to advance the research in the field of stochastic structural dynamics in general and dynamic reliability in particular.

A. H-S. Ang, NAE, Hon. Mem. ASCE

Research Professor
University of California, Irvine

Preface

As a scientific discipline, stochastic dynamics of structures has evolved from its infancy in the early 1940s to a relatively mature branch of dynamics today. In the process, basic random vibration theory is believed to have been established in the late 1950s and mainly deals with the response analysis of structures to stochastic excitations, such as the response of buildings and bridges to wind loading and earthquakes, the vibration of vehicles traveling over rough ground and the dynamic behavior of aircraft induced by atmospheric turbulence and jet noise. In the late 1960s, the importance of the effect of randomness in structural parameters on the structural response was recognized gradually and this led to stochastic structural analysis, or stochastic finite-element analysis as termed by many researchers. There has been a large amount of literature published in the past 40 years; however, careful people may find that random vibration theory and stochastic finite element analysis seem to have developed in two parallel ways. It is very hard for most engineers, even those specialists who are familiar with stochastic analysis, to organize their knowledge of the two branches of dynamics in a systematic framework. Therefore, the first aim of this book is to present a coherent and reasonably self-contained theoretical framework of the stochastic dynamics of structures which may bridge the gap between traditional random vibration theory and the stochastic finite-element analysis method. We hope such a treatment will provide a comprehensive account for stochastic dynamic response analysis, reliability evaluation and system control.

The second aim, which may be more important and seems a little bit ambitious, is to deal with the basic content of stochastic dynamics of structures in a unified new theoretical framework. We refer to this as the frame of the physical stochastic system. Most people know that, in many practical applications, the system of concern usually exhibits nonlinearities. However, it is just for nonlinear dynamical systems that the foundational stochastic dynamics theory involves huge complexities. After considerable research efforts in the field of random vibration and stochastic finite-element analysis, although some important progress has been made for simple structural models, people still cannot solve the problem of nonlinear stochastic dynamical systems rationally, especially for practical complex structures. Motivated by the need to provide a rational description of a nonlinear stochastic system and of developing appropriate analytical tools, we undertook a systematic investigation on the difficult area in the past 15 years. Tracing back to the source of the discipline, we find that there are two historical traditions in the study of stochastic dynamics: the phenomenological tradition and the physical tradition. Because of the introduction of the Wiener process, the two traditions gain an intrinsic relation. However, if we return to the physical processes themselves (that is, investigating random phenomena from a physical viewpoint), then we will be led to another possible way: approaching the stochastic system based on physics. Using this approach, we give a rational description of the relationship between the physical sample trajectories of a dynamic system and its probabilistic description and, therefore, establish a family of generalized probability density evolution equations for dynamical systems which could deal with both linear and nonlinear systems in a unified form. Furthermore, bearing in mind the physical stochastic system, we find that traditional random vibration theory and the stochastic finite-element methods can be appropriately brought into a new theoretical frame. Obviously, this provides a foundation to rearrange the content of stochastic dynamics of structures in a comprehensive framework. This book tries to present such a development, as well as pragmatic methods and algorithms where possible.

We assume that the book will be used by graduate students and professionals in civil engineering, mechanical engineering, aircraft and marine engineering, as well as in mechanics. The level of the preparation assumed of the reader corresponds to that of the bachelor's degree in science or engineering, especially those who have a basic understanding of the concept of probability theory and structural dynamics. In addition, to make the book self-sufficient, the essential concepts of random variables, stochastic processes and random fields are presented in the book.

Our sincere appreciations go first to Professor P.D. Spanos at Rice University, for his friendly encouragement, and to Professor R.G. Ghanem at the University of Southern California, for his constructive comments and fruitful discussions. For their valuable help and advice, special thanks are also due to Professor W.D. Iwan at California Institute of Technology, Professor Jinping Ou at Dalian University of Technology and Professor Yangang Zhao at Nagaya Institute of Technology. The first author would also like to take this opportunity to express his deep appreciation to Professor J.B. Roberts of Sussex University, who is greatly missed, for the generous support given for the chance to complete the investigation on stochastic analysis and modeling during 1993 and 1994 when the author spent one year as a senior visiting scholar at Sussex University, and to his colleagues Professor Xilin Lu, Professor Guoqiang Li, Professor Ming Gu, Professor Yiyi Chen and Professor Menglin Lou at Tongji University for their continuous cooperation and support.

Much of the research work of the authors was developed with support from the National Natural Science Foundation of China, over a period of decade, including the National Outstanding Young Scientist Foundation (received by the first author in 1998), the Young Scholars Foundation (received by the second author in 2004), and the Innovative Research Groups Plan. All the support is gratefully acknowledged.

Finally, a votes of thanks go to Mr James Murphy, editor at John Wiley & Sons, Ltd, for his patience and encouragement during the preparation of this book, and to Mr Roger Bullen, Project Editor, for his patience and scrutiny in the editing of this book.

Jie Li and Jianbing Chen

Shanghai
June 2008

1

Introduction

1.1 Motivations and Historical Clues

Structural dynamics deals with the problems of response analysis, reliability evaluation and system control of any given type of structure subjected to dynamic actions.[1] Structures (such as buildings, bridges, aircraft, ships and so on) refer to those bodies or systems composed of various materials in a certain way that are capable of bearing loads and actions. On the other hand, when we say an action applied on structures is dynamic, this not only indicates that the action is time varying, but also that the induced inertial effects cannot be ignored. For example, earthquakes, wind, sea waves, jet noise and turbulence in the boundary layer and the like are typical dynamic actions. The task of dynamic response analysis of structures is to capture the internal forces, deformations or other state quantities of structures when they are subjected to dynamic actions. At the same time, we may need to study whether the structural response meets some specified limit in a sense, which is generally referred to as reliability evaluation. Furthermore, to make a structure subjected to dynamic actions response in a desired way to an extent is what to be done in system control.

Most dynamic actions exhibit appreciable randomness. Actually, investigators frequently find that the results observed under almost identical conditions have obvious deviation, but simultaneously exhibit some statistical rules. In essence, the randomness results from the uncontrollability of causation of the realized phenomenon. For example, consider wind turbulence in the atmospheric boundary layer. It is well known that the observed wind speeds recorded at the same position but during different time intervals are quite different (Figure 1.1). However, if the statistics of a large number of samples are examined, then we find that the probabilistic characteristics of the wind speed are relatively stable (Figure 1.2). In fact, the randomness involved stems from a complicated physical mechanism in the wind flows, say the mechanism of turbulence. The underlying reason is the uncontrollable nature of the motion of air molecules.

In addition, the randomness involved in the physical parameters of structures is also one of the sources that induce randomness in the dynamic responses of structures. For instance, in the

[1] The dynamic properties of structures, such as the frequencies and mode shapes, are also research topics of structural dynamics. But, in a general sense, the dynamic properties of structures can be regarded as part of the dynamic analysis of structures.

Stochastic Dynamics of Structures Jie Li and Jianbing Chen

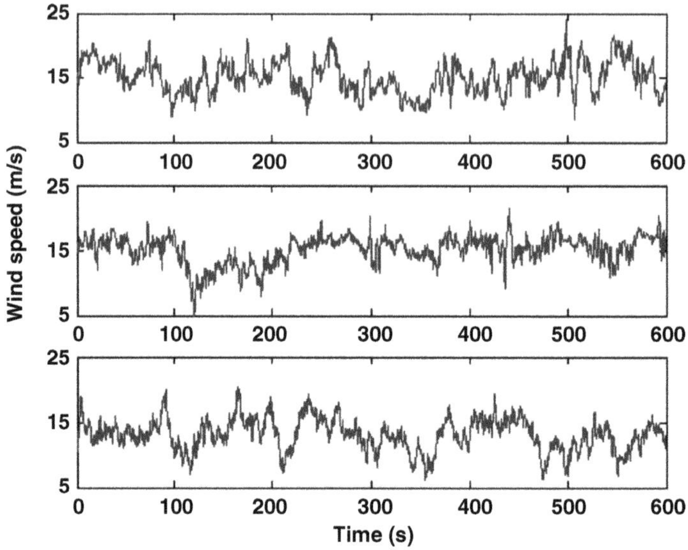

Figure 1.1 Records of wind speed.

dynamic response analysis of building structures, the soil–structure interaction is one of the basic problems where the properties of soil must be considered in the establishment of a reasonable structural analysis model. Evidently, it is impossible to measure the physical properties of soil completely at all points in the groundwork. Thus, a reasonable modeling approach is to regard the physical properties of soil, such as the shear wave speed and the damping ratio, as random variables or random fields. This will lead to the structural analysis involving random parameters, usually known as stochastic structural analysis.

Stochastic dynamic response analysis, reliability evaluation and system control compose the basic research scope of the stochastic dynamics of structures.

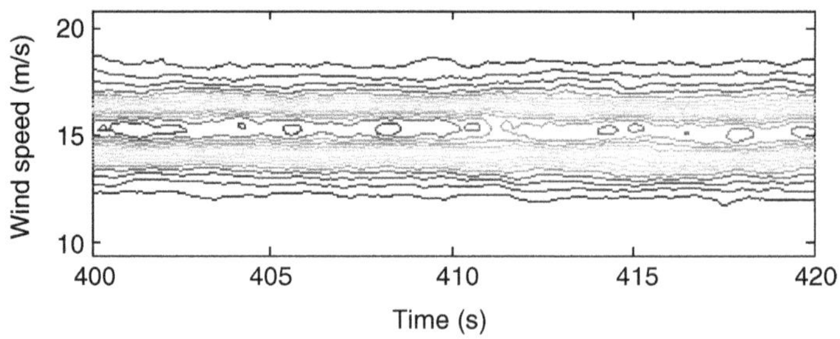

Figure 1.2 Contour of probability density of wind speed.

Although the studies on stochastic dynamical systems can be dated back to the investigations on statistical mechanics by Gibbs and Boltzmann (Gibbs, 1902; Cercignani, 1998), it is generally considered more reasonable to regard the studies on Brownian motion by Einstein (1905) as the origin of the discipline.

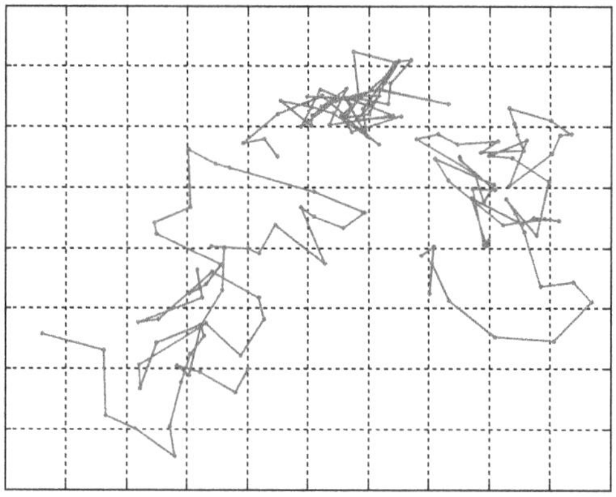

Figure 1.3 Typical trajectories of Brownian motions.

In 1905, Einstein studied the problem of the irregular motion of particles suspended in fluids, which was first observed by the Scottish botanist Robert Brown in 1827 (Figure 1.3). Einstein believed that Brownian motion of the particles was induced by the highly frequent random impacts of the fluid molecules. Based on this physical interpretation, Einstein made the following assumptions:

(a) the motion of different Brownian particles is mutually independent;
(b) the motion of Brownian particles is isotropic and no external actions except the collision of fluid molecules are applied;
(c) the collision of fluid molecules is instantaneous, such that the time of collision can be ignored (rigid collision).

Based on the above assumptions, the probability density of the particle group at two different instants of time can be derived by examining the phenomenological evolution process of the particle group; that is:

$$f(x, t+\tau) = \int_{-\infty}^{+\infty} f(x+r, t)\Phi(r)\,\mathrm{d}r \tag{1.1}$$

where $f(x, t+\tau)$ is the probability density of the position of the particles at time $t+\tau$, $f(x+r, t)$ is the probability density by transition of the particles with distance r during the time interval τ, and $\Phi(r)$ is the probability density of displacement of the particles.

Using the rigid collision assumption, expanding the functions by using the Taylor series and retaining the first-order term with respect to $f(x, t+\tau)$ and the second-order term with respect to $f(x+r, t)$ will yield

$$\frac{\partial f(x,t)}{\partial t} = D\frac{\partial^2 f(x,t)}{\partial x^2} \tag{1.2}$$

where

$$D = \frac{1}{\tau}\int_{-\infty}^{+\infty} \frac{1}{2} r^2 \Phi(r)\,\mathrm{d}r \tag{1.3}$$

Clearly, Equation (1.2) is a diffusion equation, where D is the diffusion coefficient.

In 1914 and 1917, Fokker and Planck respectively introduced the drift term for a similar physical problem, leading to the so-called Fokker–Planck equation (Fokker, 1914; Planck, 1917; Gardiner, 1983), of which the rigorous mathematical basis was later established by Kolmogorov (1931).[2]

We note that, initially, the studies on Brownian motion were based on physical concepts; however, a statistical phenomenological interpretation was subsequently introduced in the deductions. In this book, we call this historical clue the *Einstein–Fokker–Planck tradition* or *phenomenological tradition*. In this tradition, a large number of studies on the probability density evolution of stochastic dynamical systems have been done (Kozin, 1961; Lin, 1967; Roberts and Spanos, 1990; Zhu, 1992, 2003; Lin and Cai, 1995). However, for the multi-degree-of-freedom (MDOF) systems or multidimensional problems, advancement is still quite limited (Schuëller, 1997, 2001).

Soon after Einstein's work, Langevin (1908) came up with a completely different research approach. In his investigation, the physical interpretation of Brownian motion is the same as that of Einstein, but Langevin contributed to two basic aspects. He:

(a) introduced the assumption of random forces;
(b) employed Newton's equation of motion to govern the motion of the Brownian particles.

Based on this, he established the stochastic dynamics equation, which was later called the Langevin equation:

$$m\ddot{x} = -\gamma\dot{x} + \xi(t) \tag{1.4}$$

where m is the mass of the Brownian particles, $\ddot{x}$ and $\dot{x}$ are the acceleration and velocity of motion respectively, γ is the viscous damping coefficient and $\xi(t)$ is the force induced by the collision of the fluid molecules, which is randomly fluctuating.

Using the ensemble average, Langevin obtained a diffusion coefficient identical to that given by Einstein.

In contrast to the diffusion equation derived by Einstein, the Langevin equation is more direct and more physically intuitive. However, the physical features of the random forces are not completely clear in Langevin's work.

In 1923, Wiener proposed a stochastic process model for Brownian motion (Wiener, 1923). Around 20 years later, Itô introduced the Itô integral and gave the more generic Langevin equation based on the Wiener process (Itô, 1942, 1944; Itô and McKean, 1965):

$$\mathrm{d}x(t) = a[x(t), t]\,\mathrm{d}t + b[x(t), t]\,\mathrm{d}W(t) \tag{1.5}$$

where $a(\cdot)$ and $b(\cdot)$ are known deterministic functions and $W(t)$ is a Wiener process.

The form of Equation 1.5 is nowadays called the Itô stochastic differential equation. Clearly, this equation is in essence a physical equation. It is generally believed that the Itô equation

[2] Interestingly, Kolmogorov did not at first know about the work of Fokker and Planck and developed his equation independently.

provides a sample trajectory description for stochastic dynamical systems. In this book, we refer to this historical clue as the *Langevin–Itô tradition* or *physical tradition.* In this approach, the mean-square calculus theory was established, based on which correlation analysis and spectral analysis in classical random vibration analysis were well developed (Crandall, 1958, 2006; Lin, 1967; Zhu, 1992; Øksendal, 2005).

There were intrinsic and countless ties between the phenomenological tradition and the physical tradition in stochastic dynamics. As a matter of fact, upon the assumption that the system inputs are white-noise processes, it is easy to obtain the Fokker–Planck–Kolmogorov (FPK) equation via the Itô equation. This demonstrates that in the physics approach we can discover the intrinsic arcanum of the evolution of stochastic systems. Unfortunately, white noise is physically unrealizable. In other words, although mathematically it plays a fundamental role in a sense, the various singular or even ridiculous features of white noise (say, continuous but indifferentiable everywhere) are rare in the real world.

The white-noise process is, of course, an idealized model for various real physical processes. Noticing this, we naturally hope to return to the real physical processes themselves. For a specific physical dynamical process, the problem is usually easily resolvable. Thus, once further introducing the intrinsic ties between the sample trajectories and the probabilistic description, we will be led to an approach of studying stochastic systems based on physics. In this approach, we not only can establish the generalized probability density evolution equation (Li and Chen, 2003, 2006c, 2008), but also find that the nowadays available major research results, such as traditional random vibration theory and stochastic finite element methods, can be appropriately brought into the new theoretical frame (Li, 2006). In fact, correlation analysis and the spectral analysis in classical random vibration theory can be regarded as the results of combining the formal solution of physical equations and the evolution of moment characteristics of the response processes. Perturbation theory and orthogonal expansion theory in the analysis of structures with random parameters can also be reasonably interpreted in this sense. The classical FPK equation, as mentioned before, can be viewed as the result of the idealization of physical processes. In addition, the thoughts of physical stochastic system can also be used in modeling of general stochastic process, such as seismic ground motion, wind turbulence and the like (Li and Ai, 2006; Li and Zhang, 2007).

On the basis of the above thoughts on physical stochastic systems, we prefer to entitle this book *Stochastic Dynamics of Structures: a Physical Approach.*

1.2 Contents of the Book

This book deals with the basic problems of the stochastic dynamics of structures in the theoretical frame of physical stochastic systems.

In Chapter 2 the prerequisite fundamentals of probability theory are outlined, including the basic concepts of random variables, stochastic processes, random fields and the orthogonal expansion of random functions.

Chapter 3 deals with stochastic process models for typical dynamic excitations of structures, including the phenomenological and physical modeling of seismic ground motions, fluctuating wind speed and sea waves. Simultaneously, we introduce the standard orthogonal expansion of stochastic processes, which can be applied to random vibration analysis of structures.

The approaches for analysis of structures with random parameters mainly include the random simulation method, the perturbation method and the orthogonal expansion method. These approaches are discussed in detail in Chapter 4.

Chapter 5 deals with the response analysis of deterministic structures subjected to stochastic dynamic excitations, including correlation analysis, spectral analysis, the statistical linearization method and the FPK equation approach. In particular, in this chapter we introduce the pseudo-excitation method for linear systems. We believe these contents are valuable to in-depth understanding of classical random vibration theory.

Probability density evolution analysis of stochastic responses of dynamical systems is an important topic of the book. We will deal with this topic in Chapters 6 and 7. In Chapter 6 we trace in some detail the historical origin of probability density evolution analysis of stochastic dynamical systems. Using the principle of preservation of probability as a unified basis, we derive the Liouville equation, the FPK equation, the Dostupov–Pugachev equation and the generalized probability density evolution equation proposed by the authors. In Chapter 7 we study the numerical methods for probability density evolution analysis in detail, including the finite difference method, the strategy of selecting representative points via tangent spheres, lattices and the number theoretical method. For all these methods, we discuss the problems of numerical convergence and stability where possible.

The aim of structural dynamical analysis is to realize reliability-based design and performance control of structures. We discuss the problem of dynamic reliability and global reliability of structures in Chapter 8. Based on the random event description of the evolution of probability density, the absorbing boundary condition for the first-passage problem is introduced. The theory on evaluation of the extreme value distribution is elaborated through introducing a virtual stochastic process related to the extreme value of the response process. Furthermore, the principle of equivalent extreme value and its application to the global reliability evaluation of structures is discussed. It is worth pointing out that the principle of equivalent extreme value is of significance and applicable to static reliability evaluation of generic systems.

We come to the problem of the dynamic control of structures in Chapter 9. On the basis of classical dynamic control, the concept of stochastic optimal control is introduced and the approach for design of the control systems based on probability density evolution analysis is proposed. For realization of 'real' stochastic optimal control of dynamical systems, the proposed approach is undoubtedly promising.

2

Stochastic Processes and Random Fields

2.1 Random Variables

2.1.1 Introduction

By an experiment we mean taking a kind of action or operation devised to seek for a certain truth or fact. For some experiments, the results possess basic properties of deterministic phenomena once all underlying conditions are well controlled and all experimental phenomena are exactly observed. In other words, the results of these experiments are predictable. Owing to uncontrollable or immeasurable facts, however, those experiments may obtain varied results, though fundamental conditions remain invariant in some respects. This is the so-called random phenomenon. The results occurring in a set of random experiments are generally called random events, or events for simplicity. The basic property of a random event lies in that the predicted event may be observed or not when the observational conditions differ by a small amount. However, we can always identify the set whose elements consist of all the possible results for a given experiment. In other words, the union of all experimental results can be determined beforehand. This set is called the sample space and denoted by Ω. Each possible result in Ω is called a sample point, denoted by ϖ. Each event A can be understood as a subset of Ω. An event is called an elementary event if it contains only a single sample point. On the other hand, we can say that a compound event is a certain set of sample points. A family of events, denoted by $\mathcal{F}$ or termed σ-algebra, refers to the subset of A which satisfies the following statements:

(a) $\Omega \in \mathcal{F}$;
(b) $A \in \mathcal{F}$ implies $\bar{A} \in \mathcal{F}$, where $\bar{A}$ is the complement of A;
(c) $A_n \in \mathcal{F}$ $(n = 1, 2, \ldots)$ implies $\cup_{n=1}^{\infty} A_n \in \mathcal{F}$.

To measure a sample space, we need to assign a numerical value to measure the possibility of an event occurring. This gives rise to the concept of probability measure, by which every event in $\mathcal{F}$ is mapped into the unit interval $[0, 1]$. That is, the possibility of each occurrence can be represented by a nonnegative number smaller than unity. In general, we call this number the

Stochastic Dynamics of Structures Jie Li and Jianbing Chen

probability measure of Ω, or the probability of the given event A, denoted by $P(A)$ or $\Pr\{A\}$. In addition, in Kolmogorov's *Foundations of the Theory of Probability*, the triple $(\Omega, \mathcal{F}, P)$ is defined as the probability space (Kolmogorov, 1933; Loeve, 1977; Kallenberg, 2002).

We have already noted that the probability measure gives a gauge of an event's occurrence, but does not give a similar one for the sample points. In mathematics, this problem is resolved by defining on the probability space a measurable function $X(\varpi)$, which is generally called a random variable, and denoted by X for short.[1] It has two basic properties:

(a) A random variable is a single-valued real function of sample points. That is, each random variable produces a mapping from a probability space to a field of real numbers.
(b) For any real number, $\{\varpi : X(\varpi) < x\}$ is a random event.

With the concept of a random variable, we can adopt numerical values to describe the results of any random experiment. For instance, an elementary event is expressed in the form that a random variable X is equal to one deterministic number (i.e. $X = x$), while any arbitrary event can be expressed in a way that X takes values over an interval $x_1 \le X \le x_2$ and its probability of occurring is denoted by $\Pr\{x_1 \le X \le x_2\}$. There are two basic types of random variable: discrete random variable and continuous random variable. The former take values in a finite or countable infinite set, while the latter can be assigned any value in one or several intervals. When the Dirac delta function is introduced later, the discrete random variable and continuous random variable will be seen to operate in a unified way (see Appendix A), but in this book it is mainly the continuous random variables that are discussed.

In general, $F_X(x) = \Pr\{X(\varpi) < x\}$ $(-\infty < x < \infty)$ is called the cumulative distribution function (CDF) of the random variable X. It satisfies the following basic properties:

(a) $\lim_{x \to -\infty} F_X(x) = 0, \lim_{x \to \infty} F_X(x) = 1$;
(b) if $x_1 < x_2$, then $F_X(x_1) \le F_X(x_2)$;
(c) $F_X(x - 0) = F_X(x)$;
(d) $\Pr\{x_1 \le X < x_2\} = F_X(x_2) - F_X(x_1)$.

By introducing random variables, we can further deal with probability measure problems of complicated systems. This is done by the use of operations performed on random variables.

2.1.2 Operations with Random Variables

Two of the most important operations are the distributions of random variables' functions and the moments of random variables. They are both based on calculations of the probability density functions (PDFs). Thus, the concept of the PDF is introduced first.

For a continuous random variable X, the PDF is defined as the derivative of its CDF:

$$p_X(x) = \frac{\mathrm{d}}{\mathrm{d}x} F_X(x) \tag{2.1}$$

where $p_X(x)$ is a nonnegative function; that is, there always exists $p_X(x) \ge 0$.

---

[1] A random variable is usually denoted by a capital letter or Greek character, say X or ξ, while the sample value of a random variable is usually denoted by the corresponding lower case character, say x. The convention is used in the book except for special statements.

The inversion of Equation 2.1 gives

$$F_X(x) = \int_{-\infty}^{x} p_X(x)\,\mathrm{d}x \tag{2.2}$$

where the condition $F_X(-\infty) = 0$ has been used.

As the upper limit of the integral goes to infinity, we have

$$\int_{-\infty}^{\infty} p_X(x)\,\mathrm{d}x = 1 \tag{2.3}$$

Figure 2.1 depicts a typical PDF and the CDF.

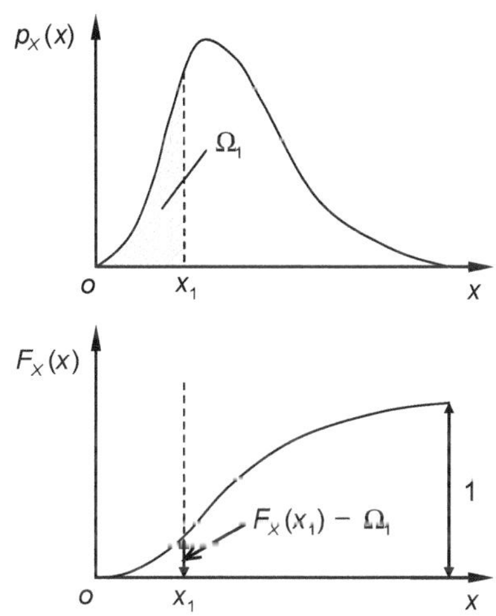

Figure 2.1 PDF and CDF.

If a random variable Y is the function of another one X, namely $Y = f(X)$, and $f(\cdot)$ only has a finite number of discontinuity points, then the CDF of Y is given by

$$F_Y(y) = \Pr\{f(X) < y\} = \int_{f(x) < y} p_X(x)\,\mathrm{d}x \tag{2.4}$$

This integral is calculated over all the segments in the x axis which satisfy the inequality below the integral symbol.

Theoretically, according to Equation 2.1, it is easy to obtain the PDF of Y from Equation 2.4. However, we may encounter difficulties when making specific operations, because sometimes $f(\cdot)$ may be a very complicated function. Thus, in general we only consider two cases as follows:

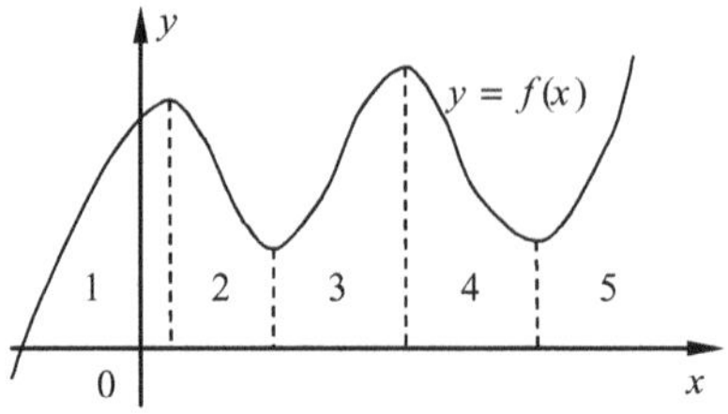

Figure 2.2 A single-valued function.

(a) Suppose $f(x)$ is a monotonic function. Then there exists $g(y)$ as the unique inverse function of $f(x)$. Using Equations 2.1 and 2.4, the PDF of Y is given by

$$p_Y(y) = p_X[g(y)]\left|\frac{\mathrm{d}g(y)}{\mathrm{d}y}\right| \tag{2.5}$$

(b) Suppose $f(x)$ is not monotonic but a single-valued function (see Figure 2.2). In this case, we can try to divide the domain of x-values into several intervals such that, over each interval, $f(x)$ is a monotonic function. Then, similar to Equation 2.4, there is

$$p_Y(y) = \sum_k p_X[g_k(y)]\left|\frac{\mathrm{d}g_k(y)}{\mathrm{d}y}\right| \tag{2.6}$$

where $g_k(y)$ is the inverse function of $f(x)$ in the kth interval.

As already noted, the CDF or PDF describes the distribution properties of random variables in a precise way. On the other hand, somewhat rough descriptions of random variables are the moments, among which two of the most useful ones are the expectation and the variance.

The expectation of a continuous random variable is defined as the first origin moment of its density function; that is,

$$\mathcal{E}[X] = \int_{-\infty}^{\infty} x p_X(x)\,\mathrm{d}x \tag{2.7}$$

Its variance is the second central moment of its density function:

$$\mathcal{D}[X] = \mathcal{E}\{(X-\mathcal{E}[X])^2\} = \int_{-\infty}^{\infty} (x-\mathcal{E}[X])^2 p_X(x)\,\mathrm{d}x \tag{2.8}$$

The basic property of the expectation is its linear superposition; that is:

$$\mathcal{E}[aX+b] = a\mathcal{E}[X]+b \tag{2.9}$$

where a and b are any two constants.

Correspondingly, the variance obeys

$$\mathcal{D}[aX+b] = a^2\mathcal{D}[X] \tag{2.10}$$

In general, we call

$$m_n = \mathcal{E}[X^n] = \int_{-\infty}^{\infty} x^n p_X(x)\,\mathrm{d}x \tag{2.11}$$

the nth origin moment of X and denote the expectation m_1 by μ.

$$K_n = \mathcal{E}[(X-\mu)^n] = \int_{-\infty}^{\infty} (x-\mu)^n p_X(x)\,\mathrm{d}x \tag{2.12}$$

is called the nth central moment of X, and σ^2 is used to denote K_2 or $\mathcal{D}[X]$. $\sigma = \sqrt{\mathcal{D}[X]}$ is usually called the standard deviation of X.

The central moments can be expressed by the linear combination of origin moments:

$$\mathcal{E}[(X-\mu)^n] = \sum_{i=0}^{n} \binom{n}{i} (-\mathcal{E}[X])^{n-i} \mathcal{E}[X^i] \tag{2.13}$$

where

$$\binom{n}{i} = \frac{n!}{i!(n-1)!}$$

Similarly, the origin moments can also be computed by the central moments.

For a continuous random variable X, the characteristic function, denoted by $f_X(\vartheta)$, is the Fourier transform of its PDF; that is:

$$f_X(\vartheta) = \int_{-\infty}^{\infty} e^{i\vartheta x} p_X(x)\,\mathrm{d}x \tag{2.14}$$

As noted, the characteristic function can serve as a mode of describing random variables like the PDF. More significantly, moment functions of a random variable can be given by derivatives of its characteristic function. In fact:

$$\frac{\mathrm{d}^n f_X(\vartheta)}{\mathrm{d}\vartheta^n} = i^n \int_{-\infty}^{\infty} e^{i\vartheta x} x^n p_X(x)\,\mathrm{d}x \tag{2.15}$$

Let $\vartheta = 0$, then

$$\left.\frac{\mathrm{d}^n f_X(\vartheta)}{\mathrm{d}\vartheta^n}\right|_{\vartheta=0} = i^n \int_{-\infty}^{\infty} x^n p_X(x)\,\mathrm{d}x = i^n \mathcal{E}[X^n] \tag{2.16}$$

where i is the imaginary number unit. Meanwhile, we obtain the Maclaurin series expansion of $f_X(\vartheta)$:

$$f_X(\vartheta) = f_X(0) + \sum_{n=1}^{\infty} \left.\frac{\mathrm{d}^n f_X}{\mathrm{d}\vartheta^n}\right|_{\vartheta=0} \frac{\vartheta^n}{n!} = 1 + \sum_{n=1}^{\infty} \frac{(i\vartheta)^n}{n!} \mathcal{E}[X^n] \tag{2.17}$$

Equation 2.17 implies that the lower-order moments contain the major parts of information about a distribution. For many practical problems, second-order statistics are enough to describe them.

2.1.3 Random Vectors

In many cases there is more than one random variable of interest. If the random variables $X_1(\varpi), X_2(\varpi), \ldots, X_n(\varpi)$ belong to the same probability space $(\Omega, \mathcal{F}, P)$, then

$$\mathbf{X} = (X_1(\varpi), X_2(\varpi), \ldots, X_n(\varpi)) \tag{2.18}$$

is an n-dimensional random vector.

The joint CDF of a random vector is defined by

$$\begin{aligned} F_{\mathbf{X}}(x_1, x_2, \ldots, x_n) &= \Pr\{X_1 < x_1, X_2 < x_2, \ldots, X_n < x_n\} \\ &= \int_{-\infty}^{x_1} \int_{-\infty}^{x_2} \cdots \int_{-\infty}^{x_n} p_{\mathbf{X}}(x_1, x_2, \ldots, x_n)\, \mathrm{d}x_1\, \mathrm{d}x_2 \ldots \mathrm{d}x_n \end{aligned} \tag{2.19}$$

where $p_{\mathbf{X}}(x_1, x_2, \ldots, x_n)$ is the joint PDF of $\mathbf{X}$. The joint density function satisfies the following properties:

$$p_{\mathbf{X}}(x_1, x_2, \ldots, x_n) \geq 0 \tag{2.20}$$

$$\int_{-\infty}^{\infty} \int_{-\infty}^{\infty} \cdots \int_{-\infty}^{\infty} p_{\mathbf{X}}(x_1, x_2, \ldots, x_n)\, \mathrm{d}x_1\, \mathrm{d}x_2 \ldots \mathrm{d}x_n = 1 \tag{2.21}$$

and there exists

$$p_{\mathbf{X}}(x_1, x_2, \ldots, x_n) = \frac{\partial^n F_{\mathbf{X}}(x_1, x_2, \ldots, x_n)}{\partial x_1 \partial x_2 \ldots \partial x_n} \tag{2.22}$$

For a certain component X_i, the marginal distribution and the marginal density function are respectively defined by

$$F_{X_i}(x_i) = \Pr\{X_i < x_i\} = F_{\mathbf{X}}(\infty, \ldots, \infty, x_i, \infty, \ldots, \infty) \tag{2.23}$$

and

$$p_{X_i}(x_i) = \int_{-\infty}^{\infty} \cdots \int_{-\infty}^{\infty} p_{\mathbf{X}}(x_1, x_2, \ldots, x_n)\, \mathrm{d}x_1\, \mathrm{d}x_2 \ldots \mathrm{d}x_{i-1}\, \mathrm{d}x_{i+1} \ldots \mathrm{d}x_n \tag{2.24}$$

Generally speaking, the marginal distribution can be uniquely determined by the joint probability distribution function, but the converse is not true. In other words, the joint PDF contains more information than each marginal density function separately, since the latter can be obtained from the former. This implies that the correlation between random variables is an important profile of a random vector.

For an n-dimensional random vector, the conditional cumulative distribution and the conditional PDF with respect to a certain component X_i are respectively defined by

$$\begin{aligned} &F_{\mathbf{X}|X_i}(x_1, \ldots, x_{i-1}, x_{i+1}, \ldots, x_n | x_i) \\ &= \Pr\{X_1 < x_1, \ldots, X_{i-1} < x_{i-1}, X_{i+1} < x_{i+1}, \ldots, X_n < x_n | X_i = x_i\} \end{aligned} \tag{2.25}$$

and

$$p_{\mathbf{X}|X_i}(x_1,\ldots,x_{i-1},x_{i+1},\ldots,x_n|x_i)=\frac{p_{\mathbf{X}}(x_1,x_2,\ldots,x_n)}{\int_{-\infty}^{\infty}p_{\mathbf{X}}(x_1,x_2,\ldots,x_n)\,\mathrm{d}x_1\ldots\,\mathrm{d}x_{i-1}\,\mathrm{d}x_{i+1}\ldots\,\mathrm{d}x_n} \tag{2.26}$$

If for all the $x_1, x_2, \ldots, x_n$ there exists

$$F_{\mathbf{X}}(x_1,x_2,\ldots,x_n)=F_{X_1}(x_1)F_{X_2}(x_2)\ldots F_{X_n}(x_n) \tag{2.27}$$

or

$$p_{\mathbf{X}}(x_1,x_2,\ldots,x_n)=p_{X_1}(x_1)p_{X_2}(x_2)\ldots p_{X_n}(x_n) \tag{2.28}$$

then we call $X_1, X_2, \ldots, X_n$ statistically independent random variables. In this case, the marginal probability distribution of a certain component is equal to the corresponding conditional probability distribution.

Let a function $Y=f(X_1, X_2, \ldots, X_n)$. Then its probability distribution function is given by

$$F_Y(y)=\Pr\{f(X_1,X_2,\ldots,X_n)<y\}=\int_{f(x_1,x_2,\ldots,x_n)<y}p_{\mathbf{X}}(x_1,x_2,\ldots,x_n)\,\mathrm{d}x_1\,\mathrm{d}x_2\ldots\,\mathrm{d}x_n \tag{2.29}$$

Similarly, for m functions $Y_i=f_i(X_1, X_2, \ldots, X_n)$, $i=1, 2, \ldots, m$, there is

$$\begin{aligned}F_{\mathbf{Y}}(y_1,y_2,\ldots,y_m)&=\Pr\{f_i(X_1,X_2,\ldots,X_n)<y_i,\ i=1,2,\ldots,m\}\\&=\int_{f_i(x_1,x_2,\ldots,x_n)<y_i,1\le i\le m}p_{\mathbf{X}}(x_1,x_2,\ldots,x_n)\,\mathrm{d}x_1\,\mathrm{d}x_2\ldots\,\mathrm{d}x_n\end{aligned} \tag{2.30}$$

Like the one-dimensional case, Equations 2.29 and 2.30 require that $f_i(\cdot)$ obey some restrictions. In particular, consider the case $m=n$. To be exact, assume that there exist the unique inverse functions of $y_i=f_i(x_1, x_2, \ldots, x_n)$, denoted by $x_i=x_i(y_1, y_2, \ldots, y_n)$, and the continuous partial derivatives $\partial x_i/\partial y_j$. Then, the PDF of the n-dimensional random vector $\mathbf{Y}$ can be given by

$$p_{\mathbf{Y}}(y_1,y_2,\ldots,y_n)=\begin{cases}p_{\mathbf{X}}(x_1,x_2,\ldots,x_n)|J| & \text{if } (y_1,y_2,\ldots,y_n)\in\Omega_{f_1,f_2,\ldots,f_n}\\0 & \text{otherwise}\end{cases} \tag{2.31}$$

where $\Omega_{f_1, f_2, \ldots, f_n}$ is the value domain of $(f_1, f_2, \ldots, f_n)$ and $|J|$ is the absolute value of the determinant of the Jacobian matrix

$$|J|=\begin{vmatrix}\dfrac{\partial x_1}{\partial y_1} & \dfrac{\partial x_2}{\partial y_1} & \cdots & \dfrac{\partial x_n}{\partial y_1}\\\dfrac{\partial x_1}{\partial y_2} & \dfrac{\partial x_2}{\partial y_2} & \cdots & \dfrac{\partial x_n}{\partial y_2}\\\cdots & \cdots & \cdots & \cdots\\\dfrac{\partial x_1}{\partial y_n} & \dfrac{\partial x_2}{\partial y_n} & \cdots & \dfrac{\partial x_n}{\partial y_n}\end{vmatrix} \tag{2.32}$$

When the inverse functions are not one-to-one, namely $y_i = f_i(x_1, x_2, \ldots, x_n)$, $i = 1, 2, \ldots, n$, have more than one solution, then one point in the space Y corresponds to multiple points in the space X. In such a case, it is necessary to partition X into several sub-domains so as to yield one-to-one transformations from Y to each sub-domain of X. Then, the probability that $\mathbf{Y}$ takes values over a certain subset of Y is equal to the sum of probabilities that $\mathbf{X}$ takes the values over the corresponding sets in each sub-domain of X; that is:

$$p_{\mathbf{Y}}(y_1, y_2, \ldots, y_n) = \begin{cases} \sum_k p_{\mathbf{X}}(x_{1,k}, x_{2,k}, \ldots, x_{n,k})|J_k| & \text{if}(y_1, y_2, \ldots, y_n) \in \Omega_{f_1, f_2, \ldots, f_n} \\ 0 & \text{otherwise} \end{cases} \tag{2.33}$$

where $\Omega_{f_1, f_2, \ldots, f_n}$ is the value domain of $(f_1, f_2, \ldots, f_n)$.

To describe the distribution of an n-dimensional random vector completely, its n-dimensional joint probability distribution function is needed. This may be difficult for most practical cases. In many practical applications, it is feasible to use a family of expected values of the random vectors.

For a random vector $(X_1, X_2, \ldots, X_n)$, the expectation is given by $(\mathcal{E}[X_1], \mathcal{E}[X_2], \ldots, \mathcal{E}[X_n])$, where

$$\mathcal{E}[X_i] = \int_{-\infty}^{\infty} x_i p_{X_i}(x_i)\,\mathrm{d}x_i \tag{2.34}$$

and its variance is $(\mathcal{D}[X_1], \mathcal{D}[X_2], \ldots, \mathcal{D}[X_n])$, where

$$\mathcal{D}[X_i] = \mathcal{E}\{(X_i - \mathcal{E}[X_i])^2\} = \int_{-\infty}^{\infty} (x_i - \mathcal{E}[X_i])^2 p_{X_i}(x_i)\,\mathrm{d}x_i \tag{2.35}$$

It is obvious that the above expectations and variances only reflect the information of each random variable itself. Apart from these, in practical problems, the correlated information between two random variables is equally worth noting. The covariance of two components X_i and X_j is defined by

$$\begin{aligned} c_{ij} &= \mathrm{cov}[X_i, X_j] = \mathcal{E}\{(X_i - \mathcal{E}[X_i])(X_j - \mathcal{E}[X_j])\} \\ &= \int_{-\infty}^{\infty}\int_{-\infty}^{\infty} (x_i - \mathcal{E}[X_i])(x_j - \mathcal{E}[X_j]) p_{X_iX_j}(x_i, x_j)\,\mathrm{d}x_i\,\mathrm{d}x_j \end{aligned} \tag{2.36}$$

The matrix

$$\mathbf{C} = \begin{bmatrix} c_{11} & c_{12} & \cdots & c_{1n} \\ c_{21} & c_{22} & \cdots & c_{2n} \\ \cdots & \cdots & \cdots & \cdots \\ c_{n1} & c_{n2} & \cdots & c_{nn} \end{bmatrix} \tag{2.37}$$

is called the covariance matrix of $(X_1, X_2, \ldots, X_n)$.

Sometimes it is convenient to define the parameter

$$\rho_{ij} = \frac{\mathrm{cov}[X_i, X_j]}{\sqrt{\mathcal{D}(X_i)\mathcal{D}(X_j)}} \tag{2.38}$$

as the *correlation coefficient* of X_i and X_j. This parameter indicates whether two random variables are linearly dependent or uncorrelated. That is, if $\rho = \pm 1$, then X_i and X_j are said to be completely correlated. In other words, they are equivalent in the sense of probability. If $\rho = 0$, then X_i and X_j are said to be completely uncorrelated. However, the uncorrelation does not imply independence. The latter usually means no function relation between the two random variables.

Note that the correlation coefficient is essentially the covariance of standardized random variables $(X_i - \mathcal{E}[X_i])/\sqrt{\mathcal{D}[X_i]}$ and $(X_j - \mathcal{E}[X_j])/\sqrt{\mathcal{D}[X_j]}$. Therefore, the matrix

$$\boldsymbol{\rho} = \begin{bmatrix} \rho_{11} & \rho_{12} & \cdots & \rho_{1n} \\ \rho_{21} & \rho_{22} & \cdots & \rho_{2n} \\ \cdots & \cdots & \cdots & \cdots \\ \rho_{n1} & \rho_{n2} & \cdots & \rho_{nn} \end{bmatrix} \tag{2.39}$$

is called the normalized covariance matrix or the correlation coefficient matrix.

The expectation of a random function $g(X_1, X_2, \ldots, X_n)$ is defined by

$$\mathcal{E}[g(X_1, X_2, \ldots, X_n)] = \int_{-\infty}^{\infty} \cdots \int_{-\infty}^{\infty} g(x_1, x_2, \ldots, x_n) p_{\mathbf{X}}(x_1, x_2, \ldots, x_n)\, \mathrm{d}x_1\, \mathrm{d}x_2 \ldots \mathrm{d}x_n \tag{2.40}$$

Comparing Equation 2.40 with Equation 2.29, we find that the expectation of a function of a random vector, $Y = g(X_1, X_2, \ldots, X_n)$, can be computed directly from Equation 2.40 without the need to first obtain its PDF $p_Y(y)$.

For more complicated functions, we can expand the random function $g(\cdot)$ in a series form:

$$Y = g(\mathcal{E}[X_1], \mathcal{E}[X_2], \ldots, \mathcal{E}[X_n]) + \sum_{i=1}^{n} \left.\frac{\partial g}{\partial X_i}\right|_{\substack{X_1=\mathcal{E}[X_1] \\ \cdots \\ X_n=\mathcal{E}[X_n]}} (X_i - \mathcal{E}[X_i])$$
$$+ \frac{1}{2}\sum_{i=1}^{n} \left.\frac{\partial^2 g}{\partial X_i \partial X_j}\right|_{\substack{X_1=\mathcal{E}[X_1] \\ \cdots \\ X_n=\mathcal{E}[X_n]}} (X_i - \mathcal{E}[X_i])(X_j - \mathcal{E}[X_j]) + \ldots \tag{2.41}$$

Retaining only the linear terms, we get the mean and the variance of Y respectively as below:

$$\mathcal{E}[Y] = g(\mathcal{E}[X_1], \mathcal{E}[X_2], \ldots, \mathcal{E}[X_n]) \tag{2.42}$$

and

$$\mathcal{D}[Y] = \sum_{i=1}^{n} \left[\left.\frac{\partial g}{\partial X_i}\right|_{\substack{X_1=\mathcal{E}[X_1] \\ \cdots \\ X_n=\mathcal{E}[X_n]}} \right]^2 \mathcal{D}[X_i] + \sum_{i=1}^{n}\sum_{j=1}^{n} \left.\frac{\partial g}{\partial X_i}\right|_{\substack{X_1=\mathcal{E}[X_1] \\ \cdots \\ X_n=\mathcal{E}[X_n]}} \left.\frac{\partial g}{\partial X_j}\right|_{\substack{X_1=\mathcal{E}[X_1] \\ \cdots \\ X_n=\mathcal{E}[X_n]}} \mathrm{cov}\left[X_i, X_j\right] \tag{2.43}$$

The expansion formula (Equation 2.41) is actually a basis for the perturbation theory of stochastic structure analysis (see Chapter 4).

2.1.4 Decomposition of Correlation Matrix

As noted in Equation 2.37, the covariance matrix is both symmetric and nonnegative definite. This fact implies that, for the covariance matrix **C**, there exists a diagonal matrix

$$\boldsymbol{\lambda} = \begin{bmatrix} \lambda_1 & 0 & \dots & 0 \\ 0 & \lambda_2 & \dots & 0 \\ 0 & \dots & \ddots & 0 \\ 0 & \dots & 0 & \lambda_n \end{bmatrix} \tag{2.44}$$

and an n-dimensional matrix

$$\boldsymbol{\phi} = [\boldsymbol{\psi}_1, \boldsymbol{\psi}_2, \dots, \boldsymbol{\psi}_n] = \begin{bmatrix} \phi_{11} & \phi_{12} & \dots & \phi_{1n} \\ \phi_{21} & \phi_{22} & \dots & \phi_{2n} \\ \dots & \dots & \dots & \dots \\ \phi_{n1} & \phi_{n2} & \dots & \phi_{nn} \end{bmatrix} \tag{2.45}$$

such that

$$\mathbf{C}\boldsymbol{\phi} = \boldsymbol{\lambda}\boldsymbol{\phi} \tag{2.46}$$

where λ_i, $i = 1, 2, \dots, n$, are the eigenvalues of **C** and $\boldsymbol{\phi}$ is the eigenvector matrix of **C**. Note that $\boldsymbol{\phi}$ is an orthogonal matrix; that is, $\boldsymbol{\phi}^{\mathrm{T}}\boldsymbol{\phi} = \mathbf{I}$, where the superscript T denotes the vector transpose.

The matrix $\boldsymbol{\phi}$ leads to a similarity transformation, through which the formerly nondiagonal matrix **C** is transformed to a diagonal matrix $\boldsymbol{\lambda}$; that is:

$$\boldsymbol{\lambda} = \boldsymbol{\phi}^{\mathrm{T}}\mathbf{C}\boldsymbol{\phi} \tag{2.47}$$

Thus, for a random vector $\mathbf{X} = (X_1, X_2, \dots, X_n)$, we are sure to obtain the eigenvalues and the eigenvectors with respect to a covariance matrix by adopting the eigenvalue theory of matrices. This process is the so-called *decomposition of the correlation matrix*, through which the former correlated random vector $(X_1, X_2, \dots, X_n)$ can be transformed into a set of uncorrelated random variables $\mathbf{Y} = (Y_1, Y_2, \dots, Y_n)$ by a linear transform:

$$\mathbf{X} = \mathbf{X}_0 + \boldsymbol{\phi}\mathbf{Y} = \mathbf{X}_0 + \sum_{i=1}^{n} \boldsymbol{\psi}_i Y_i \tag{2.48}$$

where $\mathbf{X}_0$ is the mean vector of **X**.

In fact, let $\mathbf{X}_\sigma = \boldsymbol{\phi}\mathbf{Y}$; since adding vectors with constants makes no change to the corresponding covariance, there is

$$\mathbf{C_X} = \mathbf{C_{X\sigma}} = \mathcal{E}[\mathbf{X}_\sigma \mathbf{X}_\sigma^T] \tag{2.49}$$

Combining Equations 2.49 and 2.47 gives

$$\boldsymbol{\lambda} = \boldsymbol{\phi}^T \mathcal{E}[\mathbf{X}_\sigma \mathbf{X}_\sigma^T]\boldsymbol{\phi} = \boldsymbol{\phi}^T \mathcal{E}[\boldsymbol{\phi}\mathbf{Y}\mathbf{Y}^T\boldsymbol{\phi}^T]\boldsymbol{\phi} = \mathcal{E}[\mathbf{Y}\mathbf{Y}^T] \tag{2.50}$$

This indicates that $\boldsymbol{\lambda}$ is nothing but the covariance matrix of $\mathbf{Y}$.

Let $Y_i = \sqrt{\lambda_i}\zeta_i$, then $(\zeta_1, \zeta_2, \ldots, \zeta_n)$ is a standardized sequence of uncorrelated random variables such that

$$\mathbf{X} = \mathbf{X}_0 + \sum_{i=1}^{n} \boldsymbol{\psi}_i \sqrt{\lambda_i}\zeta_i \tag{2.51}$$

This shows that a random vector can be represented in the form of a sequence of standardized uncorrelated random variables.

2.2 Stochastic Processes

2.2.1 Specification of Stochastic Processes

The so-called 'stochastic process' refers to a family of random variables defined over a parameter set. Every point of the set is associated with a random variable. In this point of view, a one-dimensional stochastic process may be understood as an extension of a random vector (Figure 2.3). Therefore, those basic concepts that hold for multidimensional random vectors are still true for one-dimensional stochastic processes.

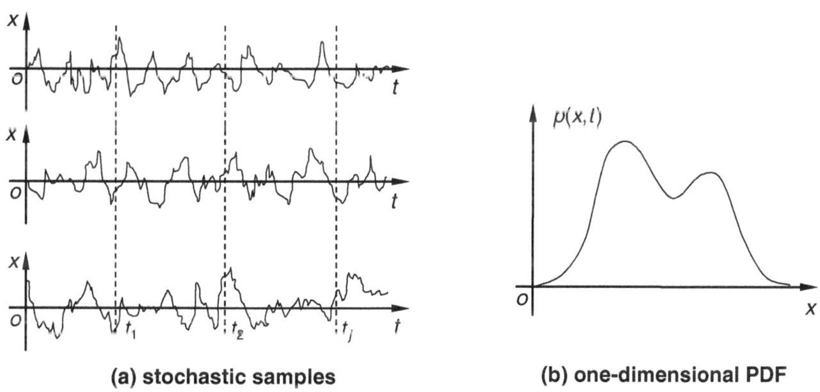

Figure 2.3 A stochastic process and the one-dimensional PDF.

Assume that $\{X(t), t \subset T\}$ is a stochastic process, where t is a time parameter that belongs to a set T (a time interval). To describe its probability properties, what first counts is the distribution of every one-time random variable (or the random variable at a certain time t, $t \in T$). This is denoted by

$$F(x,t) = \Pr\{X(t) < x\} \qquad t \in T \tag{2.52}$$

and called the one-dimensional distribution of $\{X(t), t \in T\}$.

It is obvious that the one-dimensional distribution alone is insufficient to describe a stochastic process. Then, it naturally comes into our minds to study how the random variables of a stochastic process at different time instants are correlated. For this purpose we introduce

$$F(x_1, t_1; x_2, t_2) = \Pr\{X(t_1) < x_1, X(t_2) < x_2\} \qquad t_1, t_2 \in T \tag{2.53}$$

as the two-dimensional distribution of $\{X(t), t \in T\}$.

In increasing order of completeness, for any finite $\{t_1, t_2, \ldots, t_n \in T\}$, there is

$$F(x_1, t_1; x_2, t_2; \ldots; x_n, t_n) = \Pr\{X(t_1) < x_1, X(t_2) < x_2, \ldots, X(t_n) < x_n\} \tag{2.54}$$

which is called the *n-dimensional distribution* of $\{X(t), t \in T\}$.

For a stochastic process $\{X(t), t \in T\}$, its one-, two-, ..., and n-dimensional distributions constitute its complete probabilistic structure. In fact, once this family of finite-dimensional distributions is given, we can determine the correlation of any finite random variables of the stochastic process at different time instants. That is, the probabilistic structure of $\{X(t), t \in T\}$ is obtained.

The family of finite-dimensional distributions satisfies the following three properties:

(a) Nonnegative; that is:

$$0 \le F(x_1, t_1; x_2, t_2; \ldots; x_n, t_n) \le 1 \tag{2.55}$$

(b) Symmetry; that is, for any permutation $(j_1, j_2, \ldots, j_n)$ of $(1, 2, \ldots, n)$, there is

$$F(x_{j_1}, t_{j_1}; x_{j_2}, t_{j_2}; \ldots; x_{j_n}, t_{j_n}) = F(x_1, t_1; x_2, t_2; \ldots; x_n, t_n) \tag{2.56}$$

(c) Compatibility; that is, as $m < n$, there is

$$F(x_1, t_1; \ldots; x_m, t_m; \infty, t_{m+1}; \ldots; \infty, t_n) = F(x_1, t_1; x_2, t_2; \ldots; x_m, t_m) \tag{2.57}$$

According to Equation 2.57, the lower dimensional distributions can be obtained from the higher dimensional ones.

Similar to the cases of a random variable and a random vector, for a stochastic process the finite-dimensional density functions are defined by the derivatives of the corresponding distribution functions; that is:

$$\begin{aligned} p(x, t) &= \frac{\partial F(x, t)}{\partial x} \\ p(x_1, t_1; x_2, t_2) &= \frac{\partial^2 F(x_1, t_1; x_2, t_2)}{\partial x_1 \partial x_2} \\ &\cdots \\ p(x_1, t_1; x_2, t_2; \ldots; x_n, t_n) &= \frac{\partial^n F(x_1, t_1; x_2, t_2; \ldots; x_n, t_n)}{\partial x_1 \partial x_2 \ldots \partial x_n} \end{aligned} \tag{2.58}$$

Certainly, this family can completely describe the probabilistic structures of a stochastic process as well.

A stochastic process may also be specified by a sequence of characteristic functions. In fact, performing a Fourier transform on every member of the above family yields the family of finite-dimensional characteristic functions:

$$\begin{aligned}
M(\vartheta, t) &= \mathcal{E}\{\exp[i\vartheta X(t)]\} = \int_{-\infty}^{\infty} p(x, t) e^{i\vartheta x}\, \mathrm{d}x \\
M(\vartheta_1, t_1; \vartheta_2, t_2) &= \mathcal{E}\{\exp[i\vartheta_1 X(t_1) + i\vartheta_2 X(t_2)]\} \\
&= \int_{-\infty}^{\infty} p(x_1, t_1; x_2, t_2) e^{i\vartheta_1 x_1 + i\vartheta_2 x_2}\, \mathrm{d}x_1\, \mathrm{d}x_2 \\
&\cdots \\
M(\vartheta_1, t_1; \vartheta_2, t_2; \ldots; \vartheta_n, t_n) &= \mathcal{E}\left\{\exp[i\sum_{j=1}^{n} \vartheta_j X(t_j)]\right\} \\
&= \int_{-\infty}^{\infty} p(x_1, t_1; x_2, t_2; \ldots; x_n, t_n) \exp\left(i\sum_{j=1}^{n} \vartheta_j x_j\right) \mathrm{d}x_1\, \mathrm{d}x_2 \ldots \mathrm{d}x_n
\end{aligned} \tag{2.59}$$

where ϑ_j are arbitrary real numbers.

A higher order characteristic function can be reduced to a lower order one by setting some of the ϑ_j as zero:

$$M(\vartheta_1, t_1; \vartheta_2, t_2; \ldots; \vartheta_m, t_m; 0, t_{m+1}; \ldots; 0, t_{m+k}) = M(\vartheta_1, t_1; \vartheta_2, t_2; \ldots; \vartheta_m, t_m) \tag{2.60}$$

Generally, characteristic functions are continuous and complex-valued. Since they form the Fourier transform pairs with the corresponding density functions, the family is an equivalent description of the probabilistic structure of a stochastic process.

2.2.2 *Moment Functions of a Stochastic Process*

A stochastic process can also be described by moment functions of various orders, which are defined as

$$\begin{aligned}
\mathcal{E}[X(t_1)] &= \int_{-\infty}^{\infty} x_1 p(x_1, t_1)\, \mathrm{d}x \\
\mathcal{E}[X(t_1) X(t_2)] &= \int_{-\infty}^{\infty} x_1 x_2 p(x_1, t_1; x_2, t_2)\, \mathrm{d}x_1\, \mathrm{d}x_2 \\
&\cdots
\end{aligned} \tag{2.61}$$

when these moments exist.

From the MacLaurin series expansion of a characteristic function

$$M(\vartheta_1, t_1; \vartheta_2, t_2; \ldots; \vartheta_n, t_n) = 1 + i\sum_{j=1}^{n} \vartheta_j \mathcal{E}[X(t_j)] - \frac{1}{2!}\sum_{j=1}^{n}\sum_{k=1}^{n} \vartheta_j \vartheta_k \mathcal{E}[X(t_j) X(t_k)] + \ldots \tag{2.62}$$

it is noted that all the moment functions give a complete description of a stochastic process.

The first moment function of a stochastic process $X(t)$ is called the *expectation*, which is defined by

$$m_X(t) = \mathcal{E}[X(t)] = \int_{-\infty}^{\infty} xp(x, t)\,\mathrm{d}x \tag{2.63}$$

Clearly, for a specific section of random samples, the above equation represents the first origin moment of the random variable of the stochastic process at time t. For the whole process, it represents the locus of average centers of the $X(t)$ sample functions $x_i(t)$ in the time domain.

A stochastic process is said to be stationary of first order if $m_X(t) = \text{constant}$. The stochastic process of this type can be easily reduced to a process with zero mean. With this transform, we may concentrate our attention on the deviation of the process from its expectation; that is, its variance.

The correlation function serves as a measure of interrelation of any two different states of stochastic processes. It quantifies how close the values of the random variables specified at two different time instants are in the sense of probability. There are auto- and cross-correlation functions according to whether the correlation information of one or between two stochastic processes needs to be characterized.

The *autocorrelation function* of $X(t)$ is defined for two random variables from the same process by

$$R_X(t_1, t_2) = \mathcal{E}[X(t_1)X(t_2)] = \int_{-\infty}^{\infty}\int_{-\infty}^{\infty} x_1x_2p(x_1, t_1; x_2, t_2)\,\mathrm{d}x_1\,\mathrm{d}x_2 \tag{2.64}$$

On the other hand, the *cross-correlation function* is assigned to those from two different processes. Suppose $X(t)$ and $Y(t)$ are two stochastic processes, then the cross-correlation function is defined by

$$R_{XY}(t_1, t_2) = \mathcal{E}[X(t_1)Y(t_2)] = \int_{-\infty}^{\infty}\int_{-\infty}^{\infty} x_1y_2p(x_1, t_1; y_2, t_2)\,\mathrm{d}x_1\,\mathrm{d}y_2 \tag{2.65}$$

where $p(x_1, t_1; y_2, t_2)$ is the joint PDF of $X(t_1)$ and $Y(t_2)$.

The cross-correlation function describes the interrelation between two stochastic processes in the time domain. In other words, it indicates the degree of probabilistic similarity between two stochastic processes at different time instants.

The normalized correlation functions are called as *correlation coefficient*. The *auto-correlation coefficient* is denoted by

$$r_X(t_1, t_2) = \frac{R_X(t_1, t_2)}{\sigma_X(t_1)\sigma_X(t_2)} \tag{2.66}$$

for a zero-mean stochastic process. Correspondingly, the cross-correlation coefficient is

$$r_{XY}(t_1, t_2) = \frac{R_{XY}(t_1, t_2)}{\sigma_X(t_1)\sigma_Y(t_2)} \tag{2.67}$$

As noted, the correlation is based on the second origin moments of processes, whereas the following covariance is on the second central moments.

For a stochastic process $X(t)$, the *auto-covariance function* is defined as

$$\begin{aligned}K_X(t_1, t_2) &= \mathcal{E}\{[X(t_1) - m_X(t_1)][X(t_2) - m_X(t_2)]\} \\ &= \int_{-\infty}^{\infty}\int_{-\infty}^{\infty} [x_1 - m_X(t_1)][x_2 - m_X(t_2)]p(x_1, t_1; x_2, t_2)\,\mathrm{d}x_1\,\mathrm{d}x_2\end{aligned} \tag{2.68}$$

As $t_1 = t_2 = t$, the above definition gives

$$K_X(t, t) = \mathcal{E}\{[X(t) - m_X(t)]^2\} = \mathcal{D}[X(t)] \tag{2.69}$$

where $\mathcal{D}[X(t)]$ is called the *variance* of $X(t)$, which is used to gauge how extensively $X(t)$ varies around its expectation. The standard deviation of $X(t)$ refers to the square root of $\mathcal{D}[X(t)]$; that is:

$$\sigma_X(t) = \sqrt{\mathcal{D}[X(t)]} \tag{2.70}$$

For two stochastic processes, the *cross-covariance function* is defined as

$$\begin{aligned}K_{XY}(t_1, t_2) &- \mathcal{E}\{[X(t_1) - m_X(t_1)][Y(t_2) - m_Y(t_2)]\} \\ &= \int_{-\infty}^{\infty}\int_{-\infty}^{\infty} [x_1 - m_X(t_1)][y_2 - m_Y(t_2)]p(x_1, t_1; y_2, t_2)\,\mathrm{d}x_1\,\mathrm{d}y_2\end{aligned} \tag{2.71}$$

It is easy to verify the following relationships:

$$K_X(t_1, t_2) = R_X(t_1, t_2) - m_X(t_1)m_X(t_2) \tag{2.72a}$$

$$K_{XY}(t_1, t_2) = R_{XY}(t_1, t_2) - m_X(t_1)m_Y(t_2) \tag{2.72b}$$

These two equations imply that, for stochastic processes with zero mean, the covariance function equals the correlation function.

A stochastic process is called a weakly stationary process if its expectation is a constant and its autocorrelation function depends only on the time difference $\tau = t_2 - t_1$ (independent of t_1 and t_2). In contrast, a strictly stationary process refers to one whose finite-dimensional distributions are time invariant. Generally speaking, a weakly stationary process is not necessarily a strictly stationary one, while a strictly stationary process must be a weakly stationary one. Only for a normal process, if weakly stationary, is it also strictly stationary. In practical applications, the weak stationarity is much more widely used than the strict stationarity. Thus, the stationary processes mentioned below, unless specified otherwise, mean the weakly stationary ones.

The autocorrelation function of a stationary process is expressed as

$$R_X(\tau) = R_X(t_2 - t_1) \tag{2.73}$$

and its cross-correlation function as

$$R_{XY}(\tau) = R_{XY}(t_2 - t_1) \tag{2.74}$$

Notice that, for a stationary process, the variance $\mathcal{D}[X(t)]$ in Equation 2.69 becomes

$$\mathcal{D}[X(t)] = \sigma_X^2 = K_X(0) \tag{2.75a}$$

and if $X(t)$ has zero mean, then from Equation (2.72a) there exist

$$\mathcal{D}[X(t)] = \sigma_X^2 = K_X(0) = R_X(0) \tag{2.75b}$$

The following properties hold true for the autocorrelation function:

(a) It is symmetric

$$R_X(\tau) = R_X(-\tau) \tag{2.76}$$

(b) It is nonnegative definite

$$\sum_{i=1}^{n}\sum_{j=1}^{n} R_X(t_i - t_j)h(t_i)h^*(t_j) \geq 0 \tag{2.77}$$

where $h(t)$ is any arbitrary complex function and $h^*(t)$ is the complex conjugate.
(c) It is bounded

$$|R_X(\tau)| \leq R_X(0) \tag{2.78}$$

(d) If $X(t)$ does not contain periodic components, then for a zero-mean stochastic process

$$\lim_{\tau \to \infty} R(\tau) = 0 \tag{2.79}$$

A typical autocorrelation function of a zero-mean stationary process is shown in Figure 2.4.

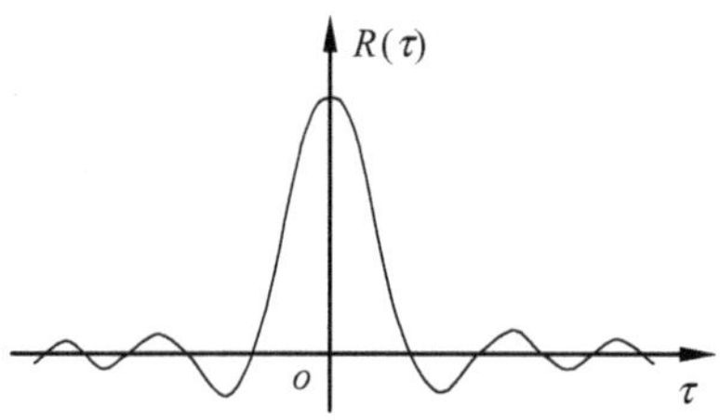

Figure 2.4 Autocorrelation function of a zero-mean stationary process.

The importance of the first- and second-order statistical properties of a stochastic process may be realized not only from the fact that the lower moments contain the major parts of information about the process (see Equation 2.62), but also from the fact that an upper-bound estimate of the probability of the event $\{|X(t) - m_X(t)| \geq \varepsilon\}$ at any t can be made from the mean and variance functions for an arbitrary stochastic process. In fact, let $\sigma_X^2(t)$ and $\sigma_{\dot{X}}^2(t)$ be the variance function of $X(t)$ and its derivative respectively; it can be verified that (Lin, 1967)

$$\Pr\{|X(t) - m_X(t)| \geq \varepsilon \text{ for } a \leq t \leq b\} \leq \frac{1}{2\varepsilon^2}\left[\sigma_X^2(a) + \sigma_X^2(b)\right] + \frac{1}{\varepsilon^2}\int_a^b \sigma_X(t)\sigma_{\dot{X}}(t)\,\mathrm{d}t \quad \varepsilon > 0 \tag{2.80}$$

2.2.3 Spectral Description of a Stochastic Process

In a general sense, the power spectral density (PSD) function is the Fourier transform of the covariance function of a stochastic process. However, it is easy to transform a stationary process to a zero-mean process. Thus, the PSD of a stationary process is defined as the Fourier transform of its correlation function.

Consider a stationary process $X(t)$. Its *auto-PSD function* is

$$S_X(\omega) = \int_{-\infty}^{\infty} R_X(\tau)e^{-i\omega\tau}\,d\tau \tag{2.81a}$$

while the inverse Fourier transform gives

$$R_X(\tau) = \frac{1}{2\pi}\int_{-\infty}^{\infty} S_X(\omega)e^{i\omega\tau}\,d\omega \tag{2.81b}$$

The above pair is the celebrated Wiener–Khintchine formula.

The auto-PSD $S_X(\omega)$ satisfies the following properties:

(a) $S_X(\omega)$ is nonnegative; that is:

$$S_X(\omega) \geq 0 \tag{2.82}$$

(b) $S_X(\omega)$ is real and even (or symmetric); that is:

$$S_X(\omega) = S_X(-\omega) \tag{2.83}$$

Using Equation 2.83, we get

$$\int_{-\infty}^{\infty} S_X(\omega)\,d\omega = 2\int_{0}^{\infty} S_X(\omega)\,d\omega \tag{2.84}$$

which leads to the definition of the *unilateral PSD*

$$G_X(\omega) = \begin{cases} 2S_X(\omega) & 0 \leq \omega < \infty \\ 0 & \text{otherwise} \end{cases} \tag{2.85}$$

Accordingly, $S_X(\omega)$ is called the *bilateral PSD*. Obviously, over the nonnegative real number field, $G_X(\omega)$ is twice that of $S_X(\omega)$.

The *cross-PSD* $S_{XY}(\omega)$ of two stochastic processes $X(t)$ and $Y(t)$ is the Fourier transform of their cross-correlation function $R_{XY}(\tau)$:

$$S_{XY}(\omega) = \int_{-\infty}^{\infty} R_{XY}(\tau)e^{-i\omega\tau}\,d\tau \tag{2.86a}$$

while the inverse Fourier transform gives

$$R_{XY}(\tau) = \frac{1}{2\pi}\int_{-\infty}^{\infty} S_{XY}(\omega)e^{i\omega\tau}\,d\omega \tag{2.86b}$$

The cross-PSD $S_{XY}(\omega)$ satisfies the following properties:

(a) $S_{XY}(\omega)$ is generally a complex function;
(b) $S_{XY}(\omega)$ satisfies

$$S_{XY}(\omega) = S^*_{YX}(\omega) = S_{XY}(-\omega) \tag{2.87}$$

(c) $S_{XY}(\omega)$ satisfies the inequality

$$|S_{XY}(\omega)|^2 \le S_X(\omega)S_Y(\omega) \tag{2.88}$$

A relationship between the PSD function of a stochastic process and the Fourier spectrum of its samples can be established. Actually, for stationary processes $X(t)$ and $Y(t)$, we have (Bendat and Piersol, 2000) (for proof, see Appendix C)

$$S_X(\omega) = \lim_{T\to\infty} \frac{1}{2T} \mathcal{E}\left[X_{\pm T}(\varpi,\omega)X^*_{\pm T}(\varpi,\omega)\right] \tag{2.89a}$$

$$S_{XY}(\omega) = \lim_{T\to\infty} \frac{1}{2T} \mathcal{E}\left[X_{\pm T}(\varpi,\omega)Y^*_{\pm T}(\varpi,\omega)\right] \tag{2.89b}$$

where $[-T, T]$ is the time interval over which the stochastic processes are specified and $X_{\pm T}(\varpi,\omega)$ and $Y_{\pm T}(\varpi,\omega)$ are the finite Fourier transform of $X(t)$ and $Y(t)$ in the sense of sample defined respectively by (see Equations C.1 and C.2 in Appendix C)

$$X_{\pm T}(\varpi,\omega) = \int_{-T}^{T} X(\varpi,t)e^{-i\omega t}\,\mathrm{d}t \tag{2.89c}$$

$$Y_{\pm T}(\varpi,\omega) = \int_{-T}^{T} Y(\varpi,t)e^{-i\omega t}\,\mathrm{d}t \tag{2.89d}$$

For a nonstationary stochastic process, the power spectrum is defined as the Fourier transform of the covariance function. As an example, consider a general stochastic process $X(t)$ and let $K_X(t_1, t_2)$ denote its covariance; the power spectrum is then given by

$$S_X(\omega_1,\omega_2) = \int_{-\infty}^{\infty}\int_{-\infty}^{\infty} K_X(t_1,t_2)e^{-i(\omega_1 t_1 - \omega_2 t_2)}\,\mathrm{d}t_1\,\mathrm{d}t_2 \tag{2.90a}$$

while the inverse Fourier transform gives

$$K_X(t_1,t_2) = \frac{1}{(2\pi)^2}\int_{-\infty}^{\infty}\int_{-\infty}^{\infty} S_X(\omega_1,\omega_2)e^{i(\omega_1 t_1 - \omega_2 t_2)}\,d\omega_1\,d\omega_2 \tag{2.90b}$$

However, one of the problems of the above pair is that the physical significance of the double-frequency spectral density function of a nonstationary stochastic process (given by Equations 2.90a and 2.90b) is not as clear as that of the PSD function of a stationary stochastic process (cf. Equations 2.81a and 2.81b). Moreover, the integral in Equation 2.90a exists provided that $K_X(t_1, t_2) \rightarrow 0$ as t_1 and t_2 tend to infinity. This condition is too strict for many idealized but widely used stochastic process models. Thus, the concept of an evolutionary PSD is preferred by many investigators. We will deal with this issue elaborately in Section 5.3.2.

2.2.4 Some Operation Rules about Expectation, Correlation and Spectrum

To deal with a stochastic process, the following computation formulae may be important in practical applications.

2.2.4.1 The Expectation Operation

The expectation is a linear operator which is homogeneous and summative. Consider stochastic processes $X_i(t)$, $i = 1, 2, \ldots, n$, if $\varphi_i(t)$ and $\nu(t)$ are deterministic functions, and

$$Y(t) = \sum_{i=1}^{n} \varphi_i(t) X_i(t) + \nu(t) \tag{2.91}$$

Then:

$$m_Y(t) = \sum_{i=1}^{n} \varphi_i(t) m_{X_i}(t) + \nu(t) \tag{2.92}$$

For an operation of expectation, the interchange order of differentiation (or integration) and expectation is allowed (Lin, 1967); that is:

$$\frac{\mathrm{d}\mathcal{E}[X(t)]}{\mathrm{d}t} = \mathcal{E}\left[\frac{\mathrm{d}X(t)}{\mathrm{d}t}\right] \tag{2.93}$$

$$\mathcal{E}\left[\int_b^a X(t)\,\mathrm{d}t\right] = \int_b^a \mathcal{E}[X(t)]\,\mathrm{d}t \tag{2.94}$$

Here, the differentiation and integration should be understood in the sense of mean-square calculus.[2]

Since both the correlation and the variance are expectation operators, these rules are the basis of the differentiation and the integration with respect to them.

2.2.4.2 The Correlation Function

Suppose $\varphi(t)$ is a deterministic function and

$$Y(t) = \varphi(t) X(t) \tag{2.95}$$

then

$$R_Y(t_1, t_2) = \varphi(t_1)\varphi(t_2) R_X(t_1, t_2) \tag{2.96}$$

Noticing that the autocovariance function of a deterministic function (say $\nu(t)$) is zero, then if $X_i(t)$'s are zero-mean and

[2] The mean-square calculus is the calculus used most in stochastic analysis. One of the advantages is that operations in the mean-square calculus are almost the same as that of ordinary calculus. For details, refer to Gardiner (1983) for example.

$$Y(t) = \sum_{i=1}^{n} \varphi_i(t) X_i(t) + \nu(t) \tag{2.97}$$

using Equation 2.96, this will lead to

$$R_Y(t_1, t_2) = \sum_{i=1}^{n} \sum_{j=1}^{n} \varphi_i(t_1)\varphi_j(t_2) R_{X_i X_j}(t_1, t_2) + \nu(t_1)\nu(t_2) \tag{2.98}$$

If $X(t)$ and $Y(t)$ are mutually independent, and let

$$Z(t) = X(t)Y(t) \tag{2.99}$$

then

$$R_Z(t_1, t_2) = R_X(t_1, t_2) R_Y(t_1, t_2) \tag{2.100}$$

According to Equation 2.93, it can be deduced that the autocorrelation functions of the derivative process of $X(t)$ equal the partial derivatives of the autocorrelation functions of $X(t)$ with respect to t_1 and t_2; that is:

$$R_{X^{(n)}X^{(m)}}(t_1, t_2) = \frac{\partial^{n+m} R_X(t_1, t_2)}{\partial t_1^n \partial t_2^m} \tag{2.101a}$$

where $X^{(n)}$ denotes the nth derivative of $X(t)$.

In particular, for a stationary process we get

$$R_{X^{(n)}}(\tau) = (-1)^n \frac{\mathrm{d}^{2n} R_X(\tau)}{\mathrm{d}\tau^{2n}} \tag{2.101b}$$

where $R_{X^{(n)}}(\tau)$ denotes the autocorrelation function of $X^{(n)}$.

Similarly, consider a stochastic process:

$$Y(t) = \int_0^t X(t)\,\mathrm{d}t \tag{2.102}$$

and note Equation 2.94; then

$$R_{YY}(t_1, t_2) = \int_0^{t_1}\int_0^{t_2} R_{XX}(t_1, t_2)\,\mathrm{d}t_1\,\mathrm{d}t_2 = \int_0^{t_1}\int_0^{t_2} R_{\dot{Y}\dot{Y}}(t_1, t_2)\,\mathrm{d}t_1\,\mathrm{d}t_2 \tag{2.103}$$

where $\dot{Y}$ denotes the first derivative of Y.

2.2.4.3 The Power Spectral Density Function

Since the PSD function of a stationary process is the Fourier transform of its autocorrelation function, its operation rules can sometimes be deduced from those of the latter.[3] Therefore, we only list two widely used equations here.

[3] A more direct and physical treatment is given in Section 5.3.1.1.

(a) The derivative equation:

$$S_{X^{(n)}}(\omega) = \omega^{2n} S_X(\omega) \tag{2.104}$$

where $S_{X^{(n)}}(\omega)$ denotes the PSD function of the nth derivative of $X(t)$.
(b) Let $X(t)$ and $Y(t)$ be two stationary processes, and

$$Z(t) = X(t) + Y(t) \tag{2.105}$$

then

$$S_Z(\omega) = S_X(\omega) + S_Y(\omega) + S_{XY}(\omega) + S_{YX}(\omega) \tag{2.106}$$

2.2.5 Karhunen–Loève Decomposition

According to the above description, a stochastic process is understood as a family of random variables with respect to the time parameter. In another point of view, however, it can also be understood as a random combination of some deterministic time functions. The Karhunen–-Loève decomposition establishes the intrinsic relationship between these two views.

Denote the mean process of $X(t)$ by $X_0(t)$; then:

$$X(t) = X_0(t) + X_\sigma(t) \tag{2.107}$$

where $X_\sigma(t)$ is a stochastic process with zero mean. Noting that the covariance of a stochastic process remains invariant when a deterministic function is added, $X_\sigma(t)$ thus has the same covariance function as $X(t)$.

Assume the covariance function of $X_\sigma(t)$ is $K_X(t_1, t_2)$. As mentioned before, it is a bounded, symmetric and nonnegative function. If

$$\int_T K_X(t_1, t_2) f_n(t_1)\, \mathrm{d}t_1 = \lambda_n f_n(t_2) \tag{2.108}$$

has nonzero solutions, then λ_n ($n = 1, 2, \ldots$) are called the eigenvalues of $K_X(t_1, t_2)$, and $f_n(t)$ ($n = 1, 2, \ldots$) are the eigenfunctions corresponding to the eigenvalues. Note that $f_n(t)$ are orthogonal, as below:

$$\int_T f_n(t) f_m(t)\, \mathrm{d}t = \delta_{nm} = \begin{cases} 1 & \text{if } n = m \\ 0 & \text{otherwise} \end{cases} \tag{2.109}$$

where T is the interval of integration.

The above property makes $f_n(t)$ satisfy conditions for forming a group of orthogonal bases. With such a set of orthogonal functions, and using the generalized Fourier expansion, $K_X(t_1, t_2)$ can be expanded as

$$K_X(t_1, t_2) = \sum_{n=1}^{\infty} \lambda_n f_n(t_1) f_n(t_2) \tag{2.110}$$

where

$$\lambda_n = \frac{1}{f_n(t_2)} \int_T K_X(t_1, t_2) f_n(t_1)\, \mathrm{d}t_1 \tag{2.111}$$

Note that Equation 2.111 is just Equation 2.108. In fact, multiplying both sides of Equation 2.110 by $f_n(t_1)$, integrating them, and noting Equation 2.109, we can obtain the expression of λ_n. Meanwhile, this deduction process shows that t_1 and t_2 on both sides of Equation 2.108 can be exchanged with one another.

On the basis of the above knowledge, Karhunen and Loève both pointed out that the stochastic process $X_\sigma(t)$ could be represented by the linear combination of $f_n(t)$ and that the combination factors are a set of uncorrelated random variables (Loève, 1977); that is:

$$X_\sigma(t) = \sum_{n=1}^{\infty} \zeta_n \sqrt{\lambda_n} f_n(t) \tag{2.112}$$

where ζ_n $(n = 1, 2, \ldots)$ are mutually uncorrelated random variables and

$$\mathcal{E}[\zeta_k \zeta_\ell] = \delta_{k\ell} = \begin{cases} 1 & \text{for } k = \ell \\ 0 & \text{otherwise} \end{cases} \tag{2.113}$$

The expression in Equation 2.112 can be proved as follows.

From the definition of the covariance and Equation 2.107, there is

$$K_{X_\sigma}(t_1, t_2) = \mathcal{E}[X_\sigma(t_1) X_\sigma(t_2)] = K_X(t_1, t_2) \tag{2.114}$$

Substituting Equation 2.112 in it yields

$$K_X(t_1, t_2) = \sum_{n=1}^{\infty} \sum_{m=1}^{\infty} \mathcal{E}[\zeta_n \zeta_m] \sqrt{\lambda_n \lambda_m} f_n(t_1) f_m(t_2) \tag{2.115}$$

Multiplying both sides of the above equation by $f_k(t_2)$, integrating them over T and noting the orthogonal relationship shown in Equation 2.109, we have

$$\int_T K_X(t_1, t_2) f_k(t_2)\, \mathrm{d}t_2 = \sum_{n=1}^{\infty} \mathcal{E}[\zeta_n \zeta_k] \sqrt{\lambda_n \lambda_k} f_n(t_1) \tag{2.116}$$

From Equation 2.108, the right-hand side of the above equation becomes

$$\sum_{n=1}^{\infty} \mathcal{E}[\zeta_n \zeta_k] \sqrt{\lambda_n \lambda_k} f_n(t_1) = \lambda_k f_k(t_1) \tag{2.117}$$

Multiplying both sides by $f_\ell(t_1)$, integrating them over T and noting the orthogonal relationship shown in Equation 2.109, we are led to

$$\sum_{n=1}^{\infty} \mathcal{E}[\zeta_n \zeta_k] \sqrt{\lambda_n \lambda_k} \delta_{n\ell} = \lambda_k \int_T f_k(t_1) f_\ell(t_1)\, \mathrm{d}t_1 = \lambda_k \delta_{k\ell} \tag{2.118}$$

Noting the basic property of the Kronecker δ symbol, there is

$$\mathcal{E}[\zeta_\ell \zeta_k]\sqrt{\lambda_\ell \lambda_k} = \lambda_k \delta_{\ell k} = \begin{cases} \lambda_k & \text{if } k = \ell \\ 0 & \text{otherwise} \end{cases} \tag{2.119}$$

Hence:

$$\mathcal{E}[\zeta_\ell \zeta_k] = \begin{cases} 1 & \text{if } k = \ell \\ 0 & \text{otherwise} \end{cases}$$

This is nothing but Equation 2.113. Thus, Equation 2.112 is proved.
Substituting Equation 2.112 in Equation 2.107 yields

$$X(t) = X_0(t) + \sum_{n=1}^{\infty} \zeta_n \sqrt{\lambda_n} f_n(t) \tag{2.120}$$

Generally, the above equation is referred to as the *Karhunen–Loève decomposition* with respect to a stochastic process.

The Karhunen–Loève decomposition implies that a stochastic process can be expanded somehow into the random superposition of a set of deterministic functions $f_n(t)$. From the viewpoint of functional analysis, Equation 2.120 is the result that a stochastic process is respectively projected on the uncorrelated variables ζ_n in terms of those orthogonal functions. Therefore, the significance of the Karhunen–Loève decomposition is that it provides the possibility of studying stochastic processes through a set of independent random variables. It is this possibility that enables us to settle many practical problems with respect to stochastic processes in a simplified way.

2.3 Random Fields

2.3.1 Basic Concepts

Extending the concept of stochastic process to a field domain, we will reach the concept of random field. What is different between the two concepts, however, is that the indexing parameter is time variable t for the stochastic processes, but space variable $\mathbf{u} = \{u, v, w\}$[4] for the random fields. Therefore, a random field is a family of random variables defined over a field-parameter set in which any point $\mathbf{u}_i$ corresponds to a random variable. In fact, the parameter sets of random fields may contain time as well as space variables, while in practice we mostly take into account random fields with space variables as indexing parameters, and denote them by $\{B(\mathbf{u}); \mathbf{u} \in D \subset \mathbb{R}^n\}$. Here, D is the field domain of $B(\mathbf{u})$, and $\mathbb{R}^n$ is the n-dimensional Euclidean space. The space coordinate $\mathbf{u}$ can have one, two or three components, corresponding to which $B(\mathbf{u})$ is called a one-, two- or three-dimensional random field respectively.

For a random field, a family of finite-dimensional probability distribution functions can be used to describe the probabilistic structure. For example, the n-dimensional probability distribution of $B(\mathbf{u})$ is given by

$$F(x_1, \mathbf{u}_1; x_2, \mathbf{u}_2; \ldots; x_n, \mathbf{u}_n) = \Pr\{B(\mathbf{u}_1) < x_1, B(\mathbf{u}_2) < x_2, \ldots, B(\mathbf{u}_n) < x_n\} \tag{2.121}$$

[4] To avoid confusion because x is used as a sampled value of the random field at a given point, here we use $\mathbf{u} = (u, v, w)$ instead of (x, y, z) to denote the space coordinates.

The family of finite-dimensional probability distribution functions is nonnegative, symmetric and consistent, too. Thus, the lower order probability distributions can be deduced from the higher order ones.

The finite-dimensional PDFs are defined as the partial derivatives of their corresponding probability distribution functions. Taking a three-dimensional scalar random field as an example, there is

$$p(x_1, \mathbf{u}_1; x_2, \mathbf{u}_2; \ldots; x_n, \mathbf{u}_n) = \frac{\partial^n F(x_1, \mathbf{u}_1; x_2, \mathbf{u}_2; \ldots; x_n, \mathbf{u}_n)}{\partial x_1 \partial x_2 \ldots \partial x_n} \tag{2.122}$$

where $\mathbf{u}_i = (u_i, v_i, w_i)$.

Obviously, it seems unfeasible to describe the probabilistic structure of a given random field with the family of finite-dimensional probability distribution functions in practical applications. Therefore, the moment functions of a random field are of great value in applications.

Let $B(\mathbf{u})$ denote a random field. Its expectation is defined by

$$m(\mathbf{u}) = \mathcal{E}[B(\mathbf{u})] = \int_{-\infty}^{\infty} x p(x, \mathbf{u})\, \mathrm{d}x \tag{2.123}$$

Geometrically, $m(\mathbf{u})$ represents the average surface centers of sample functions of $B(\mathbf{u})$ in the field domain.

The autocorrelation function of $B(\mathbf{u})$ is defined by

$$R_B(\mathbf{u}_1, \mathbf{u}_2) = \mathcal{E}[B(\mathbf{u}_1)B(\mathbf{u}_2)] = \int_{-\infty}^{\infty}\int_{-\infty}^{\infty} x_1 x_2 p(x_1, \mathbf{u}_1; x_2, \mathbf{u}_2)\, \mathrm{d}x_1\, \mathrm{d}x_2 \tag{2.124}$$

where $\mathbf{u}_1$ and $\mathbf{u}_2$ are two points in the space. The autocorrelation function here has a similar interpretation as that of a stochastic process.

The auto-covariance function of $B(\mathbf{u})$ is defined by

$$\begin{aligned} K_B(\mathbf{u}_1, \mathbf{u}_2) &= \mathcal{E}\{[B(\mathbf{u}_1) - m(\mathbf{u}_1)][B(\mathbf{u}_2) - m(\mathbf{u}_2)]\} \\ &= \int_{-\infty}^{\infty}\int_{-\infty}^{\infty} [x_1 - m(\mathbf{u}_1)][x_2 - m(\mathbf{u}_2)] p(x_1, \mathbf{u}_1; x_2, \mathbf{u}_2)\, \mathrm{d}x_1\, \mathrm{d}x_2 \end{aligned} \tag{2.125}$$

Between the autocorrelation and the auto-covariance functions, there exists

$$K_B(\mathbf{u}_1, \mathbf{u}_2) = R_B(\mathbf{u}_1, \mathbf{u}_2) - m(\mathbf{u}_1)m(\mathbf{u}_2) \tag{2.126}$$

Note that

$$K_B(\mathbf{u}, \mathbf{u}) = \sigma_B^2(\mathbf{u}) \tag{2.127}$$

is the variance function of $B(\mathbf{u})$. Hence, the normalized covariance can be defined by

$$\rho_B(\mathbf{u}_1, \mathbf{u}_2) = \frac{K_B(\mathbf{u}_1, \mathbf{u}_2)}{\sigma_B(\mathbf{u}_1)\sigma_B(\mathbf{u}_2)} \tag{2.128}$$

Corresponding to the concept of stationarity for stochastic processes, we may conceive of the analogous one for random fields, called *homogeneity*. A random field is said to be strictly homogeneous on condition that its finite-dimensional probability distribution functions remain invariant under any translation of the space coordinates. A random field can be homogeneous along a line, on a certain plate or in the whole space. In practical applications, a random field is usually just required to have homogeneity of second order. A random field, if it satisfies

$$m(\mathbf{u}) = \text{constant} \tag{2.129}$$

$$R_B(\mathbf{u}_1, \mathbf{u}_2) = R_B(\mathbf{u}_1 - \mathbf{u}_2) = R_B(\mathbf{r}) \tag{2.130}$$

where $\mathbf{r} = \mathbf{u}_1 - \mathbf{u}_2$, is called a weakly homogeneous random field. In what follows in the book the homogeneous field means this type, unless specified otherwise.

'Isotropy' is another important concept for a random field. A random field is isotropic if its finite-dimensional probability distributions make no change when the group of points $\mathbf{u}_1, \mathbf{u}_2, \ldots, \mathbf{u}_n$ takes any possible rotation around the axis passing through the origin, or a mirror reflection in any plate including the origin. Generally speaking, the so-called isotropic random fields refer to the isotropic homogeneous ones. This means their probabilistic properties are invariant under all the translations, rotations and mirror reflections of $\mathbf{u}_1, \mathbf{u}_2, \ldots,$ $\mathbf{u}_n$. Obviously, there is some trouble in understanding the 'isotropy' in a visual way. However, if we loosen the constraint conditions of the finite-dimensional probability distributions, only thinking about the isotropy of second-order characteristics of a random field, then it becomes somewhat easier to understand. A random field, if satisfying Equation 2.129, and further

$$R_B(\mathbf{r}) = R_B(|\mathbf{r}|), \tag{2.131}$$

is called a weakly isotropic random field. The term 'isotropy' clearly indicates that, for this kind of random field, the probabilistic properties are relevant only to distance, not to direction.

2.3.2 *Correlation Structures of Random Fields*

A homogeneous random field $\{B(\mathbf{u}), \mathbf{u} \in D \subset \mathbb{R}^n\}$ can be written in the form

$$B(\mathbf{u}) = B_0(\mathbf{u}) + B_\sigma(\mathbf{u}) \tag{2.132}$$

where $B_0(\mathbf{u})$ is the mean function of $B(\mathbf{u})$ and $B_\sigma(\mathbf{u})$ is a random field with zero mean.

The covariance function of $B_\sigma(\mathbf{u})$ is equal to its correlation function. Therefore, studying the correlation structure of $B_\sigma(\mathbf{u})$ usually refers to its covariance or its correlation. For most practical problems, a correlation structure is generally an empirically hypothetical model. The frequently useful patterns (expressed via normalized covariance) include the:

(a) triangular pattern

$$\rho_B(\mathbf{u}_1, \mathbf{u}_2) = \begin{cases} 1 - \dfrac{|\mathbf{u}_1 - \mathbf{u}_2|}{a} & |\mathbf{u}_1 - \mathbf{u}_2| < a \\ 0 & \text{otherwise} \end{cases} \tag{2.133}$$

(b) exponential pattern

$$\rho_B(\mathbf{u}_1, \mathbf{u}_2) = \exp\left(-\frac{|\mathbf{u}_1 - \mathbf{u}_2|}{a}\right) \tag{2.134}$$

(c) Gaussian pattern

$$\rho_B(\mathbf{u}_1, \mathbf{u}_2) = \exp\left(-\frac{|\mathbf{u}_1 - \mathbf{u}_2|^2}{a^2}\right) \tag{2.135}$$

where the constant a is called the correlation scale parameter.

For a certain physical problem, whether or not there is a widely used hypothesis about the correlation structures may serve as a mark of how developed studies of this field are.

2.3.3 *Discretization of Random Fields*

For a continuous random field, we can transform it into a set of random variables by dividing the space domain over which it is defined. This partition resembles that in the finite-element method (FEM) in some respects. The main discretization methods of random fields include the midpoint method, the shape-function method and the local average method (Vanmarcke, 1983). A two-dimensional random field will now be used to illustrate the basic concepts of these three methods.

Without loss of generality, Figure 2.5 shows a two-dimensional random field over the domain $D(x, y)$ and the form of its discretization. For a certain element, D_i denotes the area, $\mathbf{u}_{ci}$ $(i = 1, 2, \ldots, n)$ denotes the geometrical centroid, $\mathbf{u}_j$ $(j = 1, 2, \ldots, m)$ is the nodal position, and n and m are the number of elements and nodes respectively.

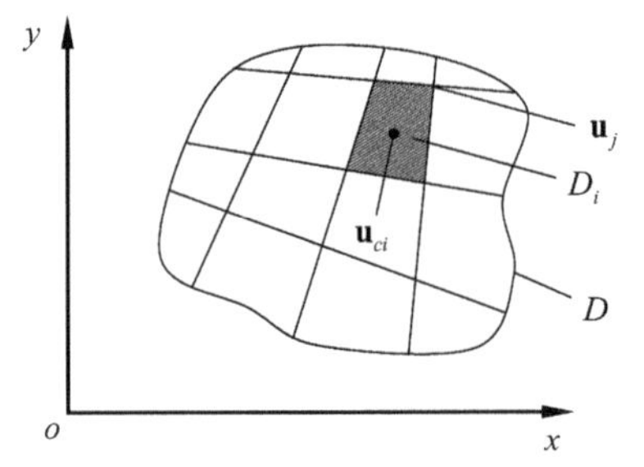

Figure 2.5 Discretization of a random field.

2.3.3.1 The Midpoint Method

The midpoint method is to substitute a set of random variables $B(\mathbf{u}_{ci})$ at geometrical centroid $\mathbf{u}_{ci}$ for the element random field $\{B(\mathbf{u}), \mathbf{u} \in D_i\}$, namely

$$B(\mathbf{u}) = B(\mathbf{u}_{ci}) \qquad \text{for } \mathbf{u} \in D_i \tag{2.136}$$

By this principle, the original random field is discretized into a random variable set $\{\xi_i = B(\mathbf{u}_{ci}), i = 1, 2, \ldots, n\}$. Then, the expectation of each variable and the correlation between variables depend on the corresponding values of ξ_i, the random variables at each element's geometrical centroid. For example, there are

$$\mathcal{E}[\xi_i] = \mathcal{E}[B(\mathbf{u}_{ci})] \tag{2.137}$$

$$\sigma^2[\xi_i] = \sigma^2[B(\mathbf{u}_{ci})] \tag{2.138}$$

$$c_{ij} = \mathrm{cov}[\xi_i, \xi_j] = \mathrm{cov}[B(\mathbf{u}_{ci}), B(\mathbf{u}_{cj})] \tag{2.139}$$

Certainly, only when the partitioned elements are very small or the original field is of little variability will the midpoint method obtain good accuracy.

2.3.3.2 The Shape Function Method

To improve the accuracy, a reasonable method is to replace the original random field with a set of random variables at nodes and then to approximate the random field inside any element by interpolation of shape functions. In other words, an element random field can be obtained from random variables at nodes by the interpolation of shape functions; that is:

$$B(\mathbf{u}) = \sum_{j=1}^{q} N_j(\mathbf{u}) B(\mathbf{u}_j) \qquad \mathbf{u} \in D_i \tag{2.140}$$

where q is the number of nodes of the given element and $N_j(\mathbf{u})$ are the shape functions. In general, the interpolation function can take the form of polynomials (Lawrence, 1987).

In this sense, the original random field is discretized into a random variable set $\{\zeta_j = B(\mathbf{u}_j), j = 1, 2, \ldots, m\}$. Then, the descriptive properties of every random variable and the correlation between random variables depend on the corresponding values of the random variables at nodes. There are expressions analogous to Equations 2.137–2.139.

For an element random field and the corresponding variables at nodes, the moment functions can be given by

$$\mathcal{E}[B(\mathbf{u})] = \sum_{j=1}^{q} N_j(\mathbf{u}) \mathcal{E}[B(\mathbf{u}_j)] \qquad \mathbf{u} \in D_i \tag{2.141}$$

$$\sigma^2[B(\mathbf{u})] = \sum_{k=1}^{q} \sum_{\ell=1}^{q} N_k(\mathbf{u}) N_\ell(\mathbf{u}) \mathrm{cov}[B(\mathbf{u}_k), B(\mathbf{u}_\ell)] \qquad \mathbf{u} \in D_i \tag{2.142}$$

If the shape functions are obtained appropriately, then the shape function method is far more accurate than the midpoint method. In the perturbed FEM, however, this method is not as concise as the local average method discussed below in generation of the stiffness matrix.

2.3.3.3 The Local Average Method

In the local average method, any element random field is represented by its local average random variable; that is:

$$\xi_i = \frac{1}{D_i}\int_{D_i} B(\mathbf{u})\,\mathrm{d}\mathbf{u} \qquad \mathbf{u} \in D_i \tag{2.143}$$

Consequently, the original random field is discretized into a random variable set $\{\xi_i, i = 1, 2, \ldots, n\}$, of which the mean is defined by

$$\mathcal{E}[\xi_i] = \frac{1}{D_i}\int_{D_i} \mathcal{E}[B(\mathbf{u})]\,\mathrm{d}\mathbf{u} \qquad \mathbf{u} \in D_i \tag{2.144}$$

while the variance is given by

$$\begin{aligned} \sigma^2[\xi_i] &= \frac{1}{D_i^2}\left(\int_{D_i} \{B(\mathbf{u}) - \mathcal{E}[B(\mathbf{u})]\}\,\mathrm{d}\mathbf{u}\right)^2 \\ &= \frac{1}{D_i^2}\int_{D_i}\int_{D_i} \operatorname{cov}[B(\mathbf{u}_1), B(\mathbf{u}_2)]\,\mathrm{d}\mathbf{u}_1\,\mathrm{d}\mathbf{u}_2 \end{aligned} \tag{2.145}$$

and the covariance between ξ_i and ξ_j is

$$c_{ij} = \operatorname{cov}[\xi_i, \xi_j] = \frac{1}{D_i D_j}\int_{D_i}\int_{D_j} \operatorname{cov}[B(\mathbf{u}_1), B(\mathbf{u}_2)]\,\mathrm{d}\mathbf{u}_1\,\mathrm{d}\mathbf{u}_2 \qquad \mathbf{u}_1 \in D_i,\ \mathbf{u}_2 \in D_j \tag{2.146}$$

Utilizing the definition of correlation coefficient (see Equation 2.128), Equation 2.146 becomes

$$c_{ij} = \frac{1}{D_i D_j}\int_{D_i}\int_{D_j} \sigma(\mathbf{u}_1)\sigma(\mathbf{u}_2)\rho(\mathbf{u}_1, \mathbf{u}_2)\,\mathrm{d}\mathbf{u}_1\,\mathrm{d}\mathbf{u}_2 \qquad \mathbf{u}_1 \in D_i,\ \mathbf{u}_2 \in D_j \tag{2.147}$$

The local average method is between the midpoint method and the shape function method in accuracy. Nevertheless, it is still broadly accepted because of its facilitating the use of general formulae of the perturbed FEM.

For the discretization of random fields, there are also some other methods, such as the weighted integral method, the optimal discretization method and so on. Readers who are interested in them can refer to the relevant literature (Takada, 1990; Li and Der Kiureghian, 1993).

2.3.4 Decomposition of Random Fields

The discretization of a random field, as noted, only finishes the transition from random field to a set of discrete random variables. However, every two of those variables may be correlated, which sometimes brings inconvenience and even difficulty in applied problems. Then, the question arises as to whether it is possible to find a set of uncorrelated random variables to replace a random field.

The answer is 'yes' for most problems. Two approaches are discussed in detail hereafter.

2.3.4.1 Karhunen–Loève Decomposition

Consider a homogeneous random field $\{B(\mathbf{u}),\ \mathbf{u} \in D \subset \mathbb{R}^n\}$. Quoting Equation 2.132, repeated here for convenience:

$$B(\mathbf{u}) = B_0(\mathbf{u}) + B_\sigma(\mathbf{u}) \tag{2.132}$$

let $K_B(\mathbf{u}_1, \mathbf{u}_2)$ denote the covariance function of $B_\sigma(\mathbf{u})$. If

$$\int_D K_B(\mathbf{u}_1, \mathbf{u}_2) f_n(\mathbf{u}_1)\, d\mathbf{u}_1 = \lambda_n f_n(\mathbf{u}_2) \tag{2.148}$$

is solvable, then according to the method of Karhunen–Loève decomposition we have

$$B(\mathbf{u}) = B_0(\mathbf{u}) + \sum_{n=0}^{\infty} \zeta_n \sqrt{\lambda_n} f_n(\mathbf{u}) \tag{2.149}$$

where ζ_n $(n = 1, 2, \ldots)$ are mutually uncorrelated random variables, satisfying

$$\mathcal{E}[\zeta_k \zeta_\ell] = \delta_{k\ell} \tag{2.150}$$

where δ is the Kronecker delta (see Appendix A).

As noted, the method of Karhunen–Loève decomposition is not counted as a method for the discretization of a random field, but one of decomposition of a random field. The distinction between 'discretization' and 'decomposition' is that the former refers to the geometrical partition of a field domain while the latter means decomposition with respect to the subspaces of a probability space.

2.3.4.2 Decomposition of Discretized Random Fields

Upon using Karhunen Loève decomposition, we need to solve the integral equation in Equation 2.148, which is not so easy in most cases. In contrast, a two-step transition method can be used without such mathematical difficulty.

In the two-step transition method, a random field is first discretized by methods described in the preceding subsection and represented by $\boldsymbol{\xi} = (\xi_1, \xi_2, \ldots, \xi_n)$, a set of correlated random variables. It is obvious that $\boldsymbol{\xi}$ can be written as

$$\boldsymbol{\xi} = \boldsymbol{\xi}_0 + \boldsymbol{\xi}_\sigma \tag{2.151}$$

where $\boldsymbol{\xi}_0$ is the mean of $\boldsymbol{\xi}$, and $\boldsymbol{\xi}_\sigma$ is a random vector with zero mean, having the same covariance matrix as $\boldsymbol{\xi}$.

Then, by decomposition of the correlation matrix in Section 2.1.4, $\boldsymbol{\xi}$ can be expressed via a sequence of normalized uncorrelated random variables:

$$\boldsymbol{\xi} = \boldsymbol{\xi}_0 + \sum_{i=1}^{n} \boldsymbol{\psi}_i \sqrt{\lambda_i} \zeta_i \tag{2.152}$$

where $\{\zeta_i, i = 1, 2, \ldots, n\}$ is the sequence of normalized uncorrelated random variables, satisfying

$$\mathcal{E}[\zeta_i] = 0 \tag{2.153}$$

$$\mathrm{Var}[\zeta_i] = 1 \tag{2.154}$$

$$\mathrm{cov}[\zeta_i, \zeta_j] = \delta_{ij} \tag{2.155}$$

where λ_i and $\boldsymbol{\psi}_i$ are respectively the eigenvalues and eigenvectors of the covariance matrix of $\boldsymbol{\xi}$, obtained from the following eigenvalue equation:

$$\mathbf{C}_{\boldsymbol{\xi}}\boldsymbol{\psi}_i = \lambda_i \boldsymbol{\psi}_i \tag{2.156}$$

2.4 Orthogonal Decomposition of Random Functions

2.4.1 Metric Spaces and Normed Linear Spaces

In mathematics, a space means a kind of set possessing certain structures. For example, a real straight-line l forms a one-dimensional space, and properties such as 'there exists distance between any two points' are structures in this space. In a given space, according to functional analysis theory, each point represents a function generally. Thus, the content of functional analysis mainly covers basic properties of spaces of continuous functions as well as the relations among the point sets that belong to these spaces.

The most basic concept in the investigation of functional spaces is the 'metric' between points. Consider a set denoted by $\mathcal{X}$. If for any two points x and y in $\mathcal{X}$, there always exists one deterministic real number $d(x, y)$, such that

(a) $d(x, y) \geq 0$, $d(x, y) = 0$ if and only if $x = y$; and
(b) $d(x, y) \leq d(x, z) + d(y, z)$ is true for any z.

Then, we call $d(x, y)$ the metric between x and y.

A space, if given a metric between any two of its points, is called a metric space, and denoted by $\mathcal{X} = (\mathcal{X}, d)$. The Euclidean space $\mathbb{R}^n$ and the space of continuous functions $C[a, b]$ are typical examples of metric spaces.

For a Euclidean space $\mathbb{R}^n$, the metric between two points

$$\mathbf{x} = (x_1, x_2, \ldots, x_n) \quad \text{and} \quad \mathbf{y} = (y_1, y_2, \ldots, y_n) \tag{2.157}$$

is given by

$$d(\mathbf{x}, \mathbf{y}) = \sqrt{\sum_{i=1}^{n} (x_i - y_i)^2} \tag{2.158}$$

For the union of continuous functions over a closed interval $[a, b]$, there is

$$d(x, y) = \max_{a \leq t \leq b} |x(t) - y(t)| \tag{2.159}$$

Assume $\{x_n\}_{n=1}^{\infty}$ is a sequence of points in the metric space $\mathcal{X}$. If, for any arbitrary number $\varepsilon > 0$, there is always a natural number $N = N(\varepsilon)$, such that as $n, m > N$, there exists

$$d(x_n, x_m) < \varepsilon \tag{2.160}$$

then we call $\{x_n\}_{n=1}^{\infty}$ the Cauchy point sequence or basic point sequence in $\mathcal{X}$.

A metric space is said to be complete if in it every Cauchy point sequence is convergent.

In functional analysis theory, the normed linear space is of great value and utility. In the normed linear spaces, elements can be added up or multiplied by scalars, and between two elements there is a metric. In addition, like a general vector, every element in the space is assigned a scalar of length that is called the norm.

Let $\mathcal{X}$ be a real (or complex) linear space. If for any $x \in \mathcal{X}$ there is a deterministic real number, denoted by $\|x\|$, such that

(a) $\|x\| \geq 0$, and $\|x\| = 0$ is equivalent to $x = 0$; and
(b) $\|ax\| = |a| \cdot \|x\|$, where a is any arbitrary real (complex) number; and
(c) $\|x + y\| \leq \|x\| + \|y\|$, for any $x, y \in \mathcal{X}$

then $\|x\|$ is called the norm of x, and $\mathcal{X}$ is thus a normed linear space with norm $\|x\|$.

The normed linear space may be illustrated with Euclidean space $\mathbb{R}^n$ and the space of continuous functions $C[a, b]$.

For any arbitrary vector $\mathbf{x} = (x_1, x_2, \ldots, x_n) \in \mathbb{R}^n$, we can define

$$\|\mathbf{x}\| = \sqrt{x_1^2 + x_2^2 + \ldots + x_n^2} \tag{2.161}$$

and for any arbitrary function $f(t) \in C[a, b]$, we can define

$$\|f\| = \max_{a \leq t \leq b} |f(t)| \tag{2.162}$$

Comparing Equations 2.158 and 2.159 with Equations 2.161 and 2.162, we find that for a normed linear space $\mathcal{X}$ the metric $d(x, y)$ can be determined from the norm by setting

$$d(x, y) = \|x - y\| \qquad (x, y \in \mathcal{X}) \tag{2.163}$$

Equations 2.162 and 2.163 clearly show us the difference between the 'norm' and 'metric': the norm is defined for a single element, whereas the metric is given to associate any two elements.

2.4.2 Hilbert Spaces and General Orthogonal Decomposition

The metric and the norm, defined in the normed linear spaces, make it possible to think over such properties as continuity and convergence for functional spaces (spaces of continuous functions). However, compared with ordinary spaces of finite-dimensional vectors, normed linear spaces still lack a geometrical property analogous to 'angle.' A functional space in which an angle is defined is called an inner product space.

Consider $\mathcal{X}$ a complex linear space. If for any two elements[5] x and y in $\mathcal{X}$ there exists a complex number $\langle x, y \rangle$, such that

[5] In linear spaces, element is equivalent to vector in concept. Therefore, they are not distinguished in this book.

(a) $\langle x, x\rangle \geq 0$ and $\langle x, x\rangle = 0 \Leftrightarrow x = 0$; and
(b) $\langle \alpha x + \beta y, z\rangle = \alpha\langle x, z\rangle + \beta\langle y, z\rangle$, for any arbitrary $z \in \mathcal{X}$, where α and β are complex numbers; and
(c) $\langle x, y\rangle = \langle y, x\rangle^*$, where $\langle\cdot\rangle^*$ is the complex conjugate,

then $\langle x, y\rangle$ is called the inner product of x, y.

For an inner product space $\mathcal{X}$, if

$$||x|| = \sqrt{\langle x, x\rangle} \tag{2.164}$$

then $||x||$ is a norm on $\mathcal{X}$. Notice that $||x||$ here is determined by the inner product. Thus, the inner product spaces are a type of special normed linear space. In other words, any inner product space, with the norm given by Equation 2.164, is a normed space.

An inner product space, if it is complete as a normed linear space, is called a *Hilbert space*.

In a Hilbert space, the inner product of two functions $f(t)$ and $g(t)$, $a \leq t \leq b$, is often defined by

$$\langle f, g\rangle = \int_a^b f(t)g(t)\,\mathrm{d}t \tag{2.165}$$

Note that the above integration is in the sense of the Lebesgue integral.

From Equation 2.165, it follows that the norm

$$||f|| = \sqrt{\int_a^b f^2(t)\,\mathrm{d}t} \tag{2.166}$$

This definition shows that the Hilbert spaces with inner products defined by Equation 2.165 is a type of special Hilbert space associated with the union of square integrable functions $L^2[a, b]$.

The metric, determined by the above norm, is in the form

$$d(f, g) = \sqrt{\int_a^b [f(t) - g(t)]^2\,\mathrm{d}t} \tag{2.167}$$

For inner product spaces, the Schwarz inequality

$$|\langle f, g\rangle| \leq ||f|| \cdot ||g|| \tag{2.168}$$

is always true.

From this inequality, in a Hilbert space there certainly exists

$$\frac{\left|\int_a^b f(t)g(t)\,\mathrm{d}t\right|}{\sqrt{\int_a^b f^2(t)\,\mathrm{d}t}\sqrt{\int_a^b g^2(t)\,\mathrm{d}t}} \leq 1 \tag{2.169}$$

Thus, the left-hand side of the above can be treated as the cosine of an angle ϑ; that is:

$$\cos\vartheta = \frac{\int_a^b f(t)g(t)\,\mathrm{d}t}{\sqrt{\int_a^b f^2(t)\,\mathrm{d}t}\sqrt{\int_a^b g^2(t)\,\mathrm{d}t}} = \frac{\langle f, g\rangle}{||f||\cdot||g||} \tag{2.170}$$

In a Hilbert space, ϑ is called the angle between $f(t)$ and $g(t)$. If $\langle f, g\rangle = 0$, then we get $\cos\vartheta = 0$, or $\vartheta = 90°$ by using Equation 2.170. Thus, f and g are considered perpendicular or orthogonal. Meanwhile:

$$\int_a^b f(t)g(t)\,\mathrm{d}t = 0 \tag{2.171}$$

Assume A and B are two subsets in the space $\mathcal{X}$. Then A and B are orthogonal if any vector in A is orthogonal to any one in B. With the concept of orthogonality, we are able to advance the concept of orthogonal decomposition in the inner product spaces.

Suppose Φ denotes a subset which includes all the nonzero points in a Hilbert space; that is:

$$\Phi = \{\varphi_1(t), \varphi_2(t), \ldots, \varphi_n(t), \ldots\} \tag{2.172}$$

If in Φ any two functions are orthogonal to each other, namely

$$\int_a^b \varphi_i(t)\varphi_j(t)\,\mathrm{d}t = 0 \qquad (i \neq j) \tag{2.173}$$

then the set Φ is called a system of orthogonal functions. Meanwhile, if the norm of every element equals unity; that is:

$$\int_a^b \varphi_k^2(t)\,\mathrm{d}t = 1 \qquad (k = 1, 2, \ldots, n, \ldots) \tag{2.174}$$

then Φ is a system of standard orthogonal functions and $\{\varphi_k(t), k = 1, 2, \ldots\}$ are orthogonal basis functions. A system of orthogonal functions is complete if there is no possibility of adding to this system a nonzero function which is orthogonal to all its functions.

The aim of introducing the system of standard orthogonal functions in an inner product space is to expand any function in the space into the series in terms of these orthogonal functions. For a nonzero Hilbert space $\mathcal{X}$ there certainly exists a complete system of standard orthogonal functions. Suppose this system is denoted by

$$\varphi_1(t), \varphi_2(t), \ldots, \varphi_n(t), \ldots$$

then any function $f(t)$ in $\mathcal{X}$ can be decomposed into a convergent series (or generalized Fourier series) in the form

$$f(t) = \lim_{i\to\infty} \sum_{i=1}^{n} a_i\varphi_i(t) = \sum_{i=1}^{\infty} a_i\varphi_i(t) \tag{2.175}$$

where the coefficients a_i are equal to the projections of $x(t)$ on $\varphi_i(t)$ and are called the Fourier coefficients of $f(t)$ associated with φ_i; that is:

$$a_i = \langle f, \varphi_i \rangle = \int_a^b f(t)\varphi_i(t)\,\mathrm{d}t \tag{2.176}$$

Equation 2.175 is an orthogonal decomposition of $f(t)$ with respect to the system of standard orthogonal functions.

It is easy to verify that

$$||f||^2 = \int_a^b f^2(t)\,\mathrm{d}t = \sum_{i=1}^{\infty} a_i^2 \tag{2.177}$$

Note that a_i are projections of $f(t)$ on $\varphi_i(t)$. Thus, we can use the partial sum of the above series

$$\tilde{f}(t) = \sum_{i=1}^{n} a_i\varphi_i(t) \tag{2.178}$$

to approximate the function $f(t)$. The error of such an approximation depends on the sum of squares of projections of $f(t)$ on the complementary set $\{\varphi_{n+1}(t), \varphi_{n+2}(t), \ldots\}$. Actually, there is an error function in the form

$$\varepsilon = ||f - \tilde{f}||^2 = \sum_{i=n+1}^{\infty} a_i^2 \tag{2.179}$$

2.4.3 *Orthogonal Decomposition of Random Functions*

When extending the functional space concepts to the probability spaces, we can deal with problems of orthogonal decomposition with respect to spaces of random functions. In these spaces, a random function can be denoted by $X(\xi, t)$, where $\xi \in \Omega$ and $t \in T$. In other words, in the space of random functions, every point in it is a function of the given random variables. Therefore, using the variables separation method for reference, there are two ways of orthogonal decomposition with respect to a random function. The first way is similar to Equation 2.175. The only difference is that, in a space of random functions, the expanding coefficients a_i should be viewed as random variables. We will discuss these situations in detail in Section 3.5. On the other hand, if we choose a system of standard orthogonal functions with respect to random variables as basic functions, then the second orthogonal decomposition method can be derived.

Consider $\mathscr{H}$, as an example, a space of random functions of random variables having standard normal distributions. Note that the PDF of the standard normal variables is given by

$$p_\xi(u) = \frac{1}{\sqrt{2\pi}} e^{-u^2/2} \tag{2.180}$$

Here, we use lower case u to represent a sample value of ξ without risk of confusion. We can define

$$H_n(\xi) = \frac{H_{e_n}(\xi)}{\sqrt{n!}} \tag{2.181}$$

where $H_{e_n}(\xi)$ are Hermite polynomials (see Appendix B). It is easy to prove that

$$\mathcal{E}[H_n(\xi)H_m(\xi)] = \int_{-\infty}^{\infty} p_\xi(u)H_n(u)H_m(u)\,\mathrm{d}u = \begin{cases} 1 & \text{if } n = m \\ 0 & \text{otherwise} \end{cases} \quad n = 0, 1, 2, \ldots \tag{2.182}$$

Thus, $H_n(\xi)$ form a system of standard orthogonal functions on $\mathscr{H}$.
The inner product of $\mathscr{H}$ is written as

$$\langle f, g\rangle = \mathcal{E}[f(\xi)g(\xi)] = \int_{-\infty}^{\infty} p_\xi(u)f(u)g(u)\,\mathrm{d}u \tag{2.183}$$

Meanwhile, we can obtain the corresponding norm, and then the metric. Finally, we find that $\mathscr{H}$ is a Hilbert space. Thus, any random function $X(\xi, t)$ in $\mathscr{H}$ can be expanded in terms of $H_n(\xi)$; that is:

$$X(\xi, t) = \sum_{i=1}^{\infty} a_i(t)H_i(\xi) \tag{2.184}$$

where the coefficients

$$a_i(t) = \langle X, H_i\rangle = \mathcal{E}[X(\xi, t)H_i(\xi)] = \int_{-\infty}^{\infty} p_\xi(u)X(u, t)H_i(u)\,\mathrm{d}u \tag{2.185}$$

are considered as the projections of $X(\xi, t)$ on the basis functions $H_n(\xi)$.

The expansion in the form of Equation 2.184 is called the second kind of orthogonal decomposition of a random function.

3

Stochastic Models of Dynamic Excitations

3.1 General Expression of Stochastic Excitations

3.1.1 Dynamic Excitations and Modeling

Most dynamic actions applied to civil engineering structures exhibit an obvious character of randomness; therefore, they are termed stochastic excitations. Typical stochastic excitations include the effects of seismic ground motions, wind gusts or turbulence in wind and ocean waves and so on. In general, these dynamic actions vary in spatial location and time. Therefore, to reflect them, the spatial random field model should be adopted as the basic model. However, owing to difficulties in observation and modeling, some simplifications must be introduced in the modeling process.

One of the most common simplifications is to neglect the variation of the dynamic actions in space and simplify the spatial random fields as a series of random processes with the same statistical characteristics. That is, use the time sequence at one point to reflect the effect of the random excitation on the structure and assume the excitation at various points to be identical within an acceptable range. A typical example is the seismic ground motion input to the bottom of building structures with a relatively small planar size. Considering the building base within the scope of the foundation as a rigid plate, the input of seismic ground motions can be reflected by using the stochastic process model at one point.

If the difference of the dynamic actions applied to different points of the structure cannot be ignored, the random field model should be adopted to reflect the spatial dynamic actions. In these cases, the homogeneous and isotropic assumption may be used to simplify the model. In the homogeneous assumption, the difference among the points in the random fields is only related to the distance between two points and has nothing to do with the location, whereas the probability distribution of the random fields is independent of the orientation according to the isotropic assumption. Owing to the lack of adequate actual measurement tests, it is usually believed that the homogeneity and isotropy are theoretical assumptions introduced to facilitate the modeling and analysis.

Stochastic Dynamics of Structures Jie Li and Jianbing Chen

The third assumption commonly used in the establishment of stochastic excitation models is the stationary assumption, which means the homogeneity of dynamic actions in the time scale. If a process meets the stationary assumption, then its variance is a constant and not changing with time. For example, it is usually assumed that wind turbulence in the atmospheric boundary layer has the characteristics of stationary stochastic processes.

In fact, when considering stochastic excitations acting on structures, one problem concerns whether the energy exchange between the structure and the external environment is taken into account. When this energy exchange is so small as to be ignored, the dynamic action on the structures may be determined only with the external environmental conditions. If this energy exchange is sufficient to impose a significant impact on the input dynamic action, then the interaction between the structure and the external media should be considered to determine the excitations affected by the structure. For example, the determination of seismic ground motions of large-scale engineering structures will encounter such a problem.

There are two basic methods for modeling stochastic dynamic excitations: phenomenology-based modeling and physics-based modeling. Owing to the difficulties in establishing finite-dimensional probability distributions, the correlation function or the PSD function is used commonly in phenomenology-based modeling. In essence, the method is based on statistical moments. In contrast, physics-based modeling focuses on giving a random function model of dynamic excitations considering the real physical background. Not only can these models give a complete mathematical description for the stochastic processes or random fields, but they also enable experimental verification of the random processes or random fields possible because of their physical significance.

This chapter will outline the mathematical models and physical models for the common types of stochastic dynamic excitations in structural analysis and design. In addition, the relevant mathematical models are introduced for modeling stochastic excitations.

3.1.2 Models of Stationary and Nonstationary Processes

If a stochastic excitation is a stationary process, particularly if it can be regarded as a Gaussian process, then, as long as the mean and the correlation function or the PSD function are known, the statistical characteristics of the excitation model can be totally determined. According to the description in Chapter 2, the mean of a stationary process $X(t)$ is a constant and the correlation function is only a function of time interval $\tau = t_2 - t_1$; that is:

$$m_X(t) = c \tag{3.1}$$

$$R_X(\tau) = R_X(t_2 - t_1) \tag{3.2}$$

If $c = 0$, then the correlation function and the PSD function of $X(t)$ have the following relationship (see Equations 2.81a and 2.81b):

$$S_X(\omega) = \int_{-\infty}^{+\infty} R_X(\tau) e^{-i\omega\tau} \, \mathrm{d}\tau \tag{3.3}$$

$$R_X(\tau) = \frac{1}{2\pi} \int_{-\infty}^{+\infty} S_X(\omega) e^{i\omega\tau} \, \mathrm{d}\omega \tag{3.4}$$

This indicates that $S_X(\omega)$ and $R_X(\tau)$ consist of a Fourier transform pair. If $c \neq 0$, then there is the same relationship as above between the covariance function $K_X(\tau)$ and the PSD function $S_X(\omega)$.

To determine a specific stationary excitation model, the ergodicity assumption is generally introduced. This indicates that, for an ergodic process, various random behaviors could appear in a long enough time-series. Therefore, the ensemble average can be replaced by the time average. For an ergodic process, the estimated values of the mean and the correlation function can be obtained using the measured samples as

$$\hat{m}_X = \frac{1}{N}\sum_{i=1}^{N} X_i \tag{3.5}$$

$$\hat{R}_X(\tau_k = k) = \frac{1}{N-k}\sum_{i=1}^{N-k} X_i X_{i+k} \tag{3.6}$$

where $X_i = X(t_i)$ is the amplitude of the sampled process at the time instant t_i, k is the lag number, N is the total number of the sampled points and the hat symbol represents the estimated value. The PSD function can be obtained from the discrete Fourier transform of $R_X(\tau_k)$. However, such spectral estimation results are usually biased. In order to obtain unbiased results, the maximum entropy spectral method can be adopted (Burg, 1967).

If a number of measured samples are available, then the sample set could be used to achieve the estimation of the PSD. In this case, the PSD is given as (see. Equation C.6, Appendix C)

$$\hat{S}_X(\omega) = \frac{1}{M}\sum_{i=1}^{M}\left[\frac{1}{T}|X_i(\omega)|^2\right] \tag{3.7}$$

where M is the sample number of the sample set, T is the observation time duration and $X_i(\omega)$ is the Fourier spectrum of the time history sample.

A stationary process is the result that is scientifically abstracted from the actual dynamic action. The majority of actual random excitations do not completely possess stationary characteristics. If there is a significant difference between the stationary assumption and the actual background, then it is necessary to adopt a nonstationary model to establish the model of random excitations. The modeling of seismic ground motions is a representative example.

For nonstationary stochastic processes, the ergodicity assumption is no longer valid. Therefore, the ensemble average cannot be replaced by the time average. In other words, the model should be established based on sample sets. In practice, a type of uniformly modulated nonstationary random process model is usually employed to establish a nonstationary random excitation model. This model can be expressed as a product of a deterministic time function $f(t)$ and a stationary process $X_s(t)$, namely

$$X(t) = f(t)X_s(t) \tag{3.8}$$

Here, we suppose the mean of $X(t) = 0$. For those processes $X'(t)$ with a nonzero mean, we can always let $X(t) = X'(t) - m_{X'}(t)$ to construct a zero-mean process.

When the correlation function or the PSD function of $X_s(t)$ is given, then it is easy to calculate the correlation function or the PSD function of the nonstationary random excitation $X(t)$; that is (see Equation 2.96):

$$R_X(t_1, t_2) = f(t_1)f(t_2)R_{X_s}(\tau) \tag{3.9}$$

$$S_X(t, \omega) = f^2(t)S_{X_s}(\omega) \tag{3.10}$$ [1]

In order to get the model of the nonstationary random excitations, the first requirement is to isolate $f(t)$ from the sample functions. For a time history sample $x_i(t)$, this can be done by a variety of approaches. For example, we can first define

$$y_i(t) = |x_i(t)| \tag{3.11}$$

and then use the technique employed in empirical mode decomposition to determine the upper and lower envelope curves in the sifting process (Huang *et al.*, 1998) to specify the upper envelope $f_i(t)$ of $y_i(t)$.

For a sample set, the specific expression of $f(t)$ can be determined by assuming the function form of $f(t)$ and using the least-squares method to fit $f(t)$ through $f_i(t)$ $(i = 1, 2, \ldots, M)$.

Once $f(t)$ is determined, it is easy to derive the sample set of the stationary process $X_s(t)$ according to Equation 3.8 and then the aforementioned modeling methods on stationary random processes can be used to complete the modeling of nonstationary random excitations.

3.1.3 Random Fourier Spectrum Model

It is very difficult to model nonstationary random excitations based on the previous classical correlation function or PSD method. In fact, only for random excitations with Gaussian normal nature can all the statistical features of the random excitations be obtained by the previous modeling methods. Unfortunately, for most practical engineering cases, the Gaussian nature cannot be fully verified.

The random Fourier spectrum model tries to build a dynamic excitation model by starting from the point of view of a random function and combining the understanding of the physical mechanism of the random excitations. For the time history sample assemble $X(t)$, the random Fourier spectrum is defined by (Li, 2006) (also see Appendix C)

$$X(\eta, \omega) = \frac{1}{\sqrt{T}} \int_0^T X(\eta, t)e^{-i\omega t}\, dt \tag{3.12}$$

where η is a random variable or random vector which affects the stochastic development process and has physical significance.

Obviously, the random Fourier spectrum is not confined just to stationary processes, but is also suitable for modeling general excitations. Using probability density evolution theory, to be

---

[1] Equation 3.10 is an evolutionary spectral density, of which the physical sense will be elaborated in Section 5.3.2.

stated in Chapter 6, the finite-dimensional probability distribution of the stochastic excitation process and its evolution with time can be observed. Thereby, the probability characteristics of the excitation can be completely determined.

When η in $X(\eta, \omega)$ is given a specific realization value z, the definite sample function $X(z, \omega)$ will be given, which provides the possibility of using the observed sample sets to complete the modeling. There are two basic modeling channels: sample-based modeling and modeling based on the statistical moments of the sample assemble.

3.1.3.1 Sample-Based Modeling

For samples in a sample assemble, the realized values of the basic random variables can be identified by using the best uniform approximation or the mean-square approximation on the basis of the observed sample values.

The best uniform approximation takes the following formula as the basic criterion to identify the parameter η:

$$J_1 = \max[\tilde{x}(\omega) - X(\eta, \omega)] \leq \varepsilon_1 \tag{3.13}$$

where x is the observed value of samples, ε_1 is the specified error bound.

The mean-square approximation takes J_2 as the basic criterion to identify the parameter η:

$$J_2 = \max(\mathcal{E}\{[\tilde{x}(\omega) - X(\eta, \omega)]^2\}) \leq \varepsilon_2 \tag{3.14}$$

where $\mathcal{E}[\cdot]$ expresses the mean of the sum of error square and ε_2 is the specified error bound.

After the sample realized values of the basic random variables are identified, common mathematical statistics can be used to obtain the probability distribution of the random variables.

3.1.3.2 Modeling Based on the Statistical Moments of the Sample Assemble

If the probability distribution function $p_\eta(\eta)$ of the random variable η is known, then the mean and the standard deviation of the random Fourier function can be obtained by

$$m_X(\omega) - \int_\Omega X(\eta, \omega) p_\eta(\eta)\, \mathrm{d}\eta \tag{3.15}$$

$$\sigma_X(\omega) = \left\{\int_\Omega [X(\eta, \omega) - m_X(\omega)]^2 p_\eta(\eta)\, \mathrm{d}\eta\right\}^{1/2} \tag{3.16}$$

where Ω is the integral domain about η.

After the mean $\hat{m}_X(\omega)$ and the standard deviation $\hat{\sigma}_X(\omega)$ of the sample set $X_i(\omega)$ $(i = 1, 2, \ldots, n)$ are obtained through mathematical statistical methods (see Equations 3.5 and 3.6), the objective random functions can be identified by using the following modeling criterion:

$$J = \alpha J_m + \beta J_\sigma \rightarrow \min \tag{3.17}$$

where α and β are weighted coefficients and

$$J_m = \frac{1}{L}\sum_{i=1}^{L}\left[m_X(\omega_i) - \hat{m}_X(\omega_i)\right]^2 \tag{3.18}$$

$$J_\sigma = \frac{1}{L}\sum_{i=1}^{L}\left[\sigma_X(\omega_i) - \hat{\sigma}_X(\omega_i)\right]^2 \tag{3.19}$$

where L is the number of scatter points within the effective frequency range.

By adjusting the distribution parameters and the probability distribution types of the basic random variables, the PDF of the identified variables can be determined.

3.2 Seismic Ground Motions

3.2.1 One-Dimensional Model

When an earthquake happens, the seismic wave produced by the seismic source is a time process. Through propagation in the Earth media, the wave shape will undergo complex changes. For a given site, the seismic ground motion at all points within a certain range of the ground surface or the vibration process under the ground surface can be characterized by the time history of ground motion displacements, velocities or accelerations. Usually, for the near-field strong ground motion records, the actual observation data is acceleration. Therefore, the seismic ground motion model usually refers to the acceleration model. The ground motion at one point usually has three spatial coordinate components. According to the studies of Penzien and coworkers (Penzien and Watabe, 1975; Kubo and Penzien, 1979), there are directions of principal axes and the components along the principal directions are uncorrelated. Therefore, for three-dimensional ground motions, only one-dimensional ground motions along the principal axis directions need to be considered.

Owing to the influence of a series of uncontrollable factors like the mechanism of the seismic source, the earthquake propagation paths and the geotechnical media distribution at the engineering site, the ground motion process is a typical stochastic process. Actual earthquake records show that the time history of the ground motion accelerations usually includes three stages of vibrations: the initial, the strong and the attenuating stages (Figure 3.1). Therefore, the ground motion is a typical nonstationary process. When the stationary process model is used to establish the ground motion models, it is usually believed that this only reflects its strong motion stage.

As a simplification, the ground motion on the surface may be regarded as a filtered white noise. In this consideration, if the ground motion on the bedrock is assumed as a zero-mean white-noise process with spectral density S_0, and the soil surface is simulated as a single-degree-of-freedom linear system, then the Kanai–Tajimi spectrum model can be obtained (Kanai, 1957; Tajimi, 1960):

$$S(\omega) = \frac{1 + 4\zeta^2(\omega/\omega_0)^2}{[1-(\omega/\omega_0)^2]^2 + 4\zeta^2(\omega/\omega_0)^2} S_0 \tag{3.20}$$

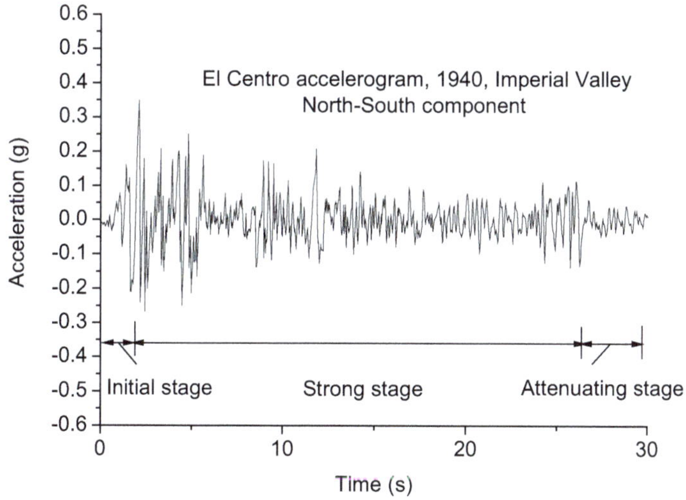

Figure 3.1 Three stages of a typical ground motion record.

where $S(\omega)$ is the PSD function of the stationary ground motion process, ζ is the damping ratio of the soil on the site and ω_0 is the natural frequency of the site.

The physical meaning of the model above is clear that the influence of the soil property on the ground motion frequency spectrum is taken into account. However, this model inappropriately exaggerates the low-frequency content of the ground motions. Meanwhile, the ground-motion velocity and displacement obtained according to the model have a singular point where the frequency is zero; therefore, a finite variance of the ground displacement and velocity cannot be achieved. To overcome these shortcomings, the following correction model has been introduced (Hu and Zhou, 1962):

$$S(\omega) = \frac{1 + 4\zeta^2(\omega/\omega_0)^2}{[1-(\omega/\omega_0)^2]^2 + 4\zeta^2(\omega/\omega_0)^2} \frac{\omega^n}{\omega^n + \omega_c^2} S_0 \tag{3.21}$$

where ω_c is the low-frequency decrease factor and $n = 4$–6.

There are many similar correction models; for example, adding a filter to the model in Equation 3.20 forms a double white-noise process (Ruiz and Penzien, 1969):

$$S(\omega) = \frac{1 + 4\zeta^2(\omega/\omega_0)^2}{[1-(\omega/\omega_0)^2]^2 + 4\zeta^2(\omega/\omega_0)^2} \frac{(\omega/\omega_1)^4}{[1-(\omega/\omega_1)^2]^2 + 4\zeta_1{}^2(\omega/\omega_1)^2} S_0 \tag{3.22}$$

where ζ_1 and ω_1 are parameters of the assumed second filter.

In contrast to the model in Equation 3.20, in which the PSD is a finite value different from zero at $\omega = 0$, both the models in Equations 3.21 and 3.22 can ensure that the PSD is zero at $\omega = 0$ (Figure 3.2).

If there is a need to reflect the rising and decaying sections of the ground-motion process – that is, to reflect the nonstationary nature of the ground motion – then a modulated nonstationary random process model can be introduced (see Equation 3.8). The modulated envelope function can be given by, say (Amin and Ang, 1968; Jennings, *et al.*, 1968).

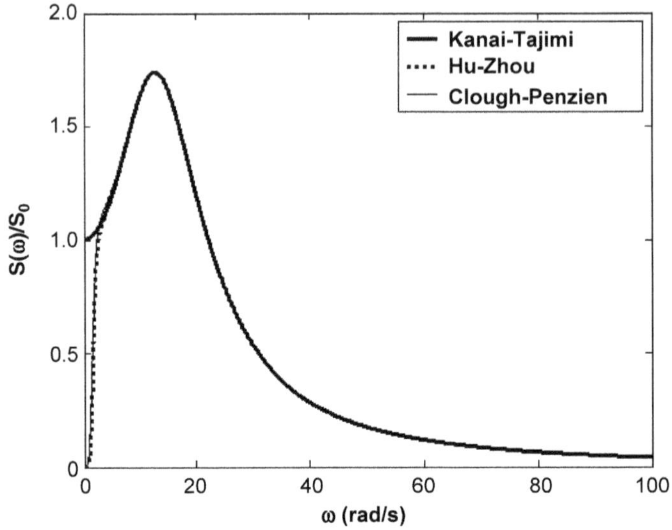

Figure 3.2 Comparison of different PSD models. Parameters: $\omega_0 = 15.71$, $\omega_c = 0.1\omega_0$, $\zeta = 0.64$, $\omega_1 = 0.1\omega_0$, $\zeta_1 = \zeta$.

$$f(t) = \begin{cases} (t/t_a)^2 & \text{for } t \leq t_a \\ 1 & \text{for } t_a < t \leq t_b \\ e^{-\alpha(t - t_b)} & \text{for } t \geq t_b \end{cases} \tag{3.23}$$

where t_a and t_b are respectively the starting time and the end time of the stationary section of the strong ground motion and α is a parameter controlling the decay speed of the attenuation section (Figure 3.3).

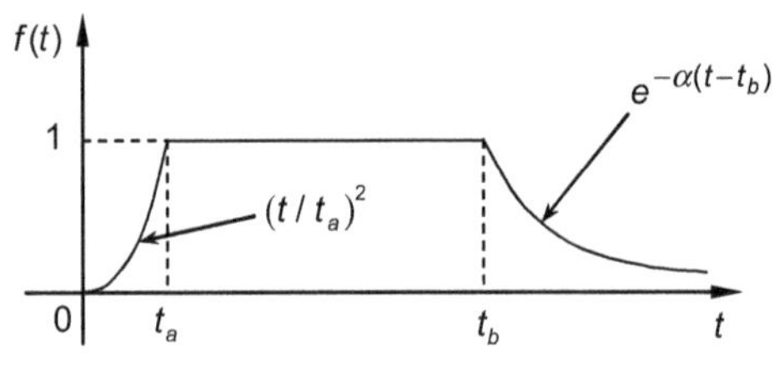

Figure 3.3 Envelope function for ground motion.

3.2.2 Random Field Model

When the difference between ground motions at two points with a certain distance cannot be ignored, then the use of random fields is required to describe seismic ground motions. Using the spatial discretization method (see Section 2.3), the continuous random field description can be transformed into a set of stochastic processes. Therefore, a homogeneous and isotropic random field B can be represented by the following PSD matrix:

$$\mathbf{S}_B(\omega) = \begin{bmatrix} S_{11}(\omega), & S_{12}(\omega), & \ldots, & S_{1m}(\omega) \\ S_{21}(\omega), & S_{22}(\omega), & \ldots, & S_{2m}(\omega) \\ \ldots & \ldots & \ldots & \ldots \\ S_{m1}(\omega), & S_{m2}(\omega), & \ldots, & S_{mm}(\omega) \end{bmatrix} \tag{3.24}$$

where m is the number of spatial points; $S_{kj}(\omega)$ is the cross-spectral density, which is a complex function and characterizes the correlation degree between the stochastic ground motions at the points k and j (see Equation 2.86a). If $k=j$, then this is the auto-spectral density of the ground motion at one point.

In the study of special earthquake ground motions, the coherency function usually represents the correlation feature between the ground motions at two different points and is defined as

$$\gamma_{kj}(\omega) = \begin{cases} \dfrac{S_{kj}(\omega)}{\sqrt{S_{kk}(\omega)S_{jj}(\omega)}} & \text{if } S_{kk}(\omega)S_{jj}(\omega) \neq 0 \\ 0 & \text{otherwise} \end{cases} \tag{3.25}$$

Obviously, the coherency function is also a complex function. Using the amplitude and phase-angle expression, there is

$$\gamma_{kj}(\omega) = |\gamma_{kj}(\omega)|\exp[i\vartheta_{kj}(\omega)] \tag{3.26}$$

The amplitude $|\gamma_{kj}(\omega)|$ of the coherency function is also called the *lagged coherency function*, and we always have (see Equation 2.88)

$$|\gamma_{kj}(\omega)| \leq 1 \tag{3.27}$$

The phase angle $\vartheta_{kj}(\omega)$ is related to the propagation speed of the harmonic wave ω and the distance between two points (Oliveira, *et al.*, 1991):

$$\vartheta_{kj}(\omega) = \frac{\omega d_{kj}^{\mathrm{L}}}{v_{\mathrm{a}}} \tag{3.28}$$

where d_{ij}^{L} is the projection of d_{ij} (the link line between points i and j) along the wave propagation direction and v_{a} is the apparent wave velocity of the ground motions. Setting a certain reference point as the starting point of the time coordinate, denoting the time instants when the earthquake waves arrives at the points k and j respectively by t_k and t_j, there is obviously

$$\frac{d_{ki}^{\mathrm{L}}}{v_{\mathrm{a}}} = t_k - t_j \tag{3.29}$$

Therefore, Equation 3.26 can also be written as

$$\gamma_{kj}(\omega) = |\gamma_{kj}(\omega)|\exp[i\omega(t_k - t_j)] \tag{3.30}$$

The preceding deduction demonstrates that $\exp[i\vartheta_{kj}(\omega)]$ represents the difference of the arrival time of ground motions at two points, which is the traveling wave effect; thus, $\exp[i\vartheta_{kj}(\omega)]$ is generally called the traveling wave effect factor, whereas $\gamma_{kj}(\omega)$ reflects the coherency effect of the ground motions between two points.

By analyzing the earthquake records at densely packed stations, an empirical expression for the lagged coherency function of ground motions can be deduced. Some typical examples follow.

3.2.2.1 Feng–Hu Model

Through analysis of the observed data from the strong earthquakes in Haicheng in 1975 in China and in Niigata in 1964 in Japan, the following formula has been proposed (Feng and Hu, 1981):

$$|\gamma(\omega, d_{kj})| = \exp[-(\rho_1\omega + \rho_2)d_{kj}] \tag{3.31}$$

where ρ_1 and ρ_2 are the coherency parameters, for which the identified values from the Haicheng and Niigata earthquakes are

$$\begin{array}{lll} \text{Haicheng earthquake}: & \rho_1 = 2 \times 10^{-5}\ \text{s/m}, & \rho_2 = 88 \times 10^{-4}\ \text{s/m} \\ \text{Niigata earthquake}: & \rho_1 = 4 \times 10^{-4}\ \text{s/m}, & \rho_2 = 19 \times 10^{-4}\ \text{s/m} \end{array}$$

3.2.2.2 Loh–Yeh Model

Through modeling using observed data from the SMART-1 Array, the following formula was obtained (Loh and Yeh, 1988):

$$|\gamma(\omega, d_{kj})| = \exp\left(-\alpha\frac{\omega d_{kj}}{2\pi v_{\mathrm{a}}}\right) \tag{3.32}$$

where α is the wave number of the ground motions; according to 40 acceleration records, $\alpha = 0.125$ is identified (Loh, 1991).

3.2.2.3 Qu–Wang Model

Through modeling according to the observed data from four earthquake observation stations including SMART-1 Array, the following lagged coherence function is suggested (Qu, *et al.*, 1996):

$$|\gamma(\omega, d_{kj})| = \exp[-a(\omega)d_{kj}^{b(\omega)}] \tag{3.33}$$

where

$$a(\omega) = (12.19 + 0.17\omega^2) \times 10^{-4} \tag{3.34}$$

$$b(\omega) = (76.74 - 0.55\omega) \times 10^{-2} \tag{3.35}$$

Figure 3.4 shows the comparison of the preceding three models as $\omega = 10\pi$.

If the auto-power spectral densities of all points are the same, then the PSD matrix of the ground motion random fields can be simplified as

$$\mathbf{S}_B(\omega) = \mathbf{G}^*\mathbf{R}\mathbf{G} \cdot S(\omega) \tag{3.36}$$

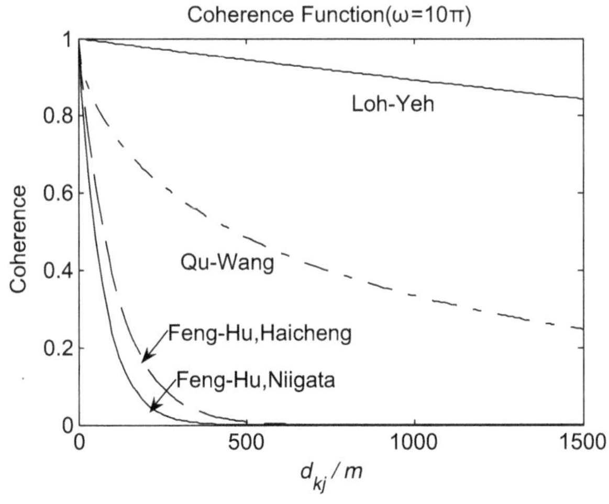

Figure 3.4 Lagged coherence functions.

where $S(\omega)$ is the PSD at each point and **G** is a diagonal matrix representing the phase-angle change of each point compared with the reference points:

$$\mathbf{G} = \mathrm{diag}\left[e^{i\omega t_1}, e^{i\omega t_2}, \ldots, e^{i\omega t_m}\right] \tag{3.37}$$

and **R** is the matrix of the lagged coherency function:

$$\mathbf{R} = \begin{bmatrix} 1 & |\gamma_{12}| & \cdots & |\gamma_{1m}| \\ |\gamma_{12}| & 1 & \cdots & |\gamma_{2m}| \\ \cdots & \cdots & \cdots & \cdots \\ |\gamma_{m1}| & |\gamma_{m2}| & \cdots & 1 \end{bmatrix} \tag{3.38}$$

3.2.3 Physical Stochastic Model

The seismic ground motion process is mainly affected by the earthquake magnitude, the seismic wave propagation distance, the site conditions and some other factors. Because most of these factors are beyond human control, a notable random nature is seen in the observed seismic ground motion processes. If the impact of earthquake magnitude and dissemination factors is not considered in the present stage and the objective is focused on the ground-motion mechanism at a specific site, then the physical relationship can be set up between the surface ground motions and the input motions at the bedrock (Li and Ai, 2006, 2007).

Without loss of generality, the actual engineering site can be simulated as an equivalent single-degree of freedom system (Figure 3.5) with the equation of motion

$$\ddot{x} + 2\zeta\omega_0\dot{x} + \omega_0^2 x = 2\zeta\omega_0\dot{u}_g + \omega_0^2 u_g \tag{3.39}$$

where $\ddot{x}$, $\dot{x}$ and x respectively represent the absolute acceleration, the absolute velocity and the displacement of a point in the fixed coordinate system, ω_0 and ζ are respectively the frequency

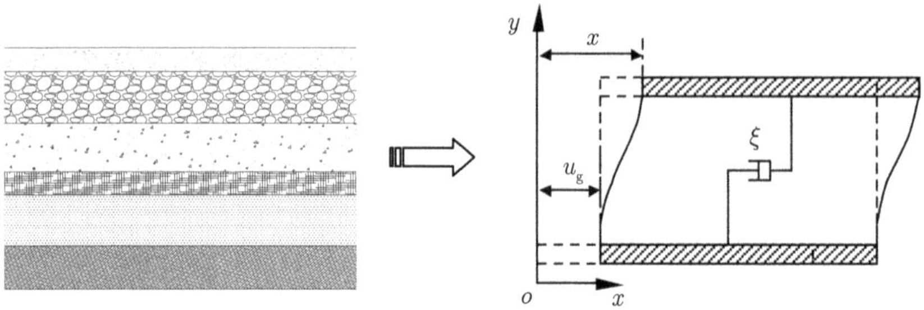

(a) (b)

Figure 3.5 Equivalent single-degree-of-freedom system.

and damping ratio of the site, and $\dot{u}_g$ and u_g are respectively the velocity and the displacement of the input seismic waves at the bedrock.

Performing a Fourier transform on both sides of Equation 3.39 and noting the relationship between displacement and acceleration, the Fourier transform of the absolute acceleration is given as

$$\ddot{X}(\omega) = \frac{\omega_0^2 + i2\zeta\omega_0\omega}{\omega_0^2 - \omega^2 + i2\zeta\omega_0\omega}\ddot{U}_g(\omega) \tag{3.40}$$

where $\ddot{U}_g(\omega)$ is the Fourier transform of the acceleration of the input seismic waves.

Introducing the concept of a random Fourier function, the equation above can be transformed as

$$F_X(\omega) = \left\{\frac{1 + 4\zeta^2(\omega/\omega_0)^2}{[1 - (\omega^2/\omega_0^2)]^2 + 4\zeta^2(\omega/\omega_0)^2}\right\}^{1/2} F_g(\eta, \omega) \tag{3.41}$$

where η is a random variable related to the amplitude of the input seismic waves.

As η, ζ and ω_0 are random variables, $F_X(\omega)$ is a random function. When the basic random variables and their probability distribution are given, the finite-dimensional probability distribution of $\ddot{x}(t)$ can be obtained.

The random Fourier function of the input seismic acceleration at the bedrock $F_g(\eta, \omega)$ may be determined according to the relationship between the physical mechanism of the source and the earthquake attenuation, or given according to the statistics of seismic records at the bedrock. When the energy density of the seismic input is assumed to be in the form in Figure 3.6, it is written as

$$F_g(\eta, \omega) = \begin{cases} \dfrac{F_0\omega}{\sqrt{T}\omega_1} & 0 \le \omega \le \omega_1 \\ \dfrac{F_0}{\sqrt{T}} & \omega_1 \le \omega \le \omega_2 \\ \dfrac{F_0}{\sqrt{T}}\dfrac{\omega_e - \omega}{\omega_e - \omega_2} & \omega_2 \le \omega \le \omega_e \end{cases} \tag{3.42}$$

where F_0 is the amplitude value of the input Fourier spectrum.

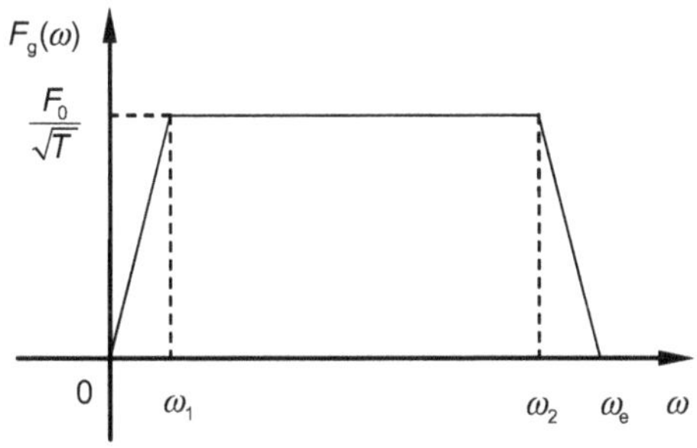

Figure 3.6 Energy distribution of the accelerations at bedrock.

According to the statistics of the acceleration records, it is found that the basic random variables obey a lognormal distribution. The identified mean value and the coefficient of variation of ω_0 and ζ are listed in Tables 3.1 and 3.2. Comparisons between the actual measurement records and the results calculated by the random Fourier spectrum are shown in Figure 3.7 (Li and Ai, 2006).

Table 3.1 Identified mean values of the random variables.

Type of soil	I	II	III	IV
ω_0	15	12	11	9
ζ	0.65	0.80	0.60	0.90

Table 3.2 Identified coefficients of variation of the random variables.

Type of soil	I	II	III	IV
ω_0	0.40	0.40	0.42	0.42
ζ	0.30	0.30	0.35	0.35

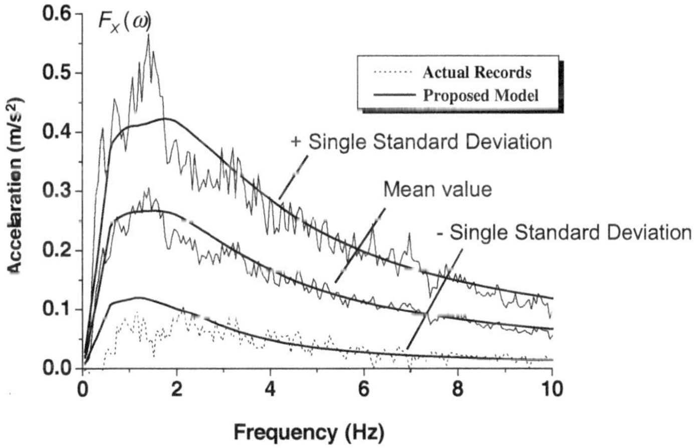

Figure 3.7 Comparison between the physical model and the measured data.

3.3 Fluctuating Wind Speed in the Boundary Layer

3.3.1 Structural Wind Pressure and Wind Speed

The movement of air forms wind. In the scope of the atmospheric boundary layer, if the flow of the wind is obstructed by structures, then there will be a lift force F_Z, a downwind force F_D and a horizontal force F_L (Figure 3.8) applied to the structures (Simiu and Scanlan, 1996)

$$F_Z = \frac{1}{2}\mu_Z \rho v^2 B \tag{3.43a}$$

$$F_D = \frac{1}{2}\mu_D \rho v^2 B \tag{3.43b}$$

$$F_L = \frac{1}{2}\mu_L \rho v^2 B \tag{3.43c}$$

where μ_Z, μ_D and μ_L are respectively the lift force coefficient, the downwind resistance coefficient and the horizontal resistance coefficient of the wind. These coefficients will change with different structure shapes and can be determined normally by wind tunnel experiments. ρ is the mass density of the air, v is the wind speed and B is the characteristic scale of the structures.

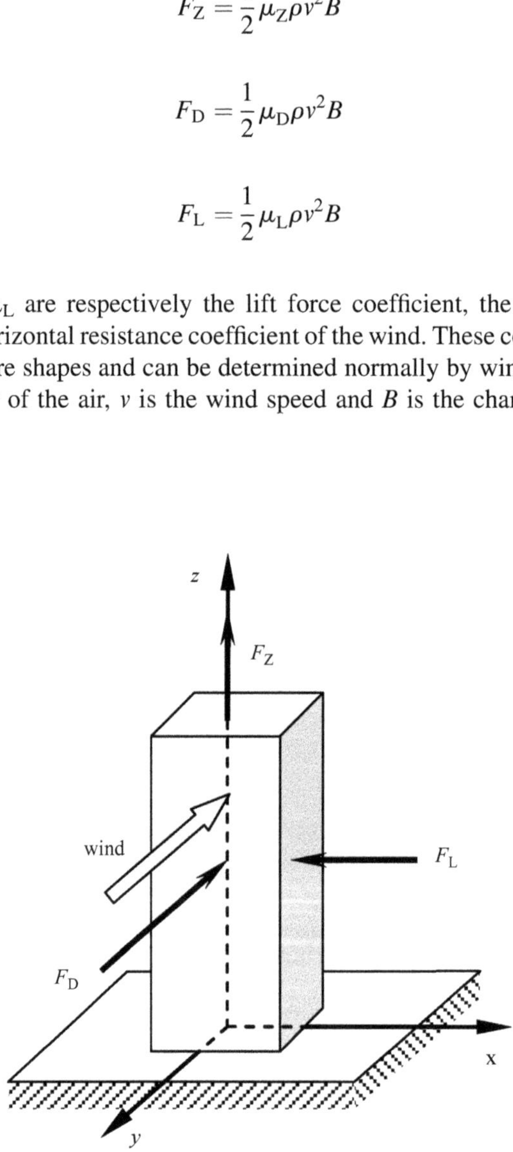

Figure 3.8 Three-component wind forces.

Dividing by B on both sides of Equations 3.43a–3.43c will give the windward pressure of the structures; that is:

$$F_i(x,z,t)=\frac{1}{2}\rho\mu_i(z)v^2(x,z,t) \qquad (i=1,2,3) \tag{3.44}$$

where $F_1=F_Z/B$, $F_2=F_D/B$, $F_3=F_L/B$, $\mu_1=\mu_Z$, $\mu_2=\mu_D$, $\mu_3=\mu_L$ and z is the height above the surface.

A large number of observations of natural wind show that, as a process of time, the wind speed can be expressed as the sum of the average wind speed and the fluctuating wind speed; namely:

$$v(x,z,t)=v_s(z)+v_D(x,z,t) \tag{3.45}$$

Introducing Equation 3.45 into Equation 3.44 and omitting the square items of the fluctuating wind speed, the wind pressure can be decomposed to an average wind pressure and a fluctuating pressure:

$$F_i(x,z,t)=F_{i,s}(x,z)+F_{i,D}(x,z,t) \qquad (i=1,2,3) \tag{3.46}$$

where

$$F_{i,s}(x,z)=\frac{1}{2}\rho\mu_i(z)v_s^2(z) \tag{3.47a}$$

$$F_{i,D}(x,z,t)=\rho\mu_i(z)v_s(z)v_D(x,z,t) \tag{3.47b}$$

Owing to the existence of a variety of uncontrollable factors, the process of the wind speed is a typical stochastic process, where the average wind speed (Figure 3.9) can be described

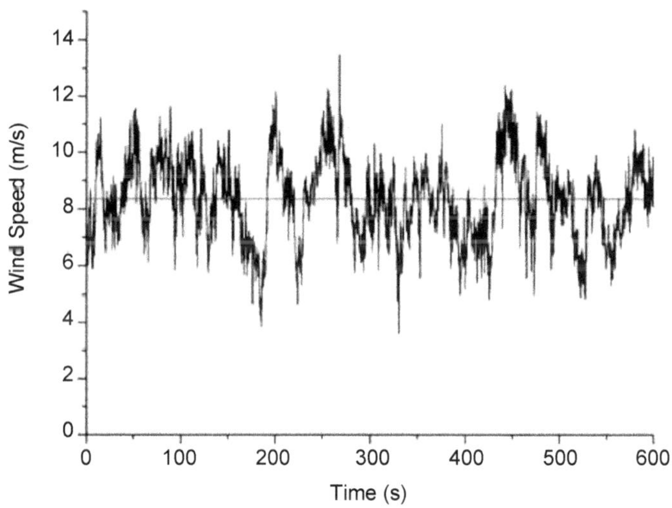

Figure 3.9 Average wind speed.

by a random variable. Generally, it is considered as the extreme-value type I distribution (Zhang, 1985):

$$p_{v_s}(v) = \exp\{-\exp[-a(v-b)]\} \tag{3.48}$$

where a and b are the distribution parameters, which can be expressed by the mean and the standard deviation of the standard wind speed (i.e. the wind speed at a height of 10 m):

$$a = \frac{\pi}{\sqrt{b}\sigma_{v_s}} \tag{3.49a}$$

$$b = \mu_{v_s} - 0.45\sigma_{v_s} \tag{3.49b}$$

Within the atmospheric boundary layer, the average speed along the height of the ground may change at the logarithm rate:

$$v_s(z) = \frac{1}{k}\mu_* \ln\frac{z}{z_0} \tag{3.50}$$

where $k \approx 0.4$, z_0 is the roughness length and μ_* is the shear velocity of the flow:

$$\mu_* = \left(\frac{\tau_0}{\rho}\right)^{1/2} \tag{3.51}$$

where τ_0 is the surface shear force.

3.3.2 Power Spectral Density of Fluctuating Wind Speed

In Equation 3.45, v_D reflects the fluctuating component of the wind speed. In essence, this component is due to the turbulence in the flowing wind. Most of the actual measured data show that fluctuating wind speed may be simulated as a zero-mean stationary Gaussian random field. Introducing this basic assumption, the longitudinal velocity fluctuations at one point can be described by the Davenport spectrum (Davenport, 1961) or the Simiu spectrum (Simiu, 1974).

The Davenport spectrum is given by

$$S_{v_D}(\omega) = 4Kv_{10}^2 \frac{f_1^2}{\omega(1+f_1^2)^{4/3}} \tag{3.52a}$$

$$f_1 = \frac{c_1\omega}{\pi v_{10}} \tag{3.52b}$$

where K is a parameter related to the ground situation (Table 3.3), v_{10} is the mean wind speed at a height of 10 m above the surface and $c_1 = 600$ m.

Table 3.3 The value of K.

Location	Height (ft)	Site description	K
Severn Bridge	100	River bend	0.003
Sale	503	Open grass land with few trees	0.005
Sale	201		
Sale	40		
Cardington	50		
Ann Arbor	25–200		
Cranfield	50	Fenced square	0.008
Brookhaven	300	Bush and 30 ft trees	0.015
London Ontario	150	Urban area	0.030

The Simiu spectrum is expressed as

$$S_{v_D}(z,\omega)=\beta\frac{2\pi\mu_*^2}{\omega}f^{-2/3} \tag{3.53a}$$

$$f=\frac{\omega}{2\pi}\frac{z}{v_s(z)} \tag{3.53b}$$

where $\beta\approx 0.26$.

Clearly, the Simiu spectrum is related to the height, while the Davenport spectrum is not. This is because Equations 3.52a and 3.52b are derived from an average of the actual measured wind speeds at different heights above the surface; therefore, the relationship between the spectrum and the height cannot be reflected. Generally, it is considered that, in the Davenport spectrum, the energy of the wind speed is overestimated in the high-frequency regions but underestimated in the low-frequency regions.

An actual measured spectrum close to the Simiu spectrum is the Kaimal spectrum given by (Kaimal *et al.*, 1972)

$$S_{v_D}(z,\omega)=\frac{2\pi\mu_*^2}{\omega}\frac{200f}{(1+50f)^{5/3}} \tag{3.54}$$

It is generally believed that Equations 3.53a and 3.54b are suitable for the low-frequency regions where $f\leq 0.2$ while Equation 3.54 is applied to the high-frequency regions where $f>0.2$. Similar to the Simiu spectrum, the Kaimal spectrum cannot meet the condition that $S(\omega)=0$ and the first-order derivative is zero as $\omega=0$. On the other hand, the Davenport spectrum can meet the condition. A comparison of the different PSD functions is shown in Figure 3.10.

The fluctuating wind speeds between two points within a certain distance at the same height or different height are correlated. Generally, the cross-PSD is used to measure the probabilistically correlated degree of two stochastic processes. Noting that the spectrum is a complex

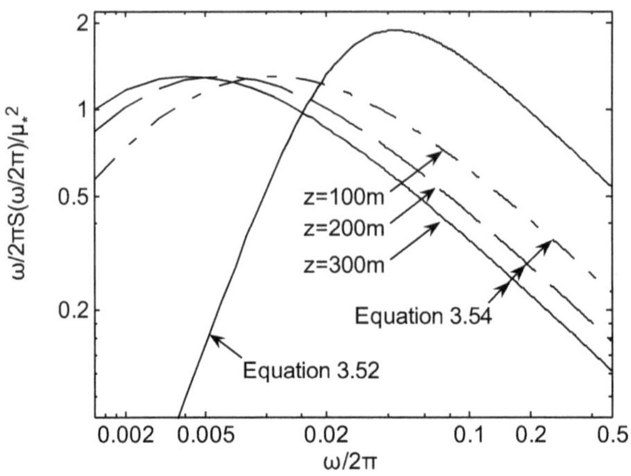

Figure 3.10 Comparison between different wind PSDs.

function, there is

$$S_{\mu_1\mu_2}(d,\omega) = S^C_{\mu_1\mu_2}(d,\omega) + iS^\alpha_{\mu_1\mu_2}(d,\omega) \tag{3.55}$$

where μ_1 and μ_2 are the wind speed records at the points M_1 and M_2 and d is the distance between the two points.

Usually, the imaginary part contributes less to the coherence function than the real part and its effect can be omitted. Therefore, the cross-PSD is expressed as (Davenport, 1968)

$$S_{\mu_1\mu_2}(d,\omega) = S^C_{\mu_1\mu_2}(d,\omega) = \sqrt{S(z_1,\omega)S(z_2,\omega)}\exp(-\hat{f}) \tag{3.56}$$

$$\hat{f} = \frac{\omega[C_Z^2(z_1-z_2)^2 + C_Y^2(x_1-x_2)^2]^{1/2}}{\pi[v_s(z_1)+v_s(z_2)]} \tag{3.57}$$

where x_1, x_2 and z_1, z_2 are the coordinates of the points M_1 and M_2 respectively. The link line between M_1 and M_2 with the mean wind direction is vertical. C_Z and C_Y are the attenuation coefficients to be determined by experience; normally, they take the values $C_Z = 10$ and $C_Y = 16$ (Simiu and Scanlan, 1996).

3.3.3 Random Fourier Spectrum of Fluctuating Wind Speed

The PSD model is essentially the second-order numerical characteristic which is adopted to express the main features of a stochastic process. If the probability distribution is Gaussian for the process, then a sequence of finite-dimensional probability distributions can be given by the first two-dimensional distributions; thereby, the statistical properties of the process can be completely determined by the first two moments of the process. Unfortunately, this condition is close to a hypothetical character, rather than to the summary of the facts observed by experience. In fact, for fluctuating wind speed, the stationary assumption is right only within a certain time-scale, while the assumption of a Gaussian normal distribution is not supported by sufficient observed facts.

The random Fourier spectrum model introduced in Section 3.1 brings about the possibility of establishing a random function model for characterizing a comprehensive probabilistic character of fluctuating wind speed (Li and Zhang, 2007). It has been said that random Fourier spectrum modeling establishes a physical model based on the physical mechanism in the process and, thus, can reflect the nature of the observed facts. A number of studies show that, in a uniform flow field, the energy produced approximately balances the energy consumed (Lumley and Panofsky, 1964). Therefore, the energy dissipation rate can be expressed as

$$\varepsilon = \frac{\tau_0}{\rho}\frac{\mathrm{d}v_{\mathrm{s}}(z)}{\mathrm{d}z} \tag{3.58}$$

where

$$v_{\mathrm{s}}(z) = \frac{1}{k}\mu_* \ln\frac{z}{z_0} \tag{3.59}$$

Noting Equation 3.51, we have

$$\varepsilon = \frac{\mu_*^3}{kz} \tag{3.60}$$

According to Kolmogorov's second hypothesis, in the inertial subrange, the eddy motion may be assumed to be independent of the viscosity, and thus determined only by the energy transfer rate. For a sufficiently high wave number κ, this is given as

$$F[E(\kappa), \kappa, \varepsilon] = 0 \tag{3.61}$$

where $E(\kappa)$ is the energy per unit wave number.

According to dimensional analysis, the above equation will give (Simiu and Scanlan, 1996)

$$E(\kappa) = a_1 \varepsilon^{2/3} \kappa^{-5/3} \tag{3.62}$$

in which a_1 is a universal constant. Note that

$$\kappa = \frac{2\pi}{\lambda} \tag{3.63}$$

and

$$\lambda = \frac{v_n}{n} \tag{3.64}$$

where λ is the wavelength and v_n is the vortex velocity with the frequency n.

For the vibration process formed by several whirlpools, it can be approximated that v_n equals the average speed $v_{\mathrm{s}}(z)$. Thus, the wave number can be written as

$$\kappa = \frac{\omega}{v_{\mathrm{s}}(z)} \tag{3.65}$$

At the same time, noting that

$$E(\kappa)\,\mathrm{d}\kappa = E(\omega)\,\mathrm{d}\omega$$

and substituting Equations 3.60 and 3.62 in Equation 3.65, this leads to

$$E(\omega) = a_1\varepsilon^{2/3}\kappa^{-5/3}\frac{1}{v_s(z)} = a\mu_*^2 z^{-2/3} v_s^{2/3}(z)\omega^{-5/3} \tag{3.66}$$

As the energy spectrum of the process and the Fourier spectrum have a square relationship, it follows immediately that

$$F(\omega) = \alpha\mu_* z^{-1/3} v_s^{1/3}(z)\omega^{-5/6} \tag{3.67}$$

Introducing the Monin coordinate

$$f = \frac{nz}{v_s(z)} \tag{3.68}$$

Equation 3.67 can be changed to

$$F(\omega) = \beta\frac{\mu_*}{\sqrt{\omega}} f^{-1/3} \tag{3.69}$$

Because μ_* and $v_s(z)$ are random variables, the above equation is a random Fourier function. In fact, the application scope of Equation 3.69 is the region of $f > 0.2$. For general cases, Equation 3.69 can be extended to

$$F(\eta, z, \omega) = \frac{\mu_*}{\sqrt{\omega}} G(f) \tag{3.70}$$

After the probability distribution of the basic random variables and the specific form of $G(\cdot)$ are identified by the measured wind speed records, the probability distribution of the random function $F(\eta, z, \omega)$ can be fully determined. As an example, assuming

$$G(f) = \frac{C_1 f^{(C_3 C_4 - 1)}}{(1 + C_2 f^{C_3})^{C_4}} \tag{3.71}$$

and taking 310 groups of measured wind speed records as the basis, the parameters can be identified and are given as: $C_1 = 4.25$, $C_2 = 0.1$, $C_3 = 0.8$, $C_4 = 0.3$. The roughness length Z_0 obeys the lognormal distribution and the 10 m high mean wind speed obeys the extreme-value type I distribution. Thus, the random Fourier spectrum of the fluctuating wind speed can be expressed as (Li and Zhang, 2007)

$$F(n) = \frac{7.02 v_{10}^{4/5} n^{-1/3}}{\ln(10/Z_0)[1 + 3.5 \times 10^4 (n/v_{10})^{9/5}]^{1/3}} \tag{3.72}$$

Figure 3.11 shows the comparison between the mean Fourier spectrum and the standard deviation spectrum of the actual measured wind-speed records with the counterparts of the random Fourier spectrum.

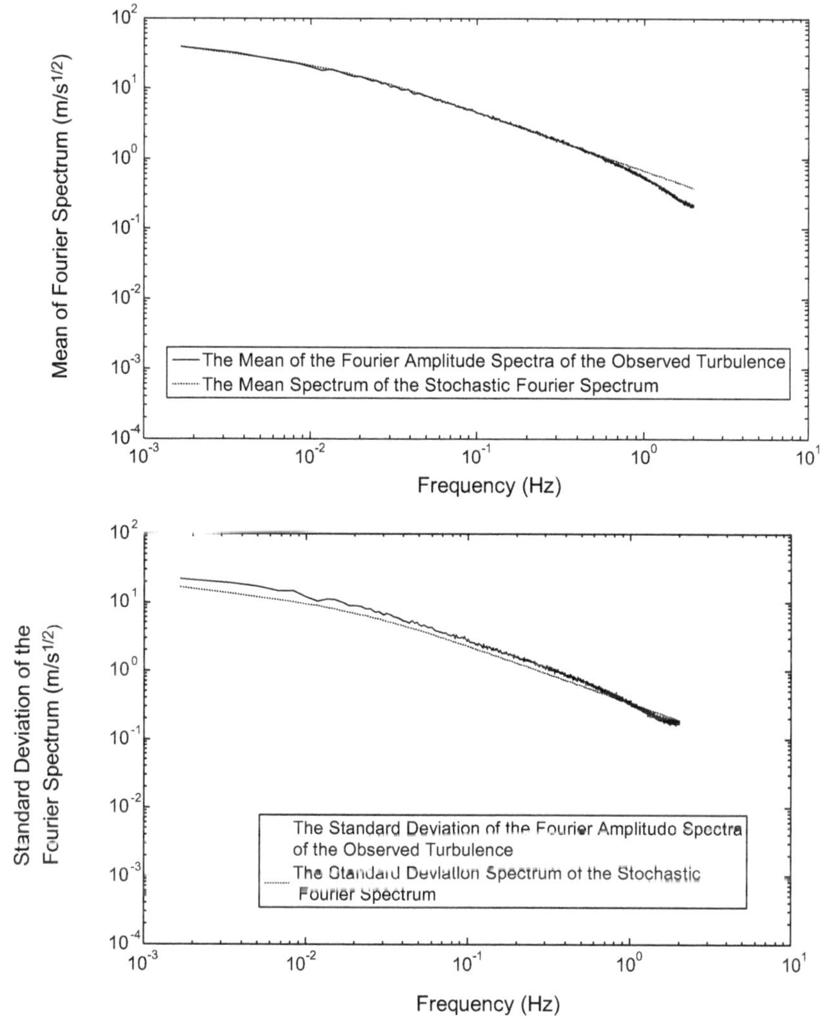

Figure 3.11 Comparison between the theoretical and measured Fourier spectrum.

3.3.4 Random Fourier Correlation Spectrum

In order to reflect the correlation characteristic of the fluctuating wind speed between two arbitrary points in space, the cross random Fourier correlation spectrum can be introduced and defined by

$$F_{\mu_1\mu_2}(\omega) = F_{\mu_1}(\omega)F_{\mu_2}(\omega)\gamma_{\mu_1\mu_2}(\omega) \tag{3.73}$$

where $F_{\mu_1}(\omega)$ and $F_{\mu_2}(\omega)$ are respectively the random Fourier spectrums at the points M_1 and M_2, and

$$\gamma_{\mu_1\mu_2} = C_1\exp(-\hat{f}) \tag{3.74}$$

is a coherence function, where C_1 is a coefficient and $\hat{f}$ is defined by Equation 3.57.

When considering only the coherence of two points at different heights in the same vertical direction, the coherence function is given as

$$\gamma_{\mu_1\mu_2}(\omega) = C_1\exp\left\{ -\frac{C_Z|z_1 - z_2|\omega}{\pi[v_s(z_1) + v_s(z_2)]}\right\} \tag{3.75}$$

In the case C_1 and C_Z are both random variables, the above equation expresses a random coherency function.

According to the statistical analysis of the actual measured data of sampled wind-speed records, it has been confirmed that C_1 and C_Z obey the normal distribution with $\mu_{C_1} = 0.492$, $\sigma_{C_1} = 0.034$, $\mu_{C_Z} = 0.03$ and $\sigma_{C_Z} = 0.042$ (Zhang and Li, 2006).

Figure 3.12 shows the comparison between the Fourier correlation spectrum of the actual measured wind-speed records and the mean and the standard deviation of random Fourier correlation spectrum calculated by Equation 3.73.

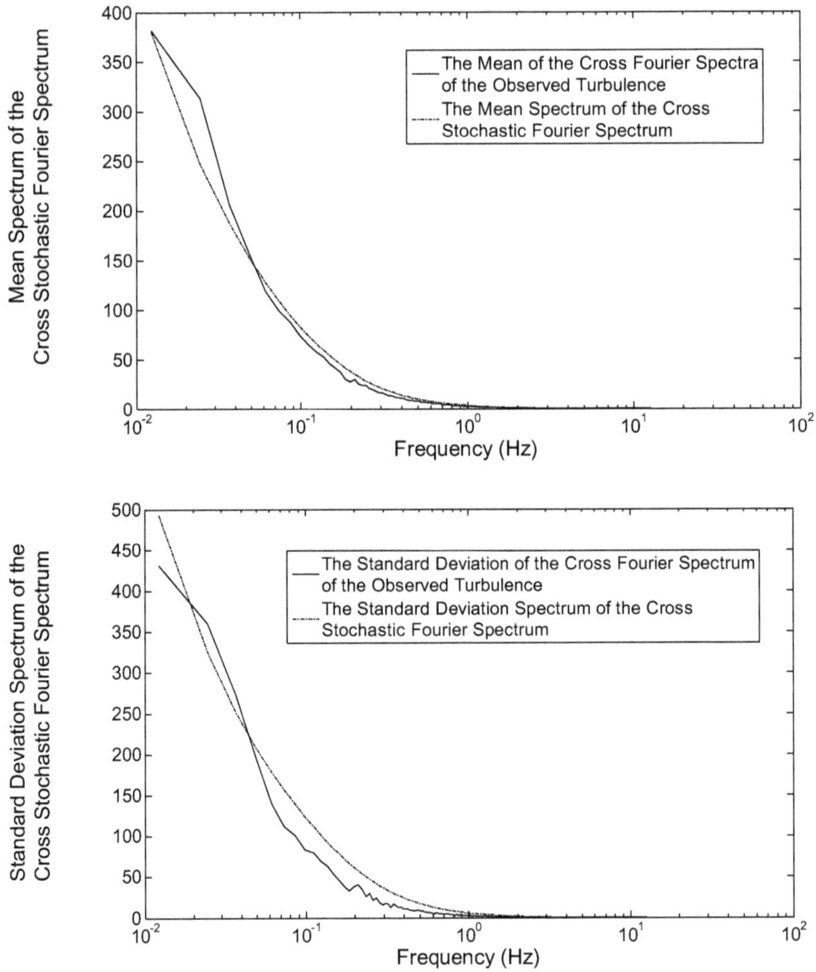

Figure 3.12 Comparison between the theoretical and measured Fourier correlation spectrum.

3.4 Wind Wave and Ocean Wave Spectrum

3.4.1 Wind Waves and Wave Forces

In the scope of its main energy distribution, an ocean wave is generated by wind power as a drive and gravity as a restoring force and thus is also termed a wind wave. In the classical theory of wave analysis, if it is assumed that the amplitude of a wave (wave height) H, the wavelength λ and the water depth h are small, then the fluid particle velocity v caused by the wave can be regarded as small. Therefore, the equation and the boundary conditions for the wave potential function φ are linear:

$$\frac{\partial^2 \varphi}{\partial x^2} + \frac{\partial^2 \varphi}{\partial z^2} = 0 \tag{3.76a}$$

$$\eta = -\frac{1}{g}\frac{\partial \varphi}{\partial t} \qquad \text{for } z = 0 \tag{3.76b}$$

$$\frac{\partial \eta}{\partial t} - \frac{\partial \varphi}{\partial z} = 0 \qquad \text{for } z = 0 \tag{3.76c}$$

$$\frac{\partial \varphi}{\partial z} = 0 \qquad \text{for } z = -h \tag{3.76d}$$

According to the above equations and the boundary conditions, the horizontal displacement of the free surface water particle of the wind wave can be obtained as

$$\eta(x, t) = \frac{H}{2}\cos(\kappa x - \omega t) \tag{3.77}$$

where κ and ω are the wave number and wave frequency respectively and can be expressed by the wavelength λ and the period T:

$$\kappa = \frac{2\pi}{\lambda} \tag{3.78a}$$

$$\omega = \frac{2\pi}{T} \tag{3.78b}$$

Figure 3.13 shows the waveform expressed in Equation 3.77. Similarly, the horizontal velocity of the water particle at a certain depth z may be obtained as

$$v(x, z, t) = \frac{\omega\cosh[\kappa(z+h)]}{\sinh(\kappa h)}\eta(x, t) \tag{3.79}$$

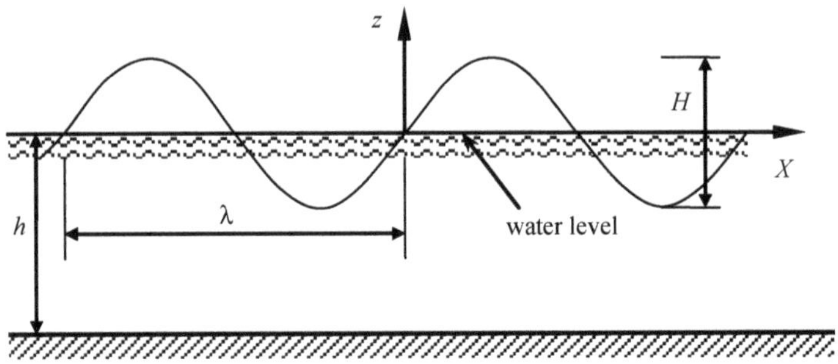

Figure 3.13 Wind wave.

and the horizontal acceleration of the water particle is

$$a(x,z,t) = \frac{\omega \cosh[\kappa(z+h)]}{\sinh(\kappa h)} \dot{\eta}(x,t) \tag{3.80}$$

The above solution is commonly known as the Airy solution (Wen and Yu, 1985). Introducing

$$H(\omega,z) = \frac{\omega \cosh[\kappa(z+h)]}{\sinh(\kappa h)} \tag{3.81}$$

Equations 3.79 and 3.80 can be simplified as follows:

$$v(x,z,t) = H(\omega,z)\eta(x,t) \tag{3.82}$$

$$a(x,z,t) = H(\omega,z)\dot{\eta}(x,t) \tag{3.83}$$

The forces generated by wind waves applied on structures in water are known as wave forces. When the ratio of the characteristic scale D of the object to the wavelength λ is comparatively small, its effect on the wave field can be neglected. In these cases, the above solution related to the wave can be applied directly.

For a vertical cylinder under the action of an ocean wave, if $D/\lambda < 0.2$, then the Morison formula can be adopted to calculate the wave forces (Morison *et al.*, 1950):

$$F(x,z,t) = \frac{1}{2}\rho C_D v(x,z,t)|v(x,z,t)| + \rho C_I \frac{\pi D^2}{4} a(x,z,t) \tag{3.84}$$

where ρ is the density of the seawater, D is the diameter of the pile, C_D is the resistance coefficient, C_I is the inertial coefficient.

The Morison formula consists of the resistance and the inertia components, among which the resistance results from the speed when the seawater flows through the piles while the inertial forces result from the acceleration by the seawater particles. The distribution of the wave forces along the height of the piles is schematically shown in Figure 3.14.

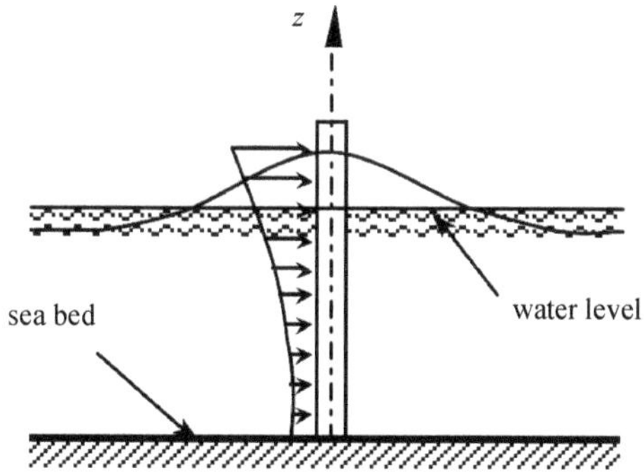

Figure 3.14 Distribution of wave forces.

If we define

$$K_{\mathrm{D}} = \frac{1}{2}\rho C_{\mathrm{D}} D \tag{3.85a}$$

$$K_{\mathrm{I}} = \rho C_{\mathrm{I}} \frac{\pi D^2}{4} \tag{3.85b}$$

then the Morison formula can be simplified to

$$F(x, z, t) = K_{\mathrm{D}} v(x, z, t)|v(x, z, t)| + K_{\mathrm{I}} a(x, z, t) \tag{3.86}$$

The formation of wind waves is affected by many uncontrollable factors. Therefore, the wave process is a typical stochastic process. However, this docs not change the physical relationships between the displacement, the velocity and the acceleration of the water particles in the wave process. For the stochastic process, Equations 3.82 and 3.83 still hold. Assuming that the waves at different points are completely correlated along the x direction, then the random field $\eta(x, t)$ can be simplified as a stochastic process $\eta(t)$.

For the random speed process $v(z, t)$, through the statistical equivalent linearization, the random resistance

$$F_v(z, t) = K_{\mathrm{D}} v(z, t)|v(z, t)| \tag{3.87}$$

can be changed to

$$F_v(z, t) = \sqrt{\frac{8}{\pi}} \sigma_v K_{\mathrm{D}} v(z, t) \tag{3.88}$$

where σ_v is the standard deviation of $v(z, t)$.

Substituting Equations 3.88 and 3.83 in Equation 3.86 and noting that the aforementioned statistical relationship assumption will give

$$F_v(z, t) = \sqrt{\frac{8}{\pi}} \sigma_v K_{\mathrm{D}} H(\omega, z)\eta(t) + K_{\mathrm{I}} H(\omega, z)\dot{\eta}(t) \tag{3.89}$$

If $\eta(t)$ is a stationary stochastic process, according to the uncorrelated nature of the stationary process and its derivative, then from Equation 3.89 it is easy to obtain the PSD of the wave force $F(z,t)$:

$$S_F(z,\omega) = \left(\frac{8}{\pi}\sigma_v^2 K_D^2 + K_I^2\omega^2\right)H^2(\omega,z)S_\eta(\omega) \tag{3.90}$$

where $S_\eta(\omega)$ is the PSD of $\eta(t)$.

3.4.2 *Power Spectral Density of Wind Waves*

A large number of actual measured data indicate that the free surface particle displacement $\eta(t)$ of the ocean wave can be reflected by a zero-mean stationary stochastic process model. So far, investigators have proposed various wind wave power spectral models, among which a significant type follows the basic form brought forward by Neumann (1952):

$$S_\eta(\omega) = \frac{A}{\omega^p}\exp\left(-B\frac{1}{\omega^q}\right) \tag{3.91}$$

where p often takes a value in the range 5–6, $q = 2$–4 and the coefficients A and B are often related to the wind speed, the wave height and some other physical parameters. In fact, in the Neumann spectrum, the parameters take the values

$$p = 6 \qquad q = 2 \qquad A = \frac{C\pi}{2} \qquad B = \frac{2g^2}{v_{7.5}^2}$$

where $C = 3.05\,\mathrm{m^2\,s^{-5}}$, g is the acceleration due to gravity and $v_{7.5}$ is the average wind speed at a height of 7.5 m over the sea surface.

In the Pierson–Moscowitz (P–M) spectrum which keeps the basic form of the Neumann spectrum, the parameters take (Pierson and Moskowitz, 1964)

$$p = 5 \qquad q = 4 \qquad A = 0.0081g^2 \qquad B = 0.74\left(\frac{g}{v_{19.5}}\right)^4$$

where $v_{19.5}$ is the average wind speed at a height of 19.5 m over the sea surface.

If we take the dimensionless constants $\alpha = 8.1 \times 10^{-3}$ and $\beta = 0.74$, then the P–M spectrum can be expressed as

$$S_\eta(\omega) = \frac{\alpha g^2}{\omega^5}\exp\left[-\beta\left(\frac{g}{v_{19.5}\omega}\right)^4\right] \tag{3.92}$$

The P–M spectrum comes from an analysis on 460 spectrums observed and recorded in the North Atlantic Ocean from 1955 to 1960. It reflects the basic characteristics of full growth of a stormy wave.

The more convincing wind wave power spectral function model is the JONSWAP spectrum (Hasselmann *et al.*, 1973), which results from the integrated spectrum analysis of 2500 observed spectra. Its form is

$$S_\eta(\omega) = \frac{\alpha g^2}{\omega^5} \exp\left[-\frac{5}{4}\left(\frac{\omega_0}{\omega}\right)^4 \right] \gamma^{\exp[-(\omega-\omega_0)^2/(2\sigma^2\omega_0^2)]} \tag{3.93}$$

where ω_0 is the peak frequency, α is the energy scale parameter, which is a function of the length $\tilde{x} = gx/v_{10}$ in the dimensionless wind region (where x is the length of the wind region and v_{10} is the mean wind speed at a height of 10 m over the sea surface). When $\tilde{x} = 10^{-1}$-10^5, the energy scale parameter is given as

$$\alpha = 0.07\tilde{x}^{-0.22} \tag{3.94}$$

γ is the peak factor, defined by

$$\gamma = \frac{S_{\eta,\max}}{S_{\eta,\max}^{\mathrm{PM}}} \tag{3.95}$$

where $S_{\eta,\max}$ is the peak of the JONSWAP spectrum and $S_{\eta,\max}^{\mathrm{PM}}$ is the peak of the P–M spectrum. The observed values of γ is between 1.5 and 6, with the average being 3.3, σ is the peak shape parameter with value

$$\sigma = \begin{cases} 0.07 & \text{for } \omega \le \omega_0 \\ 0.09 & \text{for } \omega > \omega_0 \end{cases} \tag{3.96}$$

The JONSWAP spectrum is applied to the wind wave under different stages of growth and even under a hurricane. Figure 3.15 is the comparison between the average JONSWAP spectrum and the P–M spectrum (Rye, 1974).

The preceding PSD model is, in essence, a type of observed statistical model. Using a different approach, Wen *et al.* (1994a, 1994b) attempted to draw a theoretical model of the wind wave spectrum based on an analytical viewpoint. According to their research, the spectrum is

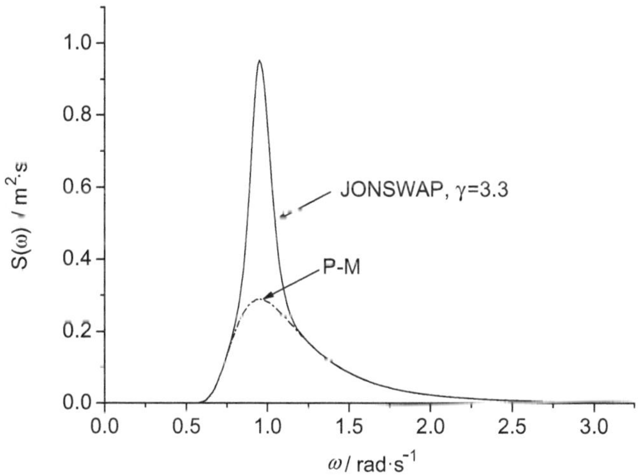

Figure 3.15 The average JONSWAP spectrum and P–M spectrum.

expressed as

$$\tilde{S}_\eta(\tilde{\omega}) = 1.01\tilde{\omega}^{-4.25}\exp[-0.773(\tilde{\omega}^{-5.5}-1)] + 1.1\exp[-41.4(\tilde{\omega}-1)^2] \tag{3.97}$$

where

$$\tilde{S}_\eta(\tilde{\omega}) = \frac{\omega}{m_0} S_\eta(\omega) \tag{3.98}$$

$$\tilde{\omega} = \frac{\omega}{\omega_0} \tag{3.99}$$

$$m_0 = \int_0^\infty S(\omega)\,\mathrm{d}\omega \tag{3.100}$$

The Wen model consists of the wind wave spectrum with full growth and the very young wind wave spectrum. In fact, Wen's model is a mathematical analysis for the wind wave spectrum curve, but is not entirely a physical interpretation. Figure 3.16 shows the comparison between the Wen spectrum and the JONSWAP spectrum.

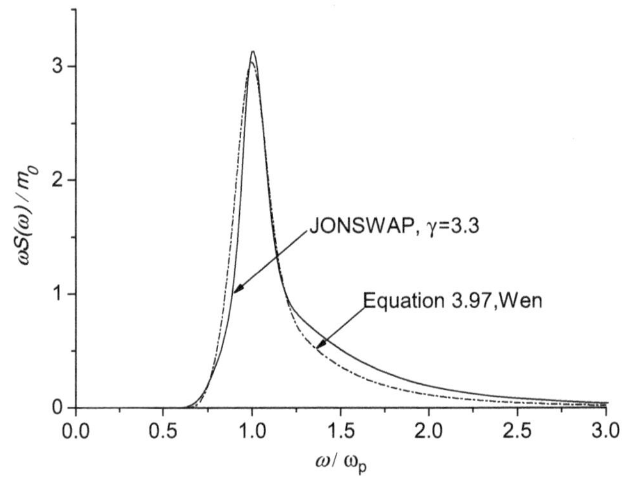

Figure 3.16 The Wen spectrum and the JONSWAP spectrum.

3.4.3 Direction Spectrum

When investigating the reaction to ocean waves by floating objects or ocean wave refraction and diffraction near large-size objects, it is essential to consider the direction distribution of the wave. The direction spectrum of the ocean wave is generally defined as

$$S(f,\vartheta) = S(f)G(f,\vartheta) \tag{3.101}$$

where $S(f)$ is the displacement frequency spectrum of the ocean wave, f is the frequency, ϑ is the angle between the oblique wave and the principal direction and $G(f,\vartheta)$ is the direction distribution function which satisfies

$$\int_{-\pi}^{\pi} G(f, \vartheta)\,\mathrm{d}\vartheta = 1 \tag{3.102}$$

Longuet-Higgins *et al.* (1963) expressed the direction distribution function as

$$G(f, \vartheta) = G_0(s)\left|\cos\frac{\vartheta}{2}\right|^{2s} \tag{3.103}$$

where

$$G_0(s) = \left(\int_{\vartheta_{\min}}^{\vartheta_{\max}} \cos^{2s}\frac{\vartheta}{2}\,\mathrm{d}\vartheta\right)^{-1} \tag{3.104}$$

Here, s is the concentration degree of the direction function and it is proper to determine it by actually measured results in different sea areas (Mitsuyasu *et al.*, 1975; Yu and Liu, 1994).

The direction distribution function advised by Donelan *et al.* (1985) is

$$G(f, \vartheta) = \frac{1}{2}\beta \mathrm{sech}^2 \beta\vartheta \tag{3.105}$$

where

$$\beta = \begin{cases} 2.61\left(\dfrac{f}{f_{\mathrm{p}}}\right)^{1.3} & \text{for } 0.56 \le \dfrac{f}{f_{\mathrm{p}}} \le 0.95 \\ 2.28\left(\dfrac{f}{f_{\mathrm{p}}}\right)^{-1.3} & \text{for } 0.95 < \dfrac{f}{f_{\mathrm{p}}} < 1.6 \\ 1.4 & \text{otherwise} \end{cases} \tag{3.106}$$

where f_{p} is the peak frequency.

Using an analytic method, Wen *et al.* (1995) gave the direction distribution function

$$G(f, \vartheta) = C(n')\cos^{n'}\vartheta \tag{3.107}$$

where

$$n' = \begin{cases} 9.91\left(\dfrac{\omega}{\omega_0}\right)^{-2} \exp(\ 0.0757\, p^{1.95}) & \text{for } \dfrac{\omega}{\omega_0} \ge 1 \\ 9.91\left(\dfrac{\omega}{\omega_0}\right)^{4.5} \exp(-0.0757\, p^{1.95}) & \text{for } \dfrac{\omega}{\omega_0} < 1 \end{cases} \tag{3.108}$$

$$p = \frac{f_{\mathrm{p}} S_\eta(f_{\mathrm{p}})}{m_0} \tag{3.109}$$

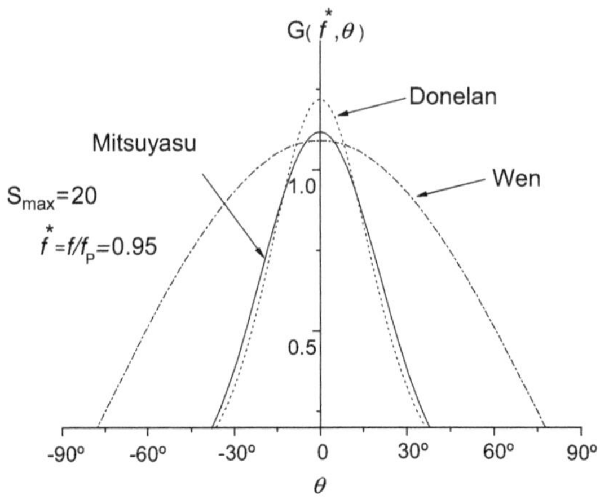

Figure 3.17 Direction spectrum.

Figure 3.17 shows the Mitsuyasu direction spectrum and the comparison between the Donelan and the Wen direction spectrum.

3.5 Orthogonal Decomposition of Random Excitations

The PSD of a random excitation describes the numerical characteristics of the stochastic process in the phenomenological sense. Correspondingly, the orthogonal decomposition of a random excitation gives the random function description of the stochastic process. The descriptions can be regarded as equivalent forms of the Karhunen–Loève decomposition described in Chapter 2.

3.5.1 Orthogonal Decomposition of a Stochastic Process

In the Karhunen–Loève decomposition of a stochastic process, to obtain the eigenvalues and the eigenvectors of the decomposed process, it is necessary to solve a Fredholm integral equation. This is usually quite difficult. To avoid the difficulty, the double expansion method of a stochastic process can be achieved based on the standard orthogonal basis (Li and Liu, 2006).

For a real, zero-mean stochastic process $\{X(t), 0 \leq t \leq T\}$ in a random functional space, introduce a standard orthogonal function set $\varphi_j(t), j = 1, 2, \ldots$, which satisfies

$$\langle \varphi_i, \varphi_j \rangle = \int_0^T \varphi_i(t)\varphi_j(t)\,\mathrm{d}t = \delta_{ij} \tag{3.110}$$

Then in the interval $[0, T]$, $X(t)$ can be expanded as

$$X(\boldsymbol{\xi}, t) = \sum_{k=1}^{\infty} \xi_k \varphi_k(t) \tag{3.111}$$

The expansion coefficients ξ_k are random variables given by

$$\xi_k = \int_0^T X(\boldsymbol{\xi}, t)\varphi_k(t)\,\mathrm{d}t \qquad k = 1, 2, \ldots \tag{3.112}$$

This integral is defined in the sense of the mean-square Riemann integral. Usually, a limited-item N can be taken as the approximation of the expansion in Equation 3.111, namely:

$$\hat{X}(\boldsymbol{\xi}, t) = \sum_{k=1}^{N} \xi_k \varphi_k(t) \tag{3.113}$$

In this case, the mean square error is

$$\varepsilon_1 = \mathcal{E}\left\{\int_0^T [X(t) - \hat{X}(t)]^2\,\mathrm{d}t\right\} = \sum_{k=N+1}^{\infty} \mathcal{E}[\xi_k^2] \tag{3.114}$$

Generally, the random variables ξ_k, $k = 1, 2, \ldots, N$, are correlated. Define the covariance matrix of the random vector $\boldsymbol{\xi} = (\xi_1, \xi_2, \ldots, \xi_N)^{\mathrm{T}}$ as

$$C = \begin{bmatrix} c_{11} & c_{12} & \ldots & c_{1N} \\ c_{21} & c_{22} & \ldots & c_{2N} \\ \ldots & \ldots & \ldots & \ldots \\ c_{N1} & c_{N2} & & c_{NN} \end{bmatrix} \tag{3.115}$$

where (see Equation 3.112)

$$\begin{aligned} c_{ij} &= \mathcal{E}[\xi_i \xi_j] \\ &= \mathcal{E}\left[\int_0^T X(\boldsymbol{\xi}, t_1)\varphi_i(t_1)\,\mathrm{d}t_1 \int_0^T X(\boldsymbol{\xi}, t_2)\varphi_j(t_2)\,\mathrm{d}t_2\right] \\ &= \int_0^T \int_0^T K_X(t_1, t_2)\varphi_i(t_1)\varphi_j(t_2)\,\mathrm{d}t_1\,\mathrm{d}t_2 \end{aligned} \tag{3.116}$$

Here, $K_X(t_1, t_2) = \mathcal{E}[X(\boldsymbol{\xi}, t_1)X(\boldsymbol{\xi}, t_2)]$ is the covariance function of the stochastic process.

According to the decomposition principle of the random vector described in Section 2.1.4, the random vector $\boldsymbol{\xi}$ has the following forms of decomposition:

$$\boldsymbol{\xi} = \sum_{j=1}^{N} \zeta_j \sqrt{\lambda_j}\boldsymbol{\psi}_j \tag{3.117}$$

where the $\boldsymbol{\psi}_j$ are the eigenvector of matrix $\mathbf{C}$, the λ_j are the corresponding eigenvalues and the ζ_j are the standardized random variables, of which the distribution is determined by the nature of the process.

Substituting Equation 3.117 in Equation 3.113 will yield

$$\hat{X}(\boldsymbol{\zeta}, t) = \sum_{k=1}^{N}\sum_{j=1}^{N} \zeta_j \sqrt{\lambda_j}\phi_{jk}\varphi_k(t) = \sum_{j=1}^{N} \zeta_j \sqrt{\lambda_j} f_j(t) \tag{3.118}$$

where ϕ_{jk} is the kth component of the eigenvector $\boldsymbol{\psi}_j$ and

$$f_j(t) = \sum_{k=1}^{N} \phi_{jk}\varphi_k(t) \tag{3.119}$$

It is easy to prove that $\{f_j(t),\ j = 1, 2, \ldots, N\}$ is a set of standard orthogonal functions; namely:

$$\langle f_i, f_j \rangle = \int_0^T f_i(t) f_j(t)\,\mathrm{d}t = \delta_{ij} \tag{3.120}$$

Equation 3.118 is termed the standard orthogonal decomposition of a stochastic process. Obviously, as $N \to \infty$, the expression of such an orthogonal expansion is equivalent to the Karhunen–Loève decomposition.

For a stochastic process with nonzero mean, it can be apparently given as

$$X(\boldsymbol{\zeta}, t) = X_0(t) + \sum_{j=1}^{\infty} \zeta_j \sqrt{\lambda_j} f_j(t) \tag{3.121}$$

Generally, for a specific random excitation, the former rth eigenvalue and the corresponding eigenvector from the largest eigenvalue in order are taken to reflect the major characteristic of the stochastic process; namely, Equation 3.118 can further be reduced to

$$\tilde{X}(\boldsymbol{\zeta}, t) = \sum_{j=1}^{r} \zeta_j \sqrt{\lambda_j} f_j(t) \tag{3.122}$$

In the sense of the mean square, the error of Equation 3.122 compared with Equation 3.118 is

$$\varepsilon_2 = \mathcal{E}\left\{ \int_0^T [\hat{X}(\boldsymbol{\zeta}, t) - \tilde{X}(\boldsymbol{\zeta}, t)]^2\,\mathrm{d}t \right\} = \sum_{j=r+1}^{N} \lambda_j \tag{3.123}$$

3.5.2 Hartley Orthogonal Basis Function

In the standard orthogonal decomposition of the stochastic process, the function $\{\varphi_j(t),\ j = 1, 2, \ldots\}$ can choose all possible orthogonal basis functions, such as the trigonometric functions and the Legendre orthogonal polynomials and the like. In comparison, using the Hartley orthogonal basis function can often obtain the best results (Li and Liu, 2006, 2008).

The Hartley transform of a real-valued function $x(t)$ is expressed as (Bracewell, 1986)

$$H_x(f) = \int_{-\infty}^{\infty} x(t)\,\mathrm{cas}(2\pi f t)\,\mathrm{d}t \tag{3.124}$$

where f is the frequency and

$$\mathrm{cas}(t) = \cos(t) + \sin(t) \tag{3.125}$$

The Hartley transform and its inverse transform have the same form of integral calculation; namely:

$$x(t) = \int_{-\infty}^{\infty} H_x(f)\text{cas}(2\pi ft)\,\text{d}f \tag{3.126}$$

It is noted that the Hartley transform of a real-valued function remains a real function. This is why the Hartley transform is simpler than the Fourier transform.

For a real time sequence $x(n)$ in a limited range, the discrete Hartley transform is

$$H_x(k) = \sum_{n=0}^{N-1} x(n)\text{cas}(2\pi k\Delta f n\Delta t) \qquad k = 0, 1, \ldots, N-1 \tag{3.127}$$

where Δf is the step of the frequency and Δt is the time step.

The corresponding inverse transform is

$$x(n) = \frac{1}{N}\sum_{k=0}^{N-1} H_x(k)\text{cas}\left(\frac{2\pi kn}{N}\right) \qquad n = 0, 1, \ldots, N-1 \tag{3.128}$$

In the interval $[0, T]$, the integrated Hartley orthogonal basis is

$$\varphi_k(t) = \frac{1}{\sqrt{T}}\text{cas}\left(\frac{2\pi kt}{T}\right) \qquad k = 0, 1, 2, \ldots \tag{3.129}$$

while the integrated trigonometric orthogonal basis is given by

$$\begin{cases} \varphi_0^s(t) & = \dfrac{1}{\sqrt{T}} \\ \varphi_{2k-1}^s(t) & = \dfrac{\sqrt{2}}{\sqrt{T}}\cos\left(\dfrac{2k\pi t}{T}\right) \qquad k = 1, 2, \ldots \\ \varphi_{2k}^s(t) & = \dfrac{\sqrt{2}}{\sqrt{T}}\sin\left(\dfrac{2k\pi t}{T}\right) \end{cases} \tag{3.130}$$

It is easy to know that

$$\varphi_{2k-1}^s(t) + \varphi_{2k}^s(t) = \sqrt{2}\varphi_k(t) \tag{3.131}$$

This indicates that when the orthogonal Hartley basis is used to substitute the trigonometric orthogonal basis as the expansion basis function of a stochastic process $x(t)$, under the same error scope, the expansion items related to the Hartley basis can be reduced by a half.

3.5.3 Orthogonal Expansion of Seismic Ground Motions

When the Hartley orthogonal basis indicated in Equation 3.129 is used to expand the process of the seismic ground displacement, there is (see Equation 3.122)

$$\tilde{X}(\boldsymbol{\zeta}, t) = \sum_{j=1}^{r} \zeta_j\sqrt{\lambda_j}f_j(t) \tag{3.132}$$

where

$$f_j(t) = \sum_{k=0}^{N-1} \phi_{j,k+1}\varphi_k(t) \tag{3.133}$$

Provided that the seismic ground motion is a Gaussian process, $\zeta_j\,(j = 1, 2, \ldots, r)$ will be a set of mutually independent standard Gaussian random variables.

Accordingly, the orthogonal expansion formula of the acceleration process is

$$\ddot{X}(\boldsymbol{\zeta}, t) = \sum_{j=1}^{r} \zeta_j \sqrt{\lambda_j} F_j(t) \tag{3.134}$$

where

$$F_j(t) = \sum_{k=0}^{N-1} \alpha_{k+1}\phi_{j,k+1}\ddot{\varphi}_k(t) \tag{3.135}$$

Here, the α_{k+1} introduced is a set of coefficients to make up the truncated error and is obtained by the principle of energy equivalence (Liu and Li, 2006).

Figure 3.18 shows the comparison between the PSD of the orthogonal expanded process of the Hu–Zhou model (Hu and Zhou, 1962) by the preceding method and the PSD of the original model. Figure 3.19 shows a typical time history sample given by the expansion function. During this process, the intensity envelope function given in Equation 3.23 is used.

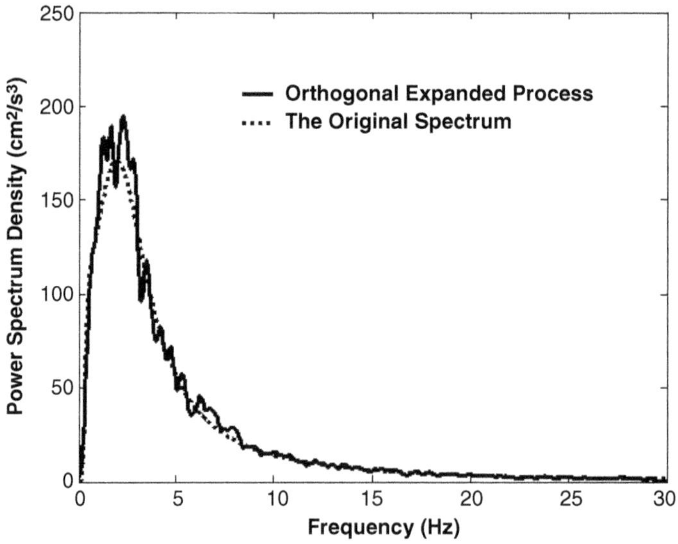

Figure 3.18 PSD of the orthogonal expanded process and the original spectrum.

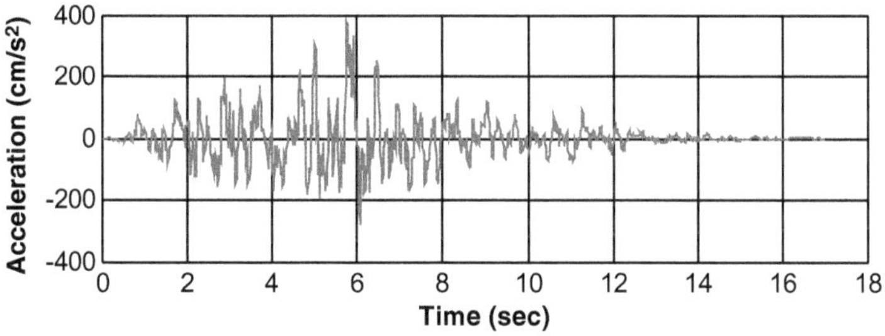

Figure 3.19 A typical acceleration sample.

3.5.4 Orthogonal Expansion of Fluctuating Wind Speed Process

Research shows that direct use of the PSD function of fluctuating wind speed for orthogonal expansion will bring more expansion items. Therefore, the concept of the *virtual wind-displacement time history*, the integration of the wind-speed time process, is introduced (Liu and Li, 2008).

For the time process of virtual wind displacement, Equation 3.122 is used for the orthogonal expansion and the Hartley orthogonal basis is taken as a standard orthogonal function set. Then the orthogonal expansion results of the fluctuating wind-speed stochastic process can be given by taking time differentiation of the above results as follows:

$$V(\boldsymbol{\zeta}, t) = \sum_{j=1}^{r} \zeta_j \sqrt{\lambda_j} G_j(t) \tag{3.136}$$

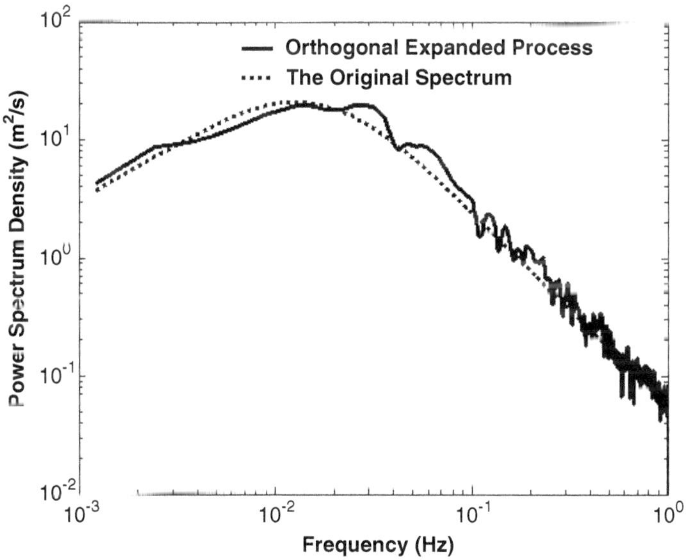

Figure 3.20 PSD of the orthogonal expanded process and the original spectrum.

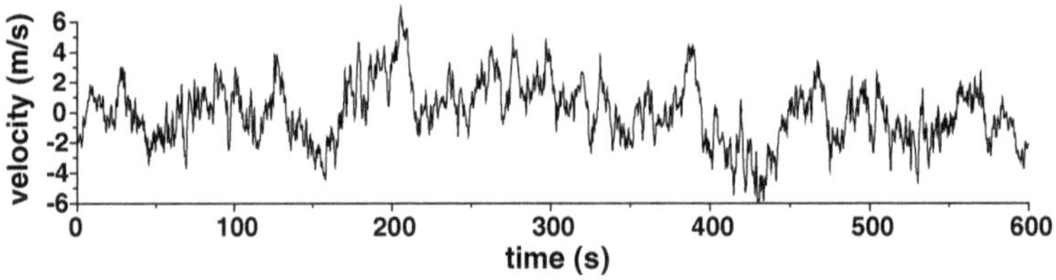

Figure 3.21 A typical sample of fluctuating wind speed.

where

$$G_j(t) = \sum_{k=0}^{N} \beta_{k+1}\phi_{j,k+1}\dot{\varphi}_k(t) \tag{3.137}$$

Here, the β_{k+1} introduced is a set of amendment coefficients to make up for the truncated error and can be given using the principle of energy equivalence (Li and Liu, 2008).

Figure 3.20 shows the comparison between the PSD obtained from the orthogonal expansion of the fluctuating wind-speed process of the Davenport spectrum and the PSD of the original model. Figure 3.21 shows a typical random fluctuating wind process sample given by the expanded function.

4

Stochastic Structural Analysis

4.1 Introductory Remarks

It has been common practice in engineering to analyze structural systems by assuming that the systems are exactly determined; for example, the system parameters and the system inputs are deterministic parameters or excitations. However, such ideal situations are rarely encountered in engineering reality. Not only the system inputs are stochastic processes (such as the fluctuating wind excitations and the earthquake ground motions), but also the structural parameters (such as the material mass density and the elastic modulus) need to be considered as uncertain variables in the design process (Vanmarcke, 1983; Li, 1996a). In this chapter, our basic concern is the uncertainties involved in the structural parameters. That is, linear differential equations with random coefficients will be dealt with in the following sections.

Three basic methodologies can be used to quantify the structural response uncertainties. The first is the Monte Carlo simulation (MCS) method (Shinozuka, 1972; Shinozuka and Jan, 1972). In such a simulated process, a set of random samples is generated first to represent the statistical uncertainties in the structures. These random samples are then substituted in the finite-element model to obtain the response of the sample structures, whereby the statistical characteristics of specified responses are analyzed to quantify the response uncertainty. The second methodology, known as perturbation technology, relies on the use of a Taylor series expansion to formulate the physical relationships between some characteristics of the random responses and the random structural parameters (Collins and Thompson, 1969; Hisada and Nakagiri, 1981, 1982; Liu *et al.*,1985, 1986; Kleiber and Hien, 1992). The third methodology, which was developed in 1990s, is the orthogonal polynomial expansion method or, as called in this book, the expanded-order system method (Spanos and Ghanem, 1989; Iwan and Jensen, 1993; Li,1995a, 1995b, 1995c, 1996a). In the approach, the responses of structures with random parameters are expanded by a suitable orthogonal polynomial in a probability space, then an expanded-order system equation can be deduced which will govern the responses of the stochastic structure. In this chapter, all these developed methodologies will be introduced.

Stochastic Dynamics of Structures Jie Li and Jianbing Chen

4.2 Fundamentals of Deterministic Structural Analysis

In any quantitative approach for analysis of stochastic structures, the deterministic analysis using the finite-element method may be an essential basis. There have been so many textbooks on the finite-element method published during the last four decades that any attempt to repeat the description would seem superfluous. However, considering that most contents of this book deal with a special aspect of the method, we at least need to set up the fundamentals of the finite-element method in such a way as to make it possible to elaborate on probabilistic issues.

4.2.1 The Basic Idea of Finite-Element Analysis

All the practical structures are essentially infinite degree-of-freedom systems. However, in studies or applications, an infinite degree-of-freedom system may be transformed into a finite degree-of-freedom system using some types of discretization method, among which the finite-element discretization method is a typical one (Bathe and Wilson, 1985; Zienkiewicz and Taylor, 2004). Using this approach, the structure is assumed to be partitioned into a system of discrete elements which are interconnected only at a finite number of nodal points. According to the nature and practical background of the problem, the connection form of element nodes may be treated, for instance, as hinged or rigid connections. Figure 4.1 is a schematic plan of the partitioning of finite elements of some typical structures.

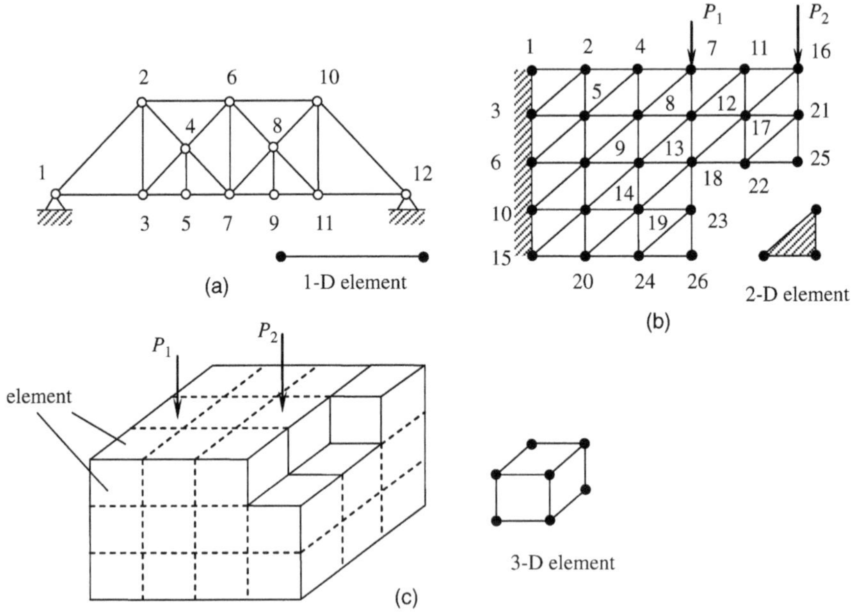

Figure 4.1 Finite-element partition of structures.

The responses of structures under external loads, such as stresses, strains, internal forces and displacements, are generally continuous functions. However, for those discretized structural systems using finite-element discretization, the previous continuous functions are replaced by approximation functions that are smooth in each element. These functions should be continuous and stepwise smooth in the whole structure. Moreover, the physical quantities at the element nodes are generally selected as the unknown variables to establish the approximate function in an element. The approximation functions in each element are given in a unified form and generally termed the shape function or interpolation function. The polynomials are usually selected as shape functions. According to the selected shape functions, combining with the stress–strain physical relationship and the boundary conditions, we can establish the expression of element energy. Then, the governing equation can be obtained by the variational principles. By solving the governing equation, the responses of the element nodes can be obtained and, accordingly, the responses of the whole structure may be computed easily through the shape functions.

According to differences in the selected physical variables and the corresponding variational principles, the finite-element method can be classified into the finite-element displacement method, the finite-element force method and the mixed finite-element method. The finite-element displacement method is based on the principle of minimum potential energy and takes the displacements of nodes as basic unknown variables. The finite-element force method is based on the principle of minimum complementary energy and takes the forces of nodes as basic unknown variables. The mixed finite-element method is based on the principle of Reissner variation and takes the nodal displacements or nodal forces as basic unknown variables in different regions (Washizu, 1975). Considering that this book is not a monograph on the finite-element method, we only briefly introduce the analysis process of the finite-element displacement method and take the common truss structure as the background.

4.2.2 Element Stiffness Matrix

Considering a typical element in a plane truss structure, let the serial numbers of its ends be i and j. The coordinate system established for the element e, the end forces and the deformations are shown in Figure 4.2. These forces and displacements can be expressed as

$$\tilde{\mathbf{F}}_{ei} = (N_{ei}, Q_{ei}, M_{ei})^{\mathrm{T}} \qquad \tilde{\mathbf{F}}_{ej} = (N_{ej}, Q_{ej}, M_{ej})^{\mathrm{T}} \tag{4.1}$$

and

$$\tilde{\Delta}_{ei} = (u_{ei}, v_{ei}, \vartheta_{ei})^{\mathrm{T}} \qquad \tilde{\Delta}_{ej} = (u_{ej}, v_{ej}, \vartheta_{ej})^{\mathrm{T}} \tag{4.2}$$

respectively.

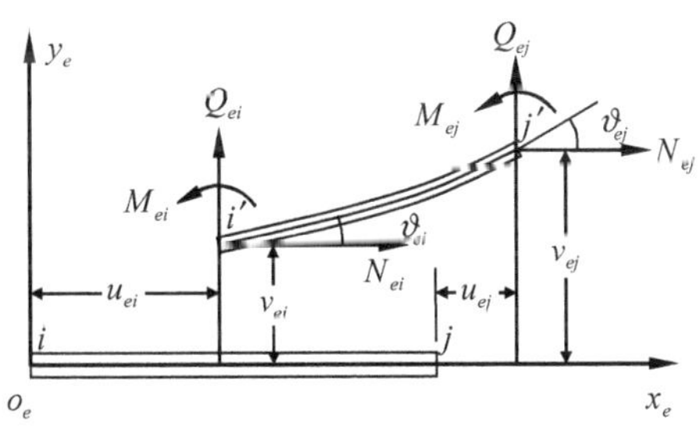

Figure 4.2 Member coordinate system.

According to material mechanics, the axial displacement u_e of an arbitrary point in a member is a linear function of the member end axial displacement; that is:

$$u_e(x_e) = u_{ei}\frac{l_e - x_e}{l_e} + u_{ej}\frac{x_e}{l_e} = u_{ei}\varphi_1(x_e) + u_{ej}\varphi_2(x_e) \tag{4.3}$$

where x_e is the horizontal coordinate of a point in the member in the local coordinate system and l_e is the length of the member.

On the other hand, the deflection v_e of a point in the member can be expressed by a cubic curve:

$$v_e(x_e) = a_1 + a_2x_e + a_3x_e^2 + a_4x_e^3 \tag{4.4}$$

According to the small deformation assumption, the following relationship exists between the rotation angle and the deflection of an arbitrary point in the member:

$$\vartheta_e(x_e) \approx \tan\vartheta_e(x_e) = \frac{\mathrm{d}v_e}{\mathrm{d}x_e} \tag{4.5}$$

and using the boundary conditions at the ends

$$\begin{cases} v_e(0) = v_{ei} \\ \vartheta_e(0) = \vartheta_{ei} \end{cases} \tag{4.6}$$

$$\begin{cases} v_e(l_e) = v_{ej} \\ \vartheta_e(l_e) = \vartheta_{ej} \end{cases} \tag{4.7}$$

there exists

$$\begin{aligned} v_e(x_e) &= v_{ei}\left(1 - 3\frac{x_e^2}{l_e^2} + 2\frac{x_e^3}{l_e^3}\right) + v_{ej}\left(3\frac{x_e^2}{l_e^2} - 2\frac{x_e^3}{l_e^3}\right) + \vartheta_{ei}\left(x_e - 2\frac{x_e^2}{l_e} + \frac{x_e^3}{l_e^2}\right) + \vartheta_{ej}\left(-\frac{x_e^2}{l_e} + \frac{x_e^3}{l_e^2}\right) \\ &= v_{ei}\varphi_3(x_e) + v_{ej}\varphi_4(x_e) + \vartheta_{ei}\varphi_5(x_e) + \vartheta_{ej}\varphi_6(x_e) \end{aligned} \tag{4.8}$$

Let

$$\mathbf{f}_e(x_e) = [u_e(x_e), v_e(x_e)]^{\mathrm{T}} \tag{4.9}$$

$$\mathbf{N}_u(x) = [\varphi_1(x_e), 0, 0, \varphi_2(x_e), 0, 0] \tag{4.10}$$

$$\mathbf{N}_v(x) = [0, \varphi_3(x_e), \varphi_5(x_e), 0, \varphi_4(x_e), \varphi_6(x_e)] \tag{4.11}$$

Then, from Equations 4.3 and 4.8, we obtain

$$\mathbf{f}_e(x_e) = \begin{bmatrix} \mathbf{N}_u(x_e) \\ \mathbf{N}_v(x_e) \end{bmatrix} \begin{bmatrix} \tilde{\Delta}_{ei} \\ \tilde{\Delta}_{ej} \end{bmatrix} = \mathbf{N}(x_e)\tilde{\Delta}_e \tag{4.12}$$

Here, $\mathbf{N}_u(x_e)$ and $\mathbf{N}_v(x_e)$ are the interpolation shape functions, which have a unified form for each structural element, and $\tilde{\Delta}_e$ is the element displacement in the local coordinate system. According to material mechanics, the strain in the element is given by

$$\boldsymbol{\varepsilon}_e = \begin{bmatrix} \dfrac{\mathrm{d}u_e}{\mathrm{d}x_e} \\ -y_e \dfrac{\mathrm{d}^2 v_e}{\mathrm{d}x_e^2} \end{bmatrix} \tag{4.13}$$

where y_e is the distance from an arbitrary point in the sections to the neutral axis of the member.

Substituting the interpolation function expression in Equation 4.13 yields

$$\boldsymbol{\varepsilon}_e = \begin{bmatrix} \dfrac{\mathrm{d}\mathbf{N}_u}{\mathrm{d}x_e} \\ -y_e \dfrac{\mathrm{d}^2 \mathbf{N}_v}{\mathrm{d}x_e^2} \end{bmatrix} \begin{bmatrix} \tilde{\Delta}_{ei} \\ \tilde{\Delta}_{ej} \end{bmatrix} = \mathbf{B}\tilde{\Delta}_e \tag{4.14}$$

where $\mathbf{B}$ is usually called the geometry matrix.

The element potential energy Π_e is composed of two parts:

$$\Pi_e = W_e + V_e \tag{4.15}$$

where W_e is the element load potential energy; in the case that only end forces are applied:

$$W_e = -[\tilde{\Delta}_{ei}^{\mathrm{T}} \quad \tilde{\Delta}_{ej}^{\mathrm{T}}] \begin{bmatrix} \tilde{\mathbf{F}}_{ei} \\ \tilde{\mathbf{F}}_{ej} \end{bmatrix} = -\tilde{\Delta}_e^{\mathrm{T}} \tilde{\mathbf{F}}_e \tag{4.16}$$

in which $\tilde{\mathbf{F}}_e$ is the element nodal force vector in the local coordinate system.

In Equation 4.15, V_e is the strain energy (deformation energy). According to elasticity mechanics, V_e is given by

$$V_e = \frac{1}{2}\int_{\Omega_e} \boldsymbol{\varepsilon}_e^{\mathrm{T}} \mathbf{E} \boldsymbol{\varepsilon}_e \, \mathrm{d}\Omega \tag{4.17}$$

where $\mathbf{E}$ is the elastic modulus matrix and Ω_e is the integration domain of the element.

Substituting Equation 4.14 in Equation 4.17, we obtain

$$\begin{aligned} V_e &= \frac{1}{2}\tilde{\Delta}_e^{\mathrm{T}} \left(\int_{\Omega_e} \mathbf{B}^{\mathrm{T}} \mathbf{E} \mathbf{B} \, \mathrm{d}\Omega \, \Delta_e \right) \\ &= \frac{1}{2}\tilde{\Delta}_e^{\mathrm{T}} \tilde{\mathbf{k}}_e \tilde{\Delta}_e \end{aligned} \tag{4.18}$$

where

$$\tilde{\mathbf{k}}_e = \int_{\Omega_e} \mathbf{B}^{\mathrm{T}} \mathbf{E} \mathbf{B} \, \mathrm{d}\Omega \tag{4.19}$$

is the element stiffness matrix in the local coordinate system. For the coordinate system determined in Figure 4.2, its explicit expression is

$$\tilde{\mathbf{k}}_e = \begin{bmatrix} \frac{EA_e}{l_e} & 0 & 0 & -\frac{EA_e}{l_e} & 0 & 0 \\ 0 & \frac{12EI_e}{l_e^3} & \frac{6EI_e}{l_e^2} & 0 & -\frac{EI_e}{l_e^3} & \frac{6EI_e}{l_e^2} \\ 0 & \frac{6EI_e}{l_e^2} & \frac{4EI_e}{l_e^2} & 0 & -\frac{6EI_e}{l_e^2} & \frac{2EI_e}{l_e} \\ -\frac{EA_e}{l_e} & 0 & 0 & \frac{EA_e}{l_e} & 0 & 0 \\ 0 & -\frac{12EI_e}{l_e^3} & -\frac{6EI_e}{l_e^2} & 0 & \frac{12EI_e}{l_e^3} & -\frac{6EI_e}{l_e^2} \\ 0 & \frac{6EI_e}{l_e^2} & \frac{2EI_e}{l_e} & 0 & -\frac{6EI_e}{l_e^2} & \frac{4EI_e}{l_e^2} \end{bmatrix} \tag{4.20}$$

where A_e is the area of the element section and I_e is the inertia moment of the element section with respect to the section central axial.

Substituting Equations 4.16 and 4.18 in Equation 4.15 yields

$$\Pi_e = \frac{1}{2}\tilde{\Delta}_e^{\mathrm{T}}\tilde{\mathbf{k}}_e\tilde{\Delta}_e - \tilde{\Delta}_e^{\mathrm{T}}\tilde{\mathbf{F}}_e \tag{4.21}$$

4.2.3 Transformation of Coordinates

The element analysis described in the preceding section is done in the local coordinate system, such as in Figure 4.2; the position of the element in the whole structure is not considered as yet. In order to derive the governing equation of the whole structures, we should transform the elements from the local coordinate system to the global coordinate system and locate the positions of elements in the whole structures. The aim of the operations is to establish an expression of the potential energy of the elements in the global coordinate system.

The relationship between the global coordinate system Oxy and the local coordinate system is shown in Figure 4.3. According to the geometric relationship, the displacement u_{ei}, v_{ei} of node i and the displacement u_i, v_i in the global coordinate system satisfy the following

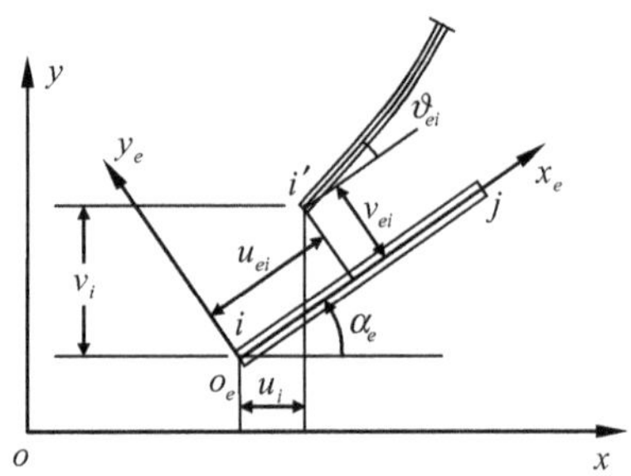

Figure 4.3 Coordinate transformation.

relationship

$$\begin{cases} u_{ei} = u_i \cos \alpha_e + v_i \sin \alpha_e \\ v_{ei} = -u_i \sin \alpha_e + v_i \cos \alpha_e \end{cases} \tag{4.22}$$

and the angles of rotation in the two coordinate systems are identical; thus

$$\vartheta_{ei} = \vartheta_i \tag{4.23}$$

According to the same principle, to consider the nodal displacement of the element, we have

$$\tilde{\Delta}_e = \mathbf{T}_\alpha \Delta_e \tag{4.24}$$

where

$$\Delta_e = (u_i, v_i, \vartheta_i, u_j, v_j, \vartheta_j)^{\mathrm{T}} \tag{4.25}$$

is the element displacement vector in the global coordinate system and

$$\mathbf{T}_\alpha = \begin{bmatrix} \cos\alpha_e & \sin\alpha_e & 0 & 0 & 0 & 0 \\ -\sin\alpha_e & \cos\alpha_e & 0 & 0 & 0 & 0 \\ 0 & 0 & 1 & 0 & 0 & 0 \\ 0 & 0 & 0 & \cos\alpha_e & \sin\alpha_e & 0 \\ 0 & 0 & 0 & -\sin\alpha_e & \cos\alpha_e & 0 \\ 0 & 0 & 0 & 0 & 0 & 1 \end{bmatrix} \tag{4.26}$$

is the element coordinate transformation matrix. This is an orthogonal matrix; that is:

$$\mathbf{T}_\alpha^{-1} = \mathbf{T}_\alpha^{\mathrm{T}} \tag{4.27}$$

Thus, the inverse transformation of Equation 4.24 gives

$$\Delta_e = \mathbf{T}_\alpha^{\mathrm{T}} \tilde{\Delta}_e \tag{4.28}$$

In a similar way, the relationship between the nodal forces of the element in the local coordinate system and that in the global coordinate system is as follows:

$$\tilde{\mathbf{F}}_e = \mathbf{T}_\alpha \mathbf{F}_e \tag{4.29}$$

$$\mathbf{F}_e = \mathbf{T}_\alpha^{\mathrm{T}} \tilde{\mathbf{F}}_e \tag{4.30}$$

where

$$\mathbf{F}_e = (N_i, Q_i, M_i, N_j, Q_j, M_j)^{\mathrm{T}} \tag{4.31}$$

is the nodal force vector of the element in the global coordinate system.

In substituting Equations 4.24 and 4.29 in Equation 4.21, the element potential energy in the global coordinate system is given by

$$\begin{aligned}\Pi_e &= \frac{1}{2}\Delta_e^{\mathrm{T}}\mathbf{T}_\alpha^{\mathrm{T}}\tilde{\mathbf{k}}_e\mathbf{T}_\alpha\Delta_e - \Delta_e^{\mathrm{T}}\mathbf{T}_\alpha^{\mathrm{T}}\mathbf{T}_\alpha\mathbf{F}_e \\ &= \frac{1}{2}\Delta_e^{\mathrm{T}}\mathbf{k}_e\Delta_e - \Delta_e^{\mathrm{T}}\mathbf{F}_e\end{aligned} \tag{4.32}$$

where

$$\mathbf{k}_e = \mathbf{T}_\alpha^{\mathrm{T}}\tilde{\mathbf{k}}_e\mathbf{T}_\alpha \tag{4.33}$$

is the element stiffness matrix in the global coordinate system.

The transformation of element coordinates only performs the transformation between the global coordinate system and the element local coordinate system. It does not resolve the problem of locating the element position in the structures. This work is realized through the concept of the position transformation matrix.

If a structure can be partitioned into N elements, then its nodal displacements column matrix can be denoted by

$$\mathbf{x} = (x_1, x_2, \ldots, x_{3n})^{\mathrm{T}} = (\mathbf{u}_1^{\mathrm{T}}, \mathbf{u}_2^{\mathrm{T}}, \ldots, \mathbf{u}_n^{\mathrm{T}})^{\mathrm{T}} \tag{4.34}$$

where $\mathbf{u}_\ell = (u_\ell, v_\ell, \vartheta_\ell)^{\mathrm{T}}(\ell = 1, 2, \ldots, n)$; n is the total number of the nodes.

The position of the eth element in $\mathbf{x}$ can be given by the following position transformation relationship:

$$\Delta_e = \begin{bmatrix}\mathbf{T}_i \\ \mathbf{T}_j\end{bmatrix}\mathbf{x} = \mathbf{T}_e\mathbf{x} \tag{4.35}$$

where

$$\mathbf{T}_i = [\mathbf{0}, \cdots, \mathbf{I}, \mathbf{0}, \cdots, \mathbf{0}] \tag{4.36}$$

in which $\mathbf{I}$ is the ith block and

$$\mathbf{T}_j = [\mathbf{0}, \cdots, \mathbf{I}, \mathbf{0}, \cdots, \mathbf{0}] \tag{4.37}$$

in which $\mathbf{I}$ is the jth block and

$$\mathbf{I} = \begin{bmatrix}1 & 0 & 0 \\ 0 & 1 & 0 \\ 0 & 0 & 1\end{bmatrix}, \quad \mathbf{0} = \begin{bmatrix}0 & 0 & 0 \\ 0 & 0 & 0 \\ 0 & 0 & 0\end{bmatrix} \tag{4.38}$$

Therefore, $\mathbf{T}_e$ is given by

$$\mathbf{T}_e = \begin{bmatrix}\mathbf{T}_i \\ \mathbf{T}_j\end{bmatrix} = \begin{bmatrix}\mathbf{0}, \cdots, \mathbf{I}, \mathbf{0}, \cdots, \mathbf{0}, \mathbf{0}, \cdots, \mathbf{0} \\ \mathbf{0}, \cdots, \mathbf{0}, \mathbf{0}, \cdots, \mathbf{I}, \mathbf{0}, \cdots, \mathbf{0}\end{bmatrix} \tag{4.39}$$

as the position transformation matrix of element e. Generally, the position transformation matrix is not an orthogonal matrix.

Equation 4.35 exists when the displacement is continuous, but there is no such relationship among element nodal forces and structural nodal loads because of equilibrium among internal nodal forces and external loads on a common node. However, by introducing the position transformation matrix, it can be used to locate the nodal forces simultaneously. In fact, if

$$\bar{\mathbf{F}}_e = \mathbf{T}_e^{\mathrm{T}}\mathbf{F}_e \tag{4.40}$$

then combining with Equation 4.35, Equation 4.32 can be rewritten as

$$\Pi_e = \frac{1}{2}\mathbf{x}^{\mathrm{T}}\mathbf{T}_e^{\mathrm{T}}\mathbf{k}_e\mathbf{T}_e\mathbf{x} - \mathbf{x}^{\mathrm{T}}\bar{\mathbf{F}}_e \tag{4.41}$$

4.2.4 Static Equations

The total potential energy of the whole structure is the sum of all the element potential energy; that is:

$$\Pi = \sum_e \Pi_e = \frac{1}{2}\mathbf{x}^{\mathrm{T}}\sum_e(\mathbf{T}_e^{\mathrm{T}}\mathbf{k}_e\mathbf{T}_e)\mathbf{x} - \mathbf{x}^{\mathrm{T}}\sum_e \bar{\mathbf{F}}_e \tag{4.42}$$

According to the equilibrium relationship between the element end forces and the nodal equivalent load, there exists

$$\sum_e \bar{\mathbf{F}}_e = \mathbf{F} \tag{4.43}$$

where **F** is the nodal force vector of the structure. For the non-nodal loads, **F** can be given by the principle of equivalence of potential energy.

Let

$$\mathbf{K} = \sum_e \mathbf{T}_e^{\mathrm{T}}\mathbf{k}_e\mathbf{T}_e \tag{4.44}$$

which is called the global stiffness matrix. Substituting this equation and Equation 4.43 in Equation 4.42 yields

$$\Pi = \frac{1}{2}\mathbf{x}^{\mathrm{T}}\mathbf{K}\mathbf{x} - \mathbf{x}^{\mathrm{T}}\mathbf{F} \tag{4.45}$$

According to the principle of minimum potential energy, for all the allowable displacement functions, the real displacements minimize the total potential energy (Washizu, 1975); that is:

$$\delta\Pi = 0 \qquad \delta^2\Pi \geq 0 \tag{4.46}$$

where δ denotes the variational symbol.

Utilizing the necessary condition of $\delta\Pi = 0$, namely

$$\frac{\partial \Pi}{\partial \mathbf{x}} = 0 \tag{4.47}$$

we have

$$\mathbf{K}\mathbf{x} = \mathbf{F} \tag{4.48}$$

This is the governing equation of deterministic structural static analysis, with $\mathbf{x}$ being the unknowns.

4.2.5 Dynamic Equations

For dynamical analysis problems, new particularity only lies in the contribution of the structural mass matrix and the damping matrix.

Still taking the previous plane truss structures as an example and using the shape functions expressed by Equations 4.10 and 4.11, the velocity of each node in the element e can be expressed by

$$\dot{\mathbf{f}}_e = \mathbf{N}(x_e)\dot{\tilde{\Delta}}_e \tag{4.49}$$

and the kinetic energy of the element e is given by

$$\begin{aligned} T_e &= \frac{1}{2}\int_{\Omega_e} \dot{\mathbf{f}}_e^{\mathrm{T}} \rho \dot{\mathbf{f}}_e \, \mathrm{d}\Omega = \frac{1}{2}\dot{\tilde{\Delta}}_e^{\mathrm{T}} \left(\int_{\Omega_e} \mathbf{N}^{\mathrm{T}} \rho \mathbf{N} \, \mathrm{d}\Omega \right) \dot{\tilde{\Delta}}_e \\ &= \frac{1}{2}\dot{\tilde{\Delta}}_e^{\mathrm{T}} \tilde{\mathbf{m}}_e \dot{\tilde{\Delta}}_e \end{aligned} \tag{4.50}$$

where ρ is the mass density of the material, Ω_e is the integration domain of the element and $\tilde{\mathbf{m}}_e$ is the element mass matrix in the local coordinate system; that is:

$$\tilde{\mathbf{m}}_e = \int_{\Omega_e} \mathbf{N}^{\mathrm{T}} \rho \mathbf{N} \, \mathrm{d}\Omega \tag{4.51}$$

Because of the effect of damping, the energy of the structural system is gradually dissipated during vibration. If the damping forces acting on a structure could be determined quantitatively, then the finite-element concept could be used again to define the element damping matrix. For example, assuming the damping is of viscosity – that is, the viscous damping force of any particle m_i is $-\eta \dot{f}_i$, where η is the viscous damping coefficient and $\dot{f}_i$ is the velocity of the particle m_i – then the dissipation function R of the element e can be given by

$$R_e = \frac{1}{2}\int_{\Omega_e} \dot{\mathbf{f}}_e^{\mathrm{T}} \eta \dot{\mathbf{f}}_e \, \mathrm{d}\Omega = \frac{1}{2}\dot{\tilde{\Delta}}_e^{\mathrm{T}} \tilde{\mathbf{c}}_e \dot{\tilde{\Delta}}_e \tag{4.52}$$

where

$$\tilde{\mathbf{c}}_e = \int_{\Omega_e} \mathbf{N}^{\mathrm{T}} \eta \mathbf{N} \, \mathrm{d}\Omega \tag{4.53}$$

is the element damping matrix in the local coordinate system.

Analogous to Equation 4.33, the element mass matrix and the element damping matrix in the global coordinate system are given by

$$\mathbf{m}_e = \mathbf{T}_\alpha^{\mathrm{T}} \tilde{\mathbf{m}}_e \mathbf{T}_\alpha \tag{4.54}$$

and

$$\mathbf{c}_e = \mathbf{T}_\alpha^{\mathrm{T}} \tilde{\mathbf{c}}_e \mathbf{T}_\alpha \tag{4.55}$$

respectively. The position transformation relationship of displacements still satisfy Equation 4.35; differentiating it will yield

$$\dot{\Delta}_e = \mathbf{T}_e \dot{\mathbf{x}} \tag{4.56}$$

where $\dot{\Delta}_e$ is the element velocity vector in the global coordinate system and $\dot{\mathbf{x}}$ is the nodal velocity vector of the structure.

Having performed the coordinate transformation and the element position transformation, the kinetic energy and the dissipation functions of the element e are

$$T_e = \frac{1}{2} \dot{\mathbf{x}}^{\mathrm{T}} \mathbf{T}_e^{\mathrm{T}} \mathbf{m}_e \mathbf{T}_e \dot{\mathbf{x}} \tag{4.57}$$

and

$$R_e = \frac{1}{2} \dot{\mathbf{x}}^{\mathrm{T}} \mathbf{T}_e^{\mathrm{T}} \mathbf{c}_e \mathbf{T}_e \dot{\mathbf{x}} \tag{4.58}$$

respectively. The total kinetic energy and dissipation functions of the whole structure are given by the sum of the corresponding values of all the elements; that is:

$$T = \frac{1}{2} \dot{\mathbf{x}}^{\mathrm{T}} \mathbf{M} \dot{\mathbf{x}} \tag{4.59}$$

$$R = \frac{1}{2} \dot{\mathbf{x}}^{\mathrm{T}} \mathbf{C} \dot{\mathbf{x}} \tag{4.60}$$

where

$$\mathbf{M} = \sum_e \mathbf{T}_e^{\mathrm{T}} \mathbf{m}_e \mathbf{T}_e \tag{4.61}$$

is the mass matrix of the structure and

$$\mathbf{C} = \sum_e \mathbf{T}_e^{\mathrm{T}} \mathbf{c}_e \mathbf{T}_e \tag{4.62}$$

is the damping matrix.

The potential energy of the structure is similar to that in the previous section (see Equation 4.45; it is denoted by V herein according to conventional notation in the Lagrange equation):

$$V = \frac{1}{2} \mathbf{x}^{\mathrm{T}} \mathbf{K} \mathbf{x} - \mathbf{x}^{\mathrm{T}} \mathbf{F} \tag{4.63}$$

It is noted that V is the potential energy with the boundary conditions having been imposed on. Hence, $\mathbf{x}$ is the nodal displacement vector of the whole structure with imposed boundary conditions, too. So is the velocity vector $\dot{\mathbf{x}}$.

The general variational principle of dynamic systems is the Hamiltonian principle. Because this principle is equivalent to the Lagrange equation, we often use the latter to establish the governing equation of dynamical systems. The Lagrange equation for viscous damping system is given by (Lanczos, 1970; Clough and Penzien, 1993)

$$\frac{\mathrm{d}}{\mathrm{d}t}\left(\frac{\partial T}{\partial \dot{\mathbf{x}}}\right) - \frac{\partial T}{\partial \mathbf{x}} + \frac{\partial V}{\partial \mathbf{x}} = -\frac{\partial R}{\partial \dot{\mathbf{x}}} \tag{4.64}$$

Substituting Equations 4.59, 4.60 and 4.63 in Equation 4.64, we can obtain the equation of motion of a deterministic structural system:

$$\mathbf{M}\ddot{\mathbf{x}} + \mathbf{C}\dot{\mathbf{x}} + \mathbf{K}\mathbf{x} = \mathbf{F} \tag{4.65}$$

This equation can be solved by various methods, such as the time-domain method, the frequency-domain method or the modal superposition technique (Clough and Penzien, 1993).

It is noted that the previously introduced mass matrix is the consistent mass matrix. In practice, when the structural system is relatively regular and simple, we may use the lumped mass matrix, and the preceding damping matrix is also generally replaced by the Rayleigh damping matrix, which is usually given by

$$\mathbf{C} = a\mathbf{M} + b\mathbf{K} \tag{4.66}$$

where a and b are coefficients that depend on the modal damping ratios.

4.3 Random Simulation Method

4.3.1 Monte Carlo Method

The Monte Carlo method is a numerical method to evaluate the approximate solutions of physical and engineering problems by means of digital simulation and statistical analysis of random variables (Robinstein, 1981; Shinozuka and Deodatis, 1991; 1996). The process of solving a problem through the Monte Carlo method may be summarized as the following three essential steps:

(a) Sampling of random variables. To generate random samples according to the known probability distributions of the basic random variables.
(b) Obtaining the sample solutions. To get the response of each sample by solving deterministic mathematical or physical equations according to the nature of the problem.
(c) Computing the statistical estimation of the response quantities. For all the sample responses, compute the mean and the variance or estimate the probability distributions of the respective output random variables.

The theoretical foundation of the Monte Carlo method is the law of large numbers in probability theory (Loève, 1977). Let n be the occurrence number of event A and $p(A)$ be the occurrence probability of the event A in N independent trials; then, according to the Bernoulli law of large numbers, for any $\varepsilon > 0$, as $N \to \infty$, the frequency n/N of event A will converge to the probability of the event with unit probability; that is:

$$\lim_{N\to\infty} \Pr\left\{\left|\frac{n}{N} - p(A)\right| < \varepsilon\right\} = 1 \tag{4.67}$$

If the random variables are independently and identically distributed – that is, if the random variable sequence $\xi_1, \xi_2, \ldots, \xi_N$ have the same distribution and mathematical expectation $\mathcal{E}[\xi_i] = \alpha, i = 1, 2, \ldots, N$ – then, according to the Kolmogorov law of large numbers, for any $\varepsilon > 0$, as $N \to \infty$, the variable $(1/N)\sum_{i=1}^{N} \xi_i$ will converge with probability unity to the expected value α; that is:

$$\lim_{N\to\infty} \Pr\left\{\left|\frac{1}{N}\sum_{i=1}^{N} \xi_i - \alpha\right| \geq \varepsilon\right\} = 0 \tag{4.68}$$

In the standard Monte Carlo method, the simple sampling method is applied to the digital simulation of random variables. Therefore, each sample is an independent random variable with the characteristic of identical distribution. According to the law of large numbers mentioned above, as the numbers of samples are large enough, the mean of the samples will converge with probability unity to the mean of the probability distribution. Meanwhile, the frequency n/N of event A will converge to the occurrence probability of event A, so the convergence of the Monte Carlo method can be ensured.

4.3.2 Sampling of Random Variables with Uniform Distribution

According to the background and characteristics of the problem to be resolved, the random variables involved in the stochastic system may belong to different probability distributions. In order to generate samples of the random variables with different types of probability distribution, a sample value of the random variable with uniform distribution on [0, 1] is usually generated first. Then the sample is converted into the desired variable according to the given type of probability distribution. Therefore, the sampling technology of uniform random variables is the basis for realization of the Monte Carlo method (Robinstein, 1981; Niederreiter, 1992).

The foundation of generating a uniform random variable is a certain mathematical recursive formula, of which the general form is

$$x_{n+1} = f(x_n, x_{n-1}, \ldots, x_{n-k}) \tag{4.69}$$

where $f(x_n, x_{n-1}, \ldots, x_{n-k})$ is a given function. According to the function, once a set of initial values $x_0, x_{-1}, \ldots, x_{-k}$ is given, the sequence $x_1, x_2, x_3, \ldots$ can be obtained one by one.

The common recursive formula is the linear congruential generator, which states that

$$y_n = (a y_{n-1} + b)(\mathrm{mod} M) \tag{4.70a}$$

$$x_n = \frac{y_n}{M} \tag{4.70b}$$

where the multiplier a, the increment b, the modulus M and the initial value y_0 are all nonnegative integers. The modulo notation (mod M) means that

$$y_n = ay_{n-1} + b - k_n M \tag{4.70c}$$

where $k_n = [(ay_{n-1} + b)/M]$ denotes the largest positive integer smaller than $(ay_{n-1} + b)/M$.

Equation 4.70a expresses such a computing process that, for a given y_{n-1}, if y_n denotes the remainder of $ay_{n-1} + b$ divided by M, then the recursive form of recursively computing y_n can be obtained by successively increasing ordinal numbers and using Equation 4.70a. Equation 4.70b shows that y_n/M gives rise to the pseudo-random numbers sequence x_n over interval [0, 1].

Example 4.1. Pseudo-Random Number Sequences Tables 4.1 and 4.2 are two sequences generated by Equations 4.70a–4.70c for different values of a, b, M and y_0.

Table 4.1 Sequence by Equations 4.70a–4.70c ($a = 2$, $b = 3$, $M = 7$, $y_0 = 1$).

n	1	2	3	4	5	6	7	8	9
y_n	5	6	1	5	6	1	5	6	1
x_n	0.7143	0.8571	0.1429	0.7143	0.8571	0.1429	0.7143	0.8571	0.1429

Table 4.2 Sequence by Equations 4.70a, 4.70b, 4.70c ($a = 3$, $b = 2$, $M = 11$, $y_0 = 5$).

n	1	2	3	4	5	6	7	8	9
y_n	6	9	7	1	5	6	9	7	1
x_n	0.5455	0.8182	0.6364	0.0909	0.4545	0.5455	0.8182	0.6364	0.0909

□

The pseudo-random number sequences generated by the previous linear congruence method have a definite period (see Tables 4.1 and 4.2; the periods are 3 and 5 respectively). However, in most MCS algorithms for practical problems, random sampling usually demands thousands or even millions of random numbers, so we hope that a simulation algorithm should have a long period and relatively perfect statistical characteristics (mainly referring to homogeneity and independence). In order to improve the property of random numbers generated by the linear congruence method, we can use two effective methods as follows.

4.3.2.1 Mixed Shuffle Method

In the method we first generate a set of random variables $v_1, v_2, \ldots, v_n$ using a generator of standard pseudo-random numbers, and then a random positive integer j uniformly distributed over $[1, n]$ is generated using another generator of random numbers. Then, take v_j as the sampled value; the vacant position originally belonging to v_j will be filled with a new random number generated by the generator of standard pseudo-random numbers. Continuing such a process will give a random variable sequence. The process is shown schematically in Figure 4.4, where Y is the random positive integer sequence. The sequences generated according to the

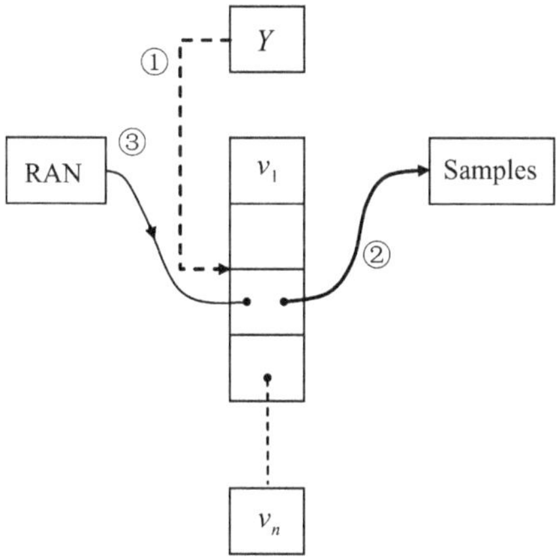

Figure 4.4 The process of the mixed shuffle method (1, 2, 3 denote the order of the mixed shuffle).

mixed shuffle method have less self correlation among sequences and better uniformity than that of sequences generated by a single generator.

4.3.2.2 Joint Sampling Method

In this method we generate random numbers with uniform distribution over [0, 1] using three linear congruential generators of random numbers. The first generator generates the maximum effective part of the random numbers, the second generator generates the minimum effective part of the random numbers, and then the third generator controls the process of the mixed shuffle process. The pseudo-random number sequence generated by the joint sampling method not only has much better independence and uniformity than that of the result of a single generator, but also has nearly infinite period in practice.

4.3.3 Sampling of Random Variables with General Probability Distribution

In principle, sampling of random variables includes the sampling of discrete random variables and continuous random variables. Because this book merely deals with continuous random variables, only this situation is discussed herein.

The sampling methods of continuous random variables mainly include two categories: the inverse transformation method and the acceptance–rejection sample method.

4.3.3.1 Inverse Transform Method

The inverse transform method is also referred to as the direct sampling method. Assume the CDF of the random variable X to be $F_X(x)$; in order to obtain the sample value of the random variable, it is required first to generate the sample value z of a random variable Z uniformly distributed over [0, 1], then the sample value of the desired random variable to be given by the inverse of the CDF; that is:

$$x = F_X^{-1}(z) \tag{4.71}$$

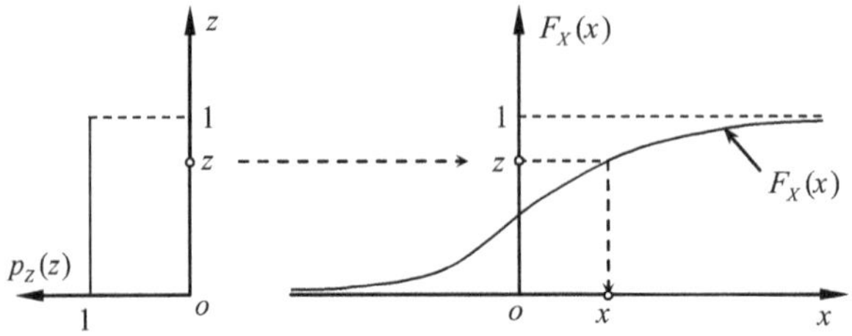

Figure 4.5 Inverse transformation method.

The principle of the inverse transform method is shown in Figure 4.5. It is thus clear that the secret of the inverse transform method is utilizing the fact that the CDF $F_X(x)$ is over [0, 1]. Thus, the samples of X can be obtained once the independent random variables of uniform distribution over [0, 1] are sampled.

4.3.3.2 Acceptance–Rejection Sampling Method

Using the inverse transformation method requires that the CDF can be expressed in an analytic form and, moreover, that its inverse function can be expressed in an explicit formula. Obviously, the limitation of this method is a bit strict for practical applications. On the other hand, the acceptance–rejection sampling method may be applied to the situation for which the inverse transformation method is not applicable.

Assume that $p_X(x)$ is the PDF of random numbers that are going to be generated; the graph of $p_X(x)$ can then be drawn in a two-dimensional coordinate system (Figure 4.6). If a two-dimensional random point (x, y) can be generated and it is scattered in the region enclosed by $p_X(x)$, then the corresponding x has the desired distribution. In order to do so, a comparative function $f(x) \geq p_X(x)$ should be introduced (in Figure 4.6 it is denoted by a rectangle surrounded by the dotted line), the domain of definition of $f(x)$ is the same as that of $p_X(x)$, and the integral of $f(x)$ in the field of definition is a finite value A. Thus, if a uniform random number $r \in (0, A)$ is taken and there exists

$$r = \int_a^x f(x)\,\mathrm{d}x = F(x) \tag{4.72}$$

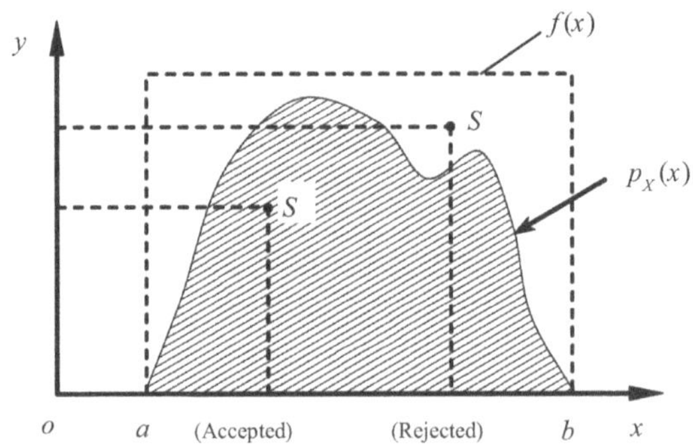

Figure 4.6 Acceptance–rejection sampling method.

and

$$x = F^{-1}(r) \tag{4.73}$$

and a uniform random number $y \in (0, f(x))$ is taken as well, then the point S decided by (x, y) must rest in the region below the curve $f(x)$. Here, if (x, y) is also below $p_X(x)$, then this point is accepted, otherwise it is rejected. By continuing such a process, the random sample under given condition $p_X(x)$ can be obtained.

Obviously, the basis of the uniformly dropped points is still the pseudo-random numbers over [0, 1], except for an affine transform.

4.3.4 Random Simulation Method

As described in Section 4.1, the random simulation method in stochastic structural analysis is very straightforward. It generally follows the steps below:

(a) Establish the dynamical analysis model of the deterministic structural system and choose the solving algorithms.
(b) Identify the basic random variables and their probability distribution functions and generate random samples according to the Monte Carlo method.
(c) Generate a random sample structure using the generated random samples and compute the structural responses by deterministic analysis methods.
(d) Compute the estimation values of the statistical characteristics of the given responses, such as the mean and the covariance of the structural responses, and so on.
(e) Stop the simulation process according to the prescribed convergence criteria.

For those stochastic structures only involving random parameters, it is enough to use the algorithm above. However, for those situations where high accuracy is required (for example, in aerospace engineering) or those problems in which the physical background cannot be briefly reflected only by using random variables (for example, in geotechnical engineering), a stochastic structural model based on random fields might be needed. In this case, it is necessary to generate the random field sample of the structural material or the geometric characteristic when the random simulation method is employed.

Random samples of a random field can be generated by a trigonometric series simulation on the basis of the spectral decomposition concept or given by random variables simulation on the basis of random field discretization.

4.3.4.1 Trigonometric Series Simulation

Let the correlation function of a homogeneous random field be $R_B(\mathbf{r})$, whose Fourier transformation may be defined as follows:

$$S_B(\boldsymbol{\kappa}) = \frac{1}{(2\pi)^n}\int_{-\infty}^{\infty} R_B(\mathbf{r})e^{-i\boldsymbol{\kappa}\cdot\mathbf{r}}\,d\mathbf{r} \tag{4.74}$$

where κ is the wave number vector. The one-dimensional wave number is defined by the reciprocal of its wave length and $\kappa \cdot \mathbf{r}$ denotes the inner product of vectors. We should specially note that the integral symbol in Equation 4.74 denotes an n-dimensional integral and n is the space dimensions of the random field.

In Equation 4.74, $S_B(\kappa)$ is called the wave number spectral density function or wave number spectrum for short. Obviously, the inverse Fourier transform of the wave number spectrum is the correlation function of the homogeneous random field; that is:

$$R_B(\mathbf{r}) = \int_{-\infty}^{\infty} S_B(\boldsymbol{\kappa}) e^{i\kappa \cdot \mathbf{r}} \, d\boldsymbol{\kappa} \tag{4.75}$$

Equation 4.74 is commonly called the spectrum decomposition of the random field correlation function. According to the concept, the following trigonometric series can be used to simulate the random field samples:

$$B(\mathbf{u}) = \sum_i A(\boldsymbol{\kappa}_i) \cos(\boldsymbol{\kappa}_i \cdot \mathbf{u} + \varphi_i) \tag{4.76}$$

where φ_i is the random phase angle uniformly distributed over $[0, 2\pi]$ and

$$A^2(\boldsymbol{\kappa}_i) = 4 S_B(\boldsymbol{\kappa}_i) \cdot |\Delta \boldsymbol{\kappa}| \tag{4.77}$$

$$\boldsymbol{\kappa}_i = (\boldsymbol{\kappa}_{1i_1}, \boldsymbol{\kappa}_{2i_2}, \cdots, \boldsymbol{\kappa}_{ni_n})^{\mathrm{T}} \tag{4.78}$$

$$\kappa_{j,i_j} = \kappa_{j,L} + \left(i_j - \frac{1}{2} \right) \Delta \kappa_j \tag{4.79}$$

$$\begin{aligned} \Delta \boldsymbol{\kappa} &= (\Delta \kappa_1, \Delta \kappa_2, \cdots, \Delta \kappa_n)^{\mathrm{T}} \\ &= \left(\frac{\kappa_{1,\mathrm{u}} - \kappa_{1,\mathrm{L}}}{N_1}, \frac{\kappa_{2,\mathrm{u}} - \kappa_{2,\mathrm{L}}}{N_2}, \cdots, \frac{\kappa_{n,\mathrm{u}} - \kappa_{n,\mathrm{L}}}{N_n} \right)^{\mathrm{T}} \end{aligned} \tag{4.80}$$

$$|\Delta \boldsymbol{\kappa}| = \prod_{i=1}^{n} \Delta \kappa_i \tag{4.81}$$

$$\sum_i = \sum_{i_1=1}^{N_1} \sum_{i_2=1}^{N_2} \cdots \sum_{i_n=1}^{N_n} \tag{4.82}$$

In the preceding formula, N_j is the sampling number along the jth wave axial; $\kappa_{j,\mathrm{u}}$ is the upper limit of the sampling wave number while $\kappa_{j,\mathrm{L}}$ is the lower limit of the sampling wave number.

In order to avoid periodic occurrences in the simulation sample, a small random wave number may be added to the cosine wave number; that is, Equation 4.76 is modified to

$$B(\mathbf{u}) = \sum_i A(\boldsymbol{\kappa}_i) \cos(\boldsymbol{\kappa}'_i \cdot \mathbf{u} + \varphi_i) \tag{4.83}$$

where

$$\boldsymbol{\kappa}'_i = \boldsymbol{\kappa}_i + \delta\boldsymbol{\kappa}_i \tag{4.84}$$

$$\delta\boldsymbol{\kappa}_i = (\delta\kappa_{1i_1}, \delta\kappa_{2i_2}, \cdots, \delta\kappa_{ni_n})^{\mathrm{T}} \tag{4.85}$$

in which $\delta\kappa_{ji_j}$ is uniformly distributed over $(-\Delta\kappa'_j/2, \Delta\kappa'_j/2)$, and $\Delta\kappa'_j \ll \Delta\kappa_j$.

4.3.4.2 Random Variables Simulation Method on the Basis of Random Field Discretization

A first-order homogeneous random field (that is, its mean is constant) can be conveniently transformed into a zero-mean random field. In such a case, the discretization of a random field can be implemented. Using the central point method, the shape function method or the local average method described in Section 2.3, the correlation coefficient matrix of the discretized random field can be obtained:

$$\mathbf{C}_{\boldsymbol{\xi}} = [c_{ij}] = [\mathrm{cov}[\xi_i, \xi_j]] \tag{4.86}$$

where ξ_i and ξ_j are the discretized basic random variables.

If the correlation coefficient matrix is symmetric and positive definite, then $\mathbf{C}\boldsymbol{\xi}$ can be decomposed into a lower triangular matrix multiplied by an upper triangular matrix using the Cholesky decomposition; that is:

$$\mathbf{C}_{\boldsymbol{\xi}} = \mathbf{L}\mathbf{L}^{\mathrm{T}} \tag{4.87}$$

where $\mathbf{L}$ is a lower triangular decomposition matrix.

Thus, if we let the discretized random vector

$$\mathbf{V} = \mathbf{L}\mathbf{Z} \tag{4.88}$$

where $\mathbf{Z} = (Z_1, Z_2, \ldots, Z_n)^{\mathrm{T}}$ is a random variable vector with zero-mean and unit variance normal distribution $\mathcal{N}(0, 1)$, then we can prove that

$$\mathcal{E}[\mathbf{V}\mathbf{V}^{\mathrm{T}}] = \mathcal{E}[\mathbf{L}\mathbf{Z}(\mathbf{L}\mathbf{Z})^{\mathrm{T}}] = \mathbf{L}\mathcal{E}[\mathbf{Z}\mathbf{Z}^{\mathrm{T}}]\mathbf{L}^{\mathrm{T}} = \mathbf{C}_{\boldsymbol{\xi}} \tag{4.89}$$

Therefore, as long as the random variables with standard normal distribution $\mathcal{N}(0, 1)$ are generated for each element, the random samples of the discretized random field can be obtained using Equation 4.88.

A two-dimensional random field sample generated according to the above method is shown schematically in Figure 4.7.

4.3.5 Accuracy of Random Simulation Method

The samples selected by the Monte Carlo method essentially belong to random variables with an independent identical distribution. Therefore, the sample responses evaluated by using a deterministic mathematical or physical equation are also random variables with the characteristic of an independent identical distribution. According to the *central limit theorem* in

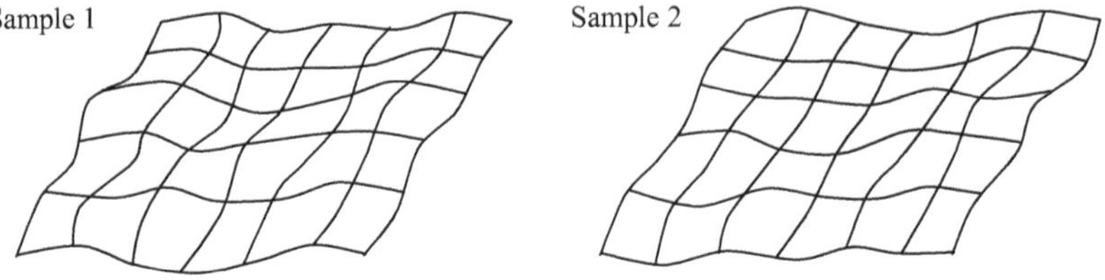

Figure 4.7 Typical random field samples.

probability theory (Loève, 1977), if the mean and variance of the random variables $\xi_1, \xi_2, \ldots$ with independent identical distribution all exist and have $\mu = \mathcal{E}[\xi], \sigma^2 = \mathcal{D}[\xi]$, then the random variable

$$\eta = \frac{\tilde{\mu} - \mu}{\sigma/\sqrt{N}} \tag{4.90}$$

will asymptotically obey the standard normal distribution; that is:

$$\lim_{N \to \infty} \Pr\left\{ \frac{\tilde{\mu} - \mu}{\sigma/\sqrt{N}} < x \right\} = \frac{1}{\sqrt{2\pi}} \int_{-\infty}^{x} e^{-x^2/2} dx \tag{4.91}$$

In the above equations, $\tilde{\mu} = (1/N) \sum_{i=1}^{N} \xi_i$ is the estimated mean of the samples.

Hence, once given a definite confidence level $1 - \alpha$, if N is large enough, then the following approximate equation exists:

$$\Pr\left(|\tilde{\mu} - \mu| < \frac{x_\alpha \sigma}{\sqrt{N}} \right) \approx \frac{2}{\sqrt{2\pi}} \int_{0}^{x_\alpha} e^{-x^2/2} dx \tag{4.92}$$

where x_α is the coordinate value of the censored bounds with the given confidence level $1 - \alpha$ (Figure 4.8). Several common sets of corresponding values are shown in Table 4.3.

According to Equation 4.92, under a given confidence level, the error between the estimated mean according to the Monte Carlo method and the real mean can be given by

$$\varepsilon = |\mu - \tilde{\mu}| < \frac{x_\alpha \sigma}{\sqrt{N}} \tag{4.93}$$

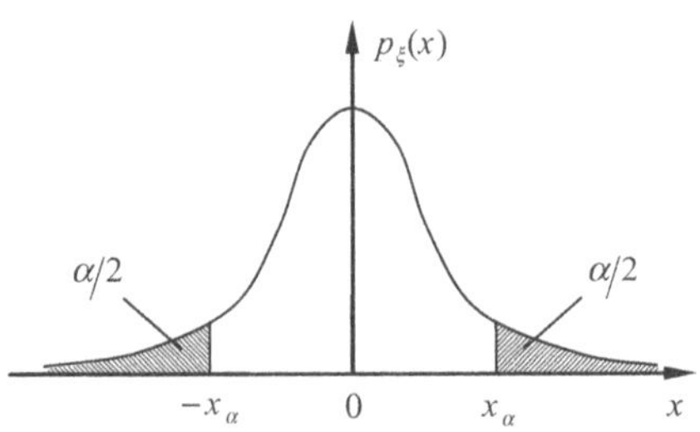

Figure 4.8 Confidence bounds.

Table 4.3 The x_α versus $1-\alpha$ relationship under a standard normal distribution.

x_α	1	2	3	4
$1-\alpha$	0.6827	0.9545	0.9973	0.9999

The previous equation is actually the estimation for the accuracy of the Monte Carlo method. It is thus clear that:

(a) the accuracy of the results can be estimated using $\sigma/\sqrt{N}$ in the trial process;
(b) the convergence speed of the Monte Carlo method is proportional to $\sqrt{N}$.

This relationship means that, if we want to increase the accuracy of the results by one digit, the computational cost of the simulation should be increased by 100 times; therefore, there is a need for several thousand (or even millions of) simulation computations when the Monte Carlo method is employed to solve a stochastic structural analysis problem.

4.4 Perturbation Approach

4.4.1 Deterministic Perturbation

The perturbation approach for stochastic structural analysis stems from the deterministic perturbation method of nonlinear analysis. In deterministic perturbation analysis, the governing equation of a physical problem is generally expressed as an equation involving small parameters; for instance:

$$\mathbf{L}(u, x, \varepsilon) = 0 \tag{4.94}$$

where $\mathbf{L}$ is a general operator, u is the solution, x is the argument and ε is a small parameter, which can naturally occur in Equation 4.94 or can be artificially introduced.

The above problem usually cannot be solved precisely. However, according to the characteristics that the solution u is a function of x and ε where ε is a small parameter, u can be expanded in an asymptotic series. For example, there exists

$$u(x, \varepsilon) = u_0(x) + \varepsilon u_1(x) + \cdots + \varepsilon^n u_n(x) + \cdots \tag{4.95a}$$

where the coefficients $u_i(x)$ are irrelevant to ε. Simultaneously, the operator $\mathbf{L}(\cdot)$ can be expanded by

$$\mathbf{L} = \mathbf{L}_0 + \varepsilon \mathbf{L}_1 + \cdots + \varepsilon^n \mathbf{L}_n + \cdots \tag{4.95b}$$

Substituting Equations 4.95a and 4.95b in Equation 4.94 and merging coefficients of the same order, we can obtain

$$(\mathbf{L}_0 u_0 - h) + (\mathbf{L}_0 u_1 + \mathbf{L}_1 u_0)\varepsilon + (\mathbf{L}_0 u_2 + \mathbf{L}_1 u_1 + \mathbf{L}_2 u_0)\varepsilon^2 + \cdots = 0 \tag{4.96}$$

where $\mathbf{L}_0, \mathbf{L}_1, \mathbf{L}_2, \ldots$ are linear operators in the space U and h is a real function of x, which can be determined according to the specific problems.

Because Equation 4.96 should be correct for any arbitrary ε and the sequence of ε is linearly independent, the coefficients of each order of ε must be zero; that is:

$$\begin{cases} \mathbf{L}_0 u_0 = h \\ \mathbf{L}_0 u_1 = -\mathbf{L}_1 u_0 \\ \mathbf{L}_0 u_2 = -\mathbf{L}_1 u_1 - \mathbf{L}_2 u_0 \\ \cdots \end{cases} \tag{4.97}$$

The above equations form a set of recursive equations of $u_i(x)$. The boundary condition and the initial condition of the problem can be obtained using similar methodology. Hereby, the above equations can be solved one by one and thus $u_i(x)$ can be obtained in sequence. Substituting the results in Equation 4.95a leads to an approximate result of $u(x, \varepsilon)$ (Nayfeh, 2000).

The preceding method is generally termed the parametric perturbation method. In this method, the expanded quantity may be a function of ε, say $\delta_i(\varepsilon)$, which is generally called the expanded asymptotic sequence and satisfies

$$\delta_i(\varepsilon) = o[\delta_{i-1}(\varepsilon)] \tag{4.98}$$

where $o(\cdot)$ denotes the infinitesimal of higher order.

This means that, in an asymptotic sequence, the latter terms of the sequence must be the high-order infinitesimal of the former terms. For example, as $\varepsilon \to 0$, the functions ε^i, $\varepsilon^{i/3}$, $(\log\varepsilon)^{-i}$ and $(\sin\varepsilon)^i$ are all asymptotic sequences.

Using the asymptotic sequence, the unknown function $u(x, \varepsilon)$ may be expanded as the following function:

$$u(x, \varepsilon) = \sum_{i=0}^{\infty} a_i(x)\delta_i(\varepsilon) \qquad (\varepsilon \to 0) \tag{4.99}$$

where a_i is a function of x and is irrelevant to ε.

For any positive integer N, there exist

$$u(x, \varepsilon) = \sum_{i=0}^{N} a_i(x)\delta_i(\varepsilon) + R_N(x, \varepsilon) \tag{4.100}$$

where $R_N(x, \varepsilon)$ is the remainder, given as

$$R_N(x, \varepsilon) = \mathcal{O}[\delta_N(\varepsilon)] \qquad (\varepsilon \to 0) \tag{4.101}$$

The left-hand side of Equation 4.100 is referred to as the nth-order asymptotic expansion equation of $u(x, \varepsilon)$.

Assume the value domain of the solution is Ω and the boundary is $\partial\Omega$. If the asymptotic expansion Equation 4.100 always holds over $\Omega + \partial\Omega$ – that is, the perturbation solutions have a uniform convergence limit with respect to x in the definition field of x – then the expansion equation is termed a consistent convergence asymptotic expansion over $\Omega + \partial\Omega$. This kind of perturbation problem is called a regular perturbation problem. Because $\delta_i(\varepsilon)$ is an asymptotic sequence, a regular perturbation means that $a_i(x)\delta_i(\varepsilon)$ is a little modification against its former value whatever x is. However, the consistent convergence condition is not always satisfied for perturbation problems. The perturbation problem with an inconsistent convergence field is referred to as a singular perturbation problem. For a deterministic perturbation problem, such problems may result from secular terms in the infinite field or the existence of singular points, and so on (Nayfeh, 2000).

4.4.2 Random Perturbation

The random perturbation method can be constructed by extending the above perturbation technology to the problem involving random parameters (Kleiber and Hien, 1992; Skorokhod et al., 2002). Herein, the random differential operator of a considered problem is given by

$$\mathbf{L}(y, x, \xi) = 0 \tag{4.102}$$

where the meanings of $\mathbf{L}$ and x are similar to those in Equation 4.94, ξ is a random variable with a given probability distribution and $y = y(x, \xi)$ is a random function.

According to Chapter 2, a random variable can be standardized by

$$\xi = \xi_0 + \sigma_\xi \zeta = \psi(\zeta) \tag{4.103}$$

where ξ_0 is the mean of ξ, σ_ξ is the standard deviation of ξ and ζ is the standardized random variable of which the mean is zero and the variance is one.

Substituting Equation 4.103 in Equation 4.102, there is

$$\mathbf{L}[y, x, \psi(\zeta)] = 0 \tag{4.104}$$

Utilizing the series expansion of the random function given in Equation 2.41, the solution $y = y(x, \xi)$ can be expanded as the series of ζ:

$$\begin{aligned} y(x, \xi) &= y[x, \psi(\zeta)] \\ &= y[x, \psi(\zeta)]|_{\zeta=0} + \left.\frac{\mathrm{d}y}{\mathrm{d}\zeta}\right|_{\zeta=0} \zeta + \left.\frac{1}{2}\frac{\mathrm{d}^2 y}{\mathrm{d}\zeta^2}\right|_{\zeta=0} \zeta^2 + \cdots \end{aligned} \tag{4.105}$$

For simplicity of notation, $y(x, \psi(\zeta))$ is written as $y(x, \zeta)$ hereafter without inducing confusion. Because y is unknown, the coefficients of $[\mathrm{d}y/\mathrm{d}\zeta]|_{\zeta=0}$ and so on are all unknown. However, the equation can be written in the equivalent form

$$y(x, \zeta) = u_0(x) + \zeta u_1(x) + \zeta^2 u_2(x) + \cdots \tag{4.106}$$

Obviously, the coefficient $u_i(x)$ is irrelevant to ζ and is a deterministic function.

Substituting Equation 4.106 in Equation 4.104 and combining the same-order coefficient of ζ after appropriate operation leads to

$$(\mathbf{L}_0 u_0 - h) + (\mathbf{L}_0 u_1 + \mathbf{L}_1 u_0)\zeta + (\mathbf{L}_0 u_2 + \mathbf{L}_1 u_1 + \mathbf{L}_2 u_0)\zeta^2 + \cdots = 0 \tag{4.107}$$

where $\mathbf{L}_0, \mathbf{L}_1, \mathbf{L}_2, \ldots$ are deterministic operators and h is a real function of x.

Because ζ is a random variable that can take any arbitrary value, the sufficient condition which makes Equation 4.107 hold is that all the coefficient terms must be zero; that is:

$$\begin{cases} \mathbf{L}_0 u_0 = h \\ \mathbf{L}_0 u_1 = -\mathbf{L}_1 u_0 \\ \mathbf{L}_0 u_2 = -\mathbf{L}_1 u_1 - \mathbf{L}_2 u_0 \\ \cdots \end{cases} \tag{4.108}$$

The equations consist of a set of deterministic operator equations. By introducing the boundary conditions and/or the initial conditions, the solutions of $u_0, u_1, \ldots$ can be obtained one by one from these equations. Substituting these solutions in Equation 4.106, the mean and variance solutions of $y(x, \zeta)$ can be given by

$$\mathcal{E}[y(x, \zeta)] = u_0(x) + u_2(x) + \cdots \tag{4.109}$$

and

$$\mathcal{D}[y(x, \zeta)] = u_1^2(x) + u_2^2(x) E[\zeta^4] + \cdots \tag{4.110}$$

respectively.

Corresponding to the deterministic parameter perturbation, in random perturbation, consistent convergence of the solution in the sense of expectation of order M must be considered. Suppose $S_M[y_N(x, \zeta)]$ is the expectation of order M of the Nth solution expanded with respect to ζ, and if there is

$$S_M[y(x, \zeta)] = S_M[y_N(x, \zeta)] + \mathcal{O}_M(x) \tag{4.111}$$

then $y_N(x, \zeta)$ is called the Nth expansion with accuracy of order N. Here, $\mathcal{O}_M(x)$ indicates that the remaining terms are infinitesimal of the same order.

4.4.3 Random Matrices

If the randomness involved in the structural parameters in a dynamical system cannot be ignored, then the corresponding dynamic matrix must be treated as a random matrix. Those parameters that render a dynamic matrix a random matrix are referred to as basic random parameters, such as the material mass density, elastic modulus, Poisson's ratio, geometric size, the damping coefficient and so on. According to the specific problem background, these parameters can be either characterized by random variables or modeled by random fields. Without loss of generality, assume the structural random field is $\{B(\mathbf{u}), \mathbf{u} \in D\}$; then, using the local average method for the random field discretization, it can be partitioned to a random variables set $\{\xi_i, i = 1, 2, \ldots, n\}$. Here, n is the partition number of the field elements. The mean and the variance of the random variable ξ_i are given by Equations 2.144 and 2.145 respectively. Transforming ξ_i to a standardized random variable, there is

$$\xi_i = \xi_{i0} + \sigma_{\xi_i} Z_i \qquad (i = 1, 2, \cdots, n) \tag{4.112}$$

where ξ_{i0} is the mean of ξ_i; σ_{ξ_i} is the standard deviation of ξ_i and Z_i is the standardized random variable of which the mean is zero and the variance is one.

On the other hand, for each element with the random parameters the property matrices can be expressed in the form of a random matrix according to the finite-element method. That is:

$$\tilde{\mathbf{m}}_i = \int_\Omega \mathbf{N}^{\mathrm{T}} \rho \mathbf{N} \, \mathrm{d}\Omega \tag{4.113}$$

$$\tilde{\mathbf{c}}_i = \int_\Omega \mathbf{N}^{\mathrm{T}} \eta \mathbf{N} \, \mathrm{d}\Omega \tag{4.114}$$

$$\tilde{\mathbf{k}}_i = \int_\Omega \mathbf{B}^{\mathrm{T}} \mathbf{E} \mathbf{B} \, \mathrm{d}\Omega \tag{4.115}$$

where ρ, η, **N** and **B** are consistent with the definition discussed in Section 4.2 and **E** is the elastic matrix that is the popularization of the elastic modulus concept in the common finite-element method. For example, for the plane-stress problem, there exists

$$\mathbf{E} = \frac{E}{1-\mu^2} \begin{bmatrix} 1 & \mu & 0 \\ \mu & 1 & 0 \\ 0 & 0 & \dfrac{1-\mu}{2} \end{bmatrix} \tag{4.116}$$

where E is the elastic modulus and μ is Poisson's ratio.

The random variable ξ_i can represent any basic variable in the element characteristic matrix, such as ρ, η, E, μ, I_i, A_i and so on, while ξ_{ij} $(j = 1, 2, \ldots)$ can also be used to represent the effect of multiple basic variables.

4.4.4 Linear Expression of Random Matrices

In the case the random variable ξ_i appears in a dynamic matrix in the form of a linear factor, the corresponding random matrix can be expressed as a linear function of the standardized random variables (Li, 1995c). Without loss of generality, we use **S** to express a general random dynamic matrix; then there exists

$$\tilde{\mathbf{S}}_i = \tilde{\mathbf{S}}_{i0} + \tilde{\mathbf{S}}_{i\sigma} Z_i \tag{4.117}$$

where $\tilde{\mathbf{S}}_i$ is the element random matrix, $\tilde{\mathbf{S}}_{i0}$ is the mean-parameter element matrix, $\tilde{\mathbf{S}}_{i\sigma}$ is the element standard deviation matrix and Z_i is a standardized random variable corresponding to the element i.

In fact, Equation 4.117 can be deduced from the series expansion of the random matrix $\mathbf{S}_i$ with respect to the standardized random variable Z_i:

$$\tilde{\mathbf{S}}_i = \tilde{\mathbf{S}}_{i0} + \left.\frac{\mathrm{d}\tilde{\mathbf{S}}_i}{\mathrm{d}Z_i}\right|_{Z_i=0} Z_i + \frac{1}{2} \left.\frac{\mathrm{d}^2\tilde{\mathbf{S}}_i}{\mathrm{d}Z_i^2}\right|_{Z_i=0} Z_i^2 + \cdots \tag{4.118}$$

Because Z_i is a linear factor in S_i, the derivatives of higher than second order are all zero. Then

$$\frac{\mathrm{d}\tilde{\mathbf{S}}_i}{\mathrm{d}Z_i} = \frac{\mathrm{d}\tilde{\mathbf{S}}_i}{\mathrm{d}\xi_i}\frac{\mathrm{d}\xi_i}{\mathrm{d}Z_i} = \frac{\mathrm{d}\tilde{\mathbf{S}}_i}{\mathrm{d}\xi_i}\sigma_{\xi_i} = \tilde{\mathbf{S}}_{i\sigma} \tag{4.119}$$

Thus, we reach Equation 4.117.

Basic variables occurring in a dynamic matrix in the form of linear factors include the mass density ρ, the damping coefficient η, the elastic modulus E, the inertial moment I and the sectional area A (see Equations 4.20, 4.51 and 4.53). For example, for the beam elements in a plane, if the axial deformation is not considered and the mass density ρ is taken as a random parameter, then there exist

$$\tilde{\mathbf{m}}_{i0} = \frac{\rho_{i0}Al}{420}\begin{bmatrix} 156 & 22l & 54 & -13l \\ 22l & 4l^2 & 13l & -3l^2 \\ 54 & 13l & 12 & -6l \\ -13l & -3l^2 & -6l & 4l^2 \end{bmatrix} \tag{4.120}$$

$$\tilde{\mathbf{m}}_{i\sigma} = \frac{\sigma_{\rho_i}Al}{420}\begin{bmatrix} 156 & 22l & 54 & -13l \\ 22l & 4l^2 & 13l & -3l^2 \\ 54 & 13l & 12 & -6l \\ -13l & -3l^2 & -6l & 4l^2 \end{bmatrix} \tag{4.121}$$

Because both the coordinate transformation and the element position transformation operation are deterministic linear transformations, the element matrix and the global dynamic matrix in the global coordinate system maintain the relationship similar to Equation 4.117; that is, there are

$$\mathbf{S}_i = \mathbf{S}_{i0} + \mathbf{S}_{i\sigma}Z_i \tag{4.122}$$

$$\mathbf{S} = \mathbf{S}_0 + \sum_i \bar{\mathbf{S}}_{i\sigma}Z_i \tag{4.123}$$

where

$$\mathbf{S}_{i0} = \mathbf{T}_\alpha^{\mathrm{T}}\tilde{\mathbf{S}}_{i0}\mathbf{T}_\alpha \tag{4.124}$$

$$\mathbf{S}_{i\sigma} = \mathbf{T}_\alpha^{\mathrm{T}}\tilde{\mathbf{S}}_{i\sigma}\mathbf{T}_\alpha \tag{4.125}$$

$$\mathbf{S}_0 = \sum_i \mathbf{T}_i^{\mathrm{T}}\tilde{\mathbf{S}}_{i0}\mathbf{T}_i \tag{4.126}$$

$$\bar{\mathbf{S}}_{i\sigma} = \mathbf{T}_i^{\mathrm{T}}\mathbf{S}_{i\sigma}\mathbf{T}_i \tag{4.127}$$

Here, $\mathbf{T}_\alpha$ is the coordinate transformation matrix of the element i (see Equation 4.26) and $\mathbf{T}_i$ is the position transformation matrix of the element i (see Equation 4.39).

Therefore, the global dynamic matrix of the system can be formed by the mean-parameter matrix $\mathbf{S}_0$ and the standard deviation matrix $\bar{\mathbf{S}}_{i\sigma}$ according to Equation 4.123. Herein, the mean-parameter matrix can be formed by adopting the mean parameter in the direct stiffness method, while the standard deviation matrix can be obtained through introducing the following virtual structures (Li, 1996a).

For a virtual structure, the parameters corresponding to the given basic variables of the elements take the value

$$\mathbf{r} = (0, 0, \cdots, \sigma_{\xi_i}, 0, 0, \cdots 0)^{\mathrm{T}} \tag{4.128}$$

Without loss of generality, suppose the element random variables in the structure can be partitioned into N subsets and the random variables (supposing the number of the variables is j_m) in each subset possess identical probability distribution, then Equation 4.123 can be expressed as

$$\mathbf{S} = \mathbf{S}_0 + \sum_{j=1}^{N} \mathbf{S}_j Z_j \tag{4.129}$$

where

$$\mathbf{S}_j = \sum_{i=1}^{j_m} \mathbf{T}_i^{\mathrm{T}} \mathbf{S}_{i\sigma} \mathbf{T}_i \tag{4.130}$$

Here, $\mathbf{S}_j$ can be formed by the direct stiffness method according to the following virtual structure:

$$\mathbf{r} = (0, \cdots, \sigma_{\xi_\ell}, 0, \cdots, \sigma_{\xi_m}, 0, \cdots, \sigma_{\xi_p}, 0, \cdots, 0)^{\mathrm{T}} \tag{4.131}$$

The above equation shows that there are a total of three elements in the jth subsets and the virtual structure is constructed by the corresponding standard deviation of these three elements and other zero elements.

Introducing the correlation decomposition technology of random vectors, the linear representation of dynamic matrices can be further simplified. In fact, according to Equation 2.152, we can obtain

$$\xi_i = \xi_{i0} + \sum_{j=1}^{n} \phi_{ij} \sqrt{\lambda_j} \zeta_j \tag{4.132}$$

Comparing Equation 4.112 with Equation 4.132, there exists

$$Z_i = \frac{1}{\sigma_{\xi_i}} \sum_{j=1}^{n} \phi_{ij} \sqrt{\lambda_j} \zeta_j \tag{4.133}$$

Substituting this in Equation 4.123 yields

$$\mathbf{S} = \mathbf{S}_0 + \sum_i \tilde{\mathbf{S}}_{i\sigma} \frac{1}{\sigma_{\xi_i}} \sum_{j=1}^{n} \phi_{ij} \sqrt{\lambda_j} \zeta_j \tag{4.134}$$

Because the random variable ξ_i occurs in the dynamic matrix in the form of a linear factor, the σ_{ξ_i} in the denominator of the above equation and the σ_{ξ_i} in $\tilde{\mathbf{S}}_{i\sigma}$ can be eliminated.

Simultaneously, the orders of the two sum operations can be exchanged and thus Equation 4.134 can be rewritten as

$$\mathbf{S} = \mathbf{S}_0 + \sum_{j=1}^{n} \mathbf{S}_j \zeta_j \tag{4.135}$$

where

$$\mathbf{S}_i = \sum_j \mathbf{T}_i^{\mathrm{T}} \mathbf{S}_{ij} \mathbf{T}_i \tag{4.136}$$

Here, $\mathbf{S}_{ij}$ is the element dynamic matrix of the element i in the global coordinate system when the virtual structure j is formed through regarding $\phi_{ij}\sqrt{\lambda_j}$ as a basic variable.

Because the variances of the independent random variables $\kappa_j = \sqrt{\lambda_j}\zeta_j$ have the characteristic of an asymptotic sequence, the original random variable set can be replaced by a subset of $q < n$; that is, there exists

$$\mathbf{S} = \mathbf{S}_0 + \sum_{j=1}^{q} \mathbf{S}_j \zeta_j \tag{4.137}$$

The standard deviation matrices $\mathbf{S}_j$ can be formed according to q virtual structures and $\mathbf{S}_0$ can be formed by using the mean parameter. Significantly, the forming ways mentioned above can all be done by the standard finite-element method and, therefore, is a great convenience.

It is worth pointing out that the normalized random variables set ζ has become an independent random variable set after undergoing correlation decomposition such as Equation 4.132. This is convenient for stochastic structural analysis. This aspect will be discussed in detail in the following section.

The random variable ξ_i may appear in dynamic matrices in the form of nonlinear factors. For example, when Poisson's ratio μ in Equation 4.116 is a basic variable, such a situation will be encountered. In these cases, the random dynamic matrix should generally be expressed by introducing a series expansion like Equation 4.118, which can be rewritten as

$$\tilde{\mathbf{S}}_i = \tilde{\mathbf{S}}_{i0} + \tilde{\mathbf{S}}_{1i} Z_i + \tilde{\mathbf{S}}_{2i} Z_i^2 + \cdots \tag{4.138}$$

where

$$\tilde{\mathbf{S}}_{ji} = \left.\frac{1}{j!}\frac{\mathrm{d}^{(j)}\tilde{\mathbf{S}}_i}{\mathrm{d}Z_i^j}\right|_{Z_i=0} \tag{4.139}$$

Through coordinate transformation and element position transformation operations, the global dynamic matrix can be given by

$$\mathbf{S} = \mathbf{S}_0 + \sum_i \bar{\mathbf{S}}_{1i} Z_i + \sum_i \bar{\mathbf{S}}_{2i} Z_i^2 + \cdots \tag{4.140}$$

where

$$\bar{\mathbf{S}}_{ji} = \mathbf{T}_i^{\mathrm{T}}\mathbf{T}_{\alpha}^{\mathrm{T}}\tilde{\mathbf{S}}_{ji}\mathbf{T}_{\alpha}\mathbf{T}_i \tag{4.141}$$

However, for the case with nonlinear variable factors, the idea of virtual structures discussed above cannot be used.

4.4.5 Dynamic Response Analysis

When one or more dynamic matrices contain random parameters, the equation of motion of a multi-degree-of-freedom (MDOF) system can be given by (see Equation 4.65)

$$\mathbf{M}\ddot{\mathbf{X}} + \mathbf{C}\dot{\mathbf{X}} + \mathbf{K}\mathbf{X} = \mathbf{F}(t) \tag{4.142}$$

Here, $\ddot{\mathbf{X}}$, $\dot{\mathbf{X}}$ and $\mathbf{X}$ are respectively the acceleration, the velocity and the displacement vectors. Note that the capital characters are used here according to the convention because they are all stochastic processes.

By introducing the series expansion of random matrices, the equation of motion of the stochastic structure can be approximately written as

$$\begin{aligned}\left(\mathbf{M}_0 + \sum_{i=1}^{n}\bar{\mathbf{M}}_{1i}\zeta_i + \sum_{i=1}^{n}\bar{\mathbf{M}}_{2i}\zeta_i^2\right)\ddot{\mathbf{X}} + \left(\mathbf{C}_0 + \sum_{i=1}^{n}\bar{\mathbf{C}}_{1i}\zeta_i + \sum_{i=1}^{n}\bar{\mathbf{C}}_{2i}\zeta_i^2\right)\dot{\mathbf{X}} \\ + \left(\mathbf{K}_0 + \sum_{i=1}^{n}\bar{\mathbf{K}}_{1i}\zeta_i + \sum_{i=1}^{n}\bar{\mathbf{K}}_{2i}\zeta_i^2\right)\mathbf{X} = \mathbf{F}(t)\end{aligned} \tag{4.143}$$

where $\mathbf{F}(t)$ is a deterministic time process vector. The matrices $\mathbf{M}_0$, $\mathbf{M}_{1i}$, $\bar{\mathbf{M}}_{2i}$, $\mathbf{K}_0$, $\bar{\mathbf{K}}_{1i}$ and $\bar{\mathbf{K}}_{2i}$ can be formed by the method described in the preceding sections. According to the Rayleigh damping assumption, the matrices $\mathbf{C}_0$, $\mathbf{C}_{1i}$ and $\bar{\mathbf{C}}_{2i}$ can be taken as

$$\mathbf{C}_0 = a\mathbf{M}_0 + b\mathbf{K}_0 \tag{4.144}$$

$$\bar{\mathbf{C}}_{1i} = a\bar{\mathbf{M}}_{1i} + b\bar{\mathbf{K}}_{1i} \tag{4.145}$$

$$\bar{\mathbf{C}}_{2i} = a\bar{\mathbf{M}}_{2i} + b\bar{\mathbf{K}}_{2i} \tag{4.146}$$

where a and b are deterministic parameters. In this treatment, we only consider the situation that the mass and the stiffness have one type of random parameter or that the mass and the stiffness have the same type of random parameters.

According to the basic idea of random perturbation analysis, the responses of structural acceleration, velocity and displacement can be expanded as a series of basic random variables ζ_i (taking second-order truncation)

$$\ddot{\mathbf{X}} = \ddot{\mathbf{X}}_0 + \sum_{i=1}^{n}\ddot{\mathbf{X}}_{1i}\zeta_i + \frac{1}{2}\sum_{i=1}^{n}\sum_{j=1}^{n}\ddot{\mathbf{X}}_{2ij}\zeta_i\zeta_j \tag{4.147}$$

$$\dot{\mathbf{X}} = \dot{\mathbf{X}}_0 + \sum_{i=1}^{n} \dot{\mathbf{X}}_{1i}\zeta_i + \frac{1}{2}\sum_{i=1}^{n}\sum_{j=1}^{n} \dot{\mathbf{X}}_{2ij}\zeta_i\zeta_j \tag{4.148}$$

$$\mathbf{X} = \mathbf{X}_0 + \sum_{i=1}^{n} \mathbf{X}_{1i}\zeta_i + \frac{1}{2}\sum_{i=1}^{n}\sum_{j=1}^{n} \mathbf{X}_{2ij}\zeta_i\zeta_j \tag{4.149}$$

Substituting them in Equation 4.143, combining the same-order terms and considering the sufficient condition of the equation obtained yields the following set of recursive equations:

$$\mathbf{M}_0\ddot{\mathbf{X}}_0 + \mathbf{C}_0\dot{\mathbf{X}}_0 + \mathbf{K}_0\mathbf{X}_0 = \mathbf{F}(t) \tag{4.150}$$

$$\mathbf{M}_0\ddot{\mathbf{X}}_{1i} + \mathbf{C}_0\dot{\mathbf{X}}_{1i} + \mathbf{K}_0\mathbf{X}_{1i} = -(\bar{\mathbf{M}}_{1i}\ddot{\mathbf{X}}_0 + \bar{\mathbf{C}}_{1i}\dot{\mathbf{X}}_0 + \bar{\mathbf{K}}_{1i}\mathbf{X}_0) \qquad (i = 1, 2, \cdots, n) \tag{4.151}$$

$$\begin{aligned}\mathbf{M}_0\ddot{\mathbf{X}}_{2ij} + \mathbf{C}_0\dot{\mathbf{X}}_{2ij} + \mathbf{K}_0\mathbf{X}_{2ij} = & -2[\bar{\mathbf{M}}_{1i}\ddot{\mathbf{X}}_{1j} + \bar{\mathbf{C}}_{1i}\dot{\mathbf{X}}_{1j} + \bar{\mathbf{K}}_{1i}\mathbf{X}_{1j} + \delta_{ij}(\bar{\mathbf{M}}_{2i}\ddot{\mathbf{X}}_0 + \bar{\mathbf{C}}_{2i}\dot{\mathbf{X}}_0 + \bar{\mathbf{K}}_{2i}\mathbf{X}_0)] \\ & (i,j = 1, 2, \cdots, n)\end{aligned} \tag{4.152}$$

where δ_{ij} is the Kronecker delta

$$\delta_{ij} = \begin{cases} 1 & \text{for } i = j \\ 0 & \text{for } i \neq j \end{cases} \tag{4.153}$$

It is thus clear that we should solve a total of $n^2 + n + 1$ equations in order to get the solution of the second-order perturbation. After introducing correlation structure decomposition technology, we only need to solve $q^2 + q + 1$ equations (see Equation 4.137). Here, q is the cardinal number of the uncorrelated random variable subset.

In principle, the time-domain analysis method, the frequency-domain analysis method or the modal superposition method can all be applied in solving the perturbation equations. However, when the degrees of freedom of the structure are relatively large, using the modal superposition method to solve the problem can greatly reduce computation effort. Noting that the left-hand side operator forms of Equations 4.150–4.152 are identical, only the following deterministic eigenvalue problem should be solved:

$$(\mathbf{K}_0 - \lambda\mathbf{M}_0)\boldsymbol{\psi} = \mathbf{0} \tag{4.154}$$

which is essentially a deterministic eigenvalue problem and, therefore, common eigenvalue algorithms can be used to get the eigenvalues λ_j and the eigenvectors $\boldsymbol{\psi}_j$ (Golub and van Loan, 1996).

After getting the specified number (for example, taking number m) of eigenvalues and eigenvectors, the responses of Equations 4.150–4.152 can be respectively approximated as follows:

$$\mathbf{X}_0 = \sum_{\ell=1}^{m} \boldsymbol{\psi}_\ell x_{0,\ell} \tag{4.155}$$

$$\mathbf{X}_{1i} = \sum_{\ell=1}^{m} \boldsymbol{\psi}_\ell x_{1i,\ell} \tag{4.156}$$

$$\mathbf{X}_{2ij} = \sum_{\ell=1}^{m} \boldsymbol{\psi}_\ell x_{2ij,\ell} \tag{4.157}$$

Substituting them in Equations 4.150–4.152 and considering the orthogonal condition of eigenvectors $\boldsymbol{\psi}_\ell$ with respect to $\mathbf{M}_0$ and $\mathbf{K}_0$ yields the following decoupled recursive equations:

$$m_\ell^* \ddot{x}_{0,\ell} + c_\ell^* \dot{x}_{0,\ell} + k_\ell^* x_{0,\ell} = f_\ell^*(t) \qquad (\ell = 1, 2, \cdots, m) \tag{4.158}$$

$$m_\ell^* \ddot{x}_{1i,\ell} + c_\ell^* \dot{x}_{1i,\ell} + k_\ell^* x_{1i,\ell} = f_{1i,\ell}^* \qquad (\ell = 1, 2, \cdots, m;\ i = 1, 2, \cdots, n) \tag{4.159}$$

$$m_\ell^* \ddot{x}_{2ij,\ell} + c_\ell^* \dot{x}_{2ij,\ell} + k_\ell^* x_{2ij,\ell} = f_{2ij,\ell}^* \qquad (\ell = 1, 2, \cdots, m;\ i,j = 1, 2, \cdots, n) \tag{4.160}$$

where

$$m_\ell^* = \boldsymbol{\psi}_\ell^{\mathrm{T}} \mathbf{M}_0 \boldsymbol{\psi}_\ell \tag{4.161}$$

$$c_\ell^* = \boldsymbol{\psi}_\ell^{\mathrm{T}} \mathbf{C}_0 \boldsymbol{\psi}_\ell \tag{4.162}$$

$$k_\ell^* = \boldsymbol{\psi}_\ell^{\mathrm{T}} \mathbf{K}_0 \boldsymbol{\psi}_\ell \tag{4.163}$$

$$f_\ell^*(t) = \boldsymbol{\psi}_\ell^{\mathrm{T}} \mathbf{F}(t) \tag{4.164}$$

$$f_{1i,\ell}^* = -\boldsymbol{\psi}_\ell^{\mathrm{T}} (\bar{\mathbf{M}}_{1i} \ddot{\mathbf{X}}_0 + \bar{\mathbf{C}}_{1i} \dot{\mathbf{X}}_0 + \bar{\mathbf{K}}_{1i} \mathbf{X}_0) \tag{4.165}$$

$$f_{2ij,\ell}^* = -2\boldsymbol{\psi}_\ell^{\mathrm{T}} [\bar{\mathbf{M}}_{1i} \ddot{\mathbf{X}}_{1i} + \bar{\mathbf{C}}_{1i} \dot{\mathbf{X}}_{1j} + \bar{\mathbf{K}}_{1i} \mathbf{X}_{1j} + \delta_{ij} (\bar{\mathbf{M}}_{2i} \ddot{\mathbf{X}}_0 + \bar{\mathbf{C}}_{2i} \dot{\mathbf{X}}_0 + \bar{\mathbf{K}}_{2i} \mathbf{X}_0)] \tag{4.166}$$

After obtaining all the zeroth-, first- and second-order solutions of the above perturbation equations, it is easy to compute the mean and covariance of the displacement responses. According to Equation 4.149, the mean of the displacement responses is given by

$$\mathcal{E}[\mathbf{X}] = \mathbf{X}_0 + \frac{1}{2} \sum_{i=1}^{n} \sum_{j=1}^{n} \mathbf{X}_{2ij} \mathcal{E}[\zeta_i \zeta_j] \tag{4.167}$$

and the covariance matrix of the displacement responses is

$$\begin{aligned} \mathrm{cov}[\mathbf{X}, \mathbf{X}] = & \sum_{i=1}^{n} \sum_{j=1}^{n} \mathbf{X}_{1i} \mathbf{X}_{1j}^{\mathrm{T}} \mathcal{E}[\zeta_i \zeta_j] + \frac{1}{2} \sum_{i=1}^{n} \sum_{j=1}^{n} \sum_{k=1}^{n} [\mathbf{X}_{1i} \mathbf{X}_{2jk}^{\mathrm{T}} + \mathbf{X}_{2ij} \mathbf{X}_{1k}^{\mathrm{T}}] \mathcal{E}[\zeta_i \zeta_j \zeta_k] \\ & + \frac{1}{4} \sum_{i=1}^{n} \sum_{j=1}^{n} \sum_{k=1}^{n} \sum_{\ell=1}^{n} \mathbf{X}_{2ij} \mathbf{X}_{2k\ell}^{\mathrm{T}} \{ \mathcal{E}[\zeta_i \zeta_\ell \zeta_j \zeta_k] - \mathcal{E}[\zeta_i \zeta_j] \mathcal{E}[\zeta_k \zeta_\ell] \end{aligned} \tag{4.168}$$

in which the component cov$[X_s, X_r]$ denotes the covariance between the displacements of the sth and the rth degrees of freedom.

By introducing the correlation decomposition technique, the above numerical eigenvalues will be further significantly simplified. In this case, Equation 4.149 can be transformed to

$$\mathbf{X} = \mathbf{X}_0 + \sum_{i=1}^{q} \mathbf{X}_{1i}\zeta_i + \frac{1}{2}\sum_{i=1}^{q}\sum_{j=1}^{q} \mathbf{X}_{2ij}\zeta_i\zeta_j + \cdots \tag{4.169}$$

Because of the uncorrelation between the variables ζ_i and ζ_j and noting that $\mathcal{E}[\zeta_i^2] = 1$, there exist

$$\mathcal{E}[\zeta_i\zeta_j] = \begin{cases} 1 & \text{for } i = j \\ 0 & \text{for } i \neq j \end{cases} \tag{4.170}$$

$$\mathcal{E}\left(\prod_{s=1}^{n} \zeta_s\right) = \prod_{s=1}^{n} \mathcal{E}[\zeta_s] \tag{4.171}$$

Hence, the mean vector and the covariance matrix of the displacement vector considering second-order truncation are respectively given as

$$\mathcal{E}[\mathbf{X}] = \mathbf{X}_0 + \frac{1}{2}\sum_{i=1}^{q} \mathbf{X}_{2ii} \tag{4.172}$$

$$\text{cov}[\mathbf{X}, \mathbf{X}] = \sum_{i=1}^{q} \mathbf{X}_i\mathbf{X}_i^{\mathrm{T}} \tag{4.173}$$

Likewise, the numerical characteristic process of other responses, such as the velocity and the acceleration and so on, can also be obtained.

In some cases, we may have interest in the correlation behavior between the responses at any two time instants of the time history. This characteristic can be depicted with the following correlation function matrix:

$$\mathbf{R} = \begin{bmatrix} \text{cov}[X_{t_1}, X_{t_1}] & \text{cov}[X_{t_1}, X_{t_2}] & \cdots & \text{cov}[X_{t_1}, X_{t_N}] \\ \text{cov}[X_{t_2}, X_{t_1}] & \text{cov}[X_{t_2}, X_{t_2}] & \cdots & \text{cov}[X_{t_2}, X_{t_N}] \\ \cdots & \cdots & \cdots & \cdots \\ \text{cov}[X_{t_N}, X_{t_1}] & \text{cov}[X_{t_N}, X_{t_2}] & \cdots & \text{cov}[X_{t_N}, X_{t_N}] \end{bmatrix} \tag{4.174}$$

where $X_{t_i} = X(t_i)$, the subscripts t_i denote the time instants and N is the number of time instants considered. The components in the above correlation matrix can be evaluated similar to Equation 4.168.

4.4.6 Secular Terms Problem

It is noted that the operators on the left-hand side of the dynamic perturbation equations are in the same form. This provides convenience for solving the dynamic perturbation equation, but meanwhile brings about the essential weakness for this kind of method. During computational practice, it was found that the mean and variance given by the perturbation method were only applicable in a short period from the beginning time instant compared with the Monte Carlo simulated results. With increasing time, the accuracy of the results will deteriorate rapidly (Liu *et al.*, 1988b). Figure 4.9 shows some results of a stochastic single-degree-of-freedom (SDOF) system. It is seen that the results of the second-order perturbation are even worse than that of the first-order perturbation. This phenomenon is the so-called secular term problem. After careful analysis, we may find that the essential reason leading to the above phenomenon lies in introducing spurious resonant inputs, which do not exist during a practical vibration process.

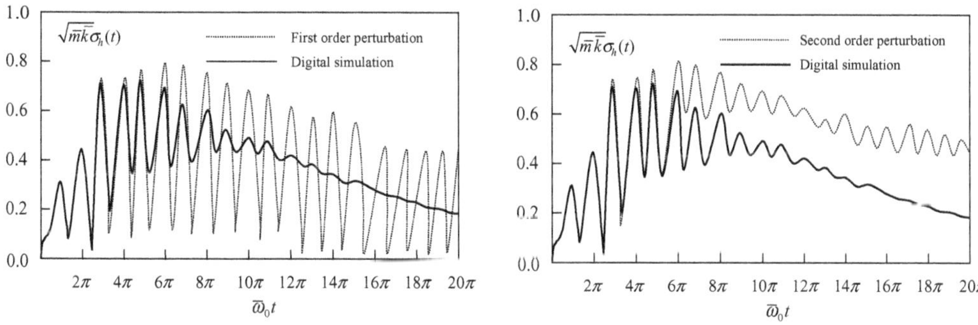

Figure 4.9 The standard deviation of a stochastic response process.

In fact, the system transfer functions[1] of Equations 4.158–4.160 are all

$$H_\ell(\omega) = \frac{1}{\omega_{0\ell}^2 - \omega^2 + 2i\xi_\ell\omega_{0\ell}\omega} \qquad (\ell = 1, 2, \ldots, m) \tag{4.175}$$

where i is the imaginary number unit and

$$\omega_{0\ell}^2 = k_\ell^*/m_\ell^* \tag{4.176}$$

$$\xi_\ell = c_\ell^*/2m_\ell^*\omega_{0\ell} \tag{4.177}$$

The amplitude of the transfer function is shown schematically in Figure 4.10.

Performing a Fourier transform for Equations 4.158–4.160 with respect to t, there exist

$$x_{0,\ell}(\omega) = H_\ell(\omega)f_\ell(\omega) \qquad (\ell = 1, 2, \cdots, m) \tag{4.178}$$

$$x_{1i,\ell}(\omega) = H_\ell(\omega)f_{1i,\ell}^*(\omega) \qquad (\ell = 1, 2, \cdots, m;\ i = 1, 2, \cdots, n) \tag{4.179}$$

$$x_{2ij,\ell}(\omega) = H_\ell(\omega)f_{2ij,\ell}^{**}(\omega) \qquad (\ell = 1, 2, \cdots, m;\ i,j = 1, 2, \cdots, n) \tag{4.180}$$

[1] The system transfer function is also referred to as the frequency response function, which bridges the relationship between the Fourier spectrum of the inputs (excitations) and that of the outputs (responses). For details, refer to say Clough and Penzien (1993). It will also be elaborated in Section 5.3.

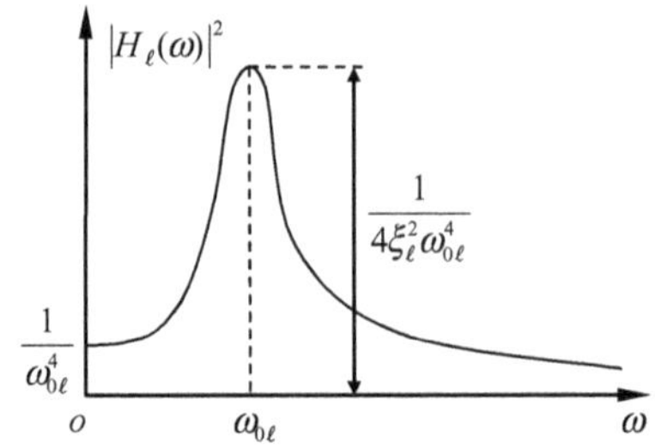

Figure 4.10 Schematic curve of $|H(\omega)|^2$.

From Figure 4.10, it is easy to find that, when the damping is small, the input components near the system frequency $\omega_{0\ell}$ will be amplified and those components which are distant from $\omega_{0\ell}$ will be suppressed. This amplifying effect is an essential reason of occurrence of secular terms.

On the basis of this analysis, some researchers suggested using a filter to restrict the influence of the secular terms (Liu *et al.*, 1988b). The key point of this modifying proposal is to transform the zeroth perturbation solution $\mathbf{X}_0(t)$ into the frequency domain after it has been obtained; that is:

$$\mathbf{X}_0(\omega) = \int_{-\infty}^{\infty} \mathbf{X}_0(t)e^{-i\omega t}\,dt \tag{4.181}$$

A retarded band function is then used to filter out each resonant components of $\mathbf{X}_0(\omega)$: that is, take

$$\hat{\mathbf{X}}_0(\omega) = \sum_{\ell=1}^{m} w_\ell(\omega)\mathbf{X}_0(\omega) \tag{4.182}$$

where the window function $w_\ell(\omega)$ can take any kind of the following types:

(a) triangle window

$$w_\ell(\omega) = \frac{|\omega_{0\ell} - \omega|}{\Delta\omega} \tag{4.183}$$

(b) cosine window

$$w_\ell(\omega) = 1 - \cos\frac{\pi(\omega_{0\ell} - \omega)}{2\Delta\omega} \tag{4.184}$$

where $\Delta\omega$ is the bandwidth of the retarded band.

Taking the inverse Fourier transformation of $\hat{\mathbf{X}}_0(\omega)$ gives

$$\hat{\mathbf{X}}_0(t) = \frac{1}{2\pi}\int_{-\infty}^{\infty} \hat{\mathbf{X}}_0(\omega)\mathbf{e}^{\mathrm{i}\omega t}\,d\omega \tag{4.185}$$

On the basis of $\hat{\mathbf{X}}_0(t), f^*_{1i;\ell}$ can be calculated and then the first-order perturbation solution can be obtained. After getting the first-order perturbation, $\mathbf{X}_1(t)$ should be filtered in a similar way and then the second-order perturbation can be solved.

In the preceding modified proposal there still exist some weaknesses as follows:

(a) There is no objective criterion for choosing the bandwidth of the retarded band and the selection of window function. In fact, different bandwidths of retarded bands might lead to completely different results.
(b) Because of filtering, the real information of the input is artificially eliminated. Thus, this will lead to perturbation input distortion for the MDOF systems, particularly for the frequency spectrum concentrated system.

Hence, as a universal method, the modifying proposal is untenable. It is this difficulty that impels investigators to find a new method for stochastic structural dynamic analysis. The orthogonal expansion approach that will be discussed in the next section is just such an achievement of these endeavors.

4.5 Orthogonal Expansion Theory

4.5.1 *Orthogonal Decomposition and Sequential Orthogonal Decomposition*

It is pointed out in Chapter 2 that, if there is a family of standard orthogonal functions in the stochastic function space, an arbitrary function in this space can be expanded as a set of generalized Fourier series in terms of this family. Generally, assume the probability measure of a stochastic function space $\mathcal{H}$ is

$$\Pr\{u \in \Omega_u\} = \int_{\Omega_u} p_\zeta(u)\,\mathrm{d}u \tag{4.186}$$

in which $p_\zeta(u)$ is the PDF of ζ, Ω_u is a given set of real variable u corresponding to ζ. Here, we use u to denote a sample value of ζ.

Suppose $\{H_i(\zeta), i=0,1,2,\ldots\}$ are standard orthogonal functions in the space $\mathcal{H}$ in which any arbitrary two functions satisfy

$$\int_\Omega p_\zeta(u)H_m(u)H_n(u)\,\mathrm{d}u = \delta_{mn} \tag{4.187}$$

where δ_{mn} is the Kronecker delta and Ω is the definition field of the real variable u.

If the inner product of two arbitrary stochastic functions in the space $\mathcal{H}$ is defined as (see Equation 2.183)

$$\langle f, g\rangle = \int_\Omega p_\zeta(u)f(u)g(u)\,\mathrm{d}u \tag{4.188}$$

then, according to this inner product, the norm and the distance of the space $\mathcal{H}$ can be given by

$$\|f(u)\| = \sqrt{\int_\Omega p_\zeta(u)f^2(u)\,\mathrm{d}x} \tag{4.189}$$

$$d(f,g) = \sqrt{\int_{\Omega} p_{\zeta}(u)[f(u) - g(u)]^2 \, \mathrm{d}u} \tag{4.190}$$

With the above definitions, if every Cauchy point sequence in the space $\mathcal{H}$ converges (this means that the stochastic functions in the space $\mathcal{H}$ satisfy the condition of mean-square convergence), then an arbitrary function in this space $\mathcal{H}$ can be expanded in terms of $H_i(\zeta)$; that is:

$$f(\zeta) = \sum_{i=0}^{\infty} a_i H_i(\zeta) \tag{4.191}$$

in which

$$a_i = \langle f, H_i \rangle = \int_{\Omega} p_{\zeta}(u) f(u) H_i(u) \, \mathrm{d}x \tag{4.192}$$

The expression of Equation 4.191 is called the orthogonal expansion of a function of a single random variable.

The orthogonal expansion can be extended to the cases of independent random variable sets. In this context, there exist two different approaches where the orthogonal polynomials as base functions are generated in different ways. The first approach employs polynomial chaos as the base functions, which has been well elaborated in the monograph by Ghanem and Spanos (1991a) and, thus, will not be detailed hereafter. The second approach is the following sequential orthogonal decomposition.

Consider a random vector

$$\boldsymbol{\zeta} = (\zeta_1, \zeta_2, \cdots, \zeta_n) \tag{4.193}$$

Noting the decomposition principle of random vectors as discussed in Chapter 2, it is reasonable to assume the random variables ζ_i and ζ_j are mutually independent. Then, the probability measure of the stochastic function space can be defined by

$$\Pr\{\mathbf{u} \in \Omega_{\mathbf{u}}\} = \int_{\Omega_{\mathbf{u}}} p_{\zeta}(\mathbf{u}) \, \mathbf{du} \tag{4.194}$$

where $\Omega_{\mathbf{u}}$ is the given set with respect to $\mathbf{u}$. In addition:

$$p_{\boldsymbol{\zeta}}(\mathbf{u}) = \prod_{i=1}^{n} p_{\zeta_i}(u_i) \tag{4.195}$$

$$\mathbf{u} = (u_1, u_2, \cdots, u_n) \tag{4.196}$$

in which $p_{\zeta_i}(u_i)$ denotes the PDFs of ζ_i.

If there is a set of functions $\{H_\ell(\boldsymbol{\zeta}), \ell = 0, 1, 2, \cdots\}$ which satisfies the relationship

$$\int_{\Omega_{\mathbf{u}}} p_{\boldsymbol{\zeta}}(\mathbf{u}) H_\ell(\mathbf{u}) H_k(\mathbf{u}) d\mathbf{u} = \delta_{\ell k} \tag{4.197}$$

in which $\Omega_{\mathbf{u}}$ is the definition field of **u**, then it is possible to regard the space $\mathcal{H}$ as a Hilbert space by introducing the inner product definition similar to Equation 4.188. An arbitrary function in this space can thus be expanded in the form of

$$Y(\boldsymbol{\zeta}) = \sum_{\ell=0}^{\infty} x_\ell H_\ell(\boldsymbol{\zeta}) \tag{4.198}$$

where

$$x_\ell = \int_{\Omega_{\mathbf{u}}} p_{\boldsymbol{\zeta}}(\mathbf{u}) Y(\mathbf{u}) H_\ell(\mathbf{u})\, \mathrm{d}\mathbf{u} \tag{4.199}$$

This equation is called the orthogonal expansion with respect to the independent random variable function set.

There are many ways to select the functions $H_\ell(\boldsymbol{\zeta})$ in Equation 4.198. For instance, the eigenvectors resulting from the correlation decomposition of the stochastic function $Y(\boldsymbol{\zeta})$ can usually be used as a set of basic functions if $Y(\boldsymbol{\zeta})$ is known. However, it is apparent that correlation decomposition cannot be implemented if this stochastic function is unknown. Nevertheless, there is a possibility of using the independence of random variables in the set of $\boldsymbol{\zeta}$ to construct the stochastic functions $H_\ell(\boldsymbol{\zeta})$ if we only need the decomposition expression with undetermined coefficients. For this purpose, decomposition with respect to the random variable ζ_1 is first considered:

$$Y(\boldsymbol{\zeta}) = \sum_{\ell_1=0}^{\infty} X_{\ell_1}(\zeta_2, \zeta_3, \ldots, \zeta_n) H_{\ell_1}(\zeta_1) \tag{4.200}$$

where

$$X_{\ell_1}(\zeta_2, \zeta_3, \ldots, \zeta_n) = \langle Y, H_{\ell_1} \rangle = \int_{\Omega_{u_1}} p_{\zeta_1}(u_1) Y(u_1, \zeta_2, \zeta_3, \ldots, \zeta_n) H_{\ell_1}(u_1)\, \mathrm{d}u_1 \tag{4.201}$$

is a stochastic function which is lower than $Y(\boldsymbol{\zeta})$ by one dimension, $\{H_{\ell_1}(\zeta_1),\ \ell_1 = 0, 1, 2, \cdots\}$ represents the orthogonal functions with respect to the random variable ζ_1 and Ω_{u_1} denotes the definition field of the real variable u_1.

Second, the decomposition of $X_{\ell_1}(\zeta_2, \zeta_3, \cdots, \zeta_n)$ over random variable ζ_2 is considered:

$$X_{\ell_1}(\zeta_2, \zeta_3, \cdots, \zeta_n) = \sum_{\ell_2=0}^{\infty} X_{\ell_1 \ell_2}(\zeta_3, \zeta_4, \ldots, \zeta_n) H_{\ell_2}(\zeta_2) \tag{4.202}$$

where

$$X_{\ell_1 \ell_2}(\zeta_3, \zeta_4, \cdots, \zeta_n) = \langle X_{\ell_1}, H_{\ell_2} \rangle = \int_{\Omega_{u_2}} p_{\zeta_2}(u_2) X_{\ell_1}(u_2, \zeta_3, \cdots, \zeta_n) H_{\ell_2}(u_2)\, \mathrm{d}u_2 \tag{4.203}$$

is a stochastic function which is lower by two dimensions than $Y(\boldsymbol{\zeta})$, $\{H_{\ell_2}(\zeta_2), \ell_2 = 0, 1, 2, \cdots\}$ are the orthogonal functions with respect to the random variable ζ_2 and, similarly, Ω_{u_2} is the definition field of the real variable u_2.

Obviously, the decomposition can be carried out in a similar way until it comes to ζ_n, then the back substitution of the items leads to

$$Y(\boldsymbol{\zeta}) = \sum_{\ell_1=0}^{\infty}\sum_{\ell_2=0}^{\infty}\cdots\sum_{l_n=0}^{\infty} x_{\ell_1\ell_2\cdots\ell_n} H_{\ell_1}(\zeta_1)H_{\ell_2}(\zeta_2)\cdots H_{\ell_n}(\zeta_n) \tag{4.204}$$

in which $x_{\ell_1\ell_2\cdots\ell_n}$ are a set of deterministic unknown coefficients; $\ell_1\ell_2\cdots\ell_n$ denotes the subscript vector.

Equation 4.204 can be approximated by a finite series as

$$\begin{aligned} Y(\boldsymbol{\zeta}) &= \sum_{\ell_1=0}^{N_1}\sum_{\ell_2=0}^{N_2}\cdots\sum_{\ell_n=0}^{N_n} x_{\ell_1\ell_2\cdots\ell_n} H_{\ell_1}(\zeta_1)H_{\ell_2}(\zeta_2)\cdots H_{\ell_n}(\zeta_n) \\ &= \sum_{\substack{0\le\ell_s\le N_s \\ 1\le s\le n}} x_{\ell_1\ell_2\cdots\ell_n} \prod_{s=1}^{n} H_{\ell_s}(\zeta_s) \end{aligned} \tag{4.205}$$

The above decomposition process is called the sequential orthogonal decomposition of the stochastic function $Y(\boldsymbol{\zeta})$ (Li, 1995a, 1996a). Obviously, the essence of sequential orthogonal decomposition is the series of orthogonal decomposition in corresponding subspaces of the stochastic function space.

4.5.2 Order-Expanded System Method

The sequential orthogonal decomposition idea provides the possibility of establishing an order-expanded system method (OEM) for stochastic dynamical analysis. In order to give a clear description of the theory, this section first discusses the problem in the frame of the static analysis of stochastic structures.

As shown in Section 4.4, the mechanical property matrices of stochastic structures can be represented in a linear form or in the form of truncated series. When only one sort of stochastic factor is considered, the structural stiffness matrix in static analysis can be expressed as (see Equation 4.117)

$$\mathbf{K} = \mathbf{K}_0 + \sum_{i=1}^{n} \mathbf{K}_i Z_i \tag{4.206}$$

in which $\mathbf{K}_i$ depends on the characteristic of the stochastic factors; that is, when the stochastic factor is in a linear form, $\mathbf{K}_i$ is the corresponding standard derivation coefficient matrix, while $\mathbf{K}_i$ is the first-order derivative matrix of $\mathbf{K}$ with respect to the stochastic factors when the factors are involved in a nonlinear form.

By introducing the correlation decomposition techniques, Equation 4.206 can be converted into (see Equation 4.137)

$$\mathbf{K} = \mathbf{K}_0 + \sum_{j=1}^{N_k} \mathbf{K}_j \zeta_j \tag{4.207}$$

The corresponding governing equation of static analysis is (see Equation 4.48)

$$\left(\mathbf{K}_0 + \sum_{j=1}^{N_k} \mathbf{K}_j \zeta_j\right)\mathbf{Y} = \mathbf{F} \tag{4.208}$$

The basic idea of sequential orthogonal decomposition is to expand sequentially the system response in an abstract space as a series of orthogonal functions by utilizing the independence of random variables ζ_i. That is, the response can be expressed as

$$\mathbf{Y}(\boldsymbol{\zeta}) = \sum_{\substack{0 \le \ell_j \le N_j \\ 1 \le j \le N_k}} \mathbf{x}_{\ell_1 \ell_2 \cdots \ell_{N_k}} \prod_{j=1}^{N_k} H_{\ell_j}(\zeta_j) \tag{4.209}$$

in which N_j represents the number of basic functions with respect to the random variable ζ_j and $H_{\ell_j}(\zeta_j)$ are the basic functions with respect to random variable ζ_j and can be selected as orthogonal polynomials in accordance with the probability distribution of the random variable. For example, weighted Hermite orthogonal polynomials are selected for random variables with a normal distribution, while Legendre polynomials are selected for random variables with a uniform distribution, and so on (see Appendix B).

Substituting Equation 4.209 in Equation 4.208 and performing a series of derivations will lead to the following equation (see the next section):

$$\sum_{p=1}^{M} (\mathbf{a}_k)_{\ell p} \mathbf{x}_p = \mathbf{f}_\ell \qquad (\ell = 1, 2, \ldots, M) \tag{4.210}$$

where

$$(\mathbf{a}_k)_{\ell p} = \mathbf{K}_0 \delta_{\ell,p} + \sum_{j=1}^{N_k} \mathbf{K}_j (\gamma_{k_j - 1} \delta_{\ell - \mu_j, p} + \beta_{k_j} \delta_{\ell,p} + \alpha_{k_j + 1} \delta_{\ell + \mu_j, p}) \qquad (0 < k_j \le N_j) \tag{4.211}$$

$$\mu_j = \begin{cases} 1 & \text{for } j = N_k \\ \prod_{i=1}^{N_k - j} (N_{N_k - i} + 1) & \text{for } j < N_k \end{cases} \tag{4.212}$$

$$\ell = 1 + \sum_{j=1}^{N_k} k_j \prod_{i=j+1}^{N_k} (N_i + 1) \tag{4.213}$$

in which α, β and γ denote the recurrence coefficients of the orthogonal polynomials that are discussed in Appendix B, and

$$M = \prod_{i=1}^{N_k} (N_i + 1) \tag{4.214}$$

$$\mathbf{f} = \mathbf{f}_{k_1 k_2 \cdots k_{N_k}} = \mathbf{F} \prod_{i=1}^{N_k} \delta_{0 k_i} \tag{4.215}$$

Equation 4.210 can be rewritten in the following concise form:

$$\mathbf{A}_{\mathrm{K}}\mathbf{X} = \mathbf{P} \tag{4.216}$$

where

$$\mathbf{A}_{\mathrm{K}} = \begin{bmatrix} a_{K,11} & a_{K,12} & \cdots & a_{K,1M} \\ a_{K,21} & a_{K,22} & \cdots & a_{K,2M} \\ \cdots & \cdots & \cdots & \cdots \\ a_{K,M1} & a_{K,M2} & \cdots & a_{K,MM} \end{bmatrix} \tag{4.217}$$

$$\mathbf{X}^{\mathrm{T}} = (\mathbf{x}_1^{\mathrm{T}}, \mathbf{x}_2^{\mathrm{T}}, \cdots, \mathbf{x}_M^{\mathrm{T}}) \tag{4.218}$$

$$\mathbf{P}^{\mathrm{T}} = (\mathbf{f}_1^{\mathrm{T}}, \mathbf{f}_2^{\mathrm{T}}, \cdots, \mathbf{f}_M^{\mathrm{T}}) \tag{4.219}$$

It is noted that Equation 4.216 has been converted into a deterministic equation and can be solved with the method for ordinary algebra equations. The degree of freedom of the original system is n_{d}, while that of the unknown variables in Equation 4.216 is enlarged to $n_{\mathrm{d}} \times M$. This is the reason why Equation 4.216 is referred to as the order-expanded equation of original system.

Note that

$$\mathbf{x}_p = \mathbf{x}_{\ell_1 \ell_2 \cdots \ell_{N_k}} \tag{4.220}$$

Then, once $\mathbf{X}$ is obtained from Equation 4.216, it can be substituted back in Equation 4.209. The mean and covariance of the stochastic system responses can thus be obtained easily according to the characteristic of orthogonal polynomials. For instance, the mean of the system response can be computed by introducing $\prod_{j=1}^{N_k} H_0(\zeta_j)$ and noticing the following characteristics:

$$H_0(\zeta_j) \equiv 1 \quad (j = 1, 2, \ldots, N_k) \tag{4.221}$$

Therefore, multiplying both sides of Equation 4.209 by $\prod_{j=1}^{N_k} H_0(\zeta_j)$ yields

$$\mathbf{Y}(\boldsymbol{\zeta}) = \sum_{\substack{0 \le \ell_j \le N_j \\ 1 \le j \le N_k}} \mathbf{x}_{\ell_1 \ell_2 \cdots \ell_{N_k}} \prod_{j=1}^{N_k} H_0(\zeta_j) \prod_{j=1}^{N_k} H_{\ell_j}(\zeta_j) \tag{4.222}$$

By taking the expectation operation on both sides of the above equation and considering the relationship

$$\mathcal{E}[H_0(\zeta_j) H_{\ell_j}(\zeta_j)] = \begin{cases} 0 & \text{for } \ell_j \neq 0 \\ 1 & \text{for } \ell_j = 0 \end{cases} \tag{4.223}$$

it then follows that

$$\mathcal{E}[\mathbf{Y}(\zeta)] = \mathbf{x}_{00\cdots 0} \tag{4.224}$$

Meanwhile, the covariance matrix of the stochastic response is given by

$$\mathrm{cov}[\mathbf{Y},\mathbf{Y}] = \mathcal{E}[\mathbf{Y}\mathbf{Y}^{\mathrm{T}}] - \mathcal{E}[\mathbf{Y}]\cdot\mathcal{E}[\mathbf{Y}^{\mathrm{T}}] \tag{4.225}$$

where

$$\mathcal{E}[\mathbf{Y}\mathbf{Y}^{\mathrm{T}}] = \mathcal{E}\left[\sum_{\substack{0\le \ell_j\le N_j\\ 1\le j\le N_k}} \mathbf{x}_{\ell_1\ell_2\cdots\ell_{N_k}} \prod_{j=1}^{N_k} H_{\ell_j}(\zeta_j) \sum_{\substack{0\le k_j\le N_j\\ 1\le j\le N_k}} \mathbf{x}^{\mathrm{T}}_{k_1k_2\cdots k_{N_k}} \prod_{j=1}^{N_k} H_{k_j}(\zeta_j)\right] \tag{4.226}$$

Since there exists

$$\mathcal{E}\left[\prod_{j=1}^{N_k} H_{\ell_j}(\zeta_j)\prod_{j=1}^{N_k} H_{k_j}(\zeta_j)\right] = \begin{cases} 1 & \text{for } \ell_j \equiv k_j \\ 0 & \text{for } \ell_j \ne k_j \end{cases} \tag{4.227}$$

Equation 4.226 becomes

$$\mathcal{E}[\mathbf{Y}\mathbf{Y}^{\mathrm{T}}] = \sum_{\substack{0\le \ell_j\le N_j\\ 1\le j\le N_k}} \mathbf{x}_{\ell_1\ell_2\cdots\ell_{N_k}} \mathbf{x}^{\mathrm{T}}_{\ell_1\ell_2\cdots\ell_{N_k}} \tag{4.228}$$

Considering Equation 4.224, substituting Equation 4.228 in Equation 4.225 yields

$$\mathrm{cov}[\mathbf{Y},\mathbf{Y}] = \sum_{\substack{0\le \ell_j\le N_j\\ 1\le j\le N_k}} \mathbf{x}_{\ell_1\ell_2\cdots\ell_{N_k}} \mathbf{x}^{\mathrm{T}}_{\ell_1\ell_2\cdots\ell_{N_k}} - \mathbf{x}_{00\cdots0}\mathbf{x}^{\mathrm{T}}_{00\cdots0} \tag{4.229}$$

It is noted that the above equation is a matrix equation in essence. The diagonal elements of the matrix can give the response variance

$$\mathrm{Var}[Y_i] = \sum_{\substack{0\le \ell_j\le N_j\\ 1\le j\le N_k}} x^2_{\ell_1\ell_2\cdots\ell_{N_k},i} - x^2_{00\cdots0,i} \qquad i = 1,2,\cdots,n_{\mathrm{d}} \tag{4.230}$$

where $x_{\ell_1\ell_2\cdots\ell_{N_k},i}$ $(i = 1,2,\ldots,n_{\mathrm{d}})$ is the ith component of $\mathbf{x}_{\ell_1\ell_2\cdots\ell_{N_k}}$.

It is shown from the preceding derivation that the numerical characteristics of the stochastic structural response can be obtained by implementing the characteristics of the orthogonal functions once the orthogonal decomposition undetermined coefficients are obtained from the deterministic order-expanded algebraic equation. The method is referred to as the *OEM* for stochastic structural analysis in this book.

Some careful readers may have already found that the expression in Equation 4.213 is relevant to the arrangement manner of the subscript vector elements. In fact, the expression is based on the rule of 'running over in an inverted order' to arrange the subscript vector. That means, if considering the arrangement order as the following matrix:

$$\begin{matrix} \ell_1 & \ell_2 & \cdots & \ell_{N_k} \end{matrix}$$
$$\begin{bmatrix} 0 & 0 & \cdots & 0 \\ 1 & 1 & \cdots & 1 \\ 2 & 2 & \cdots & 2 \\ \vdots & \vdots & \vdots & \vdots \\ N_1 & N_2 & \cdots & N_k \end{bmatrix}$$

then the order number of the corresponding subscripts can be determined in the following way: ℓ_{N_k} increases progressively while the other variables are assigned to be zero; after it is run over ℓ_{N_k}, with $\ell_{N_k-1} = 1$, it is again run progressively over ℓ_{N_k}. When running over ℓ_{N_k-1} is finished, the number 1 is assigned to ℓ_{N_k-2}. The operation is continued in this progressive way and an arrangement sequence of the $\mathbf{x}_\ell$ and $\mathbf{f}_\ell$ in the order-expanded system in Equation 4.210 can be obtained. It should be particularly noted that $\mathbf{x}_\ell$ and $\mathbf{f}_\ell$ herein are vectors with n_d elements, which is the degree of freedom of the original system, and $\mathbf{a}_\mathbf{K}$ is similarly an $n_\mathrm{d} \times n_\mathrm{d}$ matrix.

4.5.3 Proof of the Order-Expanded System Method

This section gives the proof of the order-expanded Equation 4.210 by implementing the mathematical induction method.

First, the case $N_k = 1$ is considered; that is, there is only one random variable involved in the stochastic structure. Having applied the orthogonal decomposition over ζ_1 to $\mathbf{Y}$, we have

$$\mathbf{Y} = \sum_{\ell_1=0}^{N_1} \mathbf{x}_{\ell_1} H_{\ell_1}(\zeta_1) \tag{4.231}$$

in which N_1 is the expansion order over the random variable ζ_1.

Substituting the equation in Equation 4.208 with $N_k = 1$ yields

$$\mathbf{F} = (\mathbf{K}_0 + \mathbf{K}_1\zeta_1)\left[\sum_{\ell_1=0}^{N_1} \mathbf{x}_{\ell_1} H_{\ell_1}(\zeta_1)\right] = \sum_{\ell_1=0}^{N_1} (\mathbf{K}_0 + \mathbf{K}_1\zeta_1)\mathbf{x}_{\ell_1} H_{\ell_1}(\zeta_1) \tag{4.232}$$

Multiplying the terms on both sides of the equation with $H_{k_1}(\zeta_1)$ and implementing the recurrence relationship for $\zeta_1 H_{\ell_1}(\zeta_1)$ (see Equation B.15 in Appendix B) will yield

$$\begin{aligned}\mathbf{F}H_{k_1}(\zeta_1) &= \sum_{\ell_1=0}^{N_k} \mathbf{K}_0\mathbf{x}_{\ell_1} H_{k_1}(\zeta_1) H_{\ell_1}(\zeta_1) \\ &\quad + \sum_{\ell_1=0}^{N_k} \mathbf{K}_1\mathbf{x}_{\ell_1} H_{k_1}(\zeta_1)[\alpha_{\ell_1} H_{\ell_1-1}(\zeta_1) + \beta_{\ell_1} H_{\ell_1}(\zeta_1) + \gamma_{\ell_1} H_{\ell_1+1}(\zeta_1)]\end{aligned} \tag{4.233}$$

Taking the expectation operation with respect to ζ_1 on both sides of the above equation (that is, equivalent to the inner product of Equation 4.188), utilizing Equation 4.221 and the following relationship

$$\mathcal{E}[H_{k_1} H_{\ell_1-1}] = \begin{cases} 1 & \text{for } \ell_1 - 1 = k_1 \\ 0 & \text{otherwise} \end{cases} \tag{4.234}$$

$$\mathcal{E}[H_{k_1} H_{\ell_1}] = \begin{cases} 1 & \text{for } \ell_1 = k_1 \\ 0 & \text{otherwise} \end{cases} \tag{4.235}$$

$$\mathcal{E}[H_{k_1} H_{\ell_1+1}] = \begin{cases} 1 & \text{for } \ell_1 + 1 = k_1 \\ 0 & \text{otherwise} \end{cases} \tag{4.236}$$

we are then led to

$$\begin{aligned}\mathbf{F}\delta_{0k_1} &= \mathbf{K}_0\mathbf{x}_{k_1} + \mathbf{K}_1(\alpha_{k_1+1}\mathbf{x}_{k_1+1} + \beta_{k_1}\mathbf{x}_{k_1} + \gamma_{k_1-1}\mathbf{x}_{k_1-1}) \\ &= \alpha_{k_1+1}\mathbf{K}_1\mathbf{x}_{k_1+1} + (\mathbf{K}_0 + \beta_{k_1}\mathbf{K}_1)\mathbf{x}_{k_1} + \gamma_{k_1-1}\mathbf{K}_1\mathbf{x}_{k_1-1} \qquad (k_1 = 0, 1, 2, \cdots, N_1)\end{aligned} \tag{4.237}$$

The equation can be expressed in a matrix form

$$\sum_{p=0}^{N_1+1} \mathbf{a}_{k_1 p}\mathbf{x}_p = \mathbf{f}_{k_1} \tag{4.238}$$

where

$$\mathbf{a}_{k_1 p} = \mathbf{K}_0\delta_{k_1 p} + \mathbf{K}_1(\gamma_{k_1-1}\delta_{k_1-1,p} + \beta_{k_1}\delta_{k_1,p} + \alpha_{k_1+1}\delta_{k_1+1,p}) \tag{4.239}$$

Note that the coefficient matrices of Equation 4.238 are tri-diagonal ones.

It is apparent that Equation 4.211 is equivalent to Equation 4.239 and the left-hand side of Equation 4.237 is the same as Equation 4.215 when $N_k = 1$. Therefore, the order-expanded system equation is proved to be correct when only one single random variable is involved.

Now assume that the order-expanded system equation is correct when $n-1$ independent random variables are involved ($n-1 < N_k$); that is, there exists the following order-expanded equation:

$$\sum_{s=1}^{M_{n-1}} \mathbf{a}_{rs}\mathbf{x}_s = \mathbf{f}_{k_1 k_2 \cdots k_{n-1}} \tag{4.240}$$

in which

$$M_{n-1} = \prod_{i=1}^{n-1}(N_i + 1) \tag{4.241}$$

$$\mathbf{a}_{rs} = \mathbf{K}_0\delta_{rs} + \sum_{j=1}^{n-1}\mathbf{K}_j(\gamma_{k_j-1}\delta_{r-\lambda_j,s} + \beta_{k_j}\delta_{rs} + \alpha_{k_j+1}\delta_{r+\lambda_j,s}) \tag{4.242}$$

$$\lambda_j = \begin{cases} 1 & \text{for } j = n-1 \\ \displaystyle\prod_{i=1}^{n-j-1}(N_{n-i-1} + 1) & \text{for } j < n-1 \end{cases} \tag{4.243}$$

$$r = 1 + \sum_{j-1}^{n-1} k_j \prod_{i=j+1}^{n-1}(N_j + 1) \tag{4.244}$$

$$\mathbf{x}_s = \mathbf{X}_{\ell_1\ell_2\cdots\ell_{n-1}} \tag{4.245}$$

$$\mathbf{f}_{k_1 k_2 \cdots k_{n-1}} = \mathbf{F}\prod_{j=1}^{n-1}\delta_{0k_j} \tag{4.246}$$

Then consider the case of the order-expanded system with n independent random variables, in which the orthogonal decomposition of $\mathbf{Y}$ over the previous $n-1$ random variables can be expressed as

$$\mathbf{Y}(\boldsymbol{\zeta}) = \sum_{\substack{0 \le \ell_j \le N_j \\ 1 \le j \le n-1}} \mathbf{X}_{\ell_1 \ell_2 \cdots \ell_{n-1}}(\zeta_n) \prod_{j=1}^{n-1} H_{\ell_j}(\zeta_j) \tag{4.247}$$

In accordance with Equations 4.240–4.246, the expansion yields the following order-expanded equation

$$\sum_{s=1}^{M_{n-1}} \tilde{\mathbf{a}}_{rs} \mathbf{x}_s(\zeta_n) = \mathbf{f}_{k_1 k_2 \cdots k_{n-1}} \tag{4.248}$$

in which

$$\tilde{\mathbf{a}}_{rs} = \mathbf{a}_{rs} + \mathbf{K}_n(\zeta_n)\delta_{rs} \tag{4.249}$$

$$\mathbf{x}_s(\zeta_n) = \mathbf{X}_{\ell_1 \ell_2 \cdots \ell_{n-1}}(\zeta_n) \tag{4.250}$$

The $\mathbf{x}_s(\zeta_n)$ in Equation 4.250 can be further decomposed with respect to the nth random variable by applying the idea of sequential orthogonal decomposition in the aforementioned stochastic function space; that is:

$$\mathbf{x}_s(\zeta_n) = \sum_{\ell_n=0}^{N_n} \mathbf{x}_{\ell_1 \ell_2 \cdots \ell_{n-1} \ell_n} H_{\ell_n}(\zeta_n) \tag{4.251}$$

By substituting it in Equation 4.248 and multiplying both sides by $H_{k_n}(\zeta_n)$, and implementing the recursive relationship of $\zeta_n H_{k_n}(\zeta_n)$ and the characteristics of the orthogonal basis functions, the inner product operation similar to Equation 4.188 on both sides with respect to ζ_n will give

$$\begin{aligned} \mathbf{f}_{k_1 k_2 \cdots k_{n-1}} \delta_{0k_n} = \sum_{s=1}^{M_{n-1}} [\tilde{\mathbf{a}}_{rs} \mathbf{x}_{\ell_1 \ell_2 \cdots \ell_n} + \mathbf{K}_n(\alpha_{k_n+1} \mathbf{x}_{\ell_1 \ell_2 \cdots \ell_{n-1} k_n+1} \\ + \beta_{k_n} \mathbf{x}_{\ell_1 \ell_2 \cdots \ell_{n-1} k_n} + \gamma_{k_n-1} \mathbf{x}_{\ell_1 \ell_2 \cdots \ell_{n-1} k_n-1})\delta_{rs}] \end{aligned} \tag{4.252}$$

Noting that k_n runs from 0 to N_n in the above equation, if the following notation is used

$$\mathbf{x}_{\ell_1 \ell_2 \cdots \ell_{n-1} k_n} = \mathbf{z}_m$$

and introducing

$$\mathbf{f}_{k_1 k_2 \cdots k_{n-1}} \delta_{0k_n} = \mathbf{f}_{k_1 k_2 \cdots k_n} \tag{4.253}$$

then Equation 4.252 can be expressed in the form of matrices:

$$\sum_{m=1}^{N_n+1} (\mathbf{a}_{im})_{rs} \mathbf{z}_m = \mathbf{f}_{k_1 k_2 \cdots k_n} \tag{4.254}$$

in which

$$(\mathbf{a}_{im})_{rs} = \sum_{s=1}^{M_{n-1}} [\mathbf{K}_0\delta_{rs}\delta_{im} + \mathbf{K}_n(\gamma_{k_n-1}\delta_{i-1,m} + \beta_{k_n}\delta_{im} + \alpha_{k_n+1}\delta_{i+1,m})\delta_{rs} + \sum_{j=1}^{n-1}\mathbf{K}_j(\gamma_{k_j-1}\delta_{r-\lambda_j,s} + \beta_{k_j}\delta_{rs} + \alpha_{k_j+1}\delta_{r+\lambda_j,s})\delta_{im}] \tag{4.255}$$

Let

$$\mathbf{a}_{\ell p} = (\mathbf{a}_{im})_{rs} \tag{4.256}$$

Then according to the arrangement pattern of matrices, the following relationships exist:

$$\ell = (r-1)(N_n+1) + i \tag{4.257}$$

$$p = (s-1)(N_n+1) + m \tag{4.258}$$

It can be proved that

$$\delta_{rs}\delta_{im} = \delta_{\ell p} \tag{4.259}$$

In a similar way, let

$$\delta_{i-1,m}\delta_{rs} = \delta_{\ell-\mu_n,p} \tag{4.260}$$

$$\delta_{i+1,m}\delta_{rs} = \delta_{\ell+\mu_n,p} \tag{4.261}$$

$$\delta_{im}\delta_{r-\lambda_j,s} = \delta_{\ell-\mu_j,p} \tag{4.262}$$

$$\delta_{im}\delta_{r+\lambda_j,s} = \delta_{\ell+\mu_j,p} \tag{4.263}$$

Then, from the above equalities and Equations 4.257 and 4.258, there are

$$\mu_n = 1 \tag{4.264}$$

$$\mu_j = \lambda_j(N_n+1) \tag{4.265}$$

Thus, Equation 4.254 can be rearranged as

$$\sum_{\ell=1}^{M_n} \mathbf{a}_{\ell p}\mathbf{x}_p = \mathbf{f}_\ell \tag{4.266}$$

where

$$\mathbf{a}_{\ell p} = \mathbf{K}_0\delta_{\ell p} + \sum_{j=1}^{n}\mathbf{K}_j(\gamma_{k_j-1}\delta_{\ell-\mu_j,p} + \beta_{k_j}\delta_{\ell p} + \alpha_{k_j+1}\delta_{\ell+\mu_j,p}) \tag{4.267}$$

in which

$$\mu_j = \begin{cases} 1 & \text{for} \quad j = n \\ \prod_{i=1}^{n-j}(N_{n-i}+1) & \text{for} \quad j<n \end{cases} \tag{4.268}$$

$$M_n = \prod_{i=1}^{n}(N_i+1) \tag{4.269}$$

$$\ell = 1 + \sum_{j=1}^{n} k_j \prod_{i=j+1}^{n}(N_i+1) \tag{4.270}$$

When $N_k = n$, Equation 4.267 is equivalent to Equation 4.211 and Equation 4.253 is equivalent to Equation 4.215. Therefore, it is demonstrated that the expressions of the order-expanded matrices and the loading vector are also correct when n random variables are considered.

In view of the preceding process, not only the conclusion of the order-expanded system is correct when $N_k = 1$, but also the conclusion is proved to be correct when $N_k = n$ if we assume the correctness of the order-expanded system when $N_k = n-1$. In accordance with the principle of mathematical induction theory, the order-expanded system given in the last section is correct.

4.5.4 Dynamic Analysis

When stochastic dynamic matrices are expressed in linear forms with respect to the basic random variables and these kinds of expression are simplified by applying the correlation decomposition technique, Equation 4.65 can be written as

$$\left(\mathbf{M}_0 + \sum_{j=1}^{N_\mathbf{M}} \mathbf{M}_j \zeta_j\right)\ddot{\mathbf{Y}} + \left(\mathbf{C}_0 + \sum_{j=1}^{N_\mathbf{C}} \mathbf{C}_j \zeta_j\right)\dot{\mathbf{Y}} + \left(\mathbf{K}_0 + \sum_{j=1}^{N_\mathbf{K}} \mathbf{K}_j \zeta_j\right)\mathbf{Y} = \mathbf{F}(t) \tag{4.271}$$

in which $N_{\boldsymbol{M}}$, $N_{\boldsymbol{C}}$ and $N_{\boldsymbol{K}}$ stand for the number of the independent random variables in the stochastic mass, damping and stiffness matrices respectively. According to the relevant content in this chapter, the other symbols in the above equation are not hard to understand.

For notational convenience, we introduce the following symbols

$$\mathbf{A}_{\mathbf{m}s} = \begin{cases} \mathbf{M}_j & \text{for } s \le N_\mathbf{M} \\ \mathbf{0} & \text{for } s > N_\mathbf{M} \end{cases} \qquad j = s \tag{4.272}$$

$$\mathbf{A}_{\mathbf{c}s} = \begin{cases} \mathbf{0} & \text{for } s \le N_\mathbf{M} \\ \mathbf{C}_j & \text{for } N_\mathbf{M} < s \le N_\mathbf{M} + N_\mathbf{C} \\ \mathbf{0} & \text{for } s > N_\mathbf{M} + N_\mathbf{C} \end{cases} \qquad j = s - N_\mathbf{M} \tag{4.273}$$

$$\mathbf{A}_{\mathbf{k}s} = \begin{cases} 0 & \text{for} \quad s \le N_\mathbf{M} + N_\mathbf{C} \\ \mathbf{K}_j & \text{for} \quad s > N_\mathbf{M} + N_\mathbf{C} \end{cases} \qquad j = s - (N_\mathbf{M} + N_\mathbf{C}) \tag{4.274}$$

and

$$\mathbf{A}_{\mathbf{M}_0} = \mathbf{M}_0 \qquad \mathbf{A}_{\mathbf{C}_0} = \mathbf{C}_0 \qquad \mathbf{A}_{\mathbf{K}_0} = \mathbf{K}_0 \tag{4.275}$$

Then Equation 4.271 can be further written as

$$\left(\mathbf{A}_{\mathbf{M}_0} + \sum_{s=1}^{R} \mathbf{A}_{\mathbf{M}s}\zeta_s\right)\ddot{\mathbf{Y}} + \left(\mathbf{A}_{\mathbf{C}_0} + \sum_{s=1}^{R} \mathbf{A}_{\mathbf{C}s}\zeta_s\right)\dot{\mathbf{Y}} + \left(\mathbf{A}_{\mathbf{K}_0} + \sum_{s=1}^{R} \mathbf{A}_{\mathbf{K}s}\zeta_s\right)\mathbf{Y} = \mathbf{F}(t) \tag{4.276}$$

in which $R = N_{\mathbf{M}} + N_{\mathbf{C}} + N_{\mathbf{K}}$.

According to the idea of orthogonal decomposition, if the structural response $\mathbf{Y}$ is sequentially expanded as a set of series of orthogonal basis functions in a stochastic function space, namely

$$\mathbf{Y}(\boldsymbol{\zeta}) = \sum_{0 \le \ell_s \le N_s, 1 \le s \le R} \mathbf{x}_{\ell_1 \ell_2 \ldots \ell_R}(t) \prod_{j=1}^{R} H_{\ell_j}(\zeta_j) \tag{4.277}$$

then obviously the velocity and acceleration response can be depicted as

$$\dot{\mathbf{Y}}(\boldsymbol{\zeta}) = \sum_{0 \le \ell_s \le N_s, 1 \le s \le R} \dot{\mathbf{x}}_{\ell_1 \ell_2 \ldots \ell_R}(t) \prod_{j=1}^{R} H_{\ell_j}(\zeta_j) \tag{4.278}$$

$$\ddot{\mathbf{Y}}(\boldsymbol{\zeta}) = \sum_{0 \le \ell_s \le N_s, 1 \le s \le R} \ddot{\mathbf{x}}_{\ell_1 \ell_2 \ldots \ell_R}(t) \prod_{j=1}^{R} H_{\ell_j}(\zeta_j) \tag{4.279}$$

Note that $\mathbf{x}_{\ell_1 \ell_2 \ldots \ell_R}(t)$, $\dot{\mathbf{x}}_{\ell_1 \ell_2 \ldots \ell_R}(t)$ and $\ddot{\mathbf{x}}_{\ell_1 \ell_2 \ldots \ell_R}(t)$ are all deterministic undetermined functions of time.

According to the proof for the order-expanded system of static analysis, it can be testified that through sequential orthogonal decomposition the following order-expanded system equation can be derived from Equation 4.276 (Li, 1995b):

$$\mathbf{A}_{\mathbf{M}}\ddot{\mathbf{X}} + \mathbf{A}_{\mathbf{C}}\dot{\mathbf{X}} + \mathbf{A}_{\mathbf{K}}\mathbf{X} = \mathbf{P}(t) \tag{4.280}$$

of which the component form is

$$\sum_{p=1}^{M_R} [(\mathbf{a}_{\mathbf{M}})_{\ell p}\ddot{\mathbf{x}}_p(t) + (\mathbf{a}_{\mathbf{C}})_{\ell p}\dot{\mathbf{x}}_p(t) + (\mathbf{a}_{\mathbf{K}})_{\ell p}\mathbf{x}_p(t)] = \mathbf{f}_\ell(t) \tag{4.281}$$

If the subscripts $\mathbf{M}$, $\mathbf{C}$ and $\mathbf{K}$ are skipped in the above equation, then the general expression of the partitioned matrix will be

$$\mathbf{a}_{\ell p} = \mathbf{A}_0\delta_{\ell p} + \sum_{s=1}^{R} \mathbf{A}_s(\gamma_{k_s-1}\delta_{\ell-\mu_s,p} + \beta_{k_s}\delta_{\ell p} + \alpha_{k_s+1}\delta_{\ell+\mu_s,p}) \qquad (0 \le k_s \le N_s) \tag{4.282}$$

where

$$\mu_s = \begin{cases} 1 & \text{for } s = R \\ \prod_{j=1}^{R-s}(N_{R-j}+1) & \text{for } s<R \end{cases} \tag{4.283}$$

and

$$M_R = \prod_{s=1}^{R}(N_s+1) \tag{4.284}$$

$$\ell = 1 + \sum_{s=1}^{R} k_s \prod_{j=s+1}^{R}(N_j+1) \tag{4.285}$$

$$\mathbf{f}_\ell(t) = \mathbf{f}_{k_1k_2\ldots k_R}(t) = \mathbf{F}(t)\prod_{s=1}^{R}\delta_{0k_s} \tag{4.286}$$

$$\mathbf{x}_p(t) = \mathbf{x}_{\ell_1\ell_2\ldots\ell_R}(t) \tag{4.287}$$

The coefficients α, β and γ in Equation 4.282 are from the following recursive relationship of orthogonal polynomials:

$$\zeta_s H_{\ell_s}(\zeta_s) = \alpha_{\ell_s} H_{\ell_s-1}(\zeta_s) + \beta_{\ell_s} H_{\ell_s}(\zeta_s) + \gamma_{\ell_s} H_{\ell_s+1}(\zeta_s) \tag{4.288}$$

The relationship between the matrices in both the order-expanded system Equation 4.280 and the component expression Equation 4.281 can be written as (the subscripts of the items are skipped)

$$\mathbf{A} = [\mathbf{a}_{\ell p}]_{\ell=1-M_R, p=1-M_R} \tag{4.289}$$

$$\mathbf{X} = (\mathbf{x}_p)^{\mathrm{T}}_{p=1-M_R} \tag{4.290}$$

$$\mathbf{P} = (\mathbf{f}_\ell)^{\mathrm{T}}_{\ell=1-M_R} \tag{4.291}$$

The order-expanded system equation derived from the sequential orthogonal decomposition is a dynamic equation with deterministic parameters. Using this equation, the problem of the original stochastic structural analysis is converted into a deterministic system analysis, which can be solved by any kind of algorithm for a deterministic dynamic equation. For instance, by utilizing the linear acceleration algorithm, the displacement response **X** can be expanded with respect to the time instants t_j:

$$\mathbf{X}(t_j+\tau) = \mathbf{X}(t_j) + \dot{\mathbf{X}}(t_j)\tau + \frac{\ddot{\mathbf{X}}(t_j)}{2!}\tau^2 + \frac{\dddot{\mathbf{X}}(t_j)}{3!}\tau^3 + \cdots \tag{4.292}$$

Differentiation of this expression with respect to τ yields

$$\dot{\mathbf{X}}(t_j+\tau) = \dot{\mathbf{X}}(t_j) + \ddot{\mathbf{X}}(t_j)\tau + \frac{1}{2}\dddot{\mathbf{X}}(t_j)\tau^2 + \cdots \tag{4.293}$$

According to the linear acceleration method (Clough and Penzien, 1993), the acceleration response over the time interval $\Delta t = t_{j+1} - t_j$ is assumed to be a linear function with respect to time τ; that is:

$$\dddot{\mathbf{X}}(t_j) = \frac{\ddot{\mathbf{X}}(t_{j+1}) - \ddot{\mathbf{X}}(t_j)}{\Delta t} = \text{constant} \tag{4.294}$$

Denote

$$\mathbf{X}(t_{j+1}) = \mathbf{X}_{j+1} \tag{4.295}$$

$$\mathbf{X}(t_j) = \mathbf{X}_j \tag{4.296}$$

Substituting $\tau = \Delta t$ and Equation 4.294 in Equations 4.292 and 4.293 with a third-order truncation of displacement response yields

$$\mathbf{X}_{j+1} = \mathbf{X}_j + \dot{\mathbf{X}}_j \Delta t + \frac{1}{3}\ddot{\mathbf{X}}_j (\Delta t)^2 + \frac{1}{6}\ddot{\mathbf{X}}_{j+1}(\Delta t)^2 \tag{4.297}$$

$$\dot{\mathbf{X}}_{j+1} = \dot{\mathbf{X}}_j + \frac{1}{2}\ddot{\mathbf{X}}_j \Delta t + \frac{1}{2}\ddot{\mathbf{X}}_{j+1}\Delta t \tag{4.298}$$

Substituting the above two equations in Equation 4.280 leads to

$$\tilde{\mathbf{A}}\ddot{\mathbf{X}}_{j+1} = \tilde{\mathbf{P}}_{j+1} \tag{4.299}$$

in which

$$\tilde{\mathbf{A}} = \mathbf{A}_{\mathbf{M}} + \frac{\Delta t}{2}\mathbf{A}_{\mathbf{C}} + \frac{(\Delta t)^2}{6}\mathbf{A}_{\mathbf{K}} \tag{4.300}$$

$$\tilde{\mathbf{P}}_{j+1} = \mathbf{P}_{j+1} - \mathbf{C}\left(\mathbf{X}_j + \frac{\Delta t}{2}\ddot{\mathbf{X}}_j\right) + \mathbf{K}\left(\mathbf{X}_j + \Delta t\dot{\mathbf{X}}_j + \frac{(\Delta t)^2}{3}\ddot{\mathbf{X}}_j\right) \tag{4.301}$$

Therefore, the original dynamic equation is converted into an algebraic equation at discrete time instants. Combined with the initial conditions of the problem of concern, the acceleration response can be obtained by applying the solution procedure of algebraic equations step by step. Then the displacement and velocity responses can be further obtained by implementing Equations 4.297 and 4.298.

If the initial conditions of the original stochastic system in Equation 4.271 are given by

$$\mathbf{Y}(0) = \mathbf{Y}_0, \dot{\mathbf{Y}}(0) = \dot{\mathbf{Y}}_0 \tag{4.302}$$

where $\mathbf{Y}_0$ and $\dot{\mathbf{Y}}_0$ are deterministic vectors, then the initial conditions of the order-expanded system in Equation 4.281 can be written as

$$\mathbf{x}_p(0) = \mathbf{x}_{\ell_1\ell_2\cdots\ell_R}(0) = \mathbf{Y}_0 \prod_{s=1}^{R} \delta_{0k_s} \tag{4.303}$$

$$\dot{\mathbf{x}}_p(0) = \dot{\mathbf{x}}_{\ell_1\ell_2\cdots\ell_R}(0) = \dot{\mathbf{Y}}_0 \prod_{s=1}^{R} \delta_{0k_s} \tag{4.304}$$

Once the responses of the order-expanded system are obtained, the numerical characteristics of the original structural responses can be given in a way similar to the static analysis. For example, the mean of displacement response is given as

$$\mathcal{E}[\mathbf{Y}(t)] = \mathbf{x}_{00\cdots0}(t) \tag{4.305}$$

and the correlation function matrix of the displacement responses at two arbitrary time instants can be obtained by

$$\mathbf{R}_{\mathbf{Y}}(t_1, t_2) = \sum_{\substack{0\le\ell_s\le N_s \\ 1\le s\le R}} \mathbf{x}_{\ell_1\ell_2\cdots\ell_R}(t_1)\mathbf{x}^{\mathrm{T}}_{\ell_1\ell_2\cdots\ell_R}(t_2) \tag{4.306}$$

The covariance matrix of the displacement responses at two arbitrary time instants can be evaluated further through

$$\mathbf{C}_{\mathbf{Y}}(t_1, t_2) = \sum_{\substack{0\le\ell_s\le N_s \\ 1\le s\le R}} \mathbf{x}_{\ell_1\ell_2\cdots\ell_R}(t_1)\mathbf{x}^{\mathrm{T}}_{\ell_1\ell_2\cdots\ell_R}(t_2) - \mathbf{x}_{00\cdots0}(t_1)\mathbf{x}^{\mathrm{T}}_{00\cdots0}(t_2) \tag{4.307}$$

As $t_1 = t_2 = t$, the covariance matrix is reduced to the variance matrix. In addition:

$$\mathbf{C}_{\mathbf{Y}}(t) = \sum_{\substack{0\le\ell_s\le N_s \\ 1\le s\le R}} \mathbf{x}_{\ell_1\ell_2\cdots\ell_R}(t)\mathbf{x}^{\mathrm{T}}_{\ell_1\ell_2\cdots\ell_R}(t) - \mathbf{x}_{00\cdots0}(t)\mathbf{x}^{\mathrm{T}}_{00\cdots0}(t) \tag{4.308}$$

The equation is essentially an expression in a matrix form, where the diagonal elements give the variance vector of the displacement responses for different degrees of freedom of the original systems:

$$\mathrm{Var}[Y_j] = \sum_{\substack{0\le\ell_s\le N_s \\ 1\le s\le R}} x^2_{\ell_1\ell_2\cdots\ell_R,j}(t) - x^2_{00\cdots0,j}(t) \qquad j = 1, 2, \ldots, n_{\mathrm{d}} \tag{4.309}$$

where $x_{\ell_1\ell_2\cdots\ell_R,j}$ is the jth component of $\mathbf{x}_{\ell_1\ell_2\cdots\ell_R}(t)$.

Similar to Equations 4.305–4.309, the numerical characteristics of the velocity and acceleration responses of the stochastic structures subjected to deterministic excitation can also be obtained.

Example 4.2. Stochastic Response of a Two-DOF System Consider a two DOF system subjected to deterministic excitations. The mass and stiffness are regarded as random variables with means $m_{10} = m_{20} = 1$ and $k_{10} = k_{20} = 39.48$ and the coefficients of variation respectively $\boldsymbol{\delta}_{m_1} = \boldsymbol{\delta}_{m_2} = 0.1$ and $\boldsymbol{\delta}_{k_1} = \boldsymbol{\delta}_{k_2} = 0.2$. The damping ratio takes a deterministic value of 0.05. Two types of excitation are considered. In the first case (Case 1), two sinusoid loads are applied on the lumped masses; that is, $f_1(t) = f_2(t) = \sin(\boldsymbol{\omega}_s t)$, $\boldsymbol{\omega}_s = 3.1416$. In the second case (Case 2), the El Centrol accelerogram (N–S component) is applied.

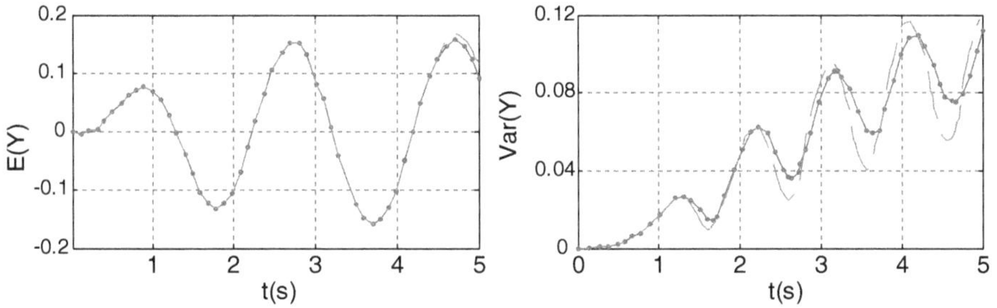

(a) Mean of the response **(b) Variance of the response**

Figure 4.11 Comparison between OEM and MCS (Case 1). (Points: 5000 MCSs; solid line: fourth OEM; dotted line: first OEM).

Stochastic response analysis of the system is performed by the OEM. Figures 4.11 and 4.12 illustrate the comparison between the means and variances of the responses by the OEM and the MCS. It is seen that the mean response can be approached by the first-order OEM, whereas the fourth order OEM is needed to approach the variance response.

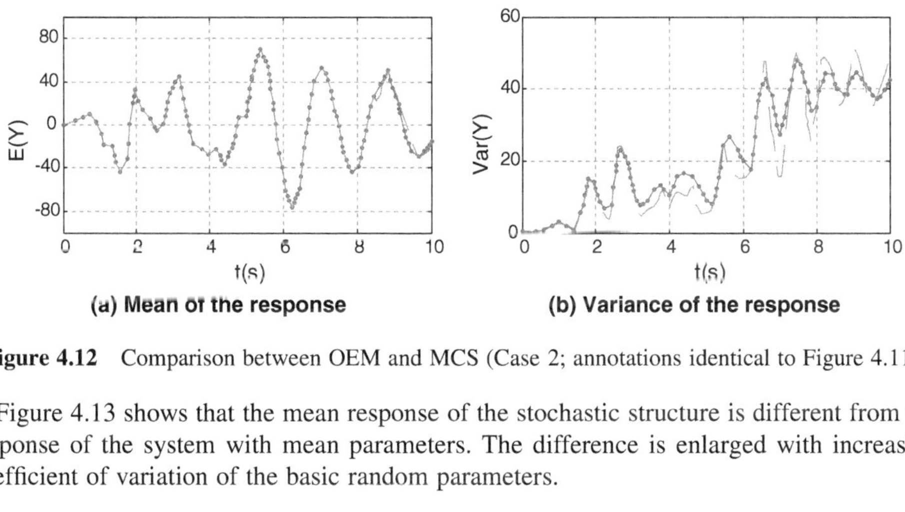

(a) Mean of the response **(b) Variance of the response**

Figure 4.12 Comparison between OEM and MCS (Case 2; annotations identical to Figure 4.11).

Figure 4.13 shows that the mean response of the stochastic structure is different from the response of the system with mean parameters. The difference is enlarged with increasing coefficient of variation of the basic random parameters. □

4.5.5 Recursive Condensation Algorithm

The recursive condensation algorithm drives the dynamic order-expanded equations by utilizing the approximate relationship among the displacement, velocity and the acceleration. For this purpose, Equation 4.280 can be rewritten as

$$\begin{bmatrix} \mathbf{m}_{uu} & \mathbf{m}_{u\vartheta} \\ \mathbf{m}_{\vartheta u} & \mathbf{m}_{\vartheta\vartheta} \end{bmatrix} \begin{Bmatrix} \ddot{\mathbf{x}}_u \\ \ddot{\mathbf{x}}_\vartheta \end{Bmatrix} + \begin{bmatrix} \mathbf{c}_{uu} & \mathbf{c}_{u\vartheta} \\ \mathbf{c}_{\vartheta u} & \mathbf{c}_{\vartheta\vartheta} \end{bmatrix} \begin{Bmatrix} \dot{\mathbf{x}}_u \\ \dot{\mathbf{x}}_\vartheta \end{Bmatrix} + \begin{bmatrix} \mathbf{k}_{uu} & \mathbf{k}_{u\vartheta} \\ \mathbf{k}_{\vartheta u} & \mathbf{k}_{\vartheta\vartheta} \end{bmatrix} \begin{Bmatrix} \mathbf{x}_u \\ \mathbf{x}_\vartheta \end{Bmatrix} = \begin{Bmatrix} \mathbf{P}_u \\ \mathbf{P}_\vartheta \end{Bmatrix} \tag{4.310}$$

in which

$$\mathbf{P}_u = \mathbf{f}_{00\cdots 0}(t) \tag{4.311}$$

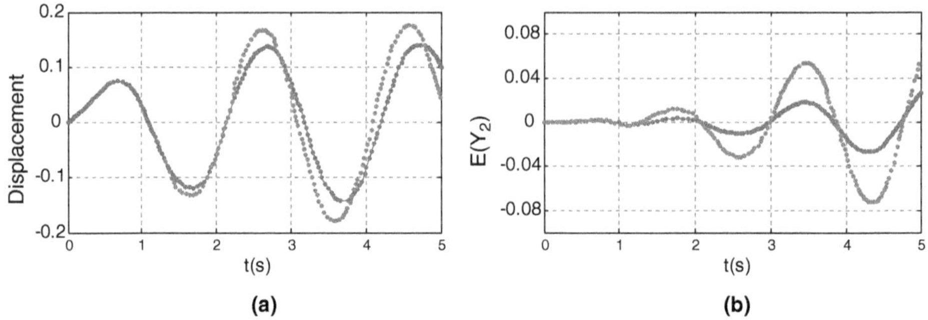

Figure 4.13 (a) Difference between the mean response and the response of the mean-parameter system (Case 1; solid line: mean response of the stochastic structure; dotted line: response of the mean-parameter system). (b) The difference when the coefficients of variation are different (solid line: COV = 0.1; dotted line: COV = 0.2).

$$\mathbf{x}_u = \mathbf{x}_{00\cdots 0}(t) \tag{4.312}$$

are the noncondensed loading vector and displacement vector. $\mathbf{P}_\vartheta$ and $\mathbf{x}_\vartheta$ are the condensed loading vector and displacement vector. The coefficient matrices are sub-matrices of the order-expanded dynamic matrices and related to the above sub-matrices of the loading vector. Considering $\mathbf{P}_\vartheta = \mathbf{0}$, the second row of Equation 4.310 is given as

$$\mathbf{m}_{\vartheta u}\ddot{\mathbf{x}}_u + \mathbf{m}_{\vartheta\vartheta}\ddot{\mathbf{x}}_\vartheta + \mathbf{c}_{\vartheta u}\dot{\mathbf{x}}_u + \mathbf{c}_{\vartheta\vartheta}\dot{\mathbf{x}}_\vartheta + \mathbf{k}_{\vartheta u}\mathbf{x}_u + \mathbf{k}_{\vartheta\vartheta}\mathbf{x}_\vartheta = \mathbf{0} \tag{4.313}$$

The equation is obviously valid for each time instant. Therefore, utilizing Equations 4.292 and 4.293 and introducing the assumption of linear acceleration will give

$$\begin{aligned}(\dot{\mathbf{x}}_u)_{j+1} &= \frac{3}{\Delta t}(\mathbf{x}_u)_{j+1} - \left[\frac{3}{\Delta t}(\mathbf{x}_u)_j + 2(\dot{\mathbf{x}}_u)_j + \frac{\Delta t}{2}(\ddot{\mathbf{x}}_u)_j\right] \\ &= \frac{3}{\Delta t}(\mathbf{x}_u)_{j+1} - \mathbf{B}_{u_j}\end{aligned} \tag{4.314}$$

$$\begin{aligned}(\ddot{\mathbf{x}}_u)_{j+1} &= \frac{6}{\Delta t^2}(\mathbf{x}_u)_{j+1} - \left[\frac{6}{\Delta t^2}(\mathbf{x}_u)_j + \frac{6}{\Delta t}(\dot{\mathbf{x}}_u)_j + 2(\ddot{\mathbf{x}}_u)_j\right] \\ &= \frac{6}{\Delta t^2}(\mathbf{x}_u)_{j+1} - \mathbf{A}_{u_j}\end{aligned} \tag{4.315}$$

$$\begin{aligned}(\dot{\mathbf{x}}_\vartheta)_{j+1} &= \frac{3}{\Delta t}(\mathbf{x}_\vartheta)_{j+1} - \left[\frac{3}{\Delta t}(\mathbf{x}_\vartheta)_j + 2(\dot{\mathbf{x}}_\vartheta)_j + \frac{\Delta t}{2}(\ddot{\mathbf{x}}_u)_j\right] \\ &= \frac{3}{\Delta t}(\mathbf{x}_\vartheta)_{j+1} - \mathbf{B}_{\vartheta_j}\end{aligned} \tag{4.316}$$

$$\begin{aligned}(\ddot{\mathbf{x}}_\vartheta)_{j+1} &= \frac{6}{\Delta t^2}(\mathbf{x}_\vartheta)_{j+1} - \left[\frac{6}{\Delta t^2}(\mathbf{x}_\vartheta)_j + \frac{6}{\Delta t}(\dot{\mathbf{x}}_\vartheta)_j + 2(\ddot{\mathbf{x}}_\vartheta)_j\right] \\ &= \frac{6}{\Delta t^2}(\mathbf{x}_\vartheta)_{j+1} - \mathbf{A}_{\vartheta_j}\end{aligned} \tag{4.317}$$

where $\Delta t = t_{j+1} - t_j$ is the given time step.

After substituting these equations in Equation 4.313 at time t_{j+1}, the following relationship can be obtained:

$$\mathbf{k}^*_{\vartheta u}\mathbf{x}_{u_{j+1}} + \mathbf{k}^*_{\vartheta\vartheta}\mathbf{x}_{\vartheta_{j+1}} - \mathbf{E}_j = 0 \tag{4.318}$$

in which

$$\mathbf{k}^*_{\vartheta u} = \frac{6}{\Delta t^2}\mathbf{m}_{\vartheta u} + \frac{3}{\Delta t}\mathbf{c}_{\vartheta u} + \mathbf{k}_{\vartheta u} \tag{4.319}$$

$$\mathbf{k}^*_{\vartheta\vartheta} = \frac{6}{\Delta t^2}\mathbf{m}_{\vartheta\vartheta} + \frac{3}{\Delta t}\mathbf{c}_{\vartheta\vartheta} + \mathbf{k}_{\vartheta\vartheta} \tag{4.320}$$

$$\mathbf{E}_j = \mathbf{m}_{\vartheta u}\mathbf{A}_{uj} + \mathbf{m}_{\vartheta\vartheta}\mathbf{A}_{\vartheta j} + \mathbf{c}_{\vartheta u}\mathbf{B}_{uj} + \mathbf{c}_{\vartheta\vartheta}\mathbf{B}_{\vartheta j} \tag{4.321}$$

From Equation 4.318 it follows that

$$\mathbf{x}_{\vartheta_{j+1}} = \mathbf{k}^{*-1}_{\vartheta\vartheta}(\mathbf{E}_j - \mathbf{k}^*_{\vartheta u}\mathbf{x}_{u_{j+1}}) \tag{4.322}$$

On the other hand, the discretized equation of the first row of Equation 4.310 can be written as

$$\mathbf{k}^*_{uu}\mathbf{x}_{u_{j+1}} + \mathbf{k}^*_{u\vartheta}\mathbf{x}_{\vartheta_{j+1}} = \mathbf{p}_{u_{j+1}} + \mathbf{F}_j \tag{4.323}$$

where

$$\mathbf{k}^*_{uu} = \frac{6}{\Delta t^2}\mathbf{m}_{uu} + \frac{3}{\Delta t}\mathbf{c}_{uu} + \mathbf{k}_{uu} \tag{4.324}$$

$$\mathbf{k}^*_{u\vartheta} = \frac{6}{\Delta t^2}\mathbf{m}_{u\vartheta} + \frac{3}{\Delta t}\mathbf{c}_{u\vartheta} + \mathbf{k}_{u\vartheta} \tag{4.325}$$

$$\mathbf{F}_j = \mathbf{m}_{uu}\mathbf{A}_{uj} + \mathbf{m}_{u\vartheta}\mathbf{A}_{\vartheta j} + \mathbf{c}_{uu}\mathbf{B}_{uj} + \mathbf{c}_{u\vartheta}\mathbf{B}_{\vartheta j} \tag{4.326}$$

Substituting Equation 4.322 in Equation 4.323 will lead to a condensation equation with respect to $\mathbf{x}_{u_{j+1}}$:

$$\tilde{\mathbf{k}}_u\mathbf{x}_{u_{j+1}} = \tilde{\mathbf{P}}_{j+1} \tag{4.327}$$

in which

$$\tilde{\mathbf{k}}_u = \mathbf{k}^*_{uu} - \mathbf{k}^*_{u\vartheta}\mathbf{k}^{*-1}_{\vartheta\vartheta}\mathbf{k}^*_{\vartheta u} \tag{4.328}$$

$$\tilde{\mathbf{P}}_{j+1} = \mathbf{P}_{u_{j+1}} + \mathbf{F}_j - \mathbf{k}^*_{u\vartheta}\mathbf{k}^{*-1}_{\vartheta\vartheta}\mathbf{E}_j \tag{4.329}$$

Obviously, the number of equations in Equation 4.327 is equal to the degrees of freedom of the original system. Substituting the solution of Equation 4.327 in Equation 4.322 step by step, the solution of the dynamic order-expanded system for the whole time history will be obtained. It is noted that, for linear structures, $\tilde{\mathbf{k}}_u$ and $\mathbf{k}^*_{u\vartheta}\mathbf{k}^{*-1}_{\vartheta\vartheta}$ are constants for different time instants,

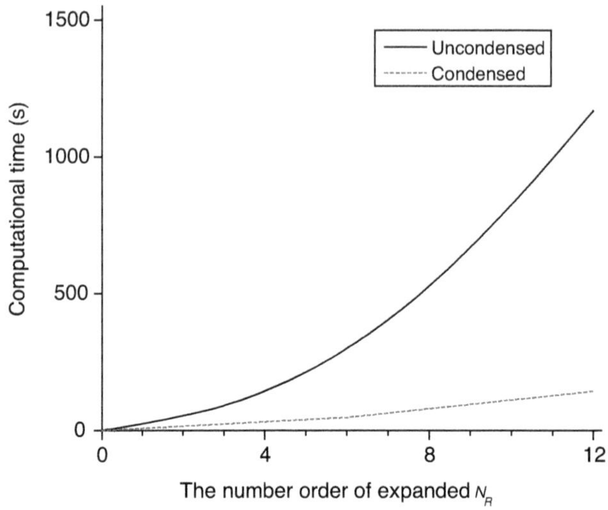

Figure 4.14 Computational efforts against the number of the expanded order.

and $\mathbf{F}_j$ and $\mathbf{E}_j$ depend on the previous structural responses. Therefore, the condensed loading term $\mathbf{P}_{u_{j+1}}$ should be modified in every step according to the results of the former step. Because the condensation equations are formed with the structural response of the former time instant, this method is called the recursive condensation algorithm (Li and Wei, 1996). Some practical calculation experiences indicate that this kind of algorithm can reduce the computational cost of the dynamic order-expanded system to that of the corresponding deterministic system. An example is shown in Figure 4.14. It can be observed that the computational cost will increase exponentially with the increment of the expansion order when the order-expanded system equation is solved directly, while the computational cost varies almost linearly with the expansion order when the recursive condensation algorithm is applied.

5

Random Vibration Analysis

5.1 Introduction

Classical random vibration theory, where the randomness involved in the excitations is taken into account but the system parameters are assumed to be exactly known and regarded as deterministic values, has experienced tens of years' development since the 1950s (Crandall, 1958). Over the last half century, a variety of approaches have been proposed and investigated extensively. These approaches can basically be classified into two categories, namely the numerical-characteristics-oriented approaches and the PDF-oriented approaches, which are logically in accordance with the two historical clues clarified in Chapter 1. The former family of approaches tries to obtain the numerical characteristics of the stochastic responses by establishing the transfer relationship from the numerical characteristics (such as the moments or the PSD functions) of the inputs to those of the responses. Here, the concept of stochastic differential equations and stochastic calculus should be introduced. For instance, the mean-square calculus is used most often. A formal solution of the stochastic differential equation or directly the stochastic differential equation is used to derive the transfer relationship or the governing differential equations in terms of the moments. In contrast, the latter family of approaches deals with the problem by transforming the random system equation to a probability density evolution equation, which is usually a deterministic multidimensional partial differential equation; for example, the Liouville equation or the FPK equation. This chapter deals with the widely used approaches among the aforementioned two families. The physical sense embedded in the approaches is particularly emphasized.

5.2 Moment Functions of the Responses

A linear transfer relationship from the moments of the inputs to those of the responses can be established for the linear structural system. Through this family of relationships, the numerical characteristics of the responses can be evaluated once the numerical characteristics, such as the mean and the convariance functions of the inputs, are known. The physical essence of this family of linear transfer relationships lies in the physical linear relationship between the excitations and the responses, which holds for any sample.

Stochastic Dynamics of Structures Jie Li and Jianbing Chen

5.2.1 Response of a Single-Degree-of-Freedom System in the Time Domain

5.2.1.1 Impulse Response Function and Duhamel's Integral

Consider the deterministic SDOF system shown in Figure 5.1 with mass m, damping c and stiffness k. If it is subjected to an excitation $f(t)$, now regarded as deterministic, then the equation of motion is

$$m\ddot{x}+c\dot{x}+kx=f(t) \tag{5.1}$$

where x is the displacement response; the overdots denote the derivative with respect to time t. Let the initial condition be $x(t_0)=x_0, \dot{x}(t_0)=\dot{x}_0$.

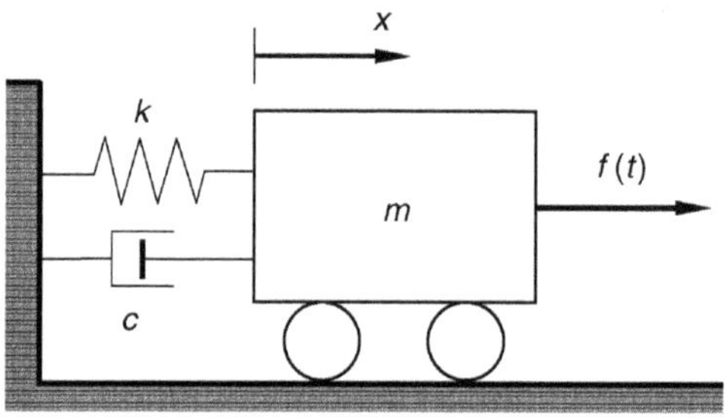

Figure 5.1 An SDOF system.

The properties of a linear dynamical system are embedded in the response of the system to some types of special excitation, although the actual excitations are usually arbitrary and irregular. One of the special excitations is the unit impulse $\delta(t)$, which is a Dirac delta function; another type of special excitation is the unit harmonic excitation, which can be expressed in a complex function as $\mathrm{e}^{\mathrm{i}\omega t}$, where $\mathrm{i}=\sqrt{-1}$ is the unit of the imaginary number. The advantage of using these special excitations as testing excitations is that any arbitrary excitation can be regarded as the linear superposition of a set of impulse functions. In fact, there is

$$f(t)=\int_{-\infty}^{\infty} f(\tau)\delta(t-\tau)\,\mathrm{d}\tau \tag{5.2}$$

In addition, the process $f(t)$ can also be regarded as the linear superposition of a set of unit harmonic excitations when employing

$$f(t)=\frac{1}{2\pi}\int_{-\infty}^{\infty} F(\omega)\mathrm{e}^{\mathrm{i}\omega t}\,\mathrm{d}\omega \tag{5.3a}$$

which is nothing but the Fourier transform, where $F(\omega)$ is the Fourier spectrum of $f(t)$. Therefore, although the unit impulse excitation and the unit harmonic excitation are both special types of excitation, they are also the basis units of any arbitrary excitations (Figure 5.2; see Appendix A). Hence, the response properties of the system to the excitations contain adequate information for gaining insight into the problem and simultaneously for practical computations of the response of the system to any arbitrary excitations. The impulse response will be discussed in this present section and the response to the unit harmonic excitation will be dealt with later.

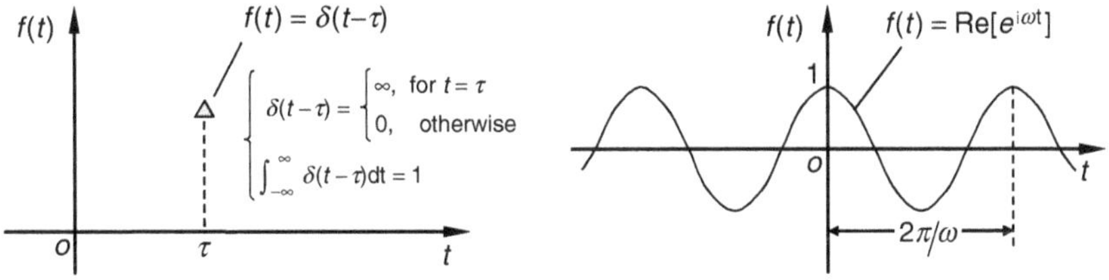

(a) Unit impulse function **(b) Unit harmonic function**

Figure 5.2 Two typical basis units.

Let the response of the SDOF system initially at rest to the unit impulse $\delta(t)$ be $h(t)$; namely:

$$m\ddot{h} + c\dot{h} + kh = \delta(t) \qquad h_0 = 0,\ \dot{h}_0 = 0 \tag{5.4}$$

Dividing by m on both sides, Equation 5.4 becomes

$$\ddot{h} + 2\zeta\omega\dot{h} + \omega^2 h = \frac{1}{m}\delta(t) \qquad h(0) = 0,\ \dot{h}(0) = 0 \tag{5.5}$$

where $\omega = \sqrt{k/m}$ is the natural circular frequency and $\zeta = c/2m\omega$ is the damping ratio. Performing integration over a small time interval $[0, \tau]$ on both sides of Equation 5.5, and noting that

$$\int_0^\tau h(t)\,\mathrm{d}t = \frac{\tau}{2}[h(0) + h(\tau)] + o(\tau) \qquad \text{and} \qquad \int_0^\tau \frac{1}{m}\delta(t)\,\mathrm{d}t = \frac{1}{m} \tag{5.6}$$

we have

$$\dot{h}(\tau) - \dot{h}(0) + 2\zeta\omega[h(\tau) - h(0)] + \frac{1}{2}\omega^2\tau[h(\tau) + h(0)] + o(\tau) = \frac{1}{m} \tag{5.7}$$

Letting $\tau \to 0$ and considering the initial conditions in Equation 5.5 together with (by the mean value theorem in calculus)

$$\lim_{\tau\to 0} h(\tau) = \lim_{\tau\to 0}\int_0^\tau \dot{h}(t)\,\mathrm{d}t = \lim_{\tau\to 0}[\dot{h}(\tilde{\tau})\tau] = 0 \tag{5.8}$$

where $\tilde{\tau}$ is some proper intermediate value over the interval $[0, \tau]$, it follows that

$$\lim_{\tau\to 0}\dot{h}(\tau) = \frac{1}{m} \tag{5.9}$$

This is essentially an initial condition for the velocity, which means that the effect of the impulse is to make a sudden finite increment of the velocity. Therefore, the response of a system initially at rest to an impulse is equivalent to the free-vibration response of the system subjected to a nonzero initial velocity from the origin. Mathematically, the solution of Equation 5.5 is equivalent to the solution of

$$\ddot{h} + 2\zeta\omega\dot{h} + \omega^2 h = 0 \qquad h(0) = 0,\ \dot{h}(0) = \frac{1}{m} \tag{5.10}$$

which is known as the impulse response function (Clough and Penzien, 1993)

$$h(t) = \frac{1}{m\omega_d} e^{-\zeta\omega t} \sin(\omega_d t) \qquad t \geq 0 \tag{5.11}$$

and $h(t)=0$ for $t<0$, where $\omega_d = \sqrt{1-\zeta^2}\omega$.

The principle of superposition holds in linear systems. For the system in Equation 5.1, if the response of the system to the excitation $f_1(t)$ is $x_1(t)$ and the response of the system to the excitation $f_2(t)$ is $x_2(t)$, then the response of the system to the linear combination $c_1 f_1(t) + c_2 f_2(t)$ is $c_1 x_1(t) + c_2 x_2(t)$. This is also true for any number of excitations. Further, if the system is subjected to the excitation

$$f(t) = \lim_{\max(\Delta\tau_j)\to 0} \sum_{j=0}^{N(t,\Delta\tau)} f_1(\tau_j) f(t,\tau_j)\Delta\tau_j = \int_0^t f_1(\tau) g(t,\tau) \mathrm{d}\tau \tag{5.12a}$$

where $g(t,\tau)$ can be regarded as some type of modulation function, $N(t,\Delta\tau)$ is the number of partitioned intervals such that $[0,t] = \cup_{j=0}^{N} \Delta\tau_j$, $\Delta\tau_j \cap \Delta\tau_k = \phi$, $\forall j \neq k$, then the response of the system is

$$x(t) = \lim_{\max(\Delta\tau_j)\to 0} \sum_{j=1}^{N(t,\Delta\tau)} x_1(\tau_j) g(t,\tau_j)\Delta\tau_j = \int_0^t x_1(\tau) g(t,\tau) \mathrm{d}\tau \tag{5.12b}$$

Introducing Equation 5.2 into Equation 5.1, the equation of motion of the system initially at rest is

$$m\ddot{x} + c\dot{x} + kx = \int_{-\infty}^{\infty} \delta(t-\tau) f(\tau)\, \mathrm{d}\tau \qquad x_0 = 0,\ \dot{x}_0 = 0 \tag{5.13}$$

Note that the response of the system initially at rest to the impulse $\delta(t-\tau)$ is $h(t-\tau)$, which is given by Equation 5.11. Then, according to the principle of superposition shown in Equations 5.12a and 5.12b, the response of the system initially at rest to $f(t)$ reads

$$x(t) = \int_0^t h(t-\tau) f(\tau)\, \mathrm{d}\tau = \int_{-\infty}^{\infty} h(t-\tau) f(\tau)\, \mathrm{d}\tau = \int_0^t h(\tau) f(t-\tau)\, \mathrm{d}\tau \tag{5.14}$$

This convolution is the well-known Duhamel integral.[1]

To include the effects of a nonresting initial condition, the free vibration of the system

$$m\ddot{x} + c\dot{x} + kx = 0 \qquad x(0) = x_0,\ \dot{x}(0) = \dot{x}_0 \tag{5.15}$$

should be considered. Again, using the principle of superposition, we know that the free vibration response of the system in Equation 5.15 is equivalent to combination of the free

[1] Because of causality – that is, $h(t)=0$ for $t<0$ – the lower integral limit in Equation can be 0 or $-\infty$ while the upper integral limit can be t or ∞; the results are equivalent.

vibration of the system with the initial conditions $x(0)=x_0$, $\dot{x}(0)=0$ and that of the system with the initial conditions $x(0)=0$, $\dot{x}(0)=\dot{x}_0$; therefore, the free vibration response of the system in Equation 5.15 is given by

$$x(t)=x_0\mathrm{e}^{-\zeta\omega t}\cos(\omega_\mathrm{d}t)+\frac{\dot{x}_0}{\omega_\mathrm{d}}\mathrm{e}^{-\zeta\omega t}\sin(\omega_\mathrm{d}t)=A_0\mathrm{e}^{-\zeta\omega t}\sin(\omega_\mathrm{d}t+\varphi_0) \tag{5.16}$$

where $A_0=\left[x_0^2+(\dot{x}_0/\omega_\mathrm{d})^2\right]^{1/2}$ is the initial amplitude and $\varphi_0=\tan^{-1}[x_0/(\dot{x}_0/\omega_\mathrm{d})]$ is the initial phase angle.

Obviously, it follows that the response $x(t)$ decays at an exponential rate and $x(t)\to 0$ as $t\to\infty$. This indicates that the effect of the initial condition can be ignored when the time duration is long enough.[2]

Again, using the principle of superposition, the total response in Equation 5.1 is the summation of the response of the corresponding system initially at rest to the excitation (given by Equation 5.14) and the free-vibration response of the system with non-resting initial conditions (given by Equation 5.16), namely

$$\begin{aligned} x(t) &= A_0\mathrm{e}^{-\zeta\omega t}\sin(\omega_\mathrm{d}t+\varphi_0)+\textstyle\int_0^t h(t-\tau)f(\tau)\,\mathrm{d}\tau \\ &= A_0\mathrm{e}^{-\zeta\omega t}\sin(\omega_\mathrm{d}t+\varphi_0)+\frac{1}{m\omega_\mathrm{d}}\int_0^t \mathrm{e}^{-\zeta\omega(t-\tau)}\sin[\omega_\mathrm{d}(t-\tau)]f(\tau)\,\mathrm{d}\tau \end{aligned} \tag{5.17}$$

This formula establishes the linear relationship between the deterministic excitation $f(t)$ and the deterministic response $x(t)$, which is illustrated in a block diagram in Figure 5.3a and expressed by the linear operator

$$x(t)=\mathscr{L}[f(t)] \tag{5.18a}$$

Now we consider the case that the input is a stochastic process $\xi(t)$. In this case, the response is undoubtedly also a stochastic process $X(t)$. However, from the point of view of the sample, or in other words from the physical point of view, the relationship in Equation 5.18a still holds, as illustrated in the upper part of Figure 5.3b, where the variable ϖ is embedded, representing the randomness involved. It is well established that a stochastic process could be probabilistically described by probabilistic information, say, a finite-dimensional PDF or moments of different orders in the time domain. Therefore, the operator in the sense of the sample in Equation 5.18a will definitely mean that a linear relationship exists between the probabilistic information of the excitation, denoted by $\mathscr{P}_{\xi(t)}$, and the probabilistic information of the response, denoted by $\mathscr{P}_{X(t)}$ (see the lower part of Figure 5.3b). Mathematically, this means that there must be some type of deterministic operator $\mathscr{L}_\mathscr{P}(\cdot)$ such that

$$\mathscr{P}_{X(t)}=\mathscr{L}_\mathscr{P}(\mathscr{P}_{\xi(t)}) \tag{5.18b}$$

One of the most important tasks in stochastic mechanics is to find the operator $\mathscr{L}_\mathscr{P}(\cdot)$ and make it feasible for analytical or numerical implementation for certain metrics of probabilistic information.

[2] Actually this is a general feature of many dynamical/iterative systems if the systems are stable. For instance, we can find the analogous feature in the iteration or evolution process of a Markov process (Lin, 1967; Gardiner, 1983), the iteration process of the Bayesian estimate (Ang and Tang, 1984), the process of iteration of a matrix (Golub and von Loan, 1996), and the Kalman filter (Stengel, 1994), and so on.

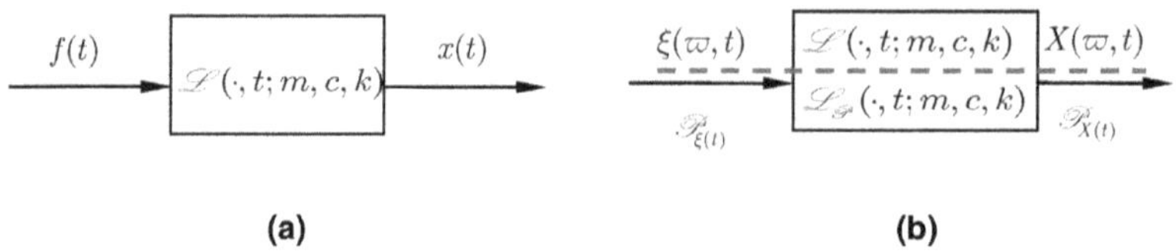

Figure 5.3 The block diagram of linear systems.

5.2.1.2 Moment Functions of the Stochastic Response of SDOF Systems in the Time Domain

Although it is more understandable to tackle the problem in a physical or sample sense as discussed above, the mathematical manipulations for a stochastic system in the sense of sample are not easy. On the other hand, to deal with the problem in the mean-square sense is more convenient, particularly because the operations of the calculus that are well established in the deterministic cases could be used directly in the mean-square sense without special revisions (Lin, 1967; Åström, 1970; Gardiner, 1983).

Denote the mean of the stochastic process $\xi(t)$ by $\mu_\xi(t)$ and the autocorrelation function by $R_\xi(t_1,t_2)$ (see Equations 2.63 and 2.64):

$$\mu_\xi(t) = \mathcal{E}[\xi(\varpi, t)] = \int_\Omega \xi(\varpi, t)P(\mathrm{d}\varpi) = \int_{\Omega_x} xp_\xi(x, t)\,\mathrm{d}x \tag{5.19}$$

$$\begin{aligned} R_\xi(t_1, t_2) &= \mathcal{E}[\xi(\varpi, t_1)\xi(\varpi, t_2)] = \int_\Omega \xi(\varpi, t_1)\xi(\varpi, t_2)P(\mathrm{d}\varpi) \\ &= \int_{\Omega_x} x_1 x_2 p_\xi(x_1, t_1; x_2, t_2)\,\mathrm{d}x_1\,\mathrm{d}x_2 \end{aligned} \tag{5.20}$$

where $P(\mathrm{d}\varpi) = \Pr\{\mathrm{d}\varpi\}$ is the probability measure, $p_\xi(x,t)$ and $p_\xi(x_1, t_1; x_2, t_2)$ are respectively the one- and two-dimensional PDFs of the stochastic process $\xi(t)$ and $\mathcal{E}[\cdot]$ stands for the ensemble average.

The response of the SDOF system to the excitation $\xi(t)$, according to Equation 5.17 when the effect of the initial condition is ignored because of the rapid attenuation, is[3]

$$X(t) = \int_0^t h(t-\tau)\xi(\tau)\,\mathrm{d}\tau \tag{5.21}$$

Taking the mathematical expectation on both sides and noting that the expectation and the integral operators are commutable in the mean-square sense will yield the mean of $X(t)$:

$$\begin{aligned} \mu_X(t) &= \mathcal{E}[X(t)] = \mathcal{E}\left[\int_0^t h(t-\tau)\xi(\tau)\,\mathrm{d}\tau\right] = \int_0^t h(t-\tau)\mathcal{E}[\xi(\tau)]\,\mathrm{d}\tau \\ &= \int_0^t h(t-\tau)\mu_\xi(\tau)\,\mathrm{d}\tau \end{aligned} \tag{5.22}$$

[3] This integral can either be understood as a sample integral or as a mean-square integral. The physical sense of the former is more direct, but the mathematical manipulation of the latter is more convenient. Fortunately, for the stochastic processes encountered most they are consistent.

Likewise, the autocorrelation function of the stochastic response $X(t)$ can be computed by

$$
\begin{aligned}
R_X(t_1, t_2) &= \mathcal{E}[X(t_1)X(t_2)] = \mathcal{E}\left[\int_0^{t_1} h(t_1-\tau)\xi(\tau)\mathrm{d}\tau \cdot \int_0^{t_2} h(t_2-\tau)\xi(\tau)\,\mathrm{d}\tau\right] \\
&= \int_0^{t_1}\int_0^{t_2} h(t_1-\tau_1)h(t_2-\tau_2)\mathcal{E}[\xi(\tau_1)\xi(\tau_2)]\,\mathrm{d}\tau_1\,\mathrm{d}\tau_2 \\
&= \int_0^{t_1}\int_0^{t_2} h(t_1-\tau_1)h(t_2-\tau_2)R_\xi(\tau_1,\tau_2)\,\mathrm{d}\tau_1\,\mathrm{d}\tau_2
\end{aligned} \tag{5.23}
$$

The variance of the response is then given by letting $t_1 = t_2 = t$:

$$
\mathrm{Var}[X(t)] = \mathcal{E}[X^2(t)] = R_X(t,t) = \int_0^t\int_0^t h(t-\tau_1)h(t-\tau_2)R_\xi(\tau_1,\tau_2)\,\mathrm{d}\tau_1\,\mathrm{d}\tau_2 \tag{5.24}
$$

In the case where the excitation is a stationary process, the steady-state response is also a stationary process. The autocorrelation function of this steady-state response is given by Equation 5.23 through letting $t_1 \to \infty$ and $t_2 \to \infty$ and noting that $R_\xi(\tau_1, \tau_2)$ is replaced by $R_\xi(\tau_1 - \tau_2)$. After the change of integral variables, we have

$$
R_X(\tau) = \int_{-\infty}^{\infty}\int_{-\infty}^{\infty} h(u_1)h(u_2)R_\xi(\tau+u_1-u_2)\,\mathrm{d}u_1\,\mathrm{d}u_2 \tag{5.25}
$$

Equations 5.22–5.25 establish the relationship between the mean and the correlation functions of the excitation and the response. We can see easily that the moments of the response are linear functionals of the corresponding moments of the excitation. In other words, the operators $\mathscr{L}_{\mathscr{P}}(\cdot)$ in Equation 5.18b and Figure 5.3b are linear; these are the expected results as the mathematical reflection of the linear physical relationship shown in the system in Equation 5.1.

Example 5.1. Response of an SDOF System to White Noise Excitation If the excitation is a white noise with $\boldsymbol{\mu}_\xi(t) = 0$, $R_\xi(t_1, t_2) = R_{\xi 0}\boldsymbol{\delta}(t_2 - t_1)$, then from Equation 5.22 it is known that the mean of the response $\boldsymbol{\mu}_X(t)$ is zero, and from Equation 5.23 the correlation function can be evaluated by

$$
\begin{aligned}
R_X(t_1,t_2) &= \int_0^{t_1}\int_0^{t_2} h(t_1-\tau_1)h(t_2-\tau_2)R_{\xi 0}\delta(\tau_2-\tau_1)\,\mathrm{d}\tau_1\,\mathrm{d}\tau_2 \\
&= R_{\xi 0}\int_0^{t_1} h(t_1-\tau_1)h(t_2-\tau_1)\,\mathrm{d}\tau_1 \\
&= \frac{R_{\xi 0}}{(m\omega_\mathrm{d})^2}\int_0^{t_1} \mathrm{e}^{-\zeta\omega(t_1+t_2-2\tau_1)}\sin\omega_\mathrm{d}(t_1-\tau_1)\sin\omega_\mathrm{d}(t_2-\tau_1)\,\mathrm{d}\tau_1 \\
&\quad - \frac{R_{\xi 0}\cos\omega_\mathrm{d}(t_2-t_1)}{4\zeta\omega(m\omega_\mathrm{d})^2}\left[\mathrm{e}^{-\zeta\omega(t_2-t_1)} - \mathrm{e}^{-\zeta\omega(t_1+t_2)}\right] \\
&\quad - \frac{R_{\xi 0}}{4\omega^2(m\omega_\mathrm{d})^2}\left[\mathrm{e}^{\zeta\omega\tau_1}(\omega_\mathrm{d}\sin\omega_\mathrm{d}\tau_1 + \zeta\omega\cos\omega_\mathrm{d}\tau_1)\right]_{\tau_1=-(t_1+t_2)}^{\tau_1=-(t_2-t_1)}
\end{aligned} \tag{5.26}
$$

Letting $\tau = t_2 - t_1$ and $t = t_1$, we then have

$$
\begin{aligned}
R_X(t,t+\tau) &= \frac{R_{\xi 0}\cos\omega_\mathrm{d}\tau}{4\zeta\omega(m\omega_\mathrm{d})^2}\left[\mathrm{e}^{-\zeta\omega\tau} - \mathrm{e}^{-\zeta\omega(\tau+2t)}\right] \\
&\quad - \frac{R_{\xi 0}}{4\omega^2(m\omega_\mathrm{d})^2}\left[\mathrm{e}^{\zeta\omega\tau_1}(\omega_\mathrm{d}\sin\omega_\mathrm{d}\tau_1 + \zeta\omega\cos\omega_\mathrm{d}\tau_1)\right]_{\tau_1=-(\tau+2t)}^{\tau_1=-\tau}
\end{aligned} \tag{5.27a}
$$

Further, as $t \to \infty$, the correlation function of the steady-state stationary process is given by

$$R_X(\tau) = \lim_{t\to\infty} R_X(t,t+\tau) = \frac{R_{\xi 0}}{4m^2\omega_d^2\omega} e^{-\zeta\omega\tau}[\cos\omega_d\tau - \zeta\omega\sin(\omega_d\tau)] \tag{5.27b}$$

From Equation 5.26 we can get the variance of the response $\text{Var}[X(t)] = R_X(t,t)$, namely

$$\begin{aligned}\text{Var}[X(t)] &= \frac{R_{\xi 0}}{4\zeta\omega(m\omega_d)^2}(1-e^{-2\zeta\omega t}) \\ &\quad - \frac{R_{\xi 0}}{4\omega^2(m\omega_d)^2}\{\zeta\omega - e^{-2\zeta\omega t}[\zeta\omega\cos(2\omega_d t) - \omega_d\sin(2\omega_d t)]\}\end{aligned} \tag{5.28a}$$

In addition, letting $t \to \infty$ yields the variance of the steady-state stationary response:

$$\lim_{t\to\infty} \text{Var}[X(t)] = R_X(0) = \frac{R_{\xi 0}}{4\zeta\omega^3 m^2} \tag{5.28b}$$

The dimensionless variance of the response with different damping ratios is shown in Figure 5.4. It is seen that the larger the damping ratio, the faster the variance approaches the steady-state variance. On the other hand, if $\zeta = 0$, then the system will be unstable because the input energy cannot be dissipated by the system. In addition, from Equations 5.27a and 5.27b it is seen that the response $X(t)$ is no longer a white noise. It is a filtered noise and often called a colored noise, which will be discussed again later. □

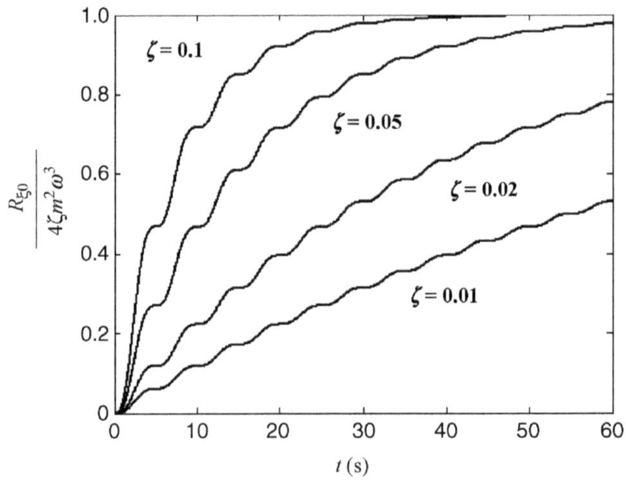

Figure 5.4 Variance function of the response to white noise ($\omega = 0.2\pi$).

5.2.2 *Response of MDOF Systems in the Time Domain*

An SDOF system not only can provide insights into the problem because of its simplicity and relatively easy mathematical manipulation, but is also applicable to a large number of engineering practical cases as a reasonable approximation, especially in the preliminary stage of analysis and design. However, most practical systems, in a more accurate sense, should be considered with infinite degrees of freedom, or at least multiple degrees of freedom. In general, with the concept of discretization, engineering structures can usually be discretized or approximated by an MDOF system, of which the equation of motion in matrix form (see Section 4.2) is

$$\mathbf{M}\ddot{\mathbf{X}} + \mathbf{C}\dot{\mathbf{X}} + \mathbf{K}\mathbf{X} = \mathbf{B}\boldsymbol{\xi}(t) \tag{5.29}$$

where $\mathbf{X} = (X_1, X_2, \ldots, X_n)^{\mathrm{T}}$ is the displacement of order n, $\mathbf{M} = [M_{ij}]_{n\times n}$, $\mathbf{C} = [C_{ij}]_{n\times n}$ and $\mathbf{K} = [K_{ij}]_{n\times n}$ are the mass, the damping and the stiffness matrices respectively, $\mathbf{B} = [B_{ij}]_{n\times r}$ is the input force influence matrix and $\boldsymbol{\xi}(t) = (\xi_1(t), \xi_2(t), \ldots, \xi_r(t))^{\mathrm{T}}$ is the stochastic excitation vector of order r.

5.2.2.1 Direct Matrix Expressions

Denote the mean vector and the correlation function matrix of the stochastic process vector $\boldsymbol{\xi}(t)$ respectively by

$$\boldsymbol{\mu}_{\boldsymbol{\xi}}(t) = \mathcal{E}[\boldsymbol{\xi}(t)] \quad \text{and} \quad \mathbf{R}_{\boldsymbol{\xi}}(t_1, t_2) = \mathcal{E}[\boldsymbol{\xi}(t_1)\boldsymbol{\xi}^{\mathrm{T}}(t_2)] \tag{5.30}$$

To obtain the moment functions of the response $\mathbf{X}(t)$ to the excitation $\boldsymbol{\xi}(t)$, the idea used in the SDOF system can be applied here directly, except for changing the scalar functions to the corresponding vector and matrix functions.

Denote the unit impulse response function matrix by $\mathbf{h}(t) = [h_{ij}(t)]_{n\times n}$, where the component $h_{ij}(t)$ of $\mathbf{h}(t)$ is the impulse response at the ith degree of freedom to the excitation acting on the jth degree of freedom. Thus, if we let $\mathbf{h}_j$ be the jth column vector of the impulse response matrix $\mathbf{h}(t)$, then the column vectors $\mathbf{h}_j$ satisfy

$$\mathbf{M}\ddot{\mathbf{h}}_j + \mathbf{C}\dot{\mathbf{h}}_j + \mathbf{K}\mathbf{h}_j = \mathbf{I}_j\delta(t) \qquad \dot{\mathbf{h}}_j(0) = \mathbf{0},\ \mathbf{h}_j(0) = \mathbf{0} \tag{5.31}$$

for $j = 1, 2, \ldots, n$. Here, $\mathrm{I}_j = (0, 0, \ldots, 0, 1, 0, \ldots, 0)^{\mathrm{T}}$ is the column vector with the components being zero except for the jth component being 1.

According to the Duhamel integral (see Equation 5.14), ignoring the effect of the initial condition of the system, we get the response of the system in Equation 5.29:

$$\mathbf{X}(t) = \int_0^t \mathbf{h}(t-\tau)\mathbf{B}\boldsymbol{\xi}(\tau)\,\mathrm{d}\tau \tag{5.32}$$

Therefore, the mean of the response is given by

$$\boldsymbol{\mu}_{\mathbf{X}}(t) = \mathcal{E}[\mathbf{X}(t)] = \int_0^t \mathbf{h}(t-\tau)\mathbf{B}\mathcal{E}[\boldsymbol{\xi}(\tau)]\,\mathrm{d}\tau = \int_0^t \mathbf{h}(t-\tau)\mathbf{B}\boldsymbol{\mu}_{\boldsymbol{\xi}}(\tau)\,\mathrm{d}\tau \tag{5.33}$$

Simultaneously, the correlation function matrix can be evaluated through

$$\begin{aligned}
\mathbf{R}_{\mathbf{X}}(t_1, t_2) &= \mathcal{E}[\mathbf{X}(t_1)\mathbf{X}^{\mathrm{T}}(t_2)] = \mathcal{E}\left[\int_0^{t_1} \mathbf{h}(t_1-\tau)\mathbf{B}\boldsymbol{\xi}(\tau)\,\mathrm{d}\tau \cdot \int_0^{t_2} \{\mathbf{h}(t_2-\tau)\mathbf{B}\boldsymbol{\xi}(\tau)\}^{\mathrm{T}}\,\mathrm{d}\tau\right] \\
&= \mathcal{E}\left[\int_0^{t_1}\int_0^{t_2} \mathbf{h}(t_1-\tau_1)\mathbf{B}\boldsymbol{\xi}(\tau_1)\boldsymbol{\xi}^{\mathrm{T}}(\tau_2)\mathbf{B}^{\mathrm{T}}\mathbf{h}^{\mathrm{T}}(t_2-\tau_2)\,\mathrm{d}\tau_1\,\mathrm{d}\tau_2\right] \\
&= \int_0^{t_1}\int_0^{t_2} \mathbf{h}(t_1-\tau_1)\mathbf{B}\mathcal{E}[\boldsymbol{\xi}(\tau_1)\boldsymbol{\xi}^{\mathrm{T}}(\tau_2)]\mathbf{B}^{\mathrm{T}}\mathbf{h}^{\mathrm{T}}(t_2-\tau_2)\,\mathrm{d}\tau_1\,\mathrm{d}\tau_2 \\
&= \int_0^{t_1}\int_0^{t_2} \mathbf{h}(t_1-\tau_1)\mathbf{B}\mathbf{R}_{\xi}(\tau_1, \tau_2)\mathbf{B}^{\mathrm{T}}\mathbf{h}^{\mathrm{T}}(t_2-\tau_2)\,\mathrm{d}\tau_1\,\mathrm{d}\tau_2
\end{aligned} \tag{5.34}$$

In the case that the excitations are stationary stochastic processes, the steady-state responses are also stationary processes. The autocorrelation function matrix of the responses can now be obtained through letting $t_1 \to \infty$, $t_2 \to \infty$ and noting that $\mathbf{R}_{\xi}(\tau_1,\tau_2)$ is reduced to $\mathbf{R}_{\xi}(\tau_1 - \tau_2)$. Therefore, similar to Equation 5.25, we have

$$\mathbf{R}_{\mathbf{X}}(\tau) = \int_{-\infty}^{\infty}\int_{-\infty}^{\infty} \mathbf{h}(u_1)\mathbf{B}\mathbf{R}_{\xi}(\tau - u_1 - u_2)\mathbf{B}^{\mathrm{T}}\mathbf{h}^{\mathrm{T}}(u_2)\,du_1\,du_2 \tag{5.35}$$

Equations 5.33–5.35 establish the relationship between the moment functions of the responses, $\boldsymbol{\mu}_{\mathbf{X}}(t)$ and $\mathbf{R}_{\mathbf{X}}(t_1, t_2)$, and those of the excitations, $\boldsymbol{\mu}_{\xi}(t)$ and $\mathbf{R}_{\xi}(t_1,t_2)$. They are just the matrix-form counterparts of the relationships in the case of SDOF systems in Equations 5.22–5.25. The underlying physical sense is clear that, as shown in Figure 5.3b, the operator $\mathscr{L}_{\mathscr{P}}(\cdot)$ in this case is still linear because of the embedded linearity in terms of the inputs and the outputs in the physical system 5.29.

Practical applications of Equations 5.33 and 5.34, however, are not convenient because (a) the closed-form solution of the impulse response matrix $\mathbf{h}(t)$ is much more difficult to reach than is that of the SDOF counterparts and (b) the computational effort involved in the implementation of Equation 5.34 is prohibitively large for the structural systems of practical interest, where the degrees of freedom might be so large that the computations involved in these equations are beyond present available computational facilities.

5.2.2.2 Modal Superposition Method

For linear systems, the modal superposition method could uncouple the original system to a set of SDOF systems and then greatly reduce the computational effort. If the damping involved is proportional damping, then the equation of motion as a second-order ordinary differential equation could be uncoupled directly through a variable separation method which employs the eigenvectors, referred to as modes, as the basis vectors (Clough and Penzien, 1993).

Consider the corresponding undamped free vibration of the system in Equation 5.29:

$$\mathbf{M}\ddot{\mathbf{X}} + \mathbf{K}\mathbf{X} = \mathbf{0} \tag{5.36}$$

Assume the free vibration is harmonic; that is, $\mathbf{X}(t) = \boldsymbol{\psi}\,\mathrm{e}^{\mathrm{i}\omega t}$. Substituting this in Equation 5.36 yields

$$[\mathbf{K} - \omega^2\mathbf{M}]\boldsymbol{\psi}\mathrm{e}^{\mathrm{i}\omega t} = \mathbf{0} \tag{5.37}$$

Because $\mathrm{e}^{\mathrm{i}\omega t}$ is not always zero, this requires

$$[\mathbf{K} - \omega^2\mathbf{M}]\boldsymbol{\psi} = \mathbf{0} \tag{5.38a}$$

which is referred to as the characteristic equation. The condition ensuring a nontrivial nonzero solution is

$$\det(\mathbf{K} - \omega^2\mathbf{M}) = 0 \tag{5.38b}$$

where $\det(\cdot)$ is the determinant of the bracketed matrix. For general structural systems, especially finite-element systems, since the stiffness $\mathbf{K}$ and mass matrix $\mathbf{M}$ are both symmetric and positive-definite, it has been proved that there exist n solutions ω_j, $j = 1, 2, \ldots, n$, to

Equation 5.38b; in turn, according to Equation 5.38a, each one possesses a corresponding vector $\boldsymbol{\psi}_j, j = 1, 2, \ldots, n$ (Golub and van Loan, 1996). In other words, there exist n characteristic pairs $(\omega_j, \boldsymbol{\psi}_j), j = 1, 2, \ldots, n$, of which ω_j^2 is the eigenvalue and $\boldsymbol{\psi}_j$ is the corresponding eigenvector, usually referred to as the mode of the structure.

The modes are weighted orthogonal with respect to the stiffness matrix **K** and the mass matrix **M**. In fact, replacing $\boldsymbol{\psi}$ in Equation 5.38a by $\boldsymbol{\psi}_j$ and pre-multiplying it by $\boldsymbol{\psi}_k^{\mathrm{T}}$ and then replacing $\boldsymbol{\psi}$ in Equation 5.38a by $\boldsymbol{\psi}_k$ and pre-multiplying it by $\boldsymbol{\psi}_j^{\mathrm{T}}$, we can get

$$\boldsymbol{\psi}_k^{\mathrm{T}}[\mathbf{K} - \omega_j^2\mathbf{M}]\boldsymbol{\psi}_j = 0 \tag{5.39a}$$

and

$$\boldsymbol{\psi}_j^{\mathrm{T}}[\mathbf{K} - \omega_k^2\mathbf{M}]\boldsymbol{\psi}_k = 0 \tag{5.39b}$$

respectively. Noting that $\mathbf{K} = \mathbf{K}^{\mathrm{T}}$ and $\mathbf{M} = \mathbf{M}^{\mathrm{T}}$, taking the transpose of the latter equation and then subtracting it from the former, considering that $\omega_k \neq \omega_j$ for $k \neq j$, we have

$$\boldsymbol{\psi}_k^{\mathrm{T}}\mathbf{M}\boldsymbol{\psi}_j = m_k\delta_{kj} \tag{5.40}$$

where $m_k = \boldsymbol{\psi}_k^{\mathrm{T}}\mathbf{M}\boldsymbol{\psi}_k$ is the kth modal mass and δ_{kj} is the Kronecker delta. Introducing Equation 5.40 into Equations 5.39a and 5.39b yields

$$\boldsymbol{\psi}_k^{\mathrm{T}}\mathbf{K}\boldsymbol{\psi}_j = \omega_j^2\boldsymbol{\psi}_k^{\mathrm{T}}\mathbf{M}\boldsymbol{\psi}_j = \omega_j^2 m_j\delta_{kj} = k_j\delta_{kj} \tag{5.41}$$

Equations 5.40 and 5.41 show that the modes are weighted orthogonal. Therefore, the modal vectors $\boldsymbol{\psi}_j, j = 1, 2, \ldots, n$, comprise a complete orthogonal basis in the n-dimensional Hilbert space; then, the response $\mathbf{X}(t)$ can be decomposed by

$$\mathbf{X}(t) = \sum_{j=1}^{n} \boldsymbol{\psi}_j u_j(t) = \boldsymbol{\phi}\mathbf{u}(t) \tag{5.42}$$

where $\boldsymbol{\phi} = (\boldsymbol{\psi}_1, \boldsymbol{\psi}_2, \ldots, \boldsymbol{\psi}_n)$ is the modal matrix and $\mathbf{u} = (u_1, u_2, \ldots, u_n)^{\mathrm{T}}$ is the modal displacement vector.

Substituting Equation 5.42 in Equation 5.29 and pre-multiplying it by $\boldsymbol{\psi}_j^{\mathrm{T}}$ on both sides yields

$$\boldsymbol{\psi}_j^{\mathrm{T}}[\mathbf{M}\boldsymbol{\phi}\ddot{\mathbf{u}}(t) + \mathbf{C}\boldsymbol{\phi}\dot{\mathbf{u}}(t) + \mathbf{K}\boldsymbol{\phi}\mathbf{u}(t)] = \boldsymbol{\psi}_j^{\mathrm{T}}\mathbf{D}\boldsymbol{\xi}(t) \qquad j = 1, 2, \ldots, n \tag{5.43}$$

Using the proportional damping, namely

$$\boldsymbol{\phi}^{\mathrm{T}}\mathbf{C}\boldsymbol{\phi} = \mathrm{diag}[c_1, c_2, \ldots, c_n] = \mathrm{diag}[2\zeta_1\omega_1 m_1, 2\zeta_2\omega_2 m_2, \ldots, 2\zeta_n\omega_n m_n] \tag{5.44}$$

where diag[·] denotes a diagonal matrix, and noting Equations 5.40 and 5.41, we have

$$m_j\ddot{u}_j + c_j\dot{u}_j + k_j u_j = \boldsymbol{\psi}_j^{\mathrm{T}}\mathbf{B}\boldsymbol{\xi}(t) = \sum_{k=1}^{n}\sum_{\ell=1}^{r} \phi_{jk}B_{k\ell}\xi_\ell(t) \qquad j = 1, 2, \ldots, n \tag{5.45}$$

where ϕ_{jk} is the kth component of $\boldsymbol{\psi}_j$ and $B_{k\ell}$ is the $k \times \ell$-th component of **B**. Equation 5.45 contains n uncoupled SDOF systems.

According to the elaboration in Section 5.2.1.1, the impulse response function of the system in Equation 5.45 $h_j(t)$ is given by Equation 5.11, where m, ζ and ω are replaced by m_j, ζ_j and ω_j respectively. Thus, it follows that (see Equation 5.14)

$$u_j = \int_0^t h_j(t-\tau)\boldsymbol{\psi}_j^{\mathrm{T}}\mathbf{B}\boldsymbol{\xi}(\tau)\,\mathrm{d}\tau \tag{5.46}$$

where the effect of the initial condition is ignored. Substituting this equation in Equation 5.42 then yields the response of the system in Equation 5.29, namely:

$$\mathbf{X}(t) = \sum_{j=1}^{n}\int_0^t h_j(t-\tau)\boldsymbol{\psi}_j\boldsymbol{\psi}_j^{\mathrm{T}}\mathbf{B}\boldsymbol{\xi}(\tau)\,\mathrm{d}\tau \tag{5.47}$$

Again, this equation could be regarded as the linear operator $\mathscr{L}(\cdot)$ which governs the responses of the system to the excitations, as shown in Figure 5.3a or the upper part of Figure 5.3b.

The mean of the responses can be given by taking the mathematical expectation on both sides of Equation 5.47:

$$\boldsymbol{\mu}_{\mathbf{X}}(t) = \mathcal{E}[\mathbf{X}(t)] = \sum_{j=1}^{n}\int_0^t h_j(t-\tau)\boldsymbol{\psi}_j\boldsymbol{\psi}_j^{\mathrm{T}}\mathbf{B}\boldsymbol{\mu}_{\boldsymbol{\xi}}(\tau)\,\mathrm{d}\tau \tag{5.48}$$

Likewise, the correlation function matrix can be evaluated through

$$\begin{aligned}
\mathbf{R}_{\mathbf{X}}(t_1,t_2) &= \mathcal{E}[\mathbf{X}(t_1)\mathbf{X}^{\mathrm{T}}(t_2)] \\
&= \mathcal{E}\left[\sum_{j=1}^{n}\int_0^{t_1} h_j(t_1-\tau_1)\boldsymbol{\psi}_j\boldsymbol{\psi}_j^{\mathrm{T}}\mathbf{B}\boldsymbol{\xi}(\tau_1)\,\mathrm{d}\tau_1\cdot\sum_{k=1}^{n}\int_0^{t_2} h_k(t_2-\tau_2)\boldsymbol{\xi}^{\mathrm{T}}(\tau_2)\mathbf{B}^{\mathrm{T}}\boldsymbol{\psi}_k\boldsymbol{\psi}_k^{\mathrm{T}}\,\mathrm{d}\tau_2\right] \\
&= \sum_{k=1}^{n}\sum_{j=1}^{n}\int_0^{t_1}\int_0^{t_2} h_j(t_1-\tau_1)h_k(t_2-\tau_2)\boldsymbol{\psi}_j\boldsymbol{\psi}_j^{\mathrm{T}}\mathbf{B}\mathcal{E}[\boldsymbol{\xi}(\tau_1)\boldsymbol{\xi}^{\mathrm{T}}(\tau_2)]\mathbf{B}^{\mathrm{T}}\boldsymbol{\psi}_k\boldsymbol{\psi}_k^{\mathrm{T}}\,\mathrm{d}\tau_1\,\mathrm{d}\tau_2 \\
&= \sum_{k=1}^{n}\sum_{j=1}^{n}\int_0^{t_1}\int_0^{t_2} h_j(t_1-\tau_1)h_k(t_2-\tau_2)\boldsymbol{\psi}_j\boldsymbol{\psi}_j^{\mathrm{T}}\mathbf{B}\mathbf{R}_{\boldsymbol{\xi}}(\tau_1,\tau_2)\mathbf{B}^{\mathrm{T}}\boldsymbol{\psi}_k\boldsymbol{\psi}_k^{\mathrm{T}}\,\mathrm{d}\tau_1\,\mathrm{d}\tau_2
\end{aligned} \tag{5.49}$$

If the excitations are stationary processes, then the steady-state response processes will also be stationary processes. In this case, Equation 5.49 reduces to

$$\begin{aligned}
\mathbf{R}_{\mathbf{X}}(\tau) &= \sum_{j=1}^{n}\sum_{k=1}^{n}\int_{-\infty}^{\infty}\int_{-\infty}^{\infty} h_k(t_1)h_j(t_2)\boldsymbol{\psi}_k\boldsymbol{\psi}_k^{\mathrm{T}}\mathbf{B}\mathbf{R}_{\boldsymbol{\xi}}(\tau+t_2-t_1)\mathbf{B}^{\mathrm{T}}\boldsymbol{\psi}_j\boldsymbol{\psi}_j^{\mathrm{T}}\,\mathrm{d}t_1\,\mathrm{d}t_2 \\
&= \int_{-\infty}^{\infty}\int_{-\infty}^{\infty}\left(\sum_{k=1}^{n}h_k(t_1)\boldsymbol{\psi}_k\boldsymbol{\psi}_k^{\mathrm{T}}\right)\mathbf{B}\mathbf{R}_{\boldsymbol{\xi}}(\tau+t_2-t_1)\mathbf{B}^{\mathrm{T}}\left(\sum_{j=1}^{n}h_j(t_2)\boldsymbol{\psi}_j\boldsymbol{\psi}_j^{\mathrm{T}}\right)\mathrm{d}t_1\,\mathrm{d}t_2
\end{aligned} \tag{5.50}$$

Again, we reach a linear relationship between the moment functions of the responses of the system and that of the excitations in Equations 5.48–5.50. However, in comparison with Equations 5.33–5.35, this set of equations is more computationally convenient because the closed form of $h_j(t)$ is known, whereas Equations 5.33–5.35 are more conceptually direct as counterparts of the SDOF systems.

Actually, these two sets of equations are mathematically equivalent. This is clear if we apply the modal superposition method to Equation 5.31 and then get

$$\mathbf{h}(t) = \sum_{j=1}^{n} h_j(t)\boldsymbol{\psi}_j\boldsymbol{\psi}_j^{\mathrm{T}} \tag{5.51}$$

Introducing Equation 5.51 into Equations 5.33 and 5.35 will immediately yield Equations 5.48 and 5.50.

When nonproportional damping is involved, the equation of motion can be changed to a state equation and again the eigenvectors could be chosen as the basis vectors in a state space; but, in this case, the eigenvalues and eigenvectors are now usually complex numbers and, therefore, this is referred to as complex modal analysis. Such an analysis for random vibrations can be found in Fang and Wang (1986) and Fang et al. (1991).

5.2.2.3 Notes on Computational Efforts

Equation 5.49 is the so-called complete quadratic combination (CQC) scheme because there are n^2 terms in total being summed. In practice, the number of participant modes q is usually chosen such that $q \ll n$ for large structural systems and, therefore, the computational efforts can be greatly reduced. However, for practical engineering systems, this is usually still too time consuming. As an approximation, the cross-terms in the summation of Equation 5.49 are sometimes ignored; therefore:

$$\mathbf{R_X}(t_1,t_2) \approx \sum_{k=1}^{q}\int_0^{t_1}\int_0^{t_2} h_k(t_1-\tau_1)h_k(t_2-\tau_2)\boldsymbol{\psi}_k\boldsymbol{\psi}_k^{\mathrm{T}}\mathbf{B}\mathbf{R}_{\boldsymbol{\xi}}(\tau_1,\tau_2)\mathbf{B}^{\mathrm{T}}\boldsymbol{\psi}_k\boldsymbol{\psi}_k^{\mathrm{T}}\,\mathrm{d}\tau_1\,\mathrm{d}\tau_2 \tag{5.52}$$

This scheme is referred to as the square root of the summation of the square (SRSS). Usually, it is believed that this approximation is acceptable when the frequencies are not densely scattered. However, it cannot be guaranteed and actually it is not the case for many large complex structural systems (Der Kiureghian, 1980).

The prohibitive computational effort of Equation 5.49 is essentially one of the greatest hinderances to the application of classical random vibration theory to problems of practical interest, since the basic theory was developed about four decades ago. Interestingly, a simple further step, although always ignored, can greatly reduce the computational effort when we rearrange Equation 5.49 by

$$\begin{aligned}\mathbf{R_X}(t_1,t_2) &= \sum_{k=1}^{n}\sum_{j=1}^{n}\int_0^{t_1}\int_0^{t_2} h_j(t_1-\tau_1)h_k(t_2-\tau_2)\boldsymbol{\psi}_j\boldsymbol{\psi}_j^{\mathrm{T}}\mathbf{B}\mathbf{R}_{\boldsymbol{\xi}}(\tau_1,\tau_2)\mathbf{B}^{\mathrm{T}}\boldsymbol{\psi}_k\boldsymbol{\psi}_k^{\mathrm{T}}\,\mathrm{d}\tau_1\,\mathrm{d}\tau_2\\ &= \int_0^{t_1}\int_0^{t_2}\left\{\sum_{j=1}^{n}[h_j(t_1-\tau_1)\boldsymbol{\psi}_j\boldsymbol{\psi}_j^{\mathrm{T}}]\right\}\mathbf{B}\mathbf{R}_{\boldsymbol{\xi}}(\tau_1,\tau_2)\mathbf{B}^{\mathrm{T}}\left\{\sum_{k=1}^{n}[h_k(t_2-\tau_2)\boldsymbol{\psi}_k\boldsymbol{\psi}_k^{\mathrm{T}}]\right\}\mathrm{d}\tau_1\,\mathrm{d}\tau_2\end{aligned} \tag{5.53}$$

We find that Equation 5.53 is in a form identical to the equation resulting from introducing Equation 5.51 into Equation 5.34. However, from the point of view of computation, the double summation in Equation 5.49 (i.e. the CQC) has been changed to two single summations; that is, the multiplication of two summations. We can view the effect of the change by examining a simple example:

$$s = \sum_{i=1}^{q}\sum_{j=1}^{q} a_i b_j = \left(\sum_{i=1}^{q} a_i\right)\left(\sum_{j=1}^{q} b_j\right) \tag{5.54}$$

Mathematically, the two equalities are equivalent. However, in the computation of the first entity $\sum_{i=1}^{q}\sum_{j=1}^{q} a_i b_j$, q^2 multiplication operations and q^2 summation operations need to be performed, whereas only one multiplication operation and $2q$ summation operations are needed in the computation of the second entity $(\sum_{i=1}^{q} a_i)(\sum_{j=1}^{q} b_j)$. Note that the multiplication operation is much more time consuming than the summation operation is. Compared with the first entity, the computational efforts in the second entity is reduced from $q^2 w_{\mathrm{m}} + q^2 w_{\mathrm{s}}$ to $w_{\mathrm{m}} + 2qw_{\mathrm{s}}$, where w_{m} and w_{s} are respectively the workload of one multiplication operation and one summation operation. If q is in the order of 10^2, then this might make the computation more efficient by times in the order of 10^8–10^4. The difference between the computational efforts of the first and second equalities in Equation 5.53 is just analogous to the situation in Equation 5.54 as discussed above. This is also true for the computation of Equation 5.50.

We will come back to this issue when the pseudo-excitation method is elaborated later.

5.3 Power Spectral Density Analysis

5.3.1 Frequency Response Function and Power Spectral Density

5.3.1.1 Response of SDOF Systems in Frequency Domain

Frequency Response Function of SDOF Systems

As discussed in Section 5.2.1.1, apart from the unit impulse, the other widely used testing excitation is the unit harmonic excitation. In this case, the response of an SDOF system, denoted by $\tilde{h}(t)$, satisfies

$$m\ddot{\tilde{h}} + c\dot{\tilde{h}} + k\tilde{h} = \mathrm{e}^{\mathrm{i}\omega t} \qquad x_0 = 0,\ \dot{x}_0 = 0 \tag{5.55}$$

Using Equation 5.14, replacing $f(t)$ by the harmonic excitation $\mathrm{e}^{\mathrm{i}\omega t}$, we have

$$\tilde{h}(t) = \int_{-\infty}^{\infty} h(t,\tau)\mathrm{e}^{\mathrm{i}\omega(t-\tau)}\,\mathrm{d}\tau = \mathrm{e}^{\mathrm{i}\omega t}\int_{-\infty}^{\infty} h(t,\tau)\mathrm{e}^{-\mathrm{i}\omega\tau}\,\mathrm{d}\tau = \tilde{H}(\omega,t)\mathrm{e}^{\mathrm{i}\omega t} \tag{5.56}$$

where

$$\tilde{H}(\omega,t) = \int_{-\infty}^{\infty} h(t,\tau)\mathrm{e}^{-\mathrm{i}\omega\tau}\,\mathrm{d}\tau \tag{5.57a}$$

and $h(t,\tau)$ is the impulse response function of a time-variant linear system. If the system is time invariant, then $h(t, \tau)$ is the function given by Equation 5.11. In this case, $\tilde{H}(\omega, t)$ reduces to

$$H(\omega) = \int_{-\infty}^{\infty} h(\tau)\mathrm{e}^{-\mathrm{i}\omega\tau}\,\mathrm{d}\tau \tag{5.57b}$$

The function $H(\omega)$ is referred to as the frequency response function. Evidently, Equation 5.57b shows that the frequency response function is the Fourier transform of the impulse response function. In turn, this indicates that the impulse response function $h(t)$ must be the inverse Fourier transform of the frequency response function:

$$h(t) = \frac{1}{2\pi}\int_{-\infty}^{\infty} H(\omega)e^{i\omega t}\,d\omega \tag{5.58}$$

If we introduce Equation 5.56, in which $\tilde{H}(\omega, t)$ is replaced by $H(\omega)$, into Equation 5.55, it follows that

$$H(\omega) = \frac{1}{k - m\omega^2 + ic\omega} = \frac{1}{m(\omega_0^2 - \omega^2 + i2\zeta\omega_0\omega)} = \frac{1}{m\omega_0^2[1 - (\omega/\omega_0)^2 + i2\zeta\omega/\omega_0]} \tag{5.59}$$

where ζ and ω_0 are the damping ratio and the circular frequency of the SDOF system respectively.

Furthermore, according to the principle of superposition, multiplying both sides of the equation of motion of the SDOF system subject to an arbitrary excitation $f(t)$ by $e^{-i\omega t}$

$$m\ddot{x} + c\dot{x} + kx = f(t) \tag{5.60}$$

and integrating with regard to t yields

$$m\int_{-\infty}^{\infty} \ddot{x}(t)e^{-i\omega t}\,dt + c\int_{-\infty}^{\infty} \dot{x}(t)e^{-i\omega t}\,dt + k\int_{-\infty}^{\infty} x(t)e^{-i\omega t}\,dt = \int_{-\infty}^{\infty} f(t)e^{-i\omega t}\,dt \tag{5.61a}$$

Denoting the Fourier transforms of $\ddot{x}(t)$, $\dot{x}(t)$, $x(t)$ and $f(t)$ by $\ddot{x}(\omega)$, $\dot{x}(\omega)$, $x(\omega)$ and $F(\omega)$ respectively, Equation 5.61a is rewritten as

$$m\ddot{x}(\omega) + c\dot{x}(\omega) + kx(\omega) = F(\omega) \tag{5.61b}$$

Note that $\ddot{x}(\omega) = i\omega\dot{x}(\omega) = -\omega^2 x(\omega)$. It follows that

$$x(\omega) = \frac{1}{k - m\omega^2 + ic\omega}F(\omega) = H(\omega)F(\omega) \tag{5.62}$$

This equation establishes the linear relationship between the input and the output in the frequency domain. We can see that all the properties of the SDOF system are contained in the frequency response function $H(\omega)$, which is irrelevant to the excitation and completely characterizes the system properties. This is one of the reasons why the unit harmonic excitation is chosen as the testing excitation. From Equation 5.62, we have the impression that the linearity embedded in the system is reflected more clearly in the frequency domain than in the time domain.

Power Spectral Density of Response of an SDOF System to Stochastic Excitation

If the excitation is a stochastic process $\xi(t)$, then the response $X(t)$ is also a stochastic process. In this case, Equation 5.62 could be interpreted in a sample sense which conforms to the physical sense; namely:

$$X(\varpi, \omega) = H(\omega)\xi(\varpi, \omega) \tag{5.63a}$$

where ϖ is the variable characterizing the embedded randomness and $X(\varpi, \omega)$ and $\xi(\varpi, \omega)$ should be understood as the Fourier spectrum of the stochastic processes $X(\varpi, t)$ and $\xi(\varpi, t)$, respectively.

Taking the complex conjugate of Equation 5.63a:

$$X^*(\varpi, \omega) = H^*(\omega) \cdot \xi^*(\varpi, \omega) \tag{5.63b}$$

Hereafter, the asterisk superscripts denote the complex conjugate.

Multiplying the left-hand side of Equation 5.63a with the left-hand side of Equation 5.63b and doing the same thing for the right-hand sides of Equations 5.63a and 5.63b, taking mathematical expectations and dividing by the duration T as $T \to \infty$, we have

$$\lim_{T \to \infty} \frac{1}{2T} \mathcal{E}[X(\varpi, \omega) X^*(\varpi, \omega)] = \lim_{T \to \infty} \frac{1}{2T} |H(\omega)|^2 \mathcal{E}[\xi(\varpi, \omega) \xi^*(\varpi, \omega)] \tag{5.64}$$

In fact, the left-hand side of Equation 5.64 is nothing but the PSD of $X(t)$, provided it is a stationary stochastic process (see Equation 2.89a and Appendix C); namely:

$$S_X(\omega) = \lim_{T \to \infty} \frac{1}{2T} \mathcal{E}[X(\varpi, \omega) X^*(\varpi, \omega)] \tag{5.65a}$$

Likewise, the terms contained on the right-hand side of Equation 5.64 involve the PSD of $\xi(t)$, provided it is a stationary stochastic process:

$$S_\xi(\omega) = \lim_{T \to \infty} \frac{1}{2T} \mathcal{E}[\xi(\varpi, \omega) \xi^*(\varpi, \omega)] \tag{5.65b}$$

Then, from Equation 5.64, it follows that

$$S_X(\omega) = |H(\omega)|^2 S_\xi(\omega) \tag{5.66}$$

where $|H(\omega)|^2$ is given from Equation 5.59 by

$$|H(\omega)|^2 = \frac{1}{(m\omega_0^2)^2} \frac{1}{[1 - (\omega/\omega_0)^2]^2 + 4\zeta^2 (\omega/\omega_0)^2} \tag{5.67}$$

which is illustrated in Figure 5.5.

Some characteristics of $|H(\omega)|^2$ are:

(a) $|H(\omega)|^2 = 1/(m\omega_0^2)^2$, or $|H(\omega)|^2 (m\omega_0^2)^2 = 1$, for $\omega/\omega_0 = 0$;
(b) $|H(\omega)|^2 = [1/(m\omega_0^2)^2][1/(4\zeta^2)]$, or $|H(\omega)|^2 (m\omega_0^2)^2 = 1/(4\zeta^2)$, for $\omega/\omega_0 = 1$;
(c) $|H(\omega)|^2 \to 0$, as $\omega/\omega_0 \to \infty$;
(d) $|H(\omega)|^2$ reaches its maximum as $\omega/\omega_0 = (1 + \zeta^4)^{1/2} - \zeta^2$. It is seen that ω/ω_0 maximizing $|H(\omega)|^2$ varies from 1 to $\sqrt{2} - 1$ monotonically as ζ varis from 0 to 1.
(e) Half-power bandwidth: for light damping, the maximum value of $|H(\omega)|^2$ occurs approximately at $\omega/\omega_0 = 1$ as shown in (d). In this case, $|H(\omega_0)|^2 (m\omega_0^2)^2 = 1/(4\zeta^2)$. Consider the point ω_1 where

$$|H(\omega_1)|^2 = \frac{1}{2} |H(\omega_0)|^2 \tag{5.68a}$$

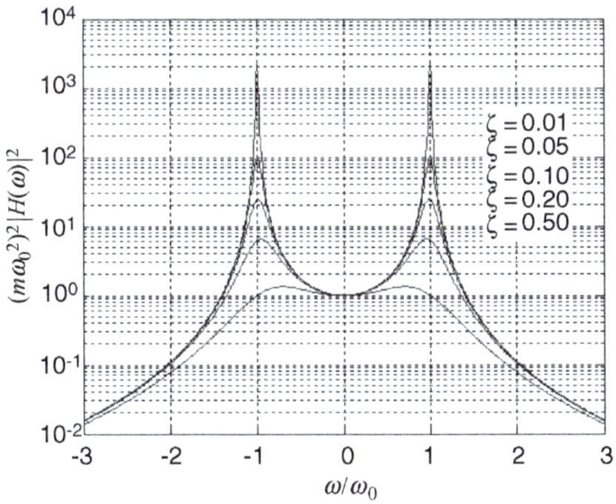

Figure 5.5 Frequency transfer function.

That is, the PSD at ω_1 is half of the maximal PSD when the excitation is a white noise; therefore, ω_1 is referred to as the half-power point. Note from Figure 5.5 that there are two half-power points ω_1 and ω_2 scattered over two sides of ω_0, $\omega_1 < \omega_0 < \omega_2$, and the differences between them are small when ζ is small; that is, $\Delta\omega = \omega_2 - \omega_1 \ll \omega_0$. Through some manipulations we have

$$\zeta \approx \frac{\Delta\omega}{2\omega_0} \tag{5.68b}$$

where $\Delta\omega$ is referred to as the half-power bandwidth.

Equation 5.68b indicates that the damping ratio can be obtained easily if the bandwidth and natural frequency can be measured through vibration testing techniques. This is the so-called half-power method for identification of damping (Clough and Penzien, 1993).

Equation 5.66 establishes a linear relationship between the PSD of the response and that of the excitation. Again, this can viewed as the linear operator $\mathscr{L}_{\mathscr{P}}(\cdot)$ in the lower part of Figure 5.3b. Obviously, the linearity exhibited here is more direct than that exhibited in the time domain, as discussed in Section 5.2.2.

Further, it is easy to get the correlation function from Equation 5.66 by the Wiener–Khintchine theorem:

$$R_X(\tau) = \frac{1}{2\pi}\int_{-\infty}^{\infty} S_Y(\omega)e^{i\omega\tau}\,d\omega = \frac{1}{2\pi}\int_{-\infty}^{\infty} |H(\omega)|^2 S_\xi(\omega)e^{i\omega\tau}\,d\omega \tag{5.69a}$$

While the variance of the response can be given by

$$\mathrm{Var}[X(t)] = R_X(0) = \frac{1}{2\pi}\int_{-\infty}^{\infty} S_X(\omega)\,d\omega = \frac{1}{2\pi}\int_{-\infty}^{\infty} |H(\omega)|^2 S_\xi(\omega)\,d\omega \tag{5.69b}$$

The derivation in Equations 5.63a–5.66 looks more intuitive than rigorous, because the Fourier transform of a stationary stochastic process does not exist when the stochastic process does not tend to zero as $|t| \to \infty$. However, this problem can be overcome by defining a Fourier transform over a finite time interval; that is (see Appendix C):

$$X_{\pm T}(\varpi,\omega)=\int_{-T}^{T}X(\varpi,t)\mathrm{e}^{-\mathrm{i}\omega t}\,\mathrm{d}t \tag{5.70a}$$

Then the Fourier transform of the derived process over the time interval reads

$$\begin{aligned}\dot{X}_{\pm T}(\varpi,\omega) &= \int_{-T}^{T}\dot{X}(\varpi,t)\mathrm{e}^{-\mathrm{i}\omega t}\,\mathrm{d}t = X(\varpi,T)\mathrm{e}^{-\mathrm{i}\omega t}-X(\varpi,-T)\mathrm{e}^{\mathrm{i}\omega t}+(\mathrm{i}\omega)\int_{-T}^{T}X(\varpi,t)\mathrm{e}^{-\mathrm{i}\omega t}\,\mathrm{d}t\\ &= X(\varpi,T)\mathrm{e}^{-\mathrm{i}\omega t}-X(\varpi,-T)\mathrm{e}^{\mathrm{i}\omega t}+(\mathrm{i}\omega)X_{\pm T}(\varpi,\omega)\end{aligned} \tag{5.70b}$$

Replacing Equation 5.63a by

$$X_{\pm T}(\varpi,\omega)=H(\omega)\xi_{\pm T}(\varpi,\omega)+H(\omega)\boldsymbol{\Psi}(\varpi,T) \tag{5.70c}$$

where

$$\begin{aligned}\boldsymbol{\Psi}(T) &= -c[X(\varpi,T)\mathrm{e}^{-\mathrm{i}\omega T}-X(\varpi,-T)\mathrm{e}^{\mathrm{i}\omega T}]\\ &= -m\{\dot{X}(\varpi,T)\mathrm{e}^{-\mathrm{i}\omega T}-\dot{X}(\varpi,-T)\mathrm{e}^{\mathrm{i}\omega T}+\mathrm{i}\omega[X(\varpi,T)\mathrm{e}^{-\mathrm{i}\omega T}-X(\varpi,-T)\mathrm{e}^{\mathrm{i}\omega T}]\}\end{aligned}$$

Multiplying Equation 5.70c on both sides by its complex conjugate, dividing it by T and letting $T\to\infty$, we find the effect of $\Psi(\varpi,T)$ vanishes because the mean and the variance of $X(\varpi,T)$ are finite; thus we get Equation 5.64. Therefore, the above derivations will lead to no problems, but will provide a much simpler and intuitive perspective. For this reason, we will use the technique extensively in the following sections.

Power Spectral Density of Derived Stochastic Processes

Sometimes we may be interested in the PSD of its derived processes, say $X^{(k)}(t)=\mathrm{d}^kX(t)/\mathrm{d}t^k$. We have given some commonly used formulae which can be proved starting with the definitions in Section 2.2.4. However, we can achieve them in a more straightforward way as shown in the preceding subsection.

Let us consider the nth derived process of $X(t)$, denoted by $X^{(n)}(t)$. Its Fourier spectrum is given by

$$X^{(n)}(\omega)=(\mathrm{i}\omega)^nX(\omega) \tag{5.71a}$$

of which the complex conjugate is

$$[X^{(n)}(\omega)]^*=[(\mathrm{i}\omega)^n]^*\cdot X^*(\omega) \tag{5.71b}$$

As with Equations 5.63a and 5.63b resulting in Equation 5.64, multiplying the left-hand side of Equation 5.71a with the left-hand side of Equation 5.71b and doing the same thing for the right-hand sides of Equations 5.71a and 5.71b, dividing by T and letting $T\to\infty$, we get

$$S_{X^{(n)}}(\omega)=\omega^{2n}S_X(\omega) \tag{5.71c}$$

which is nothing but Equation 2.104.

The same idea can be applied to Equation 2.105 and is left to the reader.

Cross-Power Spectral Density Function

Sometimes we want to get the cross-PSD, say of the response and its derived process or of the response and the excitation. Direct employment of the basic idea in the preceding section will make it quite straightforward.

First we take $S_{X\dot{X}}(\omega)$ as an example. In this case we replace Equation 5.63b by

$$\dot{X}^*(\varpi,\omega) = (-\mathrm{i}\omega)H^*(\omega)\cdot\xi^*(\varpi,\omega) \tag{5.72a}$$

where $\dot{X}(\varpi,\omega)$ is the Fourier transform of the derived process $\dot{X}(\varpi,t)$. Noting that (see Equation 2.89b and Appendix C)

$$S_{X\dot{X}}(\omega) = \lim_{T\to\infty}\frac{1}{2T}\mathcal{E}\left[X(\varpi,\omega)\dot{X}^*(\varpi,\omega)\right] \tag{5.72b}$$

then correspondingly, Equation 5.66 becomes

$$S_{X\dot{X}}(\omega) = (-\mathrm{i}\omega)|H(\omega)|^2 S_\xi(\omega) \tag{5.72c}$$

If we compare this formula with Equation 5.66, then we have

$$S_{X\dot{X}}(\omega) = (-\mathrm{i}\omega)S_X(\omega) \tag{5.72d}$$

Recalling that in harmonic motion the phase angle of the displacement is $\pi/2$ lag behind that of the velocity, the physical meaning of this relationship is quite clear.

Likewise, let us consider $S_{X\xi}(\omega)$. In this case, replacing Equation 5.63b by an equality $\xi^*(\varpi,\omega) = \xi^*(\varpi,\omega)$, and noting that (see Equation 2.89b)

$$S_{X\xi}(\omega) = \lim_{T\to\infty}\frac{1}{2T}\mathcal{E}[X(\varpi,\omega)\xi^*(\varpi,\omega)] \tag{5.73a}$$

we immediately obtain

$$S_{X\xi}(\omega) = H(\omega)S_\xi(\omega) \tag{5.73b}$$

Actually, in general, if $X(t)$ and $Y(t)$ are two stationary stochastic process determined through linear transfer operators from the input $\xi(t)$; for example:

$$X(t) = \mathcal{L}_1[\xi(t)] \tag{5.74a}$$

$$Y(t) = \mathcal{L}_2[\xi(t)] \tag{5.74b}$$

then their Fourier transform gives

$$X(\omega) = L_1(\omega)\xi(\omega) \tag{5.75a}$$

$$Y(\omega) = L_2(\omega)\xi(\omega) \tag{5.75b}$$

Thus, it follows that

$$\lim_{T\to\infty}\frac{1}{2T}\mathcal{E}[X(\omega)Y^*(\omega)] = L_1(\omega)L_2{}^*(\omega)\lim_{T\to\infty}\frac{1}{2T}\mathcal{E}[\xi(\omega)\xi^*(\omega)] \tag{5.76a}$$

That is:

$$S_{XY}(\omega) = L_1(\omega)L_2{}^*(\omega)S_\xi(\omega) \tag{5.76b}$$

5.3.1.2 Response of MDOF Systems in the Frequency Domain

Direct Matrix Expression

If $\boldsymbol{\xi}(t) = [\xi_1(t), \xi_2(t), \ldots, \xi_r(t)]^{\mathrm{T}}$ is a zero-mean weakly stationary stochastic process vector, of which the PSD matrix is $\mathbf{S}_{\boldsymbol{\xi}}(\omega) = [S_{\xi_i\xi_j}(\omega)]_{r\times r}$, then the correlation function matrix $\mathbf{R}_{\boldsymbol{\xi}}(\tau) = [R_{\xi_i\xi_j}(\tau)]_{r\times r}$ is determined by the Wiener–Khintchine formula:

$$\mathbf{S}_{\boldsymbol{\xi}}(\omega) = \int_{-\infty}^{\infty} \mathbf{R}_{\boldsymbol{\xi}}(\tau)\mathrm{e}^{-\mathrm{i}\omega\tau}\,\mathrm{d}\tau \qquad \mathbf{R}_{\boldsymbol{\xi}}(\tau) = \frac{1}{2\pi}\int_{-\infty}^{\infty} \mathbf{S}_{\boldsymbol{\xi}}(\omega)\mathrm{e}^{\mathrm{i}\omega\tau}\,\mathrm{d}\omega \tag{5.77}$$

In addition, according to Equation 5.65b, we have

$$\mathbf{S}_{\boldsymbol{\xi}}(\omega) = \lim_{T\to\infty}\frac{1}{2T}\mathcal{E}[\boldsymbol{\xi}(\varpi,\omega)\boldsymbol{\xi}^*(\varpi,\omega)] \tag{5.78}$$

Here, the asterisk denotes the transpose of the complex conjugate.

Extending the cases of the SDOF system in Section 5.3.1.1 to the MDOF system (Equation 5.29)

$$\mathbf{M}\ddot{\mathbf{X}} + \mathbf{C}\dot{\mathbf{X}} + \mathbf{K}\mathbf{X} = \mathbf{B}\boldsymbol{\xi}(\varpi, t) \tag{5.79}$$

we can get the frequency-response function matrix

$$\mathbf{H}(\omega) = (\mathbf{K} - \omega^2\mathbf{M} + \mathrm{i}\omega\mathbf{C})^{-1} \tag{5.80}$$

such that the Fourier transform of the responses, $\mathbf{X}(\varpi, \omega)$, is determined by

$$\mathbf{X}(\varpi,\omega) = \mathbf{H}(\omega)\mathbf{B}\boldsymbol{\xi}(\varpi,\omega) \tag{5.81a}$$

where $\boldsymbol{\xi}(\varpi, \omega)$ is the Fourier transform of the excitation $\boldsymbol{\xi}(\varpi, t)$.

Taking the complex conjugate of both sides of Equation 5.81a yields

$$\mathbf{X}^*(\varpi,\omega) = \boldsymbol{\xi}^*(\varpi,\omega)\mathbf{B}^{\mathrm{T}}\mathbf{H}^*(\omega) \tag{5.81b}$$

Post-multiplying both sides of Equation 5.81a by the corresponding sides of Equation 5.81b, taking the mathematical expectation, then dividing by the duration $2T$ and letting $T \to \infty$, it follows that

$$\lim_{T\to\infty}\frac{1}{2T}\mathcal{E}[\mathbf{X}(\omega)\mathbf{X}^*(\omega)] = \lim_{T\to\infty}\frac{1}{2T}\mathcal{E}\left[\mathbf{H}(\omega)\mathbf{B}\boldsymbol{\xi}(\varpi,\omega)\boldsymbol{\xi}^*(\varpi,\omega)\mathbf{B}^{\mathrm{T}}\mathbf{H}^*(\omega)\right] \tag{5.82}$$

Noting Equation 5.78 and that the PSD function matrix of $\mathbf{X}(t)$:

$$\mathbf{S}_{\mathbf{X}}(\omega) = \lim_{T\to\infty}\frac{1}{2T}\mathcal{E}[\mathbf{X}(\varpi,\omega)\mathbf{X}^*(\varpi,\omega)] \tag{5.83}$$

we reach

$$\mathbf{S}_{\mathbf{X}}(\omega) = \mathbf{H}(\omega)\mathbf{B}\mathbf{S}_{\boldsymbol{\xi}}(\omega)\mathbf{B}^{\mathrm{T}}\mathbf{H}^*(\omega) \tag{5.84}$$

Equation 5.84 is the counterpart of Equation 5.66 in the case of MDOF systems. This shows that the PSD matrix of the responses is a linear transform of the PSD matrix of the excitations. The linearity embedded here is, again, reflection of the physical linearity of the MDOF system (Equation 5.79).

The relationships in Equations 5.66 and 5.84 are algebraic, while the transfer relationships from the moments of excitations to those of the responses are integrals (see Equations 5.25 and 5.35 in the time domain); therefore, the transfer relationships in the frequency domain are more straightforward, compact and simple.

However, direct implementation of Equation 5.84 is not so easy because the computation of $\mathbf{H}(\omega)$ by Equation 5.80 might be very time consuming. This can be relieved by employing the modal superposition method.

Modal Superposition Method

According to Section 5.2.2, when employing the modal decomposition technique, an MDOF system can be uncoupled to a set of SDOF systems (see Equation 5.45):

$$m_j\ddot{u}_j + c_j\dot{u}_j + k_j u_j = \boldsymbol{\psi}_j^{\mathrm{T}}\mathbf{B}\boldsymbol{\xi}(t) \qquad j = 1, 2, \ldots, n \tag{5.85}$$

of which the frequency response functions are given by (see Equation 5.59)

$$H_j(\omega) = \frac{1}{m_j\omega_{0j}^2[1 - (\omega/\omega_{0j})^2 + \mathrm{i}2\zeta_j\omega/\omega_{0j}]} \qquad j = 1, 2, \ldots, n \tag{5.86}$$

where m_j, ζ_j and ω_{0j} are the jth modal mass, the modal damping ratio and the modal circular frequency respectively.

Using modal decomposition, Equation 5.42, the Fourier transform is

$$\mathbf{X}(\varpi, \omega) = \sum_{j=1}^{n} \boldsymbol{\psi}_j u_j(\varpi, \omega) \tag{5.87a}$$

in which $u_j\,(\varpi, \omega)$ is the Fourier transform of the modal displacement $u_j\,(\varpi, t)$ and given by

$$u_j(\varpi, \omega) = H_j(\omega)\boldsymbol{\psi}_j^{\mathrm{T}}\mathbf{B}\boldsymbol{\xi}(\omega) \tag{5.87b}$$

Employing Equation 5.83 and substituting Equation 5.87a in it, we have

$$\begin{aligned}
\mathbf{S}_{\mathbf{X}}(\omega) &= \lim_{T\to\infty}\frac{1}{2T}\mathcal{E}\left\{\left[\sum_{j=1}^{n}\boldsymbol{\psi}_j u_j(\varpi,\omega)\right]\left[\sum_{j=1}^{n}u_j^*(\varpi,\omega)\boldsymbol{\psi}_j^{\mathrm{T}}\right]\right\} \\
&= \lim_{T\to\infty}\frac{1}{2T}\mathcal{E}\left\{\left[\sum_{j=1}^{n}\boldsymbol{\psi}_j H_j(\omega)\boldsymbol{\psi}_j^{\mathrm{T}}\mathbf{B}\boldsymbol{\xi}(\omega)\right]\left[\sum_{j=1}^{n}\boldsymbol{\xi}^*(\omega)\mathbf{B}^{\mathrm{T}}\boldsymbol{\psi}_j H_j^*(\omega)\boldsymbol{\psi}_j^{\mathrm{T}}\right]\right\} \\
&= \sum_{j=1}^{n}\sum_{k=1}^{n}\boldsymbol{\psi}_k\boldsymbol{\psi}_k^{\mathrm{T}}H_k(\omega)\mathbf{B}\left\{\lim_{T\to\infty}\frac{1}{2T}\mathcal{E}\left[\boldsymbol{\xi}(\omega)\boldsymbol{\xi}^*(\omega)\right]\right\}\mathbf{B}^{\mathrm{T}}H_j^*(\omega)\boldsymbol{\psi}_j\boldsymbol{\psi}_j^{\mathrm{T}} \\
&= \sum_{j=1}^{n}\sum_{k=1}^{n}H_k(\omega)H_j^*(\omega)\boldsymbol{\psi}_k\boldsymbol{\psi}_k^{\mathrm{T}}\mathbf{B}\mathbf{S}_{\boldsymbol{\xi}}(\omega)\mathbf{B}^{T}\boldsymbol{\psi}_j\boldsymbol{\psi}_j^{\mathrm{T}}
\end{aligned} \tag{5.88}$$

Compared with Equation 5.84, the implementation of Equation 5.88 is more convenient because here the closed form of $H_j(\omega)$ is known as Equation 5.86.

On the other hand, mathematically, Equation 5.88 is equivalent to Equation 5.84. Actually, as an extension of Equation 5.62 to the matrix form, the frequency-response function matrix $\mathbf{H}(\omega)$ is the Fourier transform of the impulse-response function matrix $\mathbf{h}(t)$ (see Equation 5.31):

$$\mathbf{H}(\omega) = \int_{-\infty}^{\infty} \mathbf{h}(t)\mathrm{e}^{-\mathrm{i}\omega t}\,\mathrm{d}t \tag{5.89}$$

Introducing Equation 5.51 into this and noting $H_j(\omega)$ is the Fourier transform of $h_j(t)$ yields

$$\mathbf{H}(\omega) = \int_{-\infty}^{\infty} \mathbf{h}(t)\mathrm{e}^{-\mathrm{i}\omega t}\,\mathrm{d}t = \int_{-\infty}^{\infty} \sum_{j=1}^{n} h_j(t)\boldsymbol{\psi}_j\boldsymbol{\psi}_j^{\mathrm{T}}\mathrm{e}^{-\mathrm{i}\omega t}\,\mathrm{d}t = \sum_{j=1}^{n} H_j(\omega)\boldsymbol{\psi}_j\boldsymbol{\psi}_j^{\mathrm{T}} \tag{5.90}$$

Replacing $\mathbf{H}(\omega)$ in Equation 5.84 by the expression in Equation 5.90 immediately leads to Equation 5.88.

Notes on Computational Effort

It is seen that Equation 5.88 is a CQC scheme. Its computational workload is in the order of n^2 (or in the order of q^2 when q modes are chosen). The computational effort for a general large structural system is still prohibitive. An alternative scheme is to compute Equation 5.88 by

$$\begin{aligned}\mathbf{S}_{\mathbf{X}}(\omega) &= \sum_{j=1}^{n}\sum_{k=1}^{n} H_k(\omega)H_j^*(\omega)\boldsymbol{\psi}_k\boldsymbol{\psi}_k^{\mathrm{T}}\mathbf{B}\mathbf{S}_{\boldsymbol{\xi}}(\omega)\mathbf{B}^{\mathrm{T}}\boldsymbol{\psi}_j\boldsymbol{\psi}_j^{\mathrm{T}} \\ &= \left[\sum_{k=1}^{n} H_k(\omega)\boldsymbol{\psi}_k\boldsymbol{\psi}_k^{\mathrm{T}}\right]\mathbf{B}\mathbf{S}_{\boldsymbol{\xi}}(\omega)\mathbf{B}^{\mathrm{T}}\left[\sum_{j=1}^{n} H_j^*(\omega)\boldsymbol{\psi}_j\boldsymbol{\psi}_j^{\mathrm{T}}\right]\end{aligned} \tag{5.91}$$

Although the two equalities in Equation 5.91 are mathematically equivalent, the computational effort of the second entity is much less than that of the first entity. The reason is analogous to the discussion in Section 5.2.2.3.

Example 5.2. Response of an MDOF System to White Noise Excitation Consider an MDOF structural system subjected to earthquake ground motion modeled by a white-noise process with the PSD function $S_{\boldsymbol{\xi}}(\boldsymbol{\omega}) = S_0$. Here, $\mathbf{B} = -\mathbf{M}\{\mathbf{1}\}$, where $\mathbf{M}$ is the mass matrix and $\{\mathbf{1}\}$ is a column vector with all components 1. According to Equation 5.91:

$$\mathbf{S}_{\mathbf{X}}(\omega) = S_0\left[\sum_{k=1}^{n} H_k(\omega)\boldsymbol{\psi}_k\boldsymbol{\psi}_k^{\mathrm{T}}\right]\mathbf{M}\mathbf{B}\{\mathbf{1}\}\{\mathbf{1}\}^{\mathrm{T}}\mathbf{M}^{\mathrm{T}}\left[\sum_{j=1}^{n} H_j{}^*(\omega)\boldsymbol{\psi}_j\boldsymbol{\psi}_j^{\mathrm{T}}\right] \tag{5.92}$$

For simplicity, we assume a diagonal mass matrix $\mathbf{M} = \mathrm{diag}(M_j)$, then the component form of Equation 5.92 reads

$$\begin{aligned} S_{X_rX_s}(\omega) &= S_0\left(\sum_{k=1}^{n}\sum_{j=1}^{n}\phi_{kr}\phi_{kj}M_jH_k(\omega)\right)\left(\sum_{k=1}^{n}\sum_{j=1}^{n}\phi_{ks}\phi_{kj}M_jH_k{}^*(\omega)\right) \\ &= S_0\sum_{q=1}^{n}\sum_{p=1}^{n}\sum_{k=1}^{n}\sum_{j=1}^{n}\phi_{kr}\phi_{ps}\phi_{kj}\phi_{pq}M_jM_qH_k(\omega)H_p{}^*(\omega) \end{aligned} \tag{5.93}$$

where ϕ_{kj} is the jth component of the mode shape vector $\boldsymbol{\psi}_k$, $\boldsymbol{\psi}_k = (\phi_{k1}, \phi_{k2}, \ldots, \phi_{kn})^{\mathrm{T}}$.

In the case $r = s$, we have

$$\begin{aligned} S_{X_r}(\omega) &= S_0\left\{\sum_{k=1}^{n}\left[H_k(\omega)\sum_{j=1}^{n}\phi_{kr}\phi_{kj}M_j\right]\right\}\left\{\sum_{k=1}^{n}\left[H_k{}^*(\omega)\sum_{j=1}^{n}\phi_{kr}\phi_{kj}M_j\right]\right\} \\ &= S_0\left[\sum_{k=1}^{n}\beta_{k,r}H_k(\omega)\right]\left[\sum_{k=1}^{n}\beta_{k,r}H_k{}^*(\omega)\right] \end{aligned} \tag{5.94}$$

where $\beta_{k,r} = \sum_{j=1}^{n} \phi_{kr}\phi_{kj}M_j$.

Equation 5.94 can also be expanded by

$$S_{X_r}(\omega) = S_0\left\{\sum_{k=1}^{n}\beta_{k,r}^2|H_k(\omega)|^2 + \sum_{k=1,k\neq j}^{n}\sum_{j=1}^{n}\beta_{k,r}\beta_{j,r}\mathrm{Re}[H_k(\omega)H_j{}^*(\omega)]\right\} \tag{5.95}$$

Here, Re[·] represents the real part of the bracketed complex number.

To give a more illustrative recognition, we examine a two-DOF system with the mass, the damping and the stiffness matrices given respectively by

$$\mathbf{M} = \begin{bmatrix} 100 & 0 \\ 0 & 100 \end{bmatrix}\mathrm{kg} \quad \mathbf{C} = \begin{bmatrix} 5.758 & 4.081 \\ 4.081 & 9.839 \end{bmatrix}\mathrm{N\cdot s/m} \quad \text{and} \quad \mathbf{K} = \begin{bmatrix} 200 & -100 \\ -100 & 100 \end{bmatrix}\mathrm{N/m} \tag{5.96}$$

The modal matrix and the modal mass matrix are respectively

$$\boldsymbol{\phi} = \begin{bmatrix} 1 & 1 \\ 1.618 & -0.618 \end{bmatrix} \quad \text{and} \quad \bar{\mathbf{M}} = \begin{bmatrix} 361.8 & 0 \\ 0 & 138.2 \end{bmatrix}\mathrm{kg} \tag{5.97}$$

The modal damping ratios are $\zeta_1 = 0.10$ and $\zeta_2 = 0.01$; the frequencies are $\omega_1 = 0.618$ and $\omega_2 = 1.618$. Therefore:

$$H_1(\omega) = \frac{1}{m_1\omega_1^2}\frac{1}{[1-(\omega/\omega_1)^2]+2i\zeta_1(\omega/\omega_1)} = \frac{1}{138.2}\frac{1}{(1-\omega^2/0.382)+0.324i\omega} \tag{5.98}$$

$$H_2(\omega) = \frac{1}{m_2\omega_2^2}\frac{1}{[1-(\omega/\omega_2)^2]+2i\zeta_2(\omega/\omega_2)} = \frac{1}{361.8}\frac{1}{(1-\omega^2/2.618)+0.0124i\omega} \tag{5.99}$$

$$\beta_{1,1} = \sum_{j=1}^{2}\phi_{11}\phi_{1j}M_j = 200 \qquad \beta_{2,1} = \sum_{j=1}^{2}\phi_{21}\phi_{2j}M_j = 161.8 \tag{5.100}$$

Introducing these quantities into Equation 5.95 we have

$$S_{X_1}(\omega) = S_0\{40\,000|H_1(\omega)|^2 + 26\,179.2|H_2(\omega)|^2 + 64\,720\,\mathrm{Re}[H_1(\omega)H_2^*(\omega)]\} \quad (5.101)$$

The three terms in the above equation represent the effect of the first mode, the effect of the second mode and the cross effect respectively.

The half part of $S_{X_r}(\omega)$ as $\omega \geq 0$ is plotted in Figure 5.6.

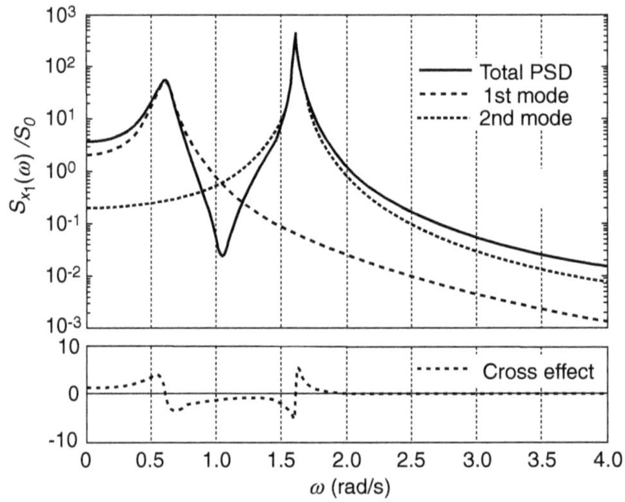

Figure 5.6 The PSD of the response of a two-DOF system.

From Figure 5.6 we see that, in the vicinity of the natural frequencies, the effect of the corresponding mode dominates the PSD of the response. While the cross effect is much smaller than the effect of the modes itself, there are points near the natural frequencies at which the cross effect is zero; the sign of the cross effect changes regularly in different intervals cut by these zero-cross-effect frequencies. Therefore, if a structure is subjected to a white noise on the base, then the natural frequencies can be easily identified and the corresponding damping ratio can also be identified by the half-power method. This is why in the shaking table tests the case of base white-noise scanning should be tested. □

5.3.2 *Evolutionary Spectral Analysis*

5.3.2.1 Evolutionary Stochastic Process

The correlation function and the PSD function of a weakly stationary stochastic process are Fourier transform pairs known as the Wiener–Khintchine formulae. However, for a nonstationary stochastic process, the traditional PSD does not exist. As mentioned in Section 2.2.3, a double-frequency PSD function and a pair of extended Wiener–Khintchine formulae exist for some kinds of nonstationary processes (see Equation 2.90). But this does not hold for all stochastic processes. In particular, it does not exist for stationary processes. In addition, the physical sense of such formulae is not as clear as the traditional PSD function.

A possible approach to modeling a nonstationary stochastic process is to regard it as a filtered stochastic process. This concept was first proposed by Priestley (1965, 1967) and then studied by a number of investigators (Liu, 1970; Fan and Ahmadi, 1990; Lin and Cai, 1995; Fang and Sun, 1997; Fang *et al.*, 2002). Let $\bar{\xi}(t)$ be a stationary stochastic process with correlation function

$$R_{\bar{\xi}}(t_1, t_2) = \mathcal{E}\left[\bar{\xi}(t_1)\bar{\xi}(t_2)\right] = \frac{1}{2\pi}\int_{-\infty}^{\infty} S_{\bar{\xi}}(\omega)\mathrm{e}^{\mathrm{i}\omega(t_2 - t_1)}\,\mathrm{d}\omega \tag{5.102}$$

Consider a linear time-dependent filter A with $\bar{\xi}(t)$ as an input. The output of the filter, $\xi(t)$, will be a nonstationary stochastic process and can be obtained by

$$\xi(\varpi, t) = \int_{-\infty}^{\infty} a(t, \tau)\bar{\xi}(\varpi, t - \tau)\,\mathrm{d}\tau \tag{5.103}$$

where $a(t, \tau)$ represents the impulse response function of the filter, namely the output of the filter A at the time instant t due to a unit impulse input at the time instant $t - \tau$. For a time-invariant system, this is $h(t - \tau)$ as given in Equation 5.11. Obviously, Equation 5.103 is an extension of the Duhamel integral in Equation 5.14.

Analogous to the treatment in Section 5.2.1.2, the correlation function of $\xi(t)$ is then

$$\begin{aligned}
R_{\xi}(t_1, t_2) &= \mathcal{E}[\xi(t_1)\xi(t_2)] \\
&= \int_{-\infty}^{\infty}\int_{-\infty}^{\infty} a(t_1, \tau_1)a(t_2, \tau_2)\mathcal{E}[\bar{\xi}(\varpi, t_1 - \tau_1)\bar{\xi}(\varpi, t_2 - \tau_2)]\,\mathrm{d}\tau_1\,\mathrm{d}\tau_2 \\
&= \frac{1}{2\pi}\int_{-\infty}^{\infty}\int_{-\infty}^{\infty}\int_{-\infty}^{\infty} a(t_1, \tau_1)a(t_2, \tau_2)S_{\bar{\xi}}(\omega)\mathrm{e}^{\mathrm{i}\omega(t_2 - \tau_2 - t_1 + \tau_1)}\,\mathrm{d}\tau_1\,\mathrm{d}\tau_2\,\mathrm{d}\omega \\
&= \frac{1}{2\pi}\int_{-\infty}^{\infty}\left[\int_{-\infty}^{\infty} a(t_1, \tau_1)\mathrm{e}^{\mathrm{i}\omega\tau_1}\,\mathrm{d}\tau_1\right]\left[\int_{-\infty}^{\infty} a(t_2, \tau_2)\mathrm{e}^{-\mathrm{i}\omega\tau_2}\,\mathrm{d}\tau_2\right]S_{\bar{\xi}}(\omega)\mathrm{e}^{\mathrm{i}\omega(t_2 - t_1)}\,\mathrm{d}\omega \\
&= \frac{1}{2\pi}\int_{-\infty}^{\infty} A(\omega, t_1)A^*(\omega, t_2)S_{\bar{\xi}}(\omega)\mathrm{e}^{\mathrm{i}\omega(t_2 - t_1)}\,\mathrm{d}\omega
\end{aligned} \tag{5.104}$$

in which

$$A(\omega, t) = \int_{-\infty}^{\infty} a(t, \tau)\mathrm{e}^{\mathrm{i}\omega\tau}\,\mathrm{d}\tau \tag{5.105}$$

We can then get the variance of the response:

$$\mathcal{E}\left[\xi^2(t)\right] = \frac{1}{2\pi}\int_{-\infty}^{\infty} |A(\omega, t)|^2 S_{\bar{\xi}}(\omega)\,\mathrm{d}\omega \tag{5.106}$$

Certainly, Equation 5.105 is an alternative form of Equation 5.57a when $h(t, \tau)$ and $\tilde{H}(\omega, t)$ are replaced by $a(t, \tau)$ and $A^*(\omega, t)$ respectively. If the filter A is time invariant, then Equation 5.105 reduces to Equation 5.57b, with $A^*(\omega, t)$ being essentially the frequency-response function $H(\omega)$. In this case, Equation 5.106 in turn reduces to Equation 5.69b. Considering this analogy, we define the evolutionary PSD by

$$S_{\xi}(\omega, t) = |A(\omega, t)|^2 S_{\bar{\xi}}(\omega) \tag{5.107}$$

Therefore:

$$\mathcal{E}\left[\xi^2(t)\right] = \frac{1}{2\pi}\int_{-\infty}^{\infty} S_{\xi}(\omega, t)\,\mathrm{d}\omega \tag{5.108}$$

This is an extension of Equation 5.69b when the process is an evolutionary stochastic process. We can see that, if $A(\omega, t) = 1$, the stochastic process is not modulated and, therefore, $\xi(t)$ is just the stationary stochastic process $\bar{\xi}(t)$. Besides the conceptual simplicity, in an evolutionary stochastic process, $A(\omega, t)$ modulates the intensity and frequency content simultaneously, which is phenomenologically the case of many stochastic processes of engineering interest, such as the ground motion of an earthquake (Liu, 1970; Fan and Ahmadi, 1990).

5.3.2.2 Evolutionary Spectral Analysis of SDOF Systems

Consider an SDOF system, when the excitation $\xi(t)$ is an evolutionary stochastic process with the evolutionary PSD in Equation 5.107:

$$m\ddot{X} + c\dot{X} + kX = \xi(t) \tag{5.109}$$

According to Equation 5.14 we have the response

$$X(\varpi, t) = \int_{-\infty}^{\infty} h(t-\tau)\xi(\varpi, \tau)\,\mathrm{d}\tau \tag{5.110}$$

The correlation function, therefore, is given by

$$\begin{aligned} R_X(t_1, t_2) &= \mathcal{E}[X(t_1)X(t_2)] \\ &= \textstyle\int_{-\infty}^{\infty}\int_{-\infty}^{\infty} h(t_1-\tau_1)h(t_2-\tau_2)\mathcal{E}[\xi(\varpi,\tau_1)\xi(\varpi,\tau_2)]\,\mathrm{d}\tau_1\,\mathrm{d}\tau_2 \end{aligned} \tag{5.111a}$$

Substituting Equation 5.104 in this yields

$$\begin{aligned} R_X(t_1,t_2) &= \textstyle\int_{-\infty}^{\infty}\int_{-\infty}^{\infty} h(t_1-\tau_1)h(t_2-\tau_2)\left(\frac{1}{2\pi}\int_{-\infty}^{\infty} A(\omega,\tau_1)A^*(\omega,\tau_2)S_{\bar{\xi}}(\omega)\mathrm{e}^{\mathrm{i}\omega(\tau_2-\tau_1)}\mathrm{d}\omega\right)\mathrm{d}\tau_1\,\mathrm{d}\tau_2 \\ &= \frac{1}{2\pi}\int_{-\infty}^{\infty}\left[\int_{-\infty}^{\infty} A(\omega,\tau_1)h(t_1-\tau_1)\mathrm{e}^{-\mathrm{i}\omega\tau_1}\mathrm{d}\tau_1\right]\left[\int_{-\infty}^{\infty} A^*(\omega,\tau_2)h(t_2-\tau_2)\mathrm{e}^{\mathrm{i}\omega\tau_2}\mathrm{d}\tau_2\right]S_{\bar{\xi}}(\omega)\,\mathrm{d}\omega \\ &= \frac{1}{2\pi}\int_{-\infty}^{\infty} H(\omega,t_1)H^*(\omega,t_2)S_{\bar{\xi}}(\omega)\,\mathrm{d}\omega \end{aligned} \tag{5.111b}$$

where we define

$$H(\omega,t) = \int_{-\infty}^{\infty} A(\omega,\tau)h(t-\tau)\mathrm{e}^{-\mathrm{i}\omega\tau}\,\mathrm{d}\tau \tag{5.112}$$

which is referred to as the evolutionary frequency-response function. Then we can get the variance of the response:

$$\mathcal{E}\left[X^2(t)\right] = \frac{1}{2\pi}\int_{-\infty}^{\infty} |H(\omega,t)|^2 S_{\bar{\xi}}(\omega)\,\mathrm{d}\omega \tag{5.113}$$

Therefore, the evolutionary PSD of the response is given by

$$S_X(\omega, t) = |H(\omega, t)|^2 S_{\bar{\xi}}(\omega) \tag{5.114}$$

This is obviously the extension of Equation 5.66. Clearly, if the excitation itself is a stationary process – that is, $A(\omega, t) = 1$ – then Equation 5.114 reduces to Equation 5.66, since Equation 5.112 reduces to Equation 5.57b.

Equation 5.114 shows that, when subjected to an evolutionary stochastic excitation, the response of the system is also an evolutionary stochastic process with time-variant intensity and simultaneously time-variant frequency content.

5.3.2.3 Evolutionary Spectral Analysis of MDOF Systems

The above principle can be extended to the case of the MDOF system:

$$\mathbf{M}\ddot{\mathbf{X}} + \mathbf{C}\dot{\mathbf{X}} + \mathbf{K}\mathbf{X} = \mathbf{B}\boldsymbol{\xi}(\varpi, t) \tag{5.115}$$

where $\boldsymbol{\xi}(\varpi, t)$ is an evolutionary stochastic process vector with the evolutionary PSD matrix

$$\mathbf{S}_{\boldsymbol{\xi}}(\omega, t) = \mathbf{A}(\omega, t)\mathbf{S}_{\bar{\boldsymbol{\xi}}}(\omega)\mathbf{A}^*(\varpi, t) \tag{5.116}$$

which is the matrix counterpart of Equation 5.107. Here, $\mathbf{A}$ is the modulating matrix and $\mathbf{A}^*$ is the transpose of the complex conjugate of $\mathbf{A}$.

Using Equation 5.32a we get

$$\mathbf{X}(t) = \int_0^t \mathbf{h}(t-\tau)\mathbf{B}\boldsymbol{\xi}(\tau)\,\mathrm{d}\tau = \int_{-\infty}^{\infty} \mathbf{h}(t-\tau)\mathbf{B}\boldsymbol{\xi}(\tau)\,\mathrm{d}\tau$$

and the similar idea in Equations 5.111a and 5.111b, the covariance function matrix is given by

$$\begin{aligned}
\mathbf{R}_{\mathbf{X}}(t_1, t_2) &= \mathcal{E}[\mathbf{X}(t_1)\mathbf{X}^{\mathrm{T}}(t_2)] \\
&= \int_{-\infty}^{\infty}\int_{-\infty}^{\infty} \mathbf{h}(t_1-\tau_1)\mathbf{B}\mathcal{E}[\boldsymbol{\xi}(\varpi,\tau_1)\xi^*(\varpi,\tau_2)]\mathbf{B}^{\mathrm{T}}\mathbf{h}^{\mathrm{T}}(t_2-\tau_2)\,\mathrm{d}\tau_1\,\mathrm{d}\tau_2 \\
&= \int_{-\infty}^{\infty}\int_{-\infty}^{\infty} \mathbf{h}(t_1-\tau_1)\mathbf{B}\left[\frac{1}{2\pi}\int_{-\infty}^{\infty} \mathbf{A}(\omega,\tau_1)\mathbf{S}_{\bar{\boldsymbol{\xi}}}(\omega)\mathrm{e}^{\mathrm{i}\omega(\tau_2-\tau_1)}\mathbf{A}^*(\omega,\tau_2)\,\mathrm{d}\omega\right]\mathbf{B}^{\mathrm{T}}\mathbf{h}^{\mathrm{T}}(t_2-\tau_2)\,\mathrm{d}\tau_1\,\mathrm{d}\tau_2 \\
&= \frac{1}{2\pi}\int_{-\infty}^{\infty}\left[\int_{-\infty}^{\infty} \mathbf{h}(t_1-\tau_1)\mathbf{B}\mathbf{A}(\omega,\tau_1)\mathrm{e}^{-\mathrm{i}\omega\tau_1}\,\mathrm{d}\tau_1\right]\mathbf{S}_{\boldsymbol{\xi}}(\omega)\left[\int_{-\infty}^{\infty} \mathbf{A}^*(\omega,\tau_2)\mathbf{B}^{\mathrm{T}}\mathbf{h}^{\mathrm{T}}(t_2-\tau_2)\mathrm{e}^{\mathrm{i}\omega\tau_2}\,\mathrm{d}\tau_2\right]\mathrm{d}\omega \\
&= \frac{1}{2\pi}\int_{-\infty}^{\infty} \mathbf{H}(\omega,t_1)\mathbf{S}_{\bar{\boldsymbol{\xi}}}(\omega)\mathbf{H}^*(\omega,t_2)\,\mathrm{d}\omega
\end{aligned} \tag{5.117}$$

where

$$\mathbf{H}(\omega,t)=\int_{-\infty}^{\infty}\mathbf{h}(t-\tau)\mathbf{B}\mathbf{A}(\omega,\tau)\mathrm{e}^{-\mathrm{i}\omega\tau}\,\mathrm{d}\tau \tag{5.118}$$

Therefore, we can get the covariance function matrix by

$$\mathcal{E}\left[\mathbf{X}(t)\mathbf{X}^{\mathrm{T}}(t)\right]=\frac{1}{2\pi}\int_{-\infty}^{\infty}\mathbf{H}(\omega,t)\mathbf{S}_{\xi}(\omega)\mathbf{H}^{*}(\omega,t)\,\mathrm{d}\omega \tag{5.119}$$

and the evolutionary PSD matrix by

$$\mathbf{S}_{\mathbf{X}}(\omega,t)=\mathbf{H}(\omega,t)\mathbf{S}_{\xi}(\omega)\mathbf{H}^{*}(\omega,t) \tag{5.120}$$

such that

$$\mathcal{E}\left[\mathbf{X}(t)\mathbf{X}^{\mathrm{T}}(t)\right]=\frac{1}{2\pi}\int_{-\infty}^{\infty}\mathbf{S}_{\mathbf{X}}(\omega,t)\,\mathrm{d}\omega \tag{5.121}$$

When $\mathbf{A}(\omega, t)$ is a unit matrix of appropriate order, Equation 5.120 will coincide with Equation 5.84.

5.3.2.4 Physical Interpretation of Evolutionary Spectral Analysis

As discussed in Section 5.3.1.2, we know that, physically, the frequency-response function $H(\omega)$ reflects the embedded properties of the structural system itself. However, the evolutionary frequency-response function $H(\omega, t)$ as defined in Equation 5.112 (or in matrix form in Equation 5.118) does not characterize the properties of the structure itself. In fact, it also includes the properties of the evolutionary stochastic excitation, since the modulation function $A(\omega, t)$ is involved.

Viewing Equation 5.112, we find that the evolutionary frequency-response function $H(\omega, t)$ is a deterministic response if the excitation in the system in Equation 5.109 is replaced by the deterministic process $A(\omega, t)\mathrm{e}^{-\mathrm{i}\omega t}$. From this point of view, the physical meaning of the modulation function $A(\omega, t)$ is clearer: it is just a filter of excitation. In other words, $A(\omega, t)$ modulates the intensity (amplitude) because $A(\omega, t)$ is time variant; simultaneously, it also modulates the frequency content because $A(\omega, t)$ is variant in terms of ω. In fact, with this understanding in mind we can develop a set of deterministic algorithms for the PSD analysis of random vibration (Fang and Sun, 1997). This is just what the pseudo-excitation method does, and will be elaborated in the following sections.

5.4 Pseudo-Excitation Method

According to the physical interpretation in Section 5.3.2.4, we have approached the position that a set of possible algorithms may be feasible for the PSD analysis of random vibration. The thoughts, referred to as the pseudo-excitation method, were first proposed by Lin in 1985, and since then the method has been systematically developed by him and his co-workers (Lin *et al.*, 1994a, 1994b, 1997; Zhong, 2004).

5.4.1 Pseudo-Excitation Method for Stationary Stochastic Response Analysis

Let us revisit the derivation in Section 5.3.1. If an SDOF system is excited by a deterministic unit harmonic excitation $e^{i\omega t}$, then the deterministic steady-state response reads

$$x(\omega, t) = H(\omega)e^{i\omega t} \tag{5.122}$$

Here, we explicitly write the arguments ω, t in $x(\cdot)$ to show the dependency of the response on the frequency of the excitation and the time.

A general deterministic, absolute integrable time history can be expressed by the superposition of harmonic components, as shown in Equation 5.3b by the inverse Fourier transform

$$\xi(t) = \frac{1}{2\pi}\int_{-\infty}^{\infty} \xi(\omega)e^{i\omega t}\,d\omega \tag{5.3b}$$

where $\xi(\omega)\,d\omega/(2\pi)$ is the amplitude of the harmonic component of frequency ω. When $\xi(t)$ is a stationary stochastic process, however, the inverse Fourier transform in Equation 5.3 does not exist. Nonetheless, the following relationship does hold (see Equation 2.89a):

$$S_\xi(\omega) = \lim_{T\to\infty}\frac{1}{2T}\mathcal{E}[\xi(\varpi,\omega)\xi^*(\varpi,\omega)] \tag{5.65b}$$

which shows that the PSD, which reflects the frequency contents of the stationary stochastic process, is in the dimension of the square of the amplitude spectrum of the sample (divided by time). Thus, it is reasonable to consider a harmonic excitation with an amplitude of $\hat{\xi} = \sqrt{S_\xi(\omega)}$, namely $\hat{\xi}e^{i\omega t} = \sqrt{S_\xi(\omega)}e^{i\omega t}$. If we use this excitation, then the steady-state response of the system will be

$$x(\omega, t) = H(\omega)\sqrt{S_\xi(\omega)}e^{i\omega t} \tag{5.123}$$

Because the excitation contains a relationship in terms of the frequency contents of the excitation – that is, the excitation $\sqrt{S_\xi(\omega)}e^{i\omega t}$ is related to the square root of the PSD, and the excitation multiplied by its complex conjugate $\sqrt{S_\xi(\omega)}e^{-i\omega t}$ will yield exactly the PSD of the excitation – we expect that the response, Equation 5.123, may be intimately related to the frequency contents of the response. This is actually the case, since when we multiply on both sides of Equation 5.123 by its complex conjugate

$$x^*(\omega, t) = H^*(\omega)\sqrt{S_\xi(\omega)}e^{-i\omega t} \tag{5.124}$$

we immediately have

$$x(\omega, t)x^*(\omega, t) = |H(\omega)|^2 S_\xi(\omega) \tag{5.125}$$

Comparing with Equation 5.66, we find that

$$S_X(\omega) = x(\omega, t)x^*(\omega, t) \tag{5.126}$$

This is an elegant formula, showing that the PSD of the response can be obtained by multiplying the deterministic response of the system to a harmonic excitation by its complex conjugate, while the excitation has an amplitude which is identical to the square root of the PSD of the stochastic excitations. Noting that the deterministic excitation used here is, in a sense, a pseudo-excitation rather than an actual excitation, the above approach is called the *pseudo-excitation method* (Lin, 1985).

The physical meaning of the pseudo-excitation is embedded in the fact that a relationship exists in the frequency contents of the sample and the ensemble properties, namely the PSD function, as shown in Equation 5.65b. Therefore, the pseudo-excitation method is essentially the embedment of the relationship between the sample and the ensemble properties.

The great advantage of the pseudo-excitation method is exhibited in random vibration analysis of MDOF systems. Actually, the preceding discussions can be extended to their matrix counterpart when an MDOF system is considered.

The PSD matrix of the stationary stochastic excitation has a relationship with the sample Fourier spectrum:

$$\mathbf{S}_{\boldsymbol{\xi}}(\omega) = \lim_{T\to\infty} \frac{1}{2T} \mathcal{E}[\boldsymbol{\xi}(\varpi, \omega)\boldsymbol{\xi}^*(\varpi, \omega)] \tag{5.127}$$

Therefore, we can expect the frequency contents of the excitation can be characterized by a set of excitation $\hat{\boldsymbol{\xi}}(\omega)e^{i\omega t}$ of which the $r \times r$ amplitude matrix $\hat{\boldsymbol{\xi}}(\omega)$ satisfies

$$\hat{\boldsymbol{\xi}}(\omega)\hat{\boldsymbol{\xi}}^*(\omega) = \mathbf{S}_{\boldsymbol{\xi}}(\omega) \tag{5.128}$$

Replacing the excitation in Equation 5.79 by $\hat{\boldsymbol{\xi}}(\omega)e^{i\omega t}$ yields

$$\mathbf{M}\ddot{\mathbf{x}} + \mathbf{C}\dot{\mathbf{x}} + \mathbf{K}\mathbf{x} = \mathbf{B}\hat{\boldsymbol{\xi}}(\omega)e^{i\omega t} \tag{5.129}$$

Clealy, the $n \times r$ steady-state response matrix is given by (see Equation 5.81a)

$$\mathbf{x}(\omega, t) = \mathbf{H}(\omega)\mathbf{B}\hat{\boldsymbol{\xi}}(\omega)e^{i\omega t} \tag{5.130}$$

Post-multiplying it by its complex conjugate yields

$$\begin{aligned}
\mathbf{x}(\omega, t)\mathbf{x}^*(\omega, t) &= [\mathbf{H}(\omega)\mathbf{B}\hat{\boldsymbol{\xi}}(\omega)e^{i\omega t}][\mathbf{H}(\omega)\mathbf{B}\hat{\boldsymbol{\xi}}(\omega)e^{i\omega t}]^* \\
&= \mathbf{H}(\omega)\mathbf{B}\left[\hat{\boldsymbol{\xi}}(\omega)\hat{\boldsymbol{\xi}}^*(\omega)\right]\mathbf{B}^{\mathrm{T}}\mathbf{H}^*(\omega) \\
&= \mathbf{H}(\omega)\mathbf{B}\mathbf{S}_{\boldsymbol{\xi}}(\omega)\mathbf{B}^{\mathrm{T}}\mathbf{H}^*(\omega)
\end{aligned} \tag{5.131}$$

Here, use has been made of Equation 5.128.

Comparing with Equation 5.84, we immediately find that

$$\mathbf{S}_{\mathbf{X}}(\omega) = \mathbf{x}(\omega, t)\mathbf{x}^*(\omega, t) \tag{5.132}$$

Again, we see that the PSD matrix of the response of an MDOF system can be obtained by a set of deterministic dynamical response analyses.

If the modal superposition method is used, noting Equation 5.90 and assuming q modes are employed, then Equation 5.131 becomes

$$\begin{aligned}
\mathbf{S}_{\mathbf{X}}(\omega) &= \mathbf{x}(\omega, t) \cdot \mathbf{x}^*(\omega, t) \\
&= \left[\sum_{j=1}^{q} H_j(\omega)\boldsymbol{\psi}_j\boldsymbol{\psi}_j^{\mathrm{T}}\mathbf{B}\hat{\boldsymbol{\xi}}(\omega)e^{i\omega t}\right]\left[\sum_{j=1}^{q} H_j(\omega)\boldsymbol{\psi}_j\boldsymbol{\psi}_j^{\mathrm{T}}\mathbf{B}\hat{\boldsymbol{\xi}}(\omega)e^{i\omega t}\right]^*
\end{aligned} \tag{5.133}$$

As is known, the PSD matrix is usually evaluated by employing Equation 5.91 (for example, see Lutes and Sarkani (2004)). It is seen that Equation 5.133 is essentially a CQC scheme, but, according to the discussion in Section 5.2.2.3, the computational effort in Equation 5.132 is greatly reduced in comparison with Equation 5.91. Hence, the pseudo-excitation method can greatly improve the efficiency of PSD analysis and makes possible the random vibration analysis of large, complex structural systems.

In addition, modal superposition is not essential in the deterministic analysis of the system in Equation 5.129. General time integration methods can also be applied. Thus, the proportional damping matrix, which is necessary for the real-mode method, is not necessary in the time integration method.

5.4.2 Pseudo-Excitation Method for Evolutionary Stochastic Response Analysis

In the case that the excitations $\boldsymbol{\xi}(t)$ are an evolutionary stochastic process vector with the modulation function matrix $\mathbf{A}(\omega, t)$ and the original stationary stochastic process $\bar{\boldsymbol{\xi}}(t)$ whose PSD matrix is $\mathbf{S}_{\bar{\xi}}(\omega)$, the pseudo-excitation corresponding to $\bar{\boldsymbol{\xi}}(t)$ is similarly chosen according to the principle discussed in the preceding section as $\hat{\boldsymbol{\xi}}(\omega)e^{i\omega t}$ of which the $r \times r$ amplitude matrix satisfies

$$\hat{\boldsymbol{\xi}}(\omega)\hat{\boldsymbol{\xi}}^*(\omega) = \mathbf{S}_{\bar{\xi}}(\omega) \tag{5.134}$$

Then it is modulated by $\mathbf{A}(\omega, t)$ such that the pseudo-excitation could be given by $\mathbf{A}(\omega, t)_{\hat{\xi}}e^{i\omega t}$. The response of the system

$$\mathbf{M}\ddot{\mathbf{x}} + \mathbf{C}\dot{\mathbf{x}} + \mathbf{K}\mathbf{x} = \mathbf{B}\mathbf{A}(\omega, t)\hat{\boldsymbol{\xi}}(\omega)e^{i\omega t} \tag{5.135}$$

is thus

$$\mathbf{x}(\omega, t) = \int_{-\infty}^{\infty} \mathbf{h}(t-\tau)\mathbf{B}\mathbf{A}(\omega, \tau)\hat{\boldsymbol{\xi}}(\omega)e^{i\omega\tau}\,d\tau = \mathbf{H}(\omega, t)\hat{\boldsymbol{\xi}}(\omega) \tag{5.136}$$

Multiplying by its complex conjugate on both sides and noting Equation 5.134 yields

$$\mathbf{x}(\omega, t)\mathbf{x}^*(\omega, t) = \mathbf{H}(\omega, t)\hat{\boldsymbol{\xi}}(\omega)\hat{\boldsymbol{\xi}}^*(\omega)\mathbf{H}^*(\omega, t) = \mathbf{H}(\omega, t)\mathbf{S}_{\bar{\xi}}(\omega)\mathbf{H}^*(\omega, t) \tag{5.137}$$

Comparing this with Equation 5.120 immediately yields

$$\mathbf{S}_{\mathbf{X}}(\omega, t) = \mathbf{x}(\omega, t)\mathbf{x}^*(\omega, t) \tag{5.138}$$

In Equations 5.136 and 5.137, use has been made of Equations 5.118 and 5.128 respectively.

We can see here that the evolutionary PSD matrix can also be obtained by a set of deterministic analyses with the deterministic excitation related to the evolutionary PSD matrix of the excitations. Recalling the discussion in Section 5.3.2.4, we can further understand how the modulation function modulates the intensity and the frequency contents of the excitations simultaneously. In the pseudo-excitation method, this modulation is embedded in modulating the deterministic harmonic excitations.

5.4.3 Notes on Sections 5.2–5.4

We note that the main results in Sections 5.2–5.4 are derived starting with a formal solution expression of the response to the excitation. In other words, we start with the linear operator bridging the functional of the excitation as a time history or Fourier spectrum and the functional of the response as a time history or Fourier spectrum, to establish the linear operator bridging the functional of probabilistic characteristics of the excitations and the functional of probabilistic characteristics of the responses. In the derivations, the physical solution is used considering the system behaviors. However, the description of the stochastic processes is phenomenological.

By the way, all previous treatments are based on describing the physical relationship by second-order differential equations. An alternative family of corresponding results can also be derived by transferring the second-order differential equations to the corresponding state equations (e.g. in Åström (1970)). Actually, in the latter description, there are some special advantages. However, this will not be elaborated in the present section, considering the aim of the book.

Besides the approaches starting with the formal solutions as elaborated in the preceding sections, we can also commence by directly tackling the random differential equations with a stochastic nonhomogeneous input to obtain the deterministic differential equations establishing the relationship between the moments of the responses and the moments of the excitations (e.g. refer to Lutes and Sarkani (2004)).

5.5 Statistical Linearization

As discussed earlier, in the analysis of linear systems, the physical solution can be obtained and used as a basis to track the propagation of the moment characteristics from the source of the randomness to the response. This, however, does not work for most nonlinear systems because the formal solutions to nonlinear systems are unavailable except some special simple cases (Nayfeh and Mook, 1995). In the analysis of deterministic nonlinear systems, one of the effective methods, mainly suitable for lightly nonlinear systems, is the perturbation method, which is first proposed by Poincare and has been extensively studied (Nayfeh, 2000). The counterpart in random analysis of nonlinear systems has also been studied by investigators (Lin, 1967; Skorokhod *et al.*, 2002) and some basic ideas have been treated in Chapter 4. On the other hand, an alternative approach is the statistical linearization method, also referred to as stochastic linearization or equivalent linearization in some of the literature (Lin, 1967; Roberts and Spanos, 1993; Crandall, 2006). This technique was first proposed independently almost simultaneously by Booton (1954), Kazakov (1954) and Caughey (1963).

5.5.1 Statistical Linearization Approximation

5.5.1.1 Nonlinear SDOF Systems

Consider the nonlinear SDOF system

$$m\ddot{X} + g(X, \dot{X}) = \xi(t) \tag{5.139}$$

where m is the mass, $g(X, \dot{X})$ includes the nonlinear damping and restoring forces and $\xi(t)$ is the stochastic excitation. The basic thought behind a statistical linearization approximation is to replace Equation 5.139 by an equivalent linear system:

$$m\ddot{Y} + c_{eq}\dot{Y} + k_{eq}Y = \xi(t) \tag{5.140}$$

where c_{eq} and k_{eq} are the equivalent damping and stiffness respectively, such that the error between the solutions of the two systems is minimized, say in the sense of the mean-square.

Comparing Equations 5.139 and 5.140, the difference is

$$\bar{e} = m\ddot{X} + g(X, \dot{X}) - (m\ddot{Y} + c_{eq}\dot{Y} + k_{eq}Y) \tag{5.141}$$

Strictly speaking, if we approximate the solution to Equation 5.139 by that of Equation 5.140, then the error should be defined by Equation 5.141 when Y is replaced by X. However, the response X of the nonlinear system is unknown; therefore, the error so defined will be intractable. In contrast, to obtain the equivalent response Y in Equation 5.140 is much easier. Therefore, Equation 5.140 is solved instead of Equation 5.139. The error can then be defined by Equation 5.141 when X is replaced by Y; namely:

$$e = g(Y, \dot{Y}) - c_{eq}\dot{Y} - k_{eq}Y \tag{5.142}$$

To choose the equivalent damping c_{eq} and the equivalent stiffness k_{eq} optimally, we should minimize the error in a statistical sense. Actually, the error defined in Equation 5.142 is a stochastic process, we naturally expect that $\mathcal{E}[e] = 0$, and the second-order moment, namely the mean-square error

$$\mathcal{E}[e^2] = \mathcal{E}\{[g(Y, \dot{Y}) - c_{eq}\dot{Y} - k_{eq}Y]^2\} \tag{5.143}$$

is minimized. This requires

$$\frac{\partial \mathcal{E}[e^2]}{\partial c_{eq}} = 0 \tag{5.144a}$$

and

$$\frac{\partial \mathcal{E}[e^2]}{\partial k_{eq}} = 0 \tag{5.144b}$$

Equations (5.144a) and (5.144b) yield two linear equations and, therefore, give the optimal values of c_{eq} and k_{eq} by the solution:

$$c_{eq} = \frac{\mathcal{E}[g(Y, \dot{Y})\dot{Y}]\mathcal{E}[Y^2] - \mathcal{E}[g(Y, \dot{Y})Y]\mathcal{E}[Y\dot{Y}]}{\mathcal{E}[\dot{Y}^2]\mathcal{E}[Y^2] - \mathcal{E}^2[Y\dot{Y}]} \tag{5.145a}$$

$$k_{eq} = \frac{\mathcal{E}[g(Y, \dot{Y})Y]\mathcal{E}[\dot{Y}^2] - \mathcal{E}[g(Y, \dot{Y})\dot{Y}]\mathcal{E}[Y\dot{Y}]}{\mathcal{E}[\dot{Y}^2]\mathcal{E}[Y^2] - \mathcal{E}^2[Y\dot{Y}]} \tag{5.145b}$$

Interestingly, if c_{eq} is replaced by k_{eq}, and simultaneously $\dot{Y}$ is replaced by Y and in turn Y is replaced by $\dot{Y}$, then Equation 5.145a becomes Equation 5.145b, and vice versa. The symmetry exhibited here lies in the fact that the terms $c_{\text{eq}}\dot{Y}$ and $k_{\text{eq}}Y$ are commutable in Equation 5.143.

We see in Equations 5.145a and 5.145b that, to obtain the optimal values of c_{eq} and k_{eq}, the joint PDF of Y and $\dot{Y}$ is needed; in turn, to solve the linear random vibration system in Equation 5.140, the values of c_{eq} and k_{eq} are needed. A circular interdependency is then formed here; therefore, an iterative algorithm is needed to break the loop. Generally, the solving flow starts with the estimated initial values of c_{eq} and k_{eq}. A cycle is illustrated in Figure 5.7, where the superscripts represent the steps of iteration and $\mathscr{P}(Y^{(j)}, \dot{Y}^{(j)})$ represents the probabilistic information (e.g. the joint statistics or PDF of Y and $\dot{Y}$ at step j). The iteration could come to an end if the error is limited within the tolerance, say by the error of c_{eq} and k_{eq}:

$$||c_{\text{eq}}^{(j)} - c_{\text{eq}}^{(j-1)}|| \leq \varepsilon_1 \qquad ||k_{\text{eq}}^{(j)} - k_{\text{eq}}^{(j-1)}|| \leq \varepsilon_2 \tag{5.146a}$$

or by the error of the probabilistic characteristics of Y and $\dot{Y}$:

$$||(\mathcal{E}[Y^2])^{(j)} - (\mathcal{E}[Y^2])^{(j-1)}|| \leq \varepsilon_1 \qquad ||(\mathcal{E}[\dot{Y}^2])^{(j)} - (\mathcal{E}[\dot{Y}^2])^{(j-1)}|| \leq \varepsilon_2 \tag{5.146b}$$

where ε_1 and ε_2 are the corresponding error tolerances.

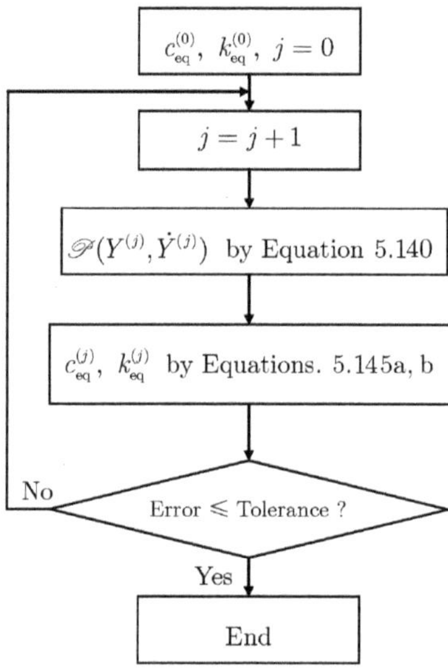

Figure 5.7 Solving flow.

We should note that such determined optimal values of c_{eq} and k_{eq} are time variant if the response is a nonstationary process; therefore, the equivalent system in Equation 5.140 is a time-variant linear system. In the case of the steady-state, stationary stochastic response, considering that $\mathcal{E}[Y\dot{Y}] = 0$, Equations 5.145a and 5.145b can be simplified to

$$c_{\text{eq}} = \frac{\mathcal{E}[g(Y, \dot{Y})\dot{Y}]\mathcal{E}[Y^2]}{\mathcal{E}[\dot{Y}^2]\mathcal{E}[Y^2]} \qquad \text{and} \qquad k_{\text{eq}} = \frac{\mathcal{E}[g(Y, \dot{Y})Y]\mathcal{E}[\dot{Y}^2]}{\mathcal{E}[\dot{Y}^2]\mathcal{E}[Y^2]} \tag{5.147}$$

More specifically, if the damping term is linear and the nonlinearity only occurs in the restoring force – that is, $g(X,\dot{X}) = c\dot{X} + g_1(X)$ – then following Equation 5.147 we have $c_{\text{eq}} = c$.

5.5.1.2 Statistical Linearization Method for MDOF Systems

The same concept can be applied to MDOF nonlinear systems. Consider an MDOF system with the equation of motion

$$\mathbf{M}\ddot{\mathbf{X}} + \mathbf{G}(\mathbf{X},\dot{\mathbf{X}}) = \mathbf{L}\boldsymbol{\xi}(t) \tag{5.148}$$

where $\mathbf{M}$ is the $n \times n$ mass matrix, $\mathbf{G}(\cdot) = (G_1, G_2, \ldots, G_n)^{\mathrm{T}}$ includes the damping and restoring forces, $\mathbf{L} = [L_{ij}]_{m\times r}$ is the $n \times r$ input force influence matrix and $\boldsymbol{\xi}(t) = (\xi_1(t), \xi_2(t), \ldots, \xi_r(t))^{\mathrm{T}}$ is an r-dimensional stochastic process vector.

Assume the system in Equation 5.148 can be replaced by the linear MDOF system

$$\mathbf{M}\ddot{\mathbf{Y}} + \mathbf{C}_{\text{eq}}\dot{\mathbf{Y}} + \mathbf{K}_{\text{eq}}\mathbf{Y} = \mathbf{L}\boldsymbol{\xi}(t) \tag{5.149}$$

where $\mathbf{C}_{\text{eq}}$ and $\mathbf{K}_{\text{eq}}$ are the $n \times n$ damping and stiffness matrixes respectively. Analogous to Equation 5.142, the error vector can be defined by

$$\mathbf{e} = \mathbf{G}(\mathbf{Y},\dot{\mathbf{Y}}) - \mathbf{C}_{\text{eq}}\dot{\mathbf{Y}} - \mathbf{K}_{\text{eq}}\mathbf{Y} \tag{5.150}$$

The optimal values of $\mathbf{C}_{\text{eq}}$ and $\mathbf{K}_{\text{eq}}$ should minimize the covariance matrix of the error; therefore:

$$\frac{\partial \mathcal{E}[\mathbf{e}\mathbf{e}^{\mathrm{T}}]}{\partial \mathbf{C}_{\text{eq}}} = 0 \quad \text{and} \quad \frac{\partial \mathcal{E}[\mathbf{e}\mathbf{e}^{\mathrm{T}}]}{\partial \mathbf{K}_{\text{eq}}} = 0 \tag{5.151}$$

This gives the equations

$$\mathbf{C}_{\text{eq}}\mathcal{E}[\dot{\mathbf{Y}}\dot{\mathbf{Y}}^{\mathrm{T}}] + \mathbf{K}_{\text{eq}}\mathcal{E}[\mathbf{Y}\dot{\mathbf{Y}}^{\mathrm{T}}] = \mathcal{E}[\mathbf{G}(\mathbf{Y},\dot{\mathbf{Y}})\dot{\mathbf{Y}}^{\mathrm{T}}] \tag{5.152a}$$

$$\mathbf{C}_{\text{eq}}\mathcal{E}[\dot{\mathbf{Y}}\mathbf{Y}^{\mathrm{T}}] + \mathbf{K}_{\text{eq}}\mathcal{E}[\mathbf{Y}\mathbf{Y}^{\mathrm{T}}] = \mathcal{E}[\mathbf{G}(\mathbf{Y},\dot{\mathbf{Y}})\mathbf{Y}^{\mathrm{T}}] \tag{5.152b}$$

These equations can be solved to give $\mathbf{C}_{\text{eq}}$ and $\mathbf{K}_{\text{eq}}$ as long as the joint PDFs of the responses are known. Likewise, a loop is formed here. Hence, an iterative algorithm should be used to solve the problem. The procedure analogous to that shown in Figure 5.7 can be used as well.

5.5.2 Random Vibrations of Hysteretic Structures

The restoring forces of structures in practical engineering are usually quite complex. Experimental studies showed that the restoring-force curves may exhibit hysteresis, degradation of strength and stiffness, and pinching and so on. The differential equation model first proposed by Bouc (1967) and Wen (1976), and then extended by other investigators (Baber and Wen, 1981; Baber and Noori, 1985, 1986), can describe the above features phenomenologically.

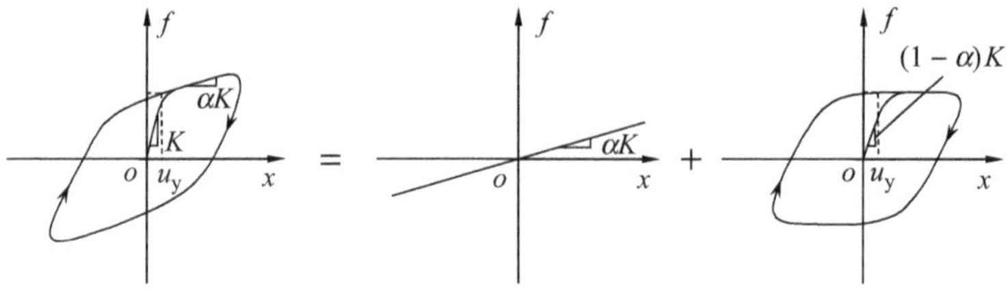

Figure 5.8 Partition of hysteretic restoring forces.

Consider the nonlinear SDOF system in Equation 5.139, where the damping is linear and the nonlinear restoring force is partitioned to the linear part and the hysteretic part; namely:

$$g(X,\dot{X}) = c\dot{X} + \alpha KX + (1-\alpha)KZ \tag{5.153}$$

where α is the ratio of the post-yield stiffness to the pre-yield stiffness (Figure 5.8).

When no effects of degradation and pinching are considered, the hysteretic component $Z(t)$ is governed by the differential equation

$$\dot{Z} = A\dot{X} - \beta|\dot{X}||Z|^{n-1}Z - \gamma\dot{X}|Z|^{n} \tag{5.154}$$

Without loss of generality, we can set $A = 1$.

If the degradation of strength and stiffness is taken into account (Figure 5.9), then this model can be extended to

$$\dot{Z} = \frac{A\dot{X} - \nu(\beta|\dot{X}||Z|^{n-1}Z + \gamma\dot{X}|Z|^{n})}{\eta} \tag{5.155}$$

in which ν and η are the factors characterizing the degradations of strength and stiffness respectively. It is seen that if $\nu = 1$ and $\eta = 1$, then Equation 5.155 reduces to Equation 5.154. In contrast, if $\nu > 1$, then the peak of $Z(t)$ will decrease and this exhibits the degradation of strength. Likewise, if $\eta > 1$, then the ratio of Z to X will decrease and this exhibits the degradation of stiffness. According to the preceding analysis, considering that the degradations of strength and stiffness will increase monotonically against increasing degree of nonlinearity,

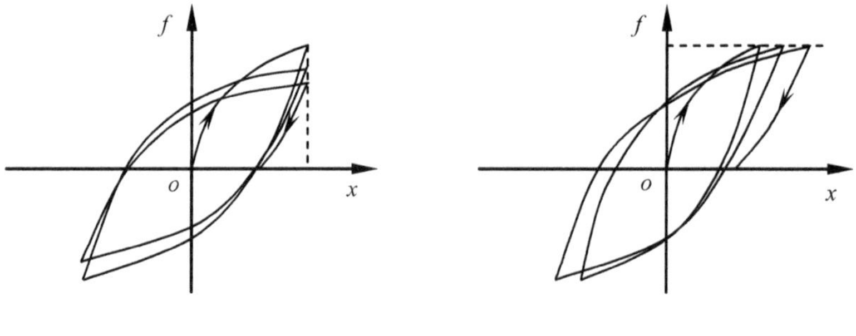

(a) Degradation of strength **(b) Degradation of stiffness**

Figure 5.9 Degradation of hysteresis.

it is reasonable to assume that degradations of the strength and the stiffness are proportional to the dissipated energy; namely:

$$\nu = 1 + d_\nu \varepsilon \qquad \text{and} \qquad \eta = 1 + d_\eta \varepsilon \tag{5.156}$$

where d_ν and d_η are parameters and ε is the dissipated energy index:

$$\varepsilon(t) = \int_0^t Z\dot{X}\,\mathrm{d}t \tag{5.157}$$

Evidently, the hysteretic dissipated energy in an element reads

$$E(t) = (1-\alpha)K\varepsilon(t) \tag{5.158}$$

Letting $\mathrm{d}Z/\mathrm{d}X = 0$, from Equation 5.155 we can get the ultimate hysteretic component

$$Z_u = \left[\frac{A}{\nu(\beta+\gamma)}\right]^{1/n} \tag{5.159}$$

Therefore, the ultimate restoring force is

$$R_\mathrm{u} = \alpha KX + (1-\alpha)KZ_\mathrm{u} \tag{5.160}$$

To take into account the effect of pinching further, Equation 5.155 can be modified by something like modulation to

$$\dot{Z} = h(Z)\frac{A\dot{X} - \nu(\beta|\dot{X}||Z|^{n-1}Z + \gamma\dot{X}|Z|^n)}{\eta} \tag{5.161}$$

If the appropriate shape is taken for $h(Z)$, then the effect of pinching will occur. For instance, we can use

$$h(Z) = 1.0 - \zeta_1 \exp\left[-\frac{Z\,\mathrm{sgn}(\dot{X}) - qZ_\mathrm{u}}{\zeta_2^2}\right] \tag{5.162}$$

where

$$\zeta_1(\varepsilon) = \zeta_\mathrm{s}(1 - e^{-p\varepsilon}) \qquad \text{and} \qquad \zeta_2(\varepsilon) = (\psi + d_\psi \varepsilon)(\lambda + \zeta_1(\varepsilon)) \tag{5.163}$$

in which ζ_s, p, q, ψ, d_ψ and λ are parameters.

In total there are 13 parameters involved. Actually, further investigation demonstrates that only 12 of the 13 parameters are independent (Ma *et al.*, 2004). If the parameters take appropriate values, then the model can phenomenologically characterize the effect of hysteresis, the degradations of stiffness and strength and the pinching. Typical hysteretic curves are shown in Figure 5.10.

Substituting Equation 5.153 in Equation 5.139 yields

$$m\ddot{X} + c\dot{X} + \alpha KX + (1-\alpha)KZ = \xi(t) \tag{5.164}$$

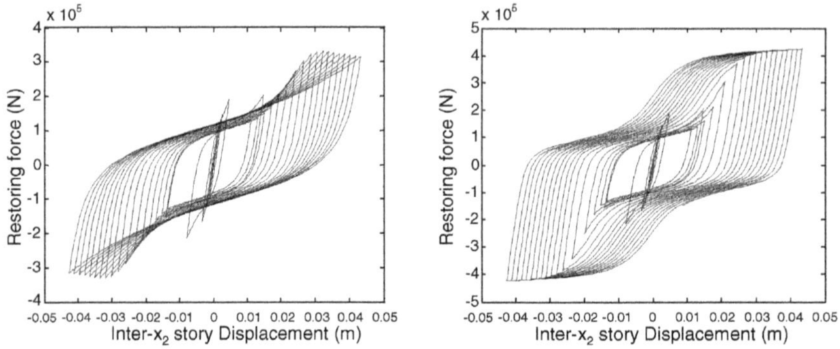

Figure 5.10 Typical hysteretic curves of the Bouc–Wen model.

Introducing an augmented state vector $(\dot{X}, X, Z)^{\mathrm{T}}$, we can change the equation set Equations 5.164 and 5.161 into a multidimensional state equation:

$$\frac{\mathrm{d}}{\mathrm{d}t}\begin{Bmatrix} \dot{X} \\ X \\ Z \end{Bmatrix} = \begin{Bmatrix} -\dfrac{c}{m}\dot{X} - \alpha\dfrac{K}{m}X - (1-\alpha)\dfrac{K}{m}Z \\ \dot{X} \\ h(Z)\dfrac{A\dot{X} - \nu(\beta|\dot{X}||Z|^{n-1}Z + \gamma\dot{X}|Z|^{n})}{\eta} \end{Bmatrix} + \begin{Bmatrix} \dfrac{1}{m} \\ 0 \\ 0 \end{Bmatrix}\xi(t) \tag{5.165}$$

Thus, by letting $\mathbf{X} = (\dot{X}, X, Z)^{\mathrm{T}}$ and $\mathbf{L} = (1/m, 0, 0)^{\mathrm{T}}$, we have

$$\dot{\mathbf{X}} = \mathbf{G}(\mathbf{X}) + \mathbf{L}\xi(t) \tag{5.166}$$

The statistical linearization technique can then be applied (Schenk and Schuëller, 2005). Actually, this nonlinear equation can be linearized through letting $\mathbf{M} = \mathbf{0}$ in Equation 5.148. Alternatively, it can also be approximated by the linear equation

$$\dot{\mathbf{Y}} = \mathbf{A}_{\mathrm{eq}}\mathbf{Y} + \mathbf{L}\xi(t) \tag{5.167}$$

Thus, the error is given by

$$\bar{\mathbf{e}} = \mathbf{G}(\mathbf{X}) - \mathbf{A}_{\mathrm{eq}}\mathbf{Y} \tag{5.168}$$

When $\mathbf{X}$ is approximated by $\mathbf{Y}$, Equation 5.168 can be replaced by

$$\mathbf{e} = \mathbf{G}(\mathbf{Y}) - \mathbf{A}_{\mathrm{eq}}\mathbf{Y} \tag{5.169}$$

Consequently, the optimal value of the coefficient matrix $\mathbf{A}_{\mathrm{eq}}$ can be determined by letting

$$\frac{\partial \mathcal{E}[\mathbf{e}\mathbf{e}^{\mathrm{T}}]}{\partial \mathbf{A}_{\mathrm{eq}}} = \frac{\partial \mathcal{E}\{[\mathbf{A}_{\mathrm{eq}}\mathbf{X} - \mathbf{G}(\mathbf{X})][\mathbf{A}_{\mathrm{eq}}\mathbf{X} - \mathbf{G}(\mathbf{X})]^{\mathrm{T}}\}}{\partial \mathbf{A}_{\mathrm{eq}}} = 0 \tag{5.170}$$

Again, this problem can be solved by an iteration method. For nonlinear MDOF systems, the same idea can be used.

5.5.3 Notes on Arguments and Some Special Issues

5.5.3.1 Probability Density Function of the Response of Statistical Linearized Systems

For simplicity, consider the steady-state, stationary stochastic response where the equivalent linearized damping and stiffness are given by Equation 5.147.

In general, when the excitation is Gaussian, the response of the linear system in Equation 5.140 is also Gaussian; namely, $Y(t)$ is Gaussian and can be characterized by the mean μ_Y and the standard deviation σ_Y. Let us introduce a stationary process $Z(t)$ and its derivative process $\dot{Z}(t)$ as subsidiary processes. If they are both Gaussian and the respective means and standard deviations are μ_Z, σ_Z and $\mu_{\dot{Z}}$, $\sigma_{\dot{Z}}$, then the PDFs are

$$p_Z(z) = \frac{1}{\sqrt{2\pi}\sigma_Z} \mathrm{e}^{-(z-\mu_Z)/2\sigma_Z^2} \tag{5.171a}$$

and

$$p_{\dot{Z}}(\dot{z}) = \frac{1}{\sqrt{2\pi}\sigma_{\dot{Z}}} \mathrm{e}^{-(\dot{z}-\mu_{\dot{Z}})/2\sigma_{\dot{Z}}^2} \tag{5.171b}$$

respectively. Replacing Y and $\dot{Y}$ by Z and $\dot{Z}$ respectively and introducing the PDF (Equation 5.171a and 5.171b into Equation 5.147, we get the expressions of c_{eq} and k_{eq} in terms of μ_Z, σ_Z and $\mu_{\dot{Z}}$, $\sigma_{\dot{Z}}$, denoted by $c_{\mathrm{eq}}(\mu_Z, \sigma_Z, \mu_{\dot{Z}}, \sigma_{\dot{Z}})$ and $k_{\mathrm{eq}}(\mu_Z, \sigma_Z, \mu_{\dot{Z}}, \sigma_{\dot{Z}})$ respectively for clarity. In sequence, substituting them in Equation 5.140 will then get μ_Y, σ_Y, $\mu_{\dot{Y}}$ and $\sigma_{\dot{Y}}$, which are functions of $c_{\mathrm{eq}}(\mu_Z, \sigma_Z, \mu_{\dot{Z}}, \sigma_{\dot{Z}})$ and $k_{\mathrm{eq}}(\mu_Z, \sigma_Z, \mu_{\dot{Z}}, \sigma_{\dot{Z}})$; namely, they are also functions of μ_Z, σ_Z, $\mu_{\dot{Z}}$ and $\sigma_{\dot{Z}}$. Because Y and $\dot{Y}$ are replaced by Z and $\dot{Z}$ respectively, certainly we require

$$\begin{cases} \mu_Y(\mu_Z, \sigma_Z, \mu_{\dot{Z}}, \sigma_{\dot{Z}}) = \mu_Z \\ \sigma_Y(\mu_Z, \sigma_Z, \mu_{\dot{Z}}, \sigma_{\dot{Z}}) = \sigma_Z \\ \mu_{\dot{Y}}(\mu_Z, \sigma_Z, \mu_{\dot{Z}}, \sigma_{\dot{Z}}) = \mu_{\dot{Z}} \\ \sigma_{\dot{Y}}(\mu_Z, \sigma_Z, \mu_{\dot{Z}}, \sigma_{\dot{Z}}) = \sigma_{\dot{Z}} \end{cases} \tag{5.172}$$

Solving these equations we will get μ_Z, σ_Z, $\mu_{\dot{Z}}$ and $\sigma_{\dot{Z}}$, and simultaneously μ_Y, σ_Y, $\mu_{\dot{Y}}$ and $\sigma_{\dot{Y}}$.

The steady-state, stationary response of the nonlinear system in Equation 5.139 to the Gaussian excitation, however, is usually non-Gaussian; therefore, employing the normal distribution (Equations 5.171a and 5.171b) will undoubtedly induce errors. Actually, this is one of the major sources of error in the statistical linearization method (Crandall, 2006). According to Caughey's theorem, which was stated in Caughey (1960) and proved in Crandall (2006), if the true distribution shape of the response is employed instead of Equations 5.171a and 5.171b, then solving Equation 5.172 will give accurate values of μ_Z, σ_Z, $\mu_{\dot{Z}}$ and $\sigma_{\dot{Z}}$. For this purpose, possible shapes of PDF are suggested; for example:

$$p_Z(z) = \frac{\exp[-(|z|/a)^m]}{\int_{-\infty}^{\infty} \exp[-(|z|/a)^m]\,\mathrm{d}z} \tag{5.173a}$$

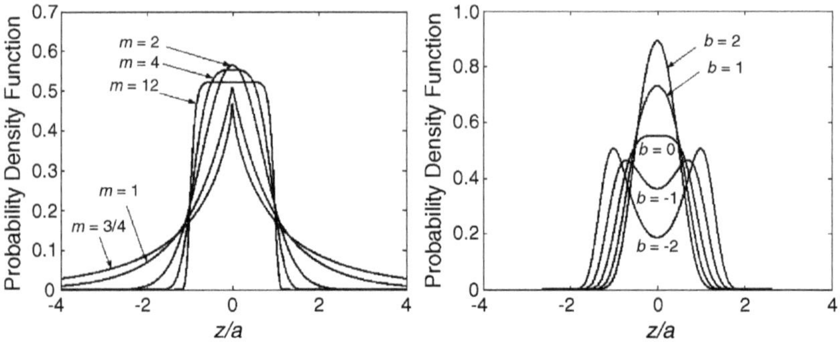

Figure 5.11 Different shapes of the PDFs.

$$p_Z(z) = \frac{\exp[-b(z^2/a^2) - (z^4/a^4)]}{\int_{-\infty}^{\infty} \exp[-b(z^2/a^2) - (z^4/a^4)]\,dz} \tag{5.173b}$$

the shapes of which are shown in Figure 5.11. For the Duffing oscillator, using the shape in Equation 5.173b will yield accurate results.

5.5.3.2 On Error-Free Linearization and Different Criteria

After the standard statistical linearization techniques as discussed in the preceding sections has been developed for nearly 40 years, Elishakoff and Colajanni (1997) thought that there were something wrong.

The problem, as pointed out, essentially lies in the deduction from Equations 5.144a–5.145b, where it was thought that the responses Y and $\dot{Y}$ are independent of c_{eq} and k_{eq}. The 'error-free' linearization, therefore, should consider Equations 5.144a and 5.144b more carefully. Because Y and $\dot{Y}$, and therefore the PDF of Y and $\dot{Y}$, are functions of c_{eq} and k_{eq}, Equation 5.143 should be first computed to give the expressions of $\mathcal{E}[e^2]$ in terms of c_{eq} and k_{eq}, then the partial derivatives in Equations 5.144a and 5.144b can be computed to yield a nonlinear equation set with c_{eq} and k_{eq} being the unknowns. Solving this equation set will give the values of c_{eq} and k_{eq} and then the statistics of Y and $\dot{Y}$ can be obtained from the equivalent linear system in Equation 5.140.

Although the above analysis is reasonable, the effect is not as good as expected. First, the deduction is much more difficult and might be impossible for complex or multidimensional problems. Second, even for simple problems, it was shown that the accuracy of the 'error-free' linearization is sometimes lower than that of standard linearization (Elishakoff and Colajanni, 1997).

The reason lies, perhaps, in the fact that in standard linearization techniques some types of iteration algorithm are used, whereas no iteration is needed in 'error-free' linearization. Actually, the iteration algorithm is one of the most effective approaches to make the loops uncoupled.

Another important issue is the criterion used in the equivalent linearization. The above criterion was usually referred to as the Kazakov II criterion (Kazakov, 1954). The major criteria include Kazakov I, Kazakov II and the energy criterion (Crandall, 2006). Research examples show that when Kazakov II is employed, Caughey's theorem described above holds, while the accuracy of the energy criterion is usually lower than that of the Kazakov II criterion.

5.5.3.3 Equivalent Stochastic Damping and Stiffness

In the statistical linearization method, the most reasonable start point might be to consider the equivalent stiffness and damping as random variables; that is, using $\mu_{C_{eq}}, \sigma_{C_{eq}}, \mu_{K_{eq}}$ and $\sigma_{K_{eq}}$ as the unknowns. Actually, for a nonlinear system, the stiffness is varying against the stochastic responses and, therefore, is essentially a stochastic process.

Set the equivalent linearized system of Equation 5.139 as

$$m\ddot{Y} + C_{eq}\dot{Y} + K_{eq}Y = \xi(t) \tag{5.174}$$

where C_{eq} and K_{eq} are random variables with means and standard deviations $\mu_{C_{eq}}, \sigma_{C_{eq}}$ and $\mu_{K_{eq}}$, $\sigma_{K_{eq}}$ respectively. The error (Equation 5.142) is now changed to

$$e = g(Y, \dot{Y}) - C_{eq}\dot{Y} - K_{eq}Y \tag{5.175}$$

Then the variance of the error is given as

$$\begin{aligned}\mathcal{E}[e^2] &= \mathcal{E}[g(Y,\dot{Y}) - C_{eq}\dot{Y} - K_{eq}Y]^2\} \\ &= \mathcal{E}\{g^2(Y,\dot{Y}) + C_{eq}^2\dot{Y}^2 + K_{eq}^2Y^2 - 2C_{eq}g(Y,\dot{Y})\dot{Y} - 2K_{eq}g(Y,\dot{Y})Y + 2C_{eq}K_{eq}\dot{Y}Y\}\end{aligned} \tag{5.176}$$

Of course, this is a function of $\mu_{C_{eq}}, \sigma_{C_{eq}}, \mu_{K_{eq}}$ and $\sigma_{K_{eq}}$. Thus, the optimal values of $\mu_{C_{eq}}, \sigma_{C_{eq}}$, $\mu_{K_{eq}}$ and $\sigma_{K_{eq}}$ should minimize $\mathcal{E}[e^2]$; that is:

$$\frac{\partial\mathcal{E}[e^2]}{\partial\mu_{C_{eq}}} = 0 \quad \frac{\partial\mathcal{E}[e^2]}{\partial\sigma_{C_{eq}}} = 0 \quad \frac{\partial\mathcal{E}[e^2]}{\partial\mu_{K_{eq}}} = 0 \quad \frac{\partial\mathcal{E}[e^2]}{\partial\sigma_{K_{eq}}} = 0 \tag{5.177}$$

To achieve the expression of $\mathcal{E}[e^2]$ with regard to $\mu_{C_{eq}}, \sigma_{C_{eq}}, \mu_{K_{eq}}$ and $\sigma_{K_{eq}}$, the stochastic structural system (Equation 5.174) should be solved by, for example, the approaches elaborated in Chapter 4.

5.5.3.4 On Applicability

With regard to applications of the statistical linearization method to problems of practical interest, we should note that:

(a) The equivalence between the original system and the linearized equivalent system is in the sense of variances; therefore, the errors in the correlation function and the PDF may be much larger. For instance, the PDF of Y in Equation 5.140 is Gaussian when the excitation is Gaussian; however, the PDF of X in Equation 5.139 might be far from Gaussian, as shown in Figure 5.11 for example.
(b) Because of the preceding reason, the statistical linearization method is usually not suitable for reliability assessment.
(c) The statistical linearization method does not work in the case where essential nonlinearity occurs, as in bifurcation, jump and limit circles for example.

5.6 Fokker–Planck–Kolmogorov Equation

The moment-characteristics-oriented approaches capture part of the probabilistic information of stochastic systems. However, this is not enough, particularly for nonlinear systems, where the probability distribution of the responses might deviate far from the normal distribution. As discussed in Chapter 1, to capture the evolution of probability densities is a dream of investigators which can be dated back to Einstein. Through the studies of Fokker and Planck and mathematically manipulated by Kolmogorov, the FPK equation governing the evolution of the joint densities of system states was established. Then the relationship between the stochastic differential equation and the FPK equation was clarified when the Itô and the Stratonovich stochastic calculus was established. These form the second thought stream in stochastic system analysis: the family of PDF-oriented approaches.

5.6.1 Stochastic Differential Equation

5.6.1.1 Itô Integral and Itô Stochastic Differential Equation

The MDOF nonlinear system in Equation 5.148 can be written alternatively in the form of the state equation

$$\dot{\mathbf{Y}} = \mathbf{A}(\mathbf{Y}, t) + \mathbf{B}(\mathbf{Y}, t)\boldsymbol{\xi}(t) \tag{5.178}$$

where

$$\mathbf{Y} = (Y_1, Y_2, \ldots, Y_m)^{\mathrm{T}} = \begin{Bmatrix} \dot{\mathbf{X}} \\ \mathbf{X} \end{Bmatrix} \quad \mathbf{A} = (A_1, A_2, \ldots, A_m)^{\mathrm{T}} = \begin{Bmatrix} -\mathbf{M}^{-1}\mathbf{G} \\ \dot{\mathbf{X}} \end{Bmatrix} \quad \mathbf{B} = [B_{ij}]_{m\times r} = \begin{bmatrix} \mathbf{M}^{-1}\mathbf{L} \\ \mathbf{0} \end{bmatrix}$$

To consider more general situations, we regard B_{ij}'s as functions of $\mathbf{Y}$ and t. $\boldsymbol{\xi}(t) = (\xi_1(t), \xi_2(t), \ldots, \xi_r(t))^{\mathrm{T}}$ is an r-dimensional stochastic process vector.

We now consider the case that $\boldsymbol{\xi}(t)$ is a Gaussian white-noise vector such that

$$\mathcal{E}[\boldsymbol{\xi}(t)] = \mathbf{0} \qquad \mathcal{E}[\boldsymbol{\xi}(t_1)\boldsymbol{\xi}^{\mathrm{T}}(t_2)] = \mathbf{D}\delta(t_1 - t_2) \tag{5.179}$$

where $\mathbf{D} = [D_{ij}]_{r\times r}$ is an $r \times r$ positive definite matrix.

The solution to Equation 5.178 is given by the integral

$$\mathbf{Y}(t) = \mathbf{Y}(t_0) + \int_{t_0}^{t} \mathbf{A}(\mathbf{Y}, t)\,\mathrm{d}t + \int_{t_0}^{t} \mathbf{B}(\mathbf{Y}, t)\boldsymbol{\xi}(t)\,\mathrm{d}t \tag{5.180}$$

or in a differential form

$$\mathrm{d}\mathbf{Y}(t) = \mathbf{A}(\mathbf{Y}, t)\,\mathrm{d}t + \mathbf{B}(\mathbf{Y}, t)\boldsymbol{\xi}(t)\,\mathrm{d}t \tag{5.181}$$

Clearly, the mean can be obtained by

$$\mathcal{E}[\mathrm{d}\mathbf{Y}(t)] = \mathcal{E}[\mathbf{A}(\mathbf{Y}, t)]\,\mathrm{d}t + \mathcal{E}[\mathbf{B}(\mathbf{Y}, t)\boldsymbol{\xi}(t)]\,\mathrm{d}t = \mathcal{E}[\mathbf{A}(\mathbf{Y}, t)]\,\mathrm{d}t \tag{5.182}$$

and the covariance matrix can be evaluated through

$$\begin{aligned}\mathcal{E}[\{\mathrm{d}\mathbf{Y}(t_1)-\mathcal{E}[\mathrm{d}\mathbf{Y}(t_1)]\}\{\mathrm{d}\mathbf{Y}(t_2)-\mathcal{E}[\mathrm{d}\mathbf{Y}(t_2)]\}^{\mathrm{T}}] &= \mathcal{E}\{[\mathbf{B}(\mathbf{Y},t_1)\boldsymbol{\xi}(t_1)\mathrm{d}t_1][\boldsymbol{\xi}^{\mathrm{T}}(t_2)\mathbf{B}^{\mathrm{T}}(\mathbf{Y},t_2)\mathrm{d}t_2]\} \\ &= \mathcal{E}[\mathbf{B}(\mathbf{Y},t_1)]\mathcal{E}[\boldsymbol{\xi}(t_1)\boldsymbol{\xi}^{\mathrm{T}}(t_2)]\mathcal{E}[\mathbf{B}^{\mathrm{T}}(\mathbf{Y},t_2)]\,\mathrm{d}t_1\,\mathrm{d}t_2 \\ &= \mathcal{E}[\mathbf{B}(\mathbf{Y},t_1)]\mathbf{D}\mathcal{E}[\mathbf{B}^{\mathrm{T}}(\mathbf{Y},t_2)]\delta(t_1-t_2)\,\mathrm{d}t_1\,\mathrm{d}t_2\end{aligned} \tag{5.183a}$$

In an alternative form, this is

$$\mathcal{E}[\{\mathrm{d}\mathbf{Y}(t)-\mathcal{E}[\mathrm{d}\mathbf{Y}(t)]\}\{\mathrm{d}\mathbf{Y}(t)-\mathcal{E}[\mathrm{d}\mathbf{Y}(t)]\}^{\mathrm{T}}] = \mathcal{E}[\mathbf{B}]\mathbf{D}\mathcal{E}[\mathbf{B}^{\mathrm{T}}]\,\mathrm{d}t \tag{5.183b}$$

The arguments in **B** are omitted for notational simplicity. Here, we use the assumption that **B** ($\mathbf{Y}$, t) and $\boldsymbol{\xi}(t)$ dt are mutually independent, the rigorous meaning of which will be clarified later.

Let $\mathbf{A} = \mathbf{0}$ and $\mathbf{B} = \mathbf{I}$, where **I** is the unit matrix. Then, from Equations 5.180–5.183b, we find that if we define the integral process

$$\mathbf{Z}(t) = \mathbf{Z}(t_0) + \int_{t_0}^{t} \boldsymbol{\xi}(t)\,\mathrm{d}t \tag{5.184}$$

then $\mathbf{Z}(t)$ is a Brownian motion process vector because the means are zero and the variances are proportional to time duration (Gardiner, 1983). However, Brownian motion is continuous but not differentiable.

The paradox that appears here can be solved in a mathematically consistent form by the Riemann–Stielgjes integral instead of Equation 5.184. By doing so, Equation 5.181 can be rewritten as

$$\mathrm{d}\mathbf{Y}(t) = \mathbf{A}(\mathbf{Y},t)\,\mathrm{d}t + \mathbf{B}(\mathbf{Y},t)\,\mathrm{d}\mathbf{W}(t) \tag{5.185}$$

where **Y**, **A** and **B** are as defined in Equation 5.178 and $\mathbf{W}(t) = (W_1(t), W_2(t), \ldots, W_r(t))^{\mathrm{T}}$ is an r-dimensional Brownian motion process vector (sometimes we will also use the terminology Wiener process later) with

$$\mathcal{E}[\mathrm{d}\mathbf{W}(t)] = \mathbf{0} \qquad \mathcal{E}[\mathrm{d}\mathbf{W}(t)\,\mathrm{d}\mathbf{W}^{\mathrm{T}}(t)] = \mathbf{D}\,\mathrm{d}t \tag{5.186}$$

in which $\mathbf{D} = [D_{ij}]_{r\times r}$ is the same as in Equation 5.179; correspondingly, Equation 5.180 is rewritten as

$$\mathbf{Y}(t) = \mathbf{Y}(t_0) + \int_{t_0}^{t} \mathbf{A}(\mathbf{Y},t)\,\mathrm{d}t + \int_{t_0}^{t} \mathbf{B}(\mathbf{Y},t)\,\mathrm{d}\mathbf{W}(t) \tag{5.187}$$

The first integral here is the common mean-square integral, while special attention should be paid to the second integral heuristically because of the highly irregular nature of the trajectory of $\mathbf{W}(t)$.

As usual, we replace the second integral by a limit of summation:

$$\mathcal{I} = \int_{t_0}^{t} \mathbf{B}(\mathbf{Y},t)\mathrm{d}\mathbf{W}(t) = \lim_{n\to\infty}\sum_{j=1}^{n} \mathbf{B}(\mathbf{Y},\tau_j)[\mathbf{W}(t_j) - \mathbf{W}(t_{j-1})] \tag{5.188}$$

where τ_j is some value over the interval $[\tau_{j-1}, \tau_j]$. In the common integral, it is known that τ_j can be anywhere over the integral $[\tau_{j-1}, \tau_j]$ and the limit in Equation 5.188 is invariant, and this invariant value is defined as the value of the integral. In the case $\mathbf{W}(t)$ is a Wiener process vector, however, this is not the case.

To make the concept clear, we first consider the expectation of a scalar integral with respect to a Wiener process $W(t)$, of which the mean and variance of increment are given by $\mathcal{E}[\mathrm{d}W(t)] = 0$ and $\mathcal{E}\{[\mathrm{d}W(t)]^2\} = D\,\mathrm{d}t$ respectively:

$$\mathcal{E}\left[\int_{t_0}^{t} W(t)\,\mathrm{d}W(t)\right] = \lim_{n\to\infty}\sum_{j=1}^{n}\mathcal{E}\{W(\tau_j)[W(t_j) - W(t_{j-1})]\} \tag{5.189}$$

Note that

$$\begin{aligned}\mathcal{E}[W(\tau_j)W(t_j)] &= \mathcal{E}\{W(\tau_j)[W(t_j) - W(\tau_j) + W(\tau_j)]\}\\ &= \mathcal{E}\{W(\tau_j)[W(t_j) - W(\tau_j)]\} + \mathcal{E}[W(\tau_j)W(\tau_j)]\\ &= D\tau_j\end{aligned} \tag{5.190a}$$

where use has been made of

$$\mathcal{E}\{W(\tau_j)[W(t_j) - W(\tau_j)]\} = 0 \tag{5.190b}$$

due to the independence of increments of the Wiener process. Likewise, $\mathcal{E}[W(\tau_j)W(t_{j-1})] = Dt_{j-1}$. Thus, we can get

$$\mathcal{E}\left[\int_{t_0}^{t} W(t)\,\mathrm{d}W(t)\right] = \lim_{n\to\infty} D\sum_{j=1}^{n}(\tau_j - t_{j-1}) \tag{5.191}$$

Somewhat surprisingly, this value is varies with different positions of τ_j. For instance, if we choose for all j

$$\tau_j = \alpha t_j + (1-\alpha)t_{j-1} \qquad (0 \le \alpha \le 1) \tag{5.192}$$

it follows that

$$\mathcal{E}\left[\int_{t_0}^{t} W(t)\,\mathrm{d}W(t)\right] = (t - t_0)\alpha D \tag{5.193}$$

Evidently, this indicates that the expectation of the integral depends on the position of the intermediate points.

This issue can be further understood. Supposing $\alpha = 0$, Equation 5.189 becomes

$$\mathcal{E}\left[\int_{t_0}^{t} W(t)\,\mathrm{d}W(t)\right] = \lim_{n\to\infty}\sum_{j=1}^{n}\mathcal{E}\{W(t_{j-1})[W(t_j) - W(t_{j-1})]\} \tag{5.194a}$$

Owing to the independence of increments, we have $\mathcal{E}\{W(t_{j-1})[W(t_j) - W(t_{j-1})]\} = 0$ for all j. Equation 5.194a immediately yields $\mathcal{E}[\int_{t_0}^{t} W(t)\,\mathrm{d}W(t)] = 0$, which is consistent with Equation 5.193 in the case $\alpha = 0$.

Now suppose $\alpha = 1$; Equation 5.189 then becomes

$$\mathcal{E}\left[\int_{t_0}^{t} W(t)\,\mathrm{d}W(t)\right] = \lim_{n\to\infty}\sum_{j=1}^{n}\mathcal{E}\{W(t_j)[W(t_j) - W(t_{j-1})]\} \tag{5.194b}$$

In this case, $W(t_j) = W(t_{j-1}) + [W(t_j) - W(t_{j-1})]$; therefore, $W(t_j)$ is not independent of the increment $[W(t_j) - W(t_{j-1})]$ because

$$\begin{aligned}\mathcal{E}\{W(t_j)[W(t_j) - W(t_{j-1})]\} &= \mathcal{E}(\{W(t_{j-1}) + [W(t_j) - W(t_{j-1})]\}[W(t_j) - W(t_{j-1})])\\ &= \mathcal{E}\{[W(t_j) - W(t_{j-1})][W(t_j) - W(t_{j-1})]\}\\ &= (t_j - t_{j-1})D\end{aligned} \tag{5.194c}$$

Thus, we now have $\mathcal{E}[\int_{t_0}^{t} W(t)\,\mathrm{d}W(t)] = (t - t_0)D$. Again, this is consistent with Equation 5.193 in the case $\alpha = 1$.

Likewise, if we use $\alpha = 1/2$ and $W[(t_j + t_{j-1})/2] = [W(t_j) + W(t_{j-1})]/2$, then Equation 5.189 is now

$$\mathcal{E}\left[\int_{t_0}^{t} W(t)\,\mathrm{d}W(t)\right] = \lim_{n\to\infty}\sum_{j=1}^{n}\mathcal{E}\left\{\frac{W(t_j) + W(t_{j-1})}{2}[W(t_j) - W(t_{j-1})]\right\} \tag{5.194d}$$

Here, $W(t_{j-1})$ is independent of the increment of $[W(t_j) - W(t_{j-1})]$ but $W(t_j)$ is not. Combining Equations 5.194a and 5.194b, we now find $\mathcal{E}[\int_{t_0}^{t} W(t)\,\mathrm{d}W(t)] = (t - t_0)D/2$, which is consistent with Equation 5.193 in the case $\alpha = 1/2$.

The preceding discussions from Equations 5.194a–5.194d show the reason that the limit of the summation in Equation 5.189 depends on the intermediate point lies in the fact that different positions of the intermediate points means different correlations between the increment $\Delta W_j = W(t_j) - W(t_{j-1})$ and the integrand. This is also true when the integrand is other types of function, say denoted by $G(Y(t),t)$, where $Y(t)$ is a stochastic process determined by a stochastic differential equation, provided at any arbitrary time t_1 the value of $G(Y(t_1),t_1)$ is independent of the value of $W(t)$ at the times $t > t_1$. Such a function $G(Y(t),t)$ is a *nonanticipating function.* Now it is obvious that, because of the nonanticipating feature, the integral defined by

$$\int_{t_0}^{t} G(Y(t),t)\,\mathrm{d}W(t) = \lim_{n\to\infty}\sum_{j=1}^{n} G(Y(t_{j-1}),t_{j-1})[W(t_j) - W(t_{j-1})] \tag{5.195}$$

is the mathematically simplest treatment, since $G(Y(t_{j-1}),t_{j-1})$ is independent of the increment ΔW_j, which will make the further mathematical manipulations much simpler than other intermediate values as demonstrated above. The definition in Equation 5.195 is the renowned *Itô integral.* Accordingly, the stochastic differential equation (Equation 5.185), when understood in this sense, is referred to as the Itô stochastic differential equation (we can now go back to Equation 5.183a, where essentially we assume that the $B_{ij}(\mathbf{Y}, t)$ are nonanticipating functions).

Here, we find that the fact of causality (that is, a present event is independent of the future events) is of critical importance in Itô calculus.

In the manipulation of Itô stochastic differentiation, because the second-order moment $\mathcal{E}\{[\mathrm{d}W(t)]^2\} = D\,\mathrm{d}t$ is in the order of $\mathrm{d}t$, if λ is a random variable with mean value $\mathcal{E}[\lambda] = 0$ and variance $\mathcal{E}[\lambda^2] = D$, then we can let $\mathrm{d}W(t) = \lambda\sqrt{\mathrm{d}t}$. Thus, $\mathrm{d}W(t)$ is in the order of $\sqrt{\mathrm{d}t}$; therefore, rather than being ignorable, the terms with regard to $\mathrm{d}W(t)$ will play important roles in computation of the second-order moments. Thus, in contrast to common differentiation, the terms of second order in the Taylor expansion should remain in Itô stochastic differentiation. For instance, for a function $f(\mathbf{Y}(t))$ where $\mathbf{Y}(t)$ is the solution of the Itô stochastic differential Equation 5.185, the differentiation of $f(\mathbf{Y}(t))$ is given as

$$\begin{aligned}
\mathrm{d}f(\mathbf{Y}(t)) &= f(\mathbf{Y}(t)+\mathrm{d}\mathbf{Y}(t)) - f(\mathbf{Y}(t)) \\
&= \sum_{\ell=1}^{m} \frac{\partial f}{\partial Y_\ell}\,\mathrm{d}Y_\ell(t) + \frac{1}{2}\sum_{k=1}^{m}\sum_{\ell=1}^{m} \frac{\partial^2 f}{\partial Y_k \partial Y_\ell}\,\mathrm{d}Y_k(t)\,\mathrm{d}Y_\ell(t) + \ldots \\
&= \sum_{\ell=1}^{m} \frac{\partial f}{\partial Y_\ell}\left[A_\ell(\mathbf{Y}(t),t)\,\mathrm{d}t + \sum_{s=1}^{r} B_{\ell s}(\mathbf{Y}(t),t)\,\mathrm{d}W_s(t)\right] \\
&\quad + \frac{1}{2}\sum_{k=1}^{m}\sum_{\ell=1}^{m} \frac{\partial^2 f}{\partial Y_k \partial Y_\ell}\left[\sum_{s=1}^{r} B_{ks}(\mathbf{Y}(t),t)\,\mathrm{d}W_s(t)\right]\left[\sum_{s=1}^{r} B_{\ell s}(\mathbf{Y}(t),t)\,\mathrm{d}W_s(t)\right] + \ldots \\
&= \sum_{l=1}^{m} \frac{\partial f}{\partial Y_\ell}\left[A_\ell(\mathbf{Y}(t),t)\,\mathrm{d}t + \sum_{s=1}^{r} B_{\ell s}(\mathbf{Y}(t),t)\,\mathrm{d}W_s(t)\right] + \frac{1}{2}\sum_{k=1}^{m}\sum_{\ell=1}^{m} (\mathbf{BDB}^{\mathrm{T}})_{k\ell}\frac{\partial^2 f}{\partial Y_k \partial Y_\ell}\,\mathrm{d}t + \ldots \\
&= \sum_{\ell=1}^{m}\left[A_\ell \frac{\partial f}{\partial Y_\ell} + \frac{1}{2}\sum_{k=1}^{m} (\mathbf{BDB}^{\mathrm{T}})_{k\ell}\frac{\partial^2 f}{\partial Y_k \partial Y_\ell}\right]\mathrm{d}t + \sum_{\ell=1}^{m}\sum_{k=1}^{r} B_{\ell k}(\mathbf{Y}(t),t)\frac{\partial f}{\partial Y_\ell}\,\mathrm{d}W_k(t)
\end{aligned} \tag{5.196}$$

where use has been made of the component form of Equation 5.185:

$$\mathrm{d}Y_\ell = A_\ell(\mathbf{Y}(t),t)\,\mathrm{d}t + \sum_{k=1}^{r} B_{\ell k}(\mathbf{Y}(t),t)\,\mathrm{d}W_k(t) \qquad \ell = 1,2,\ldots,m \tag{5.197}$$

In the case $m = 1$, $r = 1$, Equation 5.196 reduces to

$$\begin{aligned}
\mathrm{d}f(Y(t)) &= f(Y(t)+\mathrm{d}Y(t)) - f(Y(t)) = \frac{\partial f}{\partial Y}\,\mathrm{d}Y(t) + \frac{1}{2}\frac{\partial^2 f}{\partial Y^2}[\mathrm{d}Y(t)]^2 + \ldots \\
&= \frac{\partial f}{\partial Y}[A(Y(t),t)\mathrm{d}t + B(Y(t),t)\,\mathrm{d}W(t)] + \frac{1}{2}\frac{\partial^2 f}{\partial Y^2}B^2(Y(t),t)[\mathrm{d}W(t)]^2 + \ldots \\
&= \left[A(Y(t),t)\frac{\partial f}{\partial Y} + \frac{D}{2}B^2(Y(t),t)\frac{\partial^2 f}{\partial Y^2}\right]\mathrm{d}t + \frac{\partial f}{\partial Y}B(Y(t),t)\,\mathrm{d}W(t)
\end{aligned} \tag{5.198}$$

Equations 5.196 and 5.198 are usually called the *Itô lemma*. The difference between Equation 5.198 and common differentiation is that the terms of second order with respect to

$[\mathrm{d}Y(t)]^2$ (see the second equality) cannot be eliminated because they are in the order of dt and, therefore, an additional corrected term occurs in the final equality in the terms with respect to dt.

5.6.1.2 Stratonovich's Stochastic Differential Equation

The Itô integral is elegant in dealing with mathematical white noise whose correlative time is zero. In practice, however, the correlative time of the physical noise might be short, but finite and not zero. To consider this fact, the correlation should be considered; namely, α in Equation 5.192 should take values $0<\alpha<1$ rather than zero as is done in the Itô integral. Stratonovich (1963) defined the integral with $G(Y(t),t)$ being the integrand by

$$\mathcal{S}\int_{t_0}^{t} G(Y(t),t)\,\mathrm{d}W(t) = \lim_{n\to\infty}\sum_{j=1}^{n} G\left(\frac{Y(t_j)+Y(t_{j-1})}{2},t_{j-1}\right)\left[W(t_j)-W(t_{j-1})\right] \tag{5.199}$$

The corresponding stochastic differential equation is referred to as the Stratonovich's stochastic differential equation.[4]

There is no constant relationship between the Itô and Stratonovich's integrals. However, for the case where the stochastic process is related to a stochastic differential equation, we can establish a relationship. To be clear, we consider the Stratonovich's stochastic differential equation

$$\mathrm{d}\mathbf{Y}(t) = \boldsymbol{\alpha}(\mathbf{Y},t)\,\mathrm{d}t + \boldsymbol{\beta}(\mathbf{Y},t)\,\mathrm{d}\mathbf{W}(t) \tag{5.200}$$

where $\boldsymbol{\alpha} = (\alpha_1,\alpha_2,\ldots,\alpha_m)^{\mathrm{T}}$ and $\boldsymbol{\beta} = [\beta_{ij}]_{m\times r}$. Its solution is given by the Stratonovich's integral

$$\mathbf{Y}(t) = \mathbf{Y}(t_0) + \int_{t_0}^{t}\boldsymbol{\alpha}(\mathbf{Y},t)\,\mathrm{d}t + \mathcal{S}\int_{t_0}^{t}\boldsymbol{\beta}(\mathbf{Y},t)\,\mathrm{d}\mathbf{W}(t) \tag{5.201}$$

We assume that the solutions to Equations 5.201 and 5.187 are equivalent.

First, we will compute the terms of the Stratonovich's integral:

$$\mathcal{S}\int_{t_0}^{t}\boldsymbol{\beta}(\mathbf{Y},t)\,\mathrm{d}\mathbf{W}(t) = \lim_{n\to\infty}\sum_{i=1}^{n}\boldsymbol{\beta}\left(\frac{\mathbf{Y}_j+\mathbf{Y}_{j-1}}{2},t_{j-1}\right)\left[\mathbf{W}_j-\mathbf{W}_{j-1}\right] \tag{5.202}$$

Noting that $\mathbf{Y}_j = \mathbf{Y}_{j-1} + \Delta\mathbf{Y}_j$, we have

$$\boldsymbol{\beta}\left(\frac{\mathbf{Y}_j+\mathbf{Y}_{j-1}}{2},t_{j-1}\right) = \boldsymbol{\beta}\left(\mathbf{Y}_{j-1}+\frac{\Delta\mathbf{Y}_j}{2},t_{j-1}\right) - \boldsymbol{\beta}(\mathbf{Y}_{j-1},t_{j-1}) + \frac{1}{2}\sum_{\ell=1}^{m}\frac{\partial\boldsymbol{\beta}}{\partial Y_\ell}\Delta Y_{\ell,j} \tag{5.203a}$$

[4] At first glance one may be confused why the argument t in $G(Y,t)$ still takes t_{j-1} and does not get replaced by $(t_j+t_{j-1})/2$. We can try replacing t_{j-1} by $(t_j+t_{j-1})/2$ and then expanding $G(Y,t)$ with respect to t at t_{j-1}, then we find that this will lead to a term of the order $O(\Delta t)^{3/2}$ in Equation 5.199. This means that whether we replace t by $(t_j+t_{j-1})/2$ or by t_{j-1} will essentially have no influence on the results. Again, one can see from here that the influence of Y and t is different in $G(Y,t)$ because $Y(t)$ is associated with a Wiener process.

where $\Delta Y_{l,j}$ can be obtained from Equation 5.197 by Itô differentiation:

$$\Delta Y_{\ell,j} = A_\ell(\mathbf{Y}_{j-1}, t_{j-1})(t_j - t_{j-1}) + \sum_{k=1}^{r} B_{\ell k}(\mathbf{Y}_{j-1}, t_{j-1})[W_k(t_j) - W_k(t_{j-1})] \quad (5.203b)$$

Introducing Equations 5.203a and 5.203b into Equation 5.202 yields

$$\begin{aligned}
&\lim_{n\to\infty}\sum_{j=1}^{n}\boldsymbol{\beta}\left(\frac{\mathbf{Y}_j+\mathbf{Y}_{j-1}}{2}, t_{j-1}\right)\Delta\mathbf{W}_j \\
&= \lim_{n\to\infty}\sum_{j=1}^{n}\boldsymbol{\beta}(\mathbf{Y}_{j-1}, t_{j-1})\Delta\mathbf{W}_j + \frac{1}{2}\lim_{n\to\infty}\sum_{\ell=1}^{m}\sum_{j=1}^{n}\frac{\partial\boldsymbol{\beta}}{\partial Y_\ell}\Delta\mathbf{W}_j \\
&\quad\times\left[A_\ell(Y_{j-1}, t_{j-1})\Delta t_j + \sum_{k=1}^{r} B_{\ell k}(\mathbf{Y}_{j-1}, t_{j-1})\Delta W_{k,j})\right] \\
&= \int_{t_0}^{t}\boldsymbol{\beta}(\mathbf{Y}, t)\,\mathrm{d}\mathbf{W}(t) + \frac{1}{2}\lim_{n\to\infty}\sum_{\ell=1}^{m}\sum_{j=1}^{n}\frac{\partial\boldsymbol{\beta}}{\partial Y_\ell}\Delta\mathbf{W}_j \times\left[\sum_{k=1}^{r} B_{\ell k}(\mathbf{Y}_{j-1}, t_{j-1})\Delta W_{k,j}\right] \\
&= \int_{t_0}^{t}\boldsymbol{\beta}(\mathbf{Y}, t)\,\mathrm{d}\mathbf{W}(t) + \frac{1}{2}\sum_{\ell=1}^{m}\int_{t_0}^{t}\frac{\partial\boldsymbol{\beta}}{\partial Y_\ell}\mathbf{D}\mathbf{B}_{\ell,\cdot}^{\mathrm{T}}(\mathbf{Y}, t)\,\mathrm{d}t
\end{aligned} \quad (5.204)$$

where $\mathbf{B}_{\ell,\cdot}^{\mathrm{T}} = (B_{\ell 1}, B_{\ell 2}, \ldots, B_{\ell r})^{\mathrm{T}}$, $\Delta t_j = t_j - t_{j-1}$, $\Delta\mathbf{W}_j = \mathbf{W}_j - \mathbf{W}_{j-1}$ and $\Delta W_{k,j} = W_k(t_j) - W_k(t_{j-1})$.

Combining Equations 5.201, 5.202 and 5.204 and comparing with Equation 5.187, we have

$$\boldsymbol{\alpha}(\mathbf{Y}, t) = \mathbf{A}(\mathbf{Y}, t) - \frac{1}{2}\sum_{\ell=1}^{m}\frac{\partial\mathbf{B}}{\partial Y_\ell}\mathbf{D}\mathbf{B}_{\ell,\cdot}^{\mathrm{T}}(\mathbf{Y}, t) \quad (5.205a)$$

$$\boldsymbol{\beta}(\mathbf{Y}, t) = \mathbf{B}(\mathbf{Y}, t) \quad (5.205b)$$

where Equation 5.205b is used to replace $\boldsymbol{\beta}$ by $\mathbf{B}$ in Equation 5.204 to yield Equation 5.205a, which is referred to as the Wong–Zakai correction (Wong and Zakai, 1965). Note that, when $\boldsymbol{\beta}$ is independent of $\mathbf{Y}$, the Wong–Zakai correction vanishes. Conversely, we have

$$\mathbf{A}(\mathbf{Y}, t) = \boldsymbol{\alpha}(\mathbf{Y}, t) + \frac{1}{2}\sum_{\ell=1}^{m}\frac{\partial\boldsymbol{\beta}}{\partial Y_\ell}\mathbf{D}\boldsymbol{\beta}_{\ell,\cdot}^{\mathrm{T}}(Y, t) \quad (5.206a)$$

$$\mathbf{B}(\mathbf{Y}, t) = \boldsymbol{\beta}(\mathbf{Y}, t) \quad (5.206b)$$

By Equations 5.206a and 5.206b, when a practical engineering structure subjected to physical white noise was modeled by the Stratonovich's stochastic differential equation, it can be transformed into an Itô stochastic differential equation which is mathematically favorable. This is particularly preferred in establishment of the FPK equation, as discussed in the following sections.

Before we go there, the existence and uniqueness for the stochastic differential equations should be noted here. For Equation 5.185, the following theorem can be proved (Øksendal, 2005). Let $T > 0$ and $\mathbf{A}(\mathbf{y}, t)$ and $\mathbf{B}(\mathbf{y}, t)$ be measurable functions satisfying

$$|\mathbf{A}(\mathbf{y},t)| + |\mathbf{B}(\mathbf{y},t)| \le C(1+|\mathbf{y}|) \qquad \mathbf{y} \in \mathbb{R}^m,\ t \in [0,T] \tag{5.207a}$$

for some constant C (where $|\mathbf{B}|^2 = \sum |B_{ij}|^2$) and such that

$$|\mathbf{A}(\mathbf{x},t) - \mathbf{A}(\mathbf{y},t)| + |\mathbf{B}(\mathbf{x},t) - \mathbf{B}(\mathbf{y},t)| \le D|\mathbf{x}-\mathbf{y}| \qquad \mathbf{x},\mathbf{y} \in \mathbb{R}^m,\ t \in [0,T] \tag{5.207b}$$

for some constant D. Let Z be a random variable which is independent of the σ-algebra $\mathcal{F}_\infty^{(m)}$ generated by $W(s)$, $s \ge 0$ and such that

$$\mathcal{E}[|Z|^2] < \infty \tag{5.207c}$$

Then the stochastic differential Equation 5.185 has a unique t-continuous solution $\mathbf{Y}(t, \varpi)$ with the property that $\mathbf{Y}(t, \varpi)$ is adapted to the filtration $\mathcal{F}_t^Z$ generated by Z and $W(s)$, $s \ge 0$, and

$$\mathcal{E}\left[\int_0^{\mathrm{T}} |\mathbf{Y}(t)|^2\, \mathrm{d}t\right] < \infty \tag{5.207d}$$

5.6.2 Fokker–Planck–Kolmogorov Equation

Let a stochastic process vector $\mathbf{Y}(t)$ be determined by the Itô stochastic differential Equation 5.185, which is quoted here for convenience:

$$\mathrm{d}\mathbf{Y}(t) = \mathbf{A}(\mathbf{Y},t)\,\mathrm{d}t + \mathbf{B}(\mathbf{Y},t)\,\mathrm{d}\mathbf{W}(t) \tag{5.208}$$

We are interested in the function, say $f(\mathbf{Y}(t))$, with $\mathbf{Y}(t)$ being the argument. As a stochastic process, let us first examine the evolution of the mean; that is, $\mathrm{d}\mathcal{E}[f(\mathbf{Y}(t))]/\mathrm{d}t$. Because the derivative and the expectation operator are interchangeable, there is

$$\frac{\mathrm{d}\mathcal{E}[f(\mathbf{Y}(t))]}{\mathrm{d}t} = \mathcal{E}\left(\frac{\mathrm{d}f(\mathbf{Y}(t))}{\mathrm{d}t}\right) \tag{5.209a}$$

Since $\mathbf{Y}(t)$ is determined by the Itô stochastic differential Equation 5.208, then according to Equation 5.196 it follows that

$$\begin{aligned}\mathcal{E}[\mathrm{d}f(\mathbf{Y}(t))] &= \mathcal{E}\left\{\sum_{\ell=1}^{m}\left[A_\ell\frac{\partial f}{\partial Y_\ell} + \frac{1}{2}\sum_{k=1}^{m}(\mathbf{BDB}^{\mathrm{T}})_{k\ell}\frac{\partial^2 f}{\partial Y_k \partial Y_\ell}\right]\mathrm{d}t + \sum_{\ell=1}^{m}\sum_{k=1}^{r}B_{\ell k}(\mathbf{Y}(t),t)\frac{\partial f}{\partial Y_\ell}\mathrm{d}W_k(t)\right\}\\ &= \mathcal{E}\left\{\sum_{\ell=1}^{m}\left[A_\ell\frac{\partial f}{\partial Y_\ell} + \frac{1}{2}\sum_{k=1}^{m}(\mathbf{BDB}^{\mathrm{T}})_{k\ell}\frac{\partial^2 f}{\partial Y_k \partial Y_\ell}\right]\mathrm{d}t\right\}\end{aligned} \tag{5.209b}$$

Here, use has been made of the fact that the $B_{\ell k}(\mathbf{Y}(t), t)$ are nonanticipating functions so that the second term with respect to $\mathrm{d}W_k(t)$ vanishes and $(\mathbf{BDB}^{\mathrm{T}})_{\ell k}$ represents the ℓkth component of the matrix $\mathbf{BDB}^{\mathrm{T}}$.

Combination of Equations 5.209a and 5.209b yields

$$\frac{\mathrm{d}\mathcal{E}[f(\mathbf{Y}(t))]}{\mathrm{d}t} = \mathcal{E}\left\{\sum_{\ell=1}^{m} A_\ell \frac{\partial f}{\partial Y_\ell} + \frac{1}{2}\sum_{\ell=1}^{m}\sum_{k=1}^{m}(\mathbf{BDB}^{\mathrm{T}})_{k\ell}\frac{\partial^2 f}{\partial Y_k \partial Y_\ell}\right\} \tag{5.209c}$$

On the other hand, denoting the conditional probability density of $\mathbf{Y}(t)\,|\,(\mathbf{y}_0, t_0)$ as $p(\mathbf{y}, t\,|\,\mathbf{y}_0, t_0)$, we have

$$\frac{\mathrm{d}\mathcal{E}[f(\mathbf{Y}(t))]}{\mathrm{d}t} = \frac{\mathrm{d}}{\mathrm{d}t}\int_{-\infty}^{\infty} f(\mathbf{y})p_{\mathbf{Y}}(\mathbf{y}, t|\mathbf{y}_0, t_0)\,\mathrm{d}\mathbf{y} = \int_{-\infty}^{\infty} f(\mathbf{y})\frac{\partial p_{\mathbf{Y}}(\mathbf{y}, t|\mathbf{y}_0, t_0)}{\partial t}\,\mathrm{d}\mathbf{y} \tag{5.210}$$

And simultaneously, the right-hand side of Equation 5.209c yields

$$\mathcal{E}\left\{\sum_{\ell=1}^{m} A_\ell \frac{\partial f}{\partial Y_\ell} + \frac{1}{2}\sum_{\ell=1}^{m}\sum_{k=1}^{m}(\mathbf{BDB}^{\mathrm{T}})_{k\ell}\frac{\partial^2 f}{\partial Y_k \partial Y_\ell}\right\} = \int_{-\infty}^{\infty}\left\{\sum_{\ell=1}^{m} A_\ell(\mathbf{y}, t)\frac{\partial f(\mathbf{y})}{\partial y_\ell}\right.$$
$$\left. + \frac{1}{2}\sum_{\ell=1}^{m}\sum_{k=1}^{m}[\mathbf{B}(\mathbf{y}, t)\mathbf{DB}^{\mathrm{T}}(\mathbf{y}, t)]_{k\ell}\frac{\partial^2 f(\mathbf{y})}{\partial y_k \partial y_\ell}\right\} p_{\mathbf{Y}}(\mathbf{y}, t|\mathbf{y}_0, t_0)\,\mathrm{d}\mathbf{y} \tag{5.211a}$$

Integrating by parts and noting usually it holds that

$$A_\ell(\mathbf{y}, t)f(\mathbf{y})p_{\mathbf{Y}}(\mathbf{y}, t|\mathbf{y}_0, t_0)\big|_{y_\ell \to \pm\infty} = \mathbf{0} \qquad \ell = 1, 2, \ldots, m \tag{5.211b}$$

$$\mathbf{B}(\mathbf{y}, t)\mathbf{DB}^{\mathrm{T}}(\mathbf{y}, t)\frac{\partial f(\mathbf{y})}{\partial y_\ell}p_{\mathbf{Y}}(\mathbf{y}, t|\mathbf{y}_0, t_0)\bigg|_{y_\ell \to \pm\infty} = \mathbf{0} \qquad \ell = 1, 2, \ldots, m \tag{5.211c}$$

$$\mathbf{B}(\mathbf{y}, t)\mathbf{DB}^{\mathrm{T}}(\mathbf{y}, t)f(\mathbf{y})p_{\mathbf{Y}}(\mathbf{y}, t|\mathbf{y}_0, t_0)\big|_{y_\ell \to \pm\infty} = \mathbf{0} \qquad \ell = 1, 2, \ldots, m \tag{5.211d}$$

Equation 5.211a becomes

$$\mathcal{E}\left\{\sum_{\ell=1}^{m} A_\ell \frac{\partial f}{\partial Y_\ell} + \frac{1}{2}\sum_{\ell=1}^{m}\sum_{k=1}^{m}(\mathbf{BDB}^{\mathrm{T}})_{k\ell}\frac{\partial^2 f}{\partial Y_k \partial Y_\ell}\right\} = \int_{\Omega} f(\mathbf{y})\left\{-\sum_{\ell=1}^{m}\frac{\partial[A_\ell(\mathbf{y}, t)p_{\mathbf{Y}}(\mathbf{y}, t|\mathbf{y}_0, t_0)]}{\partial y_\ell}\right.$$
$$\left. + \frac{1}{2}\sum_{\ell=1}^{m}\sum_{k=1}^{m}\frac{\partial^2\{[\mathbf{B}(\mathbf{y}, t)\mathbf{DB}^{\mathrm{T}}(\mathbf{y}, t)]_{k\ell}p_{\mathbf{Y}}(\mathbf{y}, t|\mathbf{y}_0, t_0)\}}{\partial y_k \partial y_\ell}\right\}\mathrm{d}\mathbf{y} \tag{5.211e}$$

Comparing Equations 5.210 and 5.211e and noting that $f(\mathbf{y})$ is arbitrary, we must have

$$\frac{\partial p_{\mathbf{Y}}}{\partial t} = -\sum_{\ell=1}^{m}\frac{\partial[A_\ell(\mathbf{y}, t)p_{\mathbf{Y}}]}{\partial y_\ell} + \frac{1}{2}\sum_{\ell=1}^{m}\sum_{k=1}^{m}\frac{\partial^2\{[\mathbf{B}(\mathbf{y}, t)\mathbf{DB}^{\mathrm{T}}(\mathbf{y}, t)]_{k\ell}p_{\mathbf{Y}}\}}{\partial y_k \partial y_\ell} \tag{5.212}$$

where $p_{\mathbf{Y}}$ represents $p_{\mathbf{Y}}(\mathbf{y}, t\,|\,\mathbf{y}_0, t_0)$ for simplicity.

Equation 5.212 is the well-known *FPK equation.*

If $\mathbf{Q}(\mathbf{y}, t) = [Q_{\ell k}]_{r\times r}$ is an orthogonal matrix –that is, $\mathbf{QQ}^{\mathrm{T}} = \mathbf{I}_{r\times r}$, where $\mathbf{I}_{r\times r}$ is the $r \times r$ unit matrix – it can be seen that $\mathbf{QDQ}^{\mathrm{T}} = \mathbf{D}$ because $\mathbf{D}$ is an $r \times r$ diagonal matrix. Therefore, if $\mathbf{W}(t)$ is replaced by $\mathbf{QW}(t)$, then the matrix $\mathbf{B}(\mathbf{y}, t)\mathbf{DB}^{\mathrm{T}}(\mathbf{y}, t)$ is replaced by $\mathbf{B}(\mathbf{y}, t)\mathbf{Q}(\mathbf{y}, t)\mathbf{DQ}^{\mathrm{T}}(\mathbf{y}, t)\mathbf{B}^{\mathrm{T}}(\mathbf{y}, t) = \mathbf{B}(\mathbf{y}, t)\mathbf{DB}^{\mathrm{T}}(\mathbf{y}, t)$, which is essentially invariant. This means that the FPK equation is invariant against the orthogonal transformation on the Wiener process vector. In other words, the stochastic differential equation associated with an FPK equation is nonunique. This may be important to understanding the relationship between the FPK equation and the related stochastic differential equation. Actually, the physical sense of the nonuniqueness is that the description of probability density usually corresponds to an infinite set of trajectories.

In addition, Equation 5.212 also holds when $p_{\mathbf{Y}}$ represents the instantaneous probability density $p_{\mathbf{Y}}(\mathbf{y}, t)$ instead of the transition probability density. This can be achieved if we note that $p_{\mathbf{Y}}(\mathbf{y}, t) = \int p_{\mathbf{Y}}(\mathbf{y}, t|\mathbf{y}_0, t_0)p_{\mathbf{Y}}(\mathbf{y}_0, t_0)\,\mathrm{d}\mathbf{y}_0$, where $p_{\mathbf{Y}}(\mathbf{y}_0, t_0)$ is the initial joint PDF.

We now further discuss the meanings of the coefficients in the FPK equation, Equation 5.212. According to Equation 5.182 in Section 5.6.1.1, noting that, given $\{\mathbf{Y}(t) = \mathbf{y}\}$, we have

$$\mathbf{A}(\mathbf{y}, t) = \frac{\mathcal{E}[\mathrm{d}\mathbf{Y}(t)|\mathbf{Y}(t) = \mathbf{y}]}{\mathrm{d}t} = \lim_{\Delta t\to 0}\frac{\mathcal{E}[\Delta\mathbf{Y}(t)|\mathbf{Y}(t) = \mathbf{y}]}{\Delta t} \tag{5.213a}$$

where $\Delta\mathbf{Y}(t) = \mathbf{Y}(t + \Delta t) - \mathbf{Y}(t)$. Likewise, from Equation (5.183b), we have

$$\begin{aligned}\mathbf{B}(\mathbf{y}, t)\mathbf{DB}^{\mathrm{T}}(\mathbf{y}, t) &= \frac{\mathcal{E}[\mathrm{d}\mathbf{Y}(t)\,\mathrm{d}\mathbf{Y}^{\mathrm{T}}(t)|\mathbf{Y}(t) = \mathbf{y}]}{\mathrm{d}t}\\ &= \lim_{\Delta t\to 0}\frac{\mathcal{E}\{[\Delta\mathbf{Y}(t)\Delta\mathbf{Y}^{\mathrm{T}}(t)]|\mathbf{Y}(t) = \mathbf{y}\}}{\Delta t}\end{aligned} \tag{5.213b}$$

These quantities are called the *derivate moments* (Moyal, 1949). Equation 5.213a means that the coefficient $\mathbf{A}(\mathbf{y}, t)$ reflects the average tendency; therefore, it is referred to as the drift coefficient, which characterizes the mean drift velocity. In contrast, Equation 5.213b means that the matrix $\mathbf{B}(\mathbf{y}, t)\mathbf{DB}^{\mathrm{T}}(\mathbf{y}, t)$ results from the effect of Brownian motion, or a diffusion process; therefore, it is referred to as the diffusion coefficient. Now it is clear that the physical sense of Equation 5.212 is that the change of probability is due to drift and diffusion.

By the way, we point out that the derivate moments of higher order vanish if $\mathbf{Y}(t)$ is Markovian; for instance, it can be proved that (Gardiner, 1983)

$$\tilde{C}_{ijk} = \lim_{\Delta t\to 0}\frac{\mathcal{E}\{[\Delta Y_i(t)\Delta Y_j(t)\Delta Y_k(t)]|\mathbf{Y}(t) = \mathbf{y}\}}{\Delta t} = 0 \tag{5.213c}$$

This is essentially due to the fact that the transition probability density of a Markov process is determined by only two time instants; therefore, only the moment of increment of no more than second order is in the order of increment of time. More specifically, if the Markov process is related to a Brownian motion process, then we have $\mathcal{E}\{[\mathrm{d}W(t)]^N\} = 0$ for $N > 2$; therefore, the derivate moment in the form of Equation 5.213c must vanish.

In a more general sense, the FPK equation can be deduced from the Chapman–Kolmogorov equation

$$p_{\mathbf{Y}}(\mathbf{y}, t|\mathbf{y}_0, t_0) = \int_{-\infty}^{\infty} p_{\mathbf{Y}}(\mathbf{y}, t|\mathbf{z}, t_1)p_{\mathbf{Y}}(\mathbf{z}, t_1|\mathbf{y}_0, t_0)\,\mathrm{d}\mathbf{z} \tag{5.214}$$

as a specific case of differential Chapman–Kolmogorov equation when there is no jumps (Gardiner, 1983), or it can also be derived as a probability evolution equation in the specific case of a Markov process (Lin, 1967).

The initial condition for the FPK equation (Equation 5.212), when $p_{\mathbf{Y}}$ represents the transition PDF $p_{\mathbf{Y}}(\mathbf{y}, t\,|\,\mathbf{y}_0, t_0)$, is

$$p_{\mathbf{Y}}(\mathbf{y}, t_0|\mathbf{y}_0, t_0) = \delta(\mathbf{y} - \mathbf{y}_0) \tag{5.215a}$$

In the case $p_{\mathbf{Y}}$ represents the instantaneous probability density $p_{\mathbf{Y}}(\mathbf{y}, t)$, the initial condition is

$$p_{\mathbf{Y}}(\mathbf{y}, t_0) = p_{\mathbf{Y}_0}(\mathbf{y}) \tag{5.215b}$$

The boundary condition, according to different physical problems, may be different. For instance, for some systems an absorbing or reflection barrier exists in the space. For structural response analysis, the simplest but most widely used conditions for the transition PDF and the instantaneous PDF are

$$p_{\mathbf{Y}}(\mathbf{y}, t|\mathbf{y}_0, t_0)\big|_{y_\ell \to \pm\infty} = 0 \tag{5.216a}$$

and

$$p_{\mathbf{Y}}(\mathbf{y}, t)\big|_{y_\ell \to \pm\infty} = 0 \tag{5.216b}$$

respectively.

5.6.3 Solution to the Fokker–Planck–Kolmogorov Equation

5.6.3.1 Closed-Form Transient Solution to the Fokker–Planck–Kolmogorov Equation

Linear Systems

Consider the linear stochastic differential equation

$$\mathrm{d}\mathbf{Y}(t) = \tilde{\mathbf{a}}\mathbf{Y}\,\mathrm{d}t + \tilde{\mathbf{b}}\,\mathrm{d}\mathbf{W}(t) \tag{5.217}$$

where $\tilde{\mathbf{a}} = [\tilde{a}_{ij}]_{m\times m}$ and $\tilde{\mathbf{b}} = [\tilde{b}_{ij}]_{m\times r}$ are the system matrix and the input force influence matrix respectively.

According to complex modal theory, Equation 5.217 can be uncoupled (Fang and Wang, 1986). Denote the eigenmatrix of $\tilde{\mathbf{a}}$ by $\boldsymbol{\psi}$. Let $\mathbf{Y}(t) = \boldsymbol{\psi}\mathbf{Z}(t)$, where $\mathbf{Z}(t) = (Z_1(t), Z_2(t), \ldots, Z_m(t))^{\mathrm{T}}$. Introducing this into Equation 5.217, pre-multiplying it by $\boldsymbol{\psi}^{\mathrm{T}}$ and noting the orthogonality, we have

$$\mathrm{d}\mathbf{Z}(t) = \mathbf{a}\mathbf{Z}\,\mathrm{d}t + \mathbf{b}\,\mathrm{d}\mathbf{W}(t) \tag{5.218a}$$

where $\mathbf{a} = \boldsymbol{\Psi}^{\mathrm{T}}\tilde{\mathbf{a}}\boldsymbol{\Psi} = \mathrm{diag}[a_1, a_2, \ldots, a_m]$ and $\mathbf{b} = [b_{jk}]_{m\times r} = \boldsymbol{\Psi}^{\mathrm{T}}\tilde{\mathbf{b}}$. The component form is given as

$$\mathrm{d}Z_\ell(t) = a_\ell Z_\ell\,\mathrm{d}t + \sum_{k=1}^{r} b_{\ell k}\,\mathrm{d}W_k(t) \qquad \ell = 1, 2, \ldots, m \tag{5.218b}$$

Because the $W_k(t)$ are Wiener processes, the process $W_\ell^\Sigma(t) = \sum_{k=1}^r b_{\ell k} W_k(t)$ is also a Wiener process, for which the variance is

$$\mathcal{E}\{[\mathrm{d}W_\ell^\Sigma(t)]^2\} = \sum_{k=1}^r \sum_{j=1}^r b_{\ell k} b_{\ell j} D_{kj}\, \mathrm{d}t = \kappa_\ell\, \mathrm{d}t \tag{5.219}$$

where $\kappa_\ell = \sum_{k=1}^r \sum_{j=1}^r b_{\ell k} b_{\ell j} D_{kj}$. The FPK equation associated with Equation 5.218b is then

$$\frac{\partial p_{Z_\ell}(z_\ell, t|z_{\ell,0}, t_0)}{\partial t} = -\frac{\partial[a_\ell z_\ell p_{Z_\ell}(z_\ell, t|z_{\ell,0}, t_0)]}{\partial z_\ell} + \frac{1}{2}\kappa_\ell \frac{\partial^2[p_{Z_\ell}(z_\ell, t|z_{\ell,0}, t_0)]}{\partial z_\ell^2} \tag{5.220}$$

where $p_{Z_\ell}(z_\ell, t|z_{\ell,0}, t_0)$ is the transition PDF of $Z_\ell(t)$. The initial condition is $p_{Z_\ell}(z_\ell, t|z_{\ell,0}, t_0) = \delta(z_\ell - z_{\ell,0})$.

Taking the Fourier transform on both sides of Equation 5.220 and noting that

$$\phi(\vartheta, t|z_{\ell,0}, t_0) = \int_{-\infty}^{\infty} p_{Z_\ell}(z_\ell, t|z_{\ell,0}, t_0) \mathrm{e}^{-\mathrm{i}\vartheta z_\ell}\, \mathrm{d}z_\ell \tag{5.221}$$

then from Equation 5.220 we have

$$\frac{\partial \phi}{\partial t} = a_\ell \vartheta \frac{\partial \phi}{\partial \vartheta} - \frac{1}{2}\kappa_\ell \vartheta^2 \phi \tag{5.222}$$

in the deduction of which use has been made of

$$\vartheta \frac{\partial \phi(\mathrm{i}\vartheta)}{\partial \vartheta} = \int_{-\infty}^{\infty} p_{Z_\ell}(z_\ell, t|z_{\ell,0}, t_0) \mathrm{i}\vartheta z_\ell\, \mathrm{e}^{-\mathrm{i}\vartheta z_\ell}\, \mathrm{d}z_\ell \tag{5.223}$$

which is clear from Equation 5.221.

The method of characteristics can be used to solve Equation 5.222. In order to do so, a subsidiary equation is introduced as follows:

$$\frac{\mathrm{d}t}{1} = -\frac{\mathrm{d}\vartheta}{a_\ell \vartheta} = -\frac{\mathrm{d}\phi}{\frac{1}{2}\kappa_\ell \vartheta^2 \phi} \tag{5.224}$$

The integral of the first equation is given as

$$\vartheta = c_\ell \mathrm{e}^{-a_\ell(t-t_0)} \tag{5.225a}$$

while introducing it into the second equation gives

$$\phi = c \exp\left(\frac{\kappa_\ell}{4a_\ell}\vartheta^2\right) \tag{5.225b}$$

The general solution to Equation 5.222, therefore, is

$$\phi(\vartheta, t|z_{\ell,0}, t_0) = g(\vartheta \mathrm{e}^{a_\ell(t-t_0)}) \exp\left(\frac{\kappa_\ell}{4a_\ell}\vartheta^2\right) \tag{5.225c}$$

where $g(\cdot)$ is an arbitrary function.

Note from Equation 5.221 that the initial condition

$$\phi(\vartheta, t_0|z_{\ell,0}, t_0) = \int_{-\infty}^{\infty} \delta(z_\ell - z_{\ell,0}) \mathrm{e}^{-\mathrm{i}\vartheta z_\ell} \, \mathrm{d}z_\ell = \mathrm{e}^{-\mathrm{i}\vartheta z_{\ell,0}} \tag{5.226}$$

From the above two equations it follows that

$$g(\vartheta) = \exp\left(-\mathrm{i}\vartheta z_{\ell,0} - \frac{\kappa_\ell}{4a_\ell} \vartheta^2 \right) \tag{5.227}$$

Hence, we have

$$\phi(\vartheta, t|z_{\ell,0}, t_0) = \exp\left\{ -\mathrm{i}\vartheta z_{\ell,0} \, \mathrm{e}^{a_\ell(t-t_0)} + \frac{\kappa_\ell}{4a_\ell} \vartheta^2 \left[1 - \mathrm{e}^{2a_\ell(t-t_0)} \right] \right\} \tag{5.228}$$

Taking the inverse Fourier transform on both sides of it yields

$$\begin{aligned} p_{Z_\ell}(z_\ell, t|z_{\ell,0}, t_0) &= \frac{1}{2\pi} \int_{-\infty}^{\infty} \phi(\vartheta, t|z_{\ell,0}, t_0) \mathrm{e}^{\mathrm{i}\vartheta z_\ell} \, \mathrm{d}\vartheta \\ &= \frac{1}{\sqrt{2\pi}\sigma_{Z_\ell}(t)} \exp\left[\frac{z_\ell - \mu_{Z_\ell}(t)}{2\sigma_{Z_\ell}(t)} \right]^2 \end{aligned} \tag{5.229}$$

where

$$\mu_{Z_\ell}(t) = z_{\ell,0} \, \mathrm{e}^{a_\ell(t-t_0)} \tag{5.230a}$$

$$\sigma_{Z_\ell}(t) = \frac{\kappa_\ell}{2a_\ell} \left[1 - \mathrm{e}^{2a_\ell(t-t_0)} \right] \tag{5.230b}$$

Equation 5.229 shows that the stochastic process $Z_\ell(t)$ governed by the linear stochastic differential equation (Equation 5.218b) is Gaussian. Because the Wiener processes $\mathbf{W}(t)$ are Gaussian, this means that the linear stochastic differential operator will transform a Gaussian process to another Gaussian process. Actually, directly from Equation 5.218b, it follows that

$$Z_\ell(t) = Z_{\ell,0} \, \mathrm{e}^{a_\ell(t-t_0)} + \int_{t_0}^{t} \mathrm{e}^{a_\ell(t-\tau)} \, \mathrm{d}W_\ell^{\Sigma}(\tau) \tag{5.231}$$

Note there that $\sum_{k=1}^{r} b_{\ell k} W_k(t)$ is replaced by $W_\ell^{\Sigma}(t)$ for notational convenience. Under the condition $Z_{\ell,0} = z_{\ell,0}$, noting that $\mathrm{e}^{a_\ell(t-\tau)}$ is a nonanticipating function and using the Itô integral (see Section 5.6.1), we immediately get

$$\mathcal{E}[Z_\ell(t)] = z_{\ell,0} \, \mathrm{e}^{a_\ell(t-t_0)} \tag{5.232a}$$

$$\mathcal{E}(\{Z_\ell(t) - \mathcal{E}[Z_\ell(t)]\}^2) = \frac{\kappa_\ell}{2a_\ell} \left[1 - \mathrm{e}^{2a_\ell(t-t_0)} \right] \tag{5.232b}$$

which are exactly the same as Equations 5.230a and 5.230b respectively.

The preceding derivation holds for all $\ell = 1, 2, \ldots, m$; therefore, the vector process $\mathbf{Z}(t)$ is a Gaussian vector process with the component $\mu_{Z_\ell}(t)$ of the mean $\boldsymbol{\mu}_{\mathbf{Z}}(t)$ given by Equation 5.230a for $\ell = 1, 2, \ldots, m$. However, because all the Z_ℓ are dependent on the $W_\ell(t)$, it is possible that the Z_ℓ are correlated rather than independent. In other words, $\mathbf{Z}(t)$ is a joint correlated Gaussian vector. This can be demonstrated when we compute

$$\begin{aligned}\mathcal{E}(\{Z_\ell(t) - \mathcal{E}[Z_\ell(t)]\}\{Z_k(t) - \mathcal{E}[Z_k(t)]\}) &= \int_{t_0}^{t}\int_{t_0}^{t} \mathrm{e}^{a_\ell(t-\tau_1)}\mathrm{e}^{a_k(t-\tau_2)} \\ &\quad \times \mathcal{E}\left[\left(\sum_{s=1}^{r} b_{\ell s}\mathrm{d}W_s(\tau_1)\right)\left(\sum_{j=1}^{r} b_{kj}\mathrm{d}W_j(\tau_2)\right)\right] \\ &= \frac{1}{a_\ell + a_k}\left(\sum_{s=1}^{r}\sum_{j=1}^{r} b_{\ell s}b_{kj}D_{sj}\right)\left[1 - \mathrm{e}^{(a_\ell + a_k)(t - t_0)}\right] \\ &= \frac{[\mathbf{bDb}^{\mathrm{T}}]_{\ell k}}{a_\ell + a_k}\left[1 - \mathrm{e}^{(a_\ell + a_k)(t - t_0)}\right]\end{aligned} \quad (5.233)$$

Therefore, let $\mathbf{C}_{\mathbf{Z}}(t)$ be the covariance matrix, then the component is given by Equation 5.233, namely $[\mathbf{C}_{\mathbf{Z}}(t)]_{\ell k} = \mathcal{E}(\{Z_\ell(t) - \mathcal{E}[Z_\ell(t)]\}\{Z_k(t) - \mathcal{E}[Z_k(t)]\})$.

On the other hand, the FPK equation associated with Equation 5.218a is

$$\frac{\partial p_{\mathbf{Z}}}{\partial t} = -\sum_{\ell=1}^{m}\frac{\partial[a_\ell z_\ell p_{\mathbf{Z}}]}{\partial z_\ell} + \frac{1}{2}\sum_{k=1}^{m}\sum_{\ell=1}^{m}\frac{\partial^2\{[\mathbf{bDb}^{\mathrm{T}}]_{\ell k}p_Z\}}{\partial z_\ell \partial z_k} \quad (5.234)$$

where $p_{\mathbf{Z}}$ represents the transition probability density $p_{\mathbf{Z}}(\mathbf{z}, t\,|\,\mathbf{z}_0, t_0)$.

According to the preceding analysis, the joint transition probability density is

$$p_{\mathbf{Z}}(\mathbf{z}, t|\mathbf{z}_0, t_0) = (2\pi)^{-m/2}|\mathbf{C}_{\mathbf{Z}}(t)|^{-1/2}\exp\left\{-\frac{1}{2}[\mathbf{z} - \boldsymbol{\mu}_{\mathbf{Z}}(t)]^{\mathrm{T}}\mathbf{C}_{\mathbf{Z}}^{-1}(t)[\mathbf{z} - \boldsymbol{\mu}_{\mathbf{Z}}(t)]\right\} \quad (5.235)$$

where the component of $\mu_{\mathbf{Z}}(t)$ is given by Equation 5.230a and the component of $\mathbf{C}_{\mathbf{Z}}(t)$ is given by Equation 5.233. This, of course, must be the closed-form solution to the FPK equation Equation 5.234, as also can be verified by directly introducing Equation 5.235 into Equation 5.234 and knowing the uniqueness of the solution.

Likewise, the FPK equation associated directly with Equation 5.217 is

$$\frac{\partial p_{\mathbf{Y}}}{\partial t} = -\sum_{\ell=1}^{m}\frac{\partial}{\partial y_\ell}\left(\sum_{k=1}^{m}\tilde{a}_\ell y_{\ell k}p_{\mathbf{Y}}\right) + \frac{1}{2}\sum_{k=1}^{m}\sum_{\ell=1}^{m}\frac{\partial^2\{[\tilde{\mathbf{b}}\mathbf{D}\tilde{\mathbf{b}}^{\mathrm{T}}]_{\ell k}p_{\mathbf{Y}}\}}{\partial y_\ell \partial y_k} \quad (5.236)$$

where $p_{\mathbf{Y}}$ represents the transition probability density $p_{\mathbf{Y}}(\mathbf{y}, t\,|\,\mathbf{y}_0, t_0)$, $[\tilde{\mathbf{b}}\mathbf{D}\tilde{\mathbf{b}}^{\mathrm{T}}]_{\ell k}$ is the component of the matrix $\tilde{\mathbf{b}}\mathbf{D}\tilde{\mathbf{b}}^{\mathrm{T}}$ and $\tilde{a}_\ell$ is the component of $\tilde{\mathbf{a}}$. The solution of Equation 5.236 is

$$p_{\mathbf{Y}}(\mathbf{y}, t|\mathbf{y}_0, t_0) = (2\pi)^{-m/2}|\mathbf{C}_{\mathbf{Y}}(t)|^{-1/2}\exp\left\{-\frac{1}{2}[\mathbf{y} - \boldsymbol{\mu}_{\mathbf{Y}}(t)]^{\mathrm{T}}\mathbf{C}_{\mathbf{Y}}^{-1}(t)[\mathbf{y} - \boldsymbol{\mu}_{\mathbf{Y}}(t)]\right\} \quad (5.237)$$

where the mean vector $\boldsymbol{\mu}_{\mathbf{Y}}(t)$ and the covariance matrix $\mathbf{C}_{\mathbf{Y}}(t)$ can be computed directly by corresponding operators on the stochastic integral solution to Equation 5.217 under the condition $\mathbf{Y}(t_0) = \mathbf{y}_0$. Because

$$\mathbf{Y}(t) = \mathbf{Y}_0 \, e^{\tilde{\mathbf{a}}(t-t_0)} + \int_{t_0}^{t} e^{\tilde{\mathbf{a}}(t-t_0)} \tilde{\mathbf{b}} \, d\mathbf{W}(t) \tag{5.238}$$

we can get

$$\boldsymbol{\mu}_{\mathbf{Y}}(t) = \mathbf{y}_0 \, e^{\tilde{\mathbf{a}}(t-t_0)} \tag{5.239a}$$

$$\mathbf{C}_{\mathbf{Y}}(t) = \int_{t_0}^{t} e^{\tilde{\mathbf{a}}(t-t_0)} \tilde{\mathbf{b}} \mathbf{D} \tilde{\mathbf{b}}^{\mathrm{T}} e^{\tilde{\mathbf{a}}^{\mathrm{T}}(t-t_0)} \, dt \tag{5.239b}$$

Certainly, the solution equation (Equation 5.237) can also be acquired by using the linear transformation $\mathbf{Y}(t) = \boldsymbol{\psi}\mathbf{Z}(t)$, where the transition probability density of $\mathbf{Z}(t)$ has been given by Equation 5.235.

By the way, the FPK Equation 5.234 can also be solved by directly using the Fourier transform and following the like steps as employed above.

The process defined by the form of Equation 5.217 is termed the Ornstein–Uhlenbeck process, due to their initial investigations in 1930 (Uhlenbeck and Ornstein, 1930).

Understanding Nonlinear Systems

For the FPK Equation 5.212 associated with the nonlinear stochastic differential Equation 5.185, now we consider the transition probability density $p_{\mathbf{Y}}(\mathbf{y}, t + \Delta t \,|\, \mathbf{z}, t)$ with the initial condition $p_{\mathbf{Y}}(\mathbf{y}, t \,|\, \mathbf{z}, t) = \delta(\mathbf{y} - \mathbf{z})$.

In the case Δt is appropriately small, the coefficients of Equation 5.212 can be regarded as invariant and, therefore, it is approximated by

$$\frac{\partial p_{\mathbf{Y}}}{\partial t} = -\sum_{\ell=1}^{m} A_{\ell}(\mathbf{z}, t) \frac{\partial p_{\mathbf{Y}}}{\partial y_{\ell}} + \frac{1}{2} \sum_{\ell=1}^{m} \sum_{k=1}^{m} [\mathbf{B}(\mathbf{z}, t) \mathbf{D} \mathbf{B}^{\mathrm{T}}(\mathbf{z}, t)]_{\ell k} \frac{\partial^2 p_{\mathbf{Y}}}{\partial y_{\ell} \partial y_k} \tag{5.240}$$

By the same steps employed in the preceding section, the solution of the above equation is

$$p_{\mathbf{Y}}(\mathbf{y}, t + \Delta t | \mathbf{z}, t) = (2\pi)^{-m/2} |\mathbf{C}_{\mathbf{Y}|\mathbf{z}}(t, \Delta t)|^{-1/2}$$
$$\times \exp\left\{ -\frac{1}{2} [\mathbf{y} - \boldsymbol{\mu}_{\mathbf{Y}|\mathbf{z}}(t, \Delta t)]^{\mathrm{T}} \mathbf{C}_{\mathbf{Y}|\mathbf{z}}^{-1}(t, \Delta t) \left[\mathbf{y} - \boldsymbol{\mu}_{\mathbf{Y}|\mathbf{z}}(t, \Delta t)\right] \right\} \tag{5.241}$$

where the mean vector and the covariance matrix are

$$\boldsymbol{\mu}_{\mathbf{Y}|\mathbf{z}}(t, \Delta t) = \mathbf{z} + \mathbf{A}(\mathbf{z}, t) \Delta t \tag{5.242a}$$

and

$$\mathbf{C}_{\mathbf{Y}|\mathbf{z}}(t, \Delta t) = \mathbf{B}(\mathbf{z}, t) \mathbf{D} \mathbf{B}^{\mathrm{T}}(\mathbf{z}, t) \Delta t \tag{5.242b}$$

respectively.

These results demonstrate that the instantaneous transition probability density is Gaussian, while the stochastic process vector can be regarded, in an appropriate small time increment $[t, t+\Delta t]$, as the superposition of a mean (deterministic process) and the effect of a diffusion process; namely:

$$\mathbf{Y}(t+\Delta t)=\mathbf{z}+\mathbf{A}(\mathbf{z},t)\Delta t+\mathbf{B}(\mathbf{z},t)\Delta\mathbf{W}(t) \tag{5.243}$$

It is certainly quite easy to acquire Equations 5.242a and 5.242b directly from Equation 5.243, instead of solving Equation 5.240. Alternatively:

$$\mathbf{Y}(t+\Delta t)=\mathbf{z}+\mathbf{A}(\mathbf{z},t)\Delta t+\boldsymbol{\eta}(\mathbf{z},t)(\Delta t)^{1/2} \tag{5.244}$$

where $\boldsymbol{\eta}(\mathbf{z}, t)$ is an m-dimensional zero-mean Gaussian stochastic process vector with covariance matrix $\mathcal{E}[\boldsymbol{\eta}(\mathbf{z},t)\boldsymbol{\eta}^T(\mathbf{z},t)]=\mathbf{B}(\mathbf{z},t)\mathbf{D}\mathbf{B}^{\mathrm{T}}(\mathbf{z},t)$. Clearly, Equation 5.244 is consistent with Equation 5.243. The above equation shows that the trajectory of $\mathbf{Y}(t)$ is continuous but is quite irregular due to the terms of $(\Delta t)^{1/2}$. More specifically, if $\mathbf{A}(\mathbf{z}, t)=0$ and $\boldsymbol{\eta}(\mathbf{z}, t)$ do not depend on $\mathbf{z}$ and t, then the trajectory of $\mathbf{Y}(t)$ is so irregular that it is undifferentiable at any time.

Moreover, from the understanding that the process $\mathbf{Y}$ can be regarded as the combination of a drift process and the effect of a diffusion process (see Equations 5.243 and 5.244), we can derive the FPK equation using the principle of preservation of probability as a physical basis. This issue will be discussed in detail in Section 6.3.2.

5.6.3.2 Notes on Solution to General Fokker–Planck–Kolmogorov Equation

In past decades, great endeavors have been devoted to seeking solutions to the FPK equation. As demonstrated in the preceding section, the closed-form solution to the FPK equation associated with linear MDOF systems is known. Moreover, the closed-form solution to the FPK equation associated with some specific SDOF nonlinear systems, say, the Duffing oscillator, is available (Caughey, 1971; Zhu and Wu, 1990). However, as far as the solution to the FPK equation associated with general MDOF nonlinear systems is concerned, little is available so far in spite of the decades of effort.

The problem with less difficulty is the steady-state, stationary solution to the FPK equation; that is, the solution as $t\rightarrow\infty$ is time independent provided specified conditions are satisfied. In this case, $\partial p_{\mathbf{Y}}/\partial t=0$ and the FPK equation, Equation 5.212, reduces to

$$-\sum_{j=1}^{m}\frac{\partial[A_j(\mathbf{y})p_{\mathbf{Y}}]}{\partial y_j}+\frac{1}{2}\sum_{j=1}^{m}\sum_{k=1}^{m}\frac{\partial^2}{\partial y_j\partial y_k}\left\{[\mathbf{B}(\mathbf{y})\,\mathbf{D}\mathbf{B}^{\mathrm{T}}(\mathbf{y})]_{jk}p_{\mathbf{Y}}\right\}=0 \tag{5.245}$$

In the past decade, a family of new approaches in the framework of a Hamiltonian formulation has been developed and the availability of stationary solutions is greatly extended (Zhu, 2003, 2006). Unfortunately, although in many problems of practical interest the stationary solution is meaningful or sometimes provides enough information, this is not the case in many problems in, say, earthquake engineering, where the transient response, or nonstationary response, is, of course, what really matters.

Meanwhile, a number of numerical approaches for the solution of the FPK equation have also been investigated: for example, the path-integral method (Wehner and Wolf, 1983; Naess and Johnsen, 1993; Naess and Moe, 2000), the finite-element method (Spencer and Bergman,

1993), the random walk method, the cell-mapping method and expansion solutions and so on (Schuëller, 1997). However, no approach works in the FPK equation of large dimension, for $m \geq 6$ for instance, while in the problems of practical interest the dimension is usually in the order from hundreds to millions. The gap is so huge that in the near future few approaches can tackle such problems in the theoretical frame of the FPK equation.

6

Probability Density Evolution Analysis: Theory

6.1 Introduction

In his investigations on differential equations, Liouville proved a theorem in 1838, the alternative form of which in the Hamiltonian systems was later called the celebrated Liouville theorem on the volume of phase space (see Lützen (1990)). This theory was later extensively elaborated and developed by Gibbs (1902) in statistical mechanics and finally resulted in what is now called the Liouville equation. The equation could reflect the evolution of the joint probability density of states of systems with randomness involved only in the initial conditions (Kozin, 1961; Syski, 1967; Soong, 1973; Arnold, 1978).

On the other hand, as described in Chapter 1, Einstein derived a diffusion equation in terms of the probability density of the position of a particle in Brownian motion in 1905 (Einstein, 1905). The sequent investigations on more general cases where the effects of drift and diffusion occur simultaneously by Fokker (1914) and Planck (1917) led to the celebrated equation nowadays attributed to their names in physicist circles. Without knowing their pioneering work, Kolmogorov independently established the same partial differential equation in his investigations on Markov processes (Kolmogorov, 1931). In addition to what is now called the FPK equation, he also came up with a backward equation in the same paper. This was not only very elegant result, but also of methodological sense, because it indicates that a stochastic system can be treated by a deterministic equation. Since then, with the explosive development of theory on stochastic processes and stochastic differential equations, particularly the establishment of the Itô and Stratonovich's stochastic calculus and the straightforward relation between the stochastic differential equation and the FPK equation, as discussed in detail in Chapter 5, the FPK equation has become one of the major tools in a wide range of science and engineering disciplines (Itô, 1957, Stratonovich, 1963; Lin, 1967; Gihman and Skorohod, 1975). Representing the stochastic excitations in a different way from Kolmogorov, Dostupov and Pugachev (1957) transformed a system with stochastic excitations to a system involving random parameters and reached a partial differential equation, which is in a form like the Liouville equation containing parameters.

Obviously, the Liouville equation, the FPK equation and the Dostupov–Pugachev equation are different types of probability density evolution equations. Had these equations been solvable, we could capture the evolution of the probability density of a stochastic system. Unfortunately, this is not the case for most problems of practical interest, because these equations are usually high-dimensional partial differential equations and usually strong nonlinearity is involved in the coefficients. Despite great endeavors, the available solutions to the above equations are still very limited (Soong, 1973; Risken, 1984; Zhu, 2003, 2006).

Since the celebrated work of Kolmogorov (1931), the rigorous mathematical aspect has been increasingly stressed. While great advancement has been made, the physical fundamental aspect seems to have been somewhat ignored by many investigators. In 2003 to 2006, on the grounds of a physical fundamental point of view, the Dostupov–Pugachev equation was first uncoupled for the linear systems by Li and Chen (2003, 2004a) and then a generalized density evolution equation (GDEE) was established in a unified way for linear and nonlinear systems (Li and Chen, 2004b, 2006a, 2006c). This family of density evolution equations threw a new light upon the feasibility of probability density evolution analysis for stochastic dynamical systems.

In this chapter, thoughts on the evolution of densities are elaborately investigated. The principle of preservation of probability is adopted as a unified foundation to derive different types of probability density evolution equations. Rather than mathematics-oriented derivations, a direct physical treatment cooperating with the state space description of the principle of preservation of probability is adopted to re-establish the Liouville equation and the FPK equation. In addition, the Dostupov–Pugachev equation is found to be the result when a hybrid treatment of the random event description of the principle of preservation of probability and coupling physical equations of the system is adopted. As a logically spontaneous result, when the problem is viewed from the random event description and uncoupling physical equations, the generalized probability density evolution equation is reached.

For brevity, in this chapter we will sometimes just use the term 'density' to represent 'probability density' or 'PDF.'

6.2 The Principle of Preservation of Probability

6.2.1 Functions of Random Variables and their Probability Density Function Revisited

Let $X(\varpi)$ be a continuous random variable with PDF $p_X(x)$; namely:

$$\Pr\{X(\varpi) \in (x, x+\mathrm{d}x)\} = \mathrm{d}\Pr\{\varpi\} = p_X(x)\,\mathrm{d}x \tag{6.1}$$

where $\Pr\{\cdot\}$ is the probability measure and ϖ represents a basic random event.

If a map $\mathcal{G}$ exists from X to Y such that

$$\mathcal{G}: X \to Y \quad \text{or} \quad Y = g(X) \tag{6.2}$$

then Y is a random variable. Denoting the density of Y by $p_Y(y)$, our present task is to obtain $p_Y(y)$ through the known density $p_X(x)$.

Suppose the CDF of Y is differentiable; we then have

$$p_Y(y)\,\mathrm{d}y = \Pr\{Y(\varpi) \in (y, y+\mathrm{d}y)\} \tag{6.3a}$$

or

$$p_Y(y) = \frac{\mathrm{d}}{\mathrm{d}y}\Pr\{Y(\varpi) \in (y, y+\mathrm{d}y)\} \tag{6.3b}$$

Because

$$\Pr\{Y(\varpi) \in (y, y+\mathrm{d}y)\} = \mathrm{d}\Pr\{\varpi\} \tag{6.4}$$

and noting Equations 6.4 and 6.1, it follows that

$$\Pr\{Y(\varpi) \in (y, y+\mathrm{d}y)\} = \Pr\{X(\varpi) \in (x, x+\mathrm{d}x)\} = \mathrm{d}\Pr\{\varpi\} \tag{6.5}$$

Namely:

$$p_Y(y)\,\mathrm{d}y = p_X(x)\,\mathrm{d}x \tag{6.6}$$

For the same set of ϖ, the relationship between $X(\varpi)$ and $Y(\varpi)$ is given by Equation 6.2; therefore, the relationship

$$y = g(x) \tag{6.7}$$

exists for x and y in Equation 6.6. Consequently, if $g(\cdot)$ has an inverse, denoting the inverse function by $g^{-1}(\cdot)$, then from Equation 6.6 for the monotonic functions we have[1]

$$p_Y(y) = p_X\left[g^{-1}(y)\right]\frac{\mathrm{d}x}{\mathrm{d}y} = |J|p_X\left[g^{-1}(y)\right] \tag{6.8a}$$

where J is the Jacobian:

$$J = \frac{\mathrm{d}x}{\mathrm{d}y} = \left[\frac{1}{\mathrm{d}g(x)/\mathrm{d}x}\right]_{x=g^{-1}(y)} = \frac{\mathrm{d}g^{-1}(y)}{\mathrm{d}y} \tag{6.8b}$$

In fact, Equations 6.5 and 6.6 are more fundamental than Equation 6.8a, in that no constraints are imposed on the attributes of $g(\cdot)$ in the former two equations. Likewise, the same thoughts hold in the case of random vectors.

Denote the joint density of the random vector $\mathbf{X} = (X_1, X_2, \ldots, X_n)^{\mathrm{T}}$ by $p_{\mathbf{X}}(\mathbf{x})$, where $\mathbf{x} = (x_1, x_2, \ldots, x_n)^{\mathrm{T}}$; namely:

$$\Pr\{\mathbf{X}(\varpi) \in (\mathbf{x}, \mathbf{x}+\mathrm{d}\mathbf{x})\} = \mathrm{d}\Pr\{\varpi\} = p_{\mathbf{X}}(\mathbf{x})\,\mathrm{d}\mathbf{x} \tag{6.9}$$

Suppose there is a map determining a vector $\mathbf{Y} = (Y_1, Y_2, \ldots, Y_m)^{\mathrm{T}}$ by $\mathbf{X}$ through

$$\mathcal{G}:\mathbf{X} \to \mathbf{Y} \quad \text{or} \quad \mathbf{Y} = g(\mathbf{X}) \tag{6.10}$$

Here, n and m are respectively the dimension of $\mathbf{X}$ and $\mathbf{Y}$.

Denoting the joint density of $\mathbf{Y}$ by $p_{\mathbf{Y}}(\mathbf{y})$, where $\mathbf{y} = (y_1, y_2, \ldots, y_m)^{\mathrm{T}}$, we of course have

$$\Pr\{\mathbf{Y}(\varpi) \in (\mathbf{y}, \mathbf{y}+\mathrm{d}\mathbf{y})\} = \mathrm{d}\Pr\{\varpi\} = p_{\mathbf{Y}}(\mathbf{y})\,\mathrm{d}\mathbf{y} \tag{6.11}$$

[1] Here, note that, because the probability densities are nonnegative, $\mathrm{d}x$ and $\mathrm{d}y$ should both be positive. Thus, an absolute value of the Jacobian is adopted here.

Considering Equation 6.11 in conjunction with Equation 6.9 yields[2]

$$p_{\mathbf{Y}}(\mathbf{y})\,\mathrm{d}\mathbf{y} = p_{\mathbf{X}}(\mathbf{x})\,\mathrm{d}\mathbf{x} \tag{6.12}$$

where, from Equation 6.10:

$$\mathbf{y} = g(\mathbf{x}) \tag{6.13}$$

Equations 6.12 and 6.13 mean that when there exists a map from the space of **X** to the space of **Y**, then if we can find **X** in an element domain $\mathrm{d}\mathbf{x}$ in the space of **X** with a prescribed probability, then we must be able to find **Y** in a corresponding element domain $\mathrm{d}\mathbf{y}$ in the space of **Y**, which is determined by the map from **X** to **Y**, with the same probability. In this sense, the probability is preserved in a map, as shown schematically in Figure 6.1. The principle can be called the *principle of preservation of probability*.

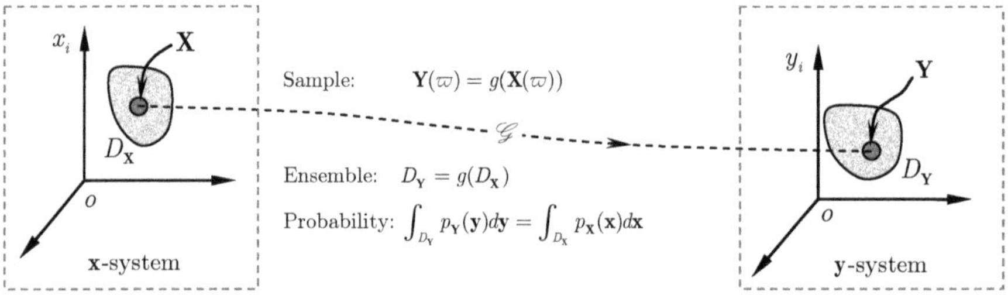

Figure 6.1 Map and preservation of probability.

The map in Equation 6.10 is essentially in the sense of a sample. Consequently, if an arbitrary ensemble domain $D_{\mathbf{X}}$ is considered in the **x**-system, then a corresponding ensemble domain $D_{\mathbf{Y}}$ is determined by the map

$$\mathcal{G}: D_{\mathbf{X}} \to D_{\mathbf{Y}} \quad \text{or} \quad D_{\mathbf{Y}} = g(D_{\mathbf{X}}) \tag{6.14}$$

Because the probability is preserved in the map of any arbitrary element, we then have

$$\int_{D_{\mathbf{Y}}} p_{\mathbf{Y}}(\mathbf{y})\,\mathrm{d}\mathbf{y} = \int_{D_{\mathbf{X}}} p_{\mathbf{X}}(\mathbf{x})\,\mathrm{d}\mathbf{x} \tag{6.15}$$

This can be understood as an integral form of preservation of probability. From Equation 6.15, it is known that $p_{\mathbf{Y}}(\mathbf{y})$ can be determined by $p_{\mathbf{X}}(\mathbf{x})$. Formally, we have

$$p_{\mathbf{Y}}(\mathbf{y}) = \mathcal{F}[p_{\mathbf{X}}(\mathbf{x})] \tag{6.16}$$

where $\mathcal{F}$ is called the *Frobenius–Perron operator* (Lasota and Mackey, 1994).

[2] A more rigorous and general treatment is to use the form of Riemann–Stieltjes integral instead of the Riemann integral, which is widely used in many monographs on probability theory because it can cover the case when discontinuity occurs in the distribution functions (Loève, 1977).

With Equations 6.12 and 6.13 in mind, for a one-to-one map and $n = m$, we have

$$p_{\mathbf{Y}}(\mathbf{y}) = |J| p_{\mathbf{X}}[\mathbf{x} = g^{-1}(\mathbf{y})]$$

where $J = |\partial\mathbf{x}/\partial\mathbf{y}|$ is the Jacobian.

The preceding elaboration indicates that the principle of preservation of probability results from the fact that the probability will be preserved as long as the random event, which can be represented in a different but equivalent form and is the argument of the associated random variables, is retained in the map. Here, the concept is essentially important that the random variable itself is not an independent argument but a function of basic random events. Insight is provided in Equations 6.5, 6.9 and 6.11 where the term $\mathrm{d}\,\Pr\{\varpi\}$ is explicitly written, which is rooted in the measure theory of probability (Chung, 1974).

6.2.2 *The Principle of Preservation of Probability*

In a general sense, the principle of preservation of probability can be stated as: if the random factors involved in a stochastic system are retained – in other words, if no new random factors arise nor existing factors vanish in a physical process – then the probability will be preserved in the evolution process of the system.

We will now investigate the principle both from the random event description and the state space description.

6.2.2.1 Random Event Description of the Principle of Preservation of Probability

The analysis in Section 6.2.1 on the functions of random variables via a map holds in a very general sense, because any function, transformation, or operator can be regarded as a map. Specifically, a dynamical system can also be taken into account in this framework.

For example, consider a dynamical system with the state equation

$$\dot{\mathbf{Y}} = \mathbf{A}(\mathbf{Y}, t) \qquad \mathbf{Y}(t_0) = \mathbf{Y}_0 \tag{6.17}$$

where $\mathbf{Y} = (Y_1, Y_2, \ldots, Y_m)^{\mathrm{T}}$ is the m-dimensional state vector, $\mathbf{Y}_0 = (Y_{1,0}, Y_{2,0}, \ldots, Y_{m,0})^{\mathrm{T}}$ is the initial value vector and $\mathbf{A} = (A_1, A_2, \ldots, A_m)^{\mathrm{T}}$ is the m-dimensional operator vector.

This system establishes a map $\mathcal{G}_t$ from $\mathbf{Y}(t_0)$ to $\mathbf{Y}(t)$; namely:

$$\mathcal{G}_t : (Y_1(t_0), Y_2(t_0), \ldots, Y_m(t_0)) \to (Y_1(t), Y_2(t), \ldots, Y_m(t)) \quad \text{or} \quad \mathbf{Y}(t) = g[t, \mathbf{Y}(t_0)] \tag{6.18a}$$

Here, $\mathbf{Y}(t)$ is the solution of Equation 6.17, or, in other words, the Lagrangian description of the system in Equation 6.17.

Compared with the map in Equation 6.10, $\mathcal{G}$, $\mathbf{X}$ and $\mathbf{Y}$ here are respectively replaced by $\mathcal{G}_t$, $\mathbf{Y}(t_0)$ and $\mathbf{Y}(t)$. Accordingly, Figure 6.1 becomes Figure 6.2.

Examine an arbitrary ensemble domain D_{t_0} in the state space, which belongs to the definition domain. The map in Equation 6.18a simultaneously determines a corresponding domain D_t; namely:

$$D_t = g(t, D_{t_0}) \tag{6.18b}$$

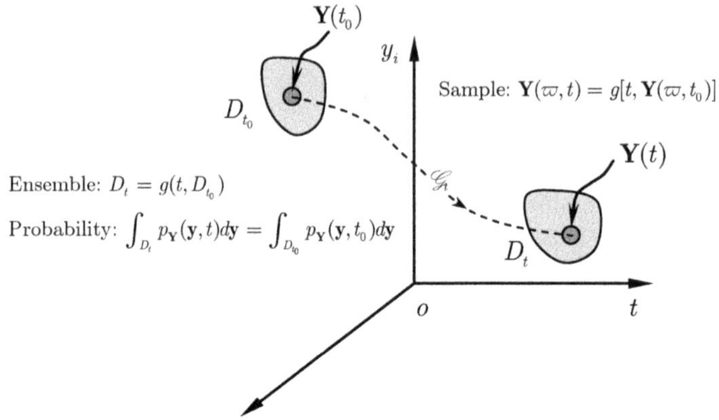

Figure 6.2 Dynamical system, map and probability.

According to Equation 6.15, it follows that

$$\int_{D_t} p_{\mathbf{Y}}(\mathbf{y}, t)\,\mathrm{d}\mathbf{y} = \int_{g(t, D_{t_0})} p_{\mathbf{Y}}(\mathbf{y}, t)\,\mathrm{d}\mathbf{y} = \int_{D_{t_0}} p_{\mathbf{Y}}(\mathbf{y}, t_0)\,\mathrm{d}\mathbf{y} \tag{6.19}$$

where $p_{\mathbf{Y}}(\mathbf{y}, t)$ is the joint PDF of $\mathbf{Y}(t)$.

This is essentially what the Liouville theorem 'on the volume in phase space' states, and the latter is extensively elaborated by Gibbs in statistical mechanics (Gibbs, 1902; Syski, 1967; Arnold, 1978; Lützen, 1990; Lasota and Mackey, 1994).

Certainly, Equation 6.19 is equivalent to

$$\frac{\mathrm{D}}{\mathrm{D}t}\int_{D_t} p_{\mathbf{Y}}(\mathbf{y}, t)\,\mathrm{d}\mathbf{y} = 0 \tag{6.20}$$

where $\mathrm{D}(\cdot)/\mathrm{D}t$ denotes the total derivative, which is usually also called the substantial or material derivative (Fung and Tong, 2001). Equation 6.20 should be understood as

$$\frac{\mathrm{D}}{\mathrm{D}t}\int_{D_t} p_{\mathbf{Y}}(y, t)\,\mathrm{d}\mathbf{y} = \lim_{\Delta t \to 0}\frac{1}{\Delta t}\left[\int_{D_{t+\Delta t}} p_{\mathbf{Y}}(\mathbf{y}, t+\Delta t)\,\mathrm{d}\mathbf{y} - \int_{D_t} p_{\mathbf{Y}}(\mathbf{y}, t)\,\mathrm{d}\mathbf{y}\right] = 0 \tag{6.21}$$

In the previous discussions, the principle of preservation of probability is understood by rooting on the samples and then ensembles. That is, tracing the trajectory of a given random event ϖ, which is shown clearly schematically in Figure 6.2. We refer to this point of view as the *random event description* of the principle of preservation of probability. Recalling the motion of particles in continuum physics, this is of course the counterpart of the *Lagrangian description* (Fung, 1994; Dafermos, 2000; Li and Chen, 2008; Chen and Li, 2009).

6.2.2.2 State Space Description of the Principle of Preservation of Probability

Instead of tracing the trajectories, a dynamical system can also be examined by considering the change of quantities in an arbitrary fixed domain D_{fixed}; that is, using the domain D_{fixed} as a window to figure out what happens there.

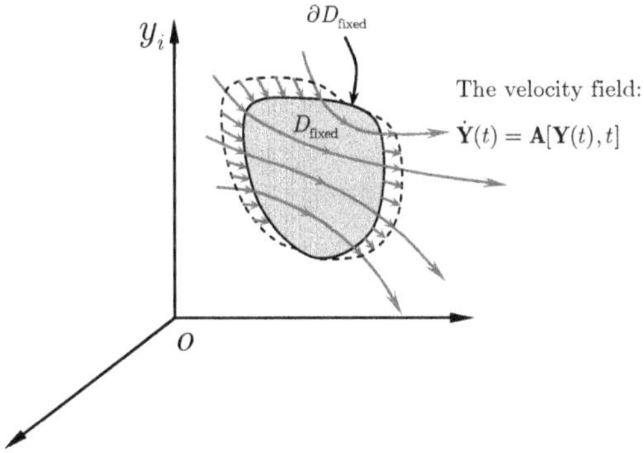

Figure 6.3 Fixed domain and transition of probability.

Focusing attention now on the fixed domain D_{fixed} with boundary $\partial D_{\text{fixed}}$ in the state space, because this domain is fixed in a time-variant velocity field determined by Equation 6.17, the phase of an ensemble will change with time. As time goes, through the boundary $\partial D_{\text{fixed}}$ some of the phase particles will enter this window whereas some other phase particles might go out of the window (see Figure 6.3). Adherent to the phase particles is the probability. Therefore, according to the principle of preservation of probability, during an arbitrary time interval $[t_1, t_2]$, the incremental change of probability in the domain D_{fixed} is due to the transition of probability through the boundary $\partial D_{\text{fixed}}$. Mathematically, this means that

$$\Delta_{[t_1,t_2]} P_{D_{\text{fixed}}} = \Delta_{[t_1,t_2]} P_{\partial D_{\text{fixed}}} \tag{6.22}$$

where

$$\Delta_{[t_1,t_2]} P_{D_{\text{fixed}}} = \int_{D_{\text{fixed}}} p_{\mathbf{Y}}(\mathbf{y}, t_2)\, \mathrm{d}\mathbf{y} - \int_{D_{\text{fixed}}} p_{\mathbf{Y}}(\mathbf{y}, t_1)\, \mathrm{d}\mathbf{y} \tag{6.23}$$

is the incremental change of probability in the domain D_{fixed} during the time interval $[t_1, t_2]$ and

$$\Delta_{[t_1,t_2]} P_{\partial D_{\text{fixed}}} = -\int_{t_1}^{t_2} \int_{\partial D_{\text{fixed}}} p_{\mathbf{Y}}(\mathbf{y}, t)(\mathbf{v}\mathrm{d}t) \cdot \mathbf{n}\, \mathrm{d}S = -\int_{t_1}^{t_2} \int_{\partial D_{\text{fixed}}} p_{\mathbf{Y}}(\mathbf{y}, t)\mathbf{A}(\mathbf{y}, t) \cdot \mathbf{n}\, \mathrm{d}S\, \mathrm{d}t \tag{6.24}$$

is the probability transiting through the boundary $\partial D_{\text{fixed}}$ during the time interval $[t_1, t_2]$, where $\mathbf{n}$ is the unit outward normal vector of the boundary surface $\partial D_{\text{fixed}}$; the negative sign exists because when the probability transits outward the retained probability in the domain decreases.

Here, the principle of preservation of probability is understood in the state space by examining the change of the probability density at a fixed position in the space. This shows that, because of the preservation of probability, the increment of probability in any arbitrary fixed domain is equivalent to the probability imported through the boundary.

6.3 Markovian Systems and State Space Description: Liouville and Fokker–Planck–Kolmogorov Equations

6.3.1 The Liouville Equation

6.3.1.1 Derivation of the Liouville Equation

In many systems of engineering interest, the initial conditions might not be known exactly. This uncertainty can usually be described by random variables with known probability distributions. Without loss of generality, the state equation of the system and the initial condition read

$$\dot{\mathbf{Y}} = \mathbf{A}(\mathbf{Y}, t) \qquad \mathbf{Y}(t_0) = \mathbf{Y}_0 \tag{6.25}$$

where $\mathbf{Y} = (Y_1, Y_2, \cdots, Y_m)^{\mathrm{T}}$ is the m-dimensional state vector, $\mathbf{A} = (A_1, A_2, \ldots, A_m)^{\mathrm{T}}$ is a deterministic m-dimensional operator vector and $\mathbf{Y}_0 = (Y_{1,0}, Y_{2,0}, \ldots, Y_{m,0})^{\mathrm{T}}$ is the initial value vector which is random with known joint density $p_{\mathbf{Y}_0}(\mathbf{y}_0)$ in which $\mathbf{y}_0$ denotes $(y_{1,0}, y_{2,0}, \ldots, y_{m,0})$.

Being a first-order ordinary differential equation, if $\mathbf{A}$ is well behaved, the solution process $\mathbf{Y}(t)$ is completely determined once the initial vector is known. Further, at any time $t_1 \geq t_0$, $\mathbf{Y}(t_1)$ is completely determined; therefore, it can be regarded as the new initial condition for $t \geq t_1$ if t_1 is regarded as the new initial time. Thus, once $\mathbf{Y}(t_1)$ is known, the solution process $\mathbf{Y}(t)$ for $t \geq t_1$ is completely determined without knowledge at time $t < t_1$; namely:

$$\{\mathbf{Y}(t), t > t_1 | \mathbf{Y}(\tau), \tau \leq t_1\} = \{\mathbf{Y}(t), t > t_1 | \mathbf{Y}(t_1)\} \tag{6.26}$$

The above discussions hold true no matter whether $\mathbf{Y}_0$ is a deterministic or random vector. Consequently, the stochastic process $\mathbf{Y}(t)$ determined by the system in Equation 6.25 is a Markov process.

Because all the randomness comes from the initial condition without other random factors involved, the system in Equation 6.25 is a probability preserved system. According to the state space description of the principle of preservation of probability elaborated in Section 6.2.2.2, when the behavior of the phase particles in an arbitrary fixed domain D_{fixed} during an arbitrary time duration $[t_1, t_2]$ is examined, Equations 6.22–6.24 hold.

Further, the incremental change of probability in D_{fixed} during the time interval $[t_1, t_2]$, given by Equation 6.23, can be rewritten as

$$\begin{aligned} \Delta_{[t_1,t_2]} P_{D_{\text{fixed}}} &= \int_{D_{\text{fixed}}} p_{\mathbf{Y}}(\mathbf{y}, t_2)\, \mathrm{d}\mathbf{y} - \int_{D_{\text{fixed}}} p_{\mathbf{Y}}(\mathbf{y}, t_1)\, \mathrm{d}\mathbf{y} \\ &= \int_{D_{\text{fixed}}} \int_{t_1}^{t_2} \frac{\partial p_{\mathbf{Y}}(\mathbf{y}, t)}{\partial t}\, \mathrm{d}\mathbf{y}\, \mathrm{d}t \end{aligned} \tag{6.27}$$

Simultaneously, the probability of transiting through the boundary $\partial D_{\text{fixed}}$ during the time interval $[t_1, t_2]$, given by Equation 6.24, can be rearranged into

$$\begin{aligned} \Delta_{[t_1,t_2]} P_{\partial D_{\text{fixed}}} &= -\int_{t_1}^{t_2} \int_{\partial D_{\text{fixed}}} p_{\mathbf{Y}}(\mathbf{y}, t)(\mathbf{v}\, \mathrm{d}t) \cdot \mathbf{n}\, \mathrm{d}S = -\int_{t_1}^{t_2} \int_{\partial D_{\text{fixed}}} p_{\mathbf{Y}}(\mathbf{y}, t) \mathbf{A}(\mathbf{y}, t) \cdot \mathbf{n}\, \mathrm{d}S\, \mathrm{d}t \\ &= -\int_{t_1}^{t_2} \int_{D_{\text{fixed}}} \sum_{\ell=1}^{m} \frac{\partial [p_{\mathbf{Y}}(\mathbf{y}, t) A_\ell(\mathbf{y}, t)]}{\partial y_\ell}\, \mathrm{d}\mathbf{y}\, \mathrm{d}t \end{aligned} \tag{6.28}$$

where use has been made of the *divergence theorem* (Korn and Korn,1968).

The probability being preserved requires that $\Delta_{[t_1,t_2]}P_{D_{\text{fixed}}} = \Delta_{[t_1,t_2]}P_{\partial D_{\text{fixed}}}$ (see Equation 6.22). Substituting Equations 6.27 and 6.28 in this and noting the arbitrariness of D_{fixed} and $[t_1, t_2]$, the integrands must be equal; therefore:

$$\frac{\partial p_{\mathbf{Y}}(\mathbf{y},t)}{\partial t} + \sum_{\ell=1}^{m} \frac{\partial[p_{\mathbf{Y}}(\mathbf{y},t)A_\ell(\mathbf{y},t)]}{\partial y_\ell} = 0 \tag{6.29}$$

This is the *Liouville equation.*

To provide more insight into the equation, we may consider the problem from the *total probability formula* as well. If we consider the transition probability density $p_{\mathbf{Y}}(\mathbf{y}, t \mid \mathbf{y}_0, t_0)$, for a general stochastic process we have

$$p_{\mathbf{Y}}(\mathbf{y}_3, t_3|\mathbf{y}_1, t_1) = \int p_{\mathbf{Y}}(\mathbf{y}_3, t_3|\mathbf{y}_2, t_2; \mathbf{y}_1, t_1) \cdot p_{\mathbf{Y}}(\mathbf{y}_2, t_2|\mathbf{y}_1, t_1)\, d\mathbf{y}_2 \tag{6.30}$$

which can be verified when we note that $p_{\mathbf{Y}}(\mathbf{y}_3, t_3|\mathbf{y}_1, t_1) = p_{\mathbf{Y}}(\mathbf{y}_3, t_3; \mathbf{y}_1, t_1)/p_{\mathbf{Y}}(\mathbf{y}_1, t_1)$ and the like relationship. Equation 6.30 is of course essentially the total probability formula. Here, we stress that Equation 6.30 is also an embedment of the preservation of probability.

For a Markov process at time instants $t_3 > t_2 > t_1$, Equation 6.30 reduces to

$$p_{\mathbf{Y}}(\mathbf{y}_3, t_3|\mathbf{y}_1, t_1) = \int p_{\mathbf{Y}}(\mathbf{y}_3, t_3|\mathbf{y}_2, t_2) \cdot p_{\mathbf{Y}}(\mathbf{y}_2, t_2|\mathbf{y}_1, t_1)\, d\mathbf{y}_2 \tag{6.31a}$$

This is the *Chapman–Kolmogorov equation* for Markov processes. Clearly, it is an embedment of the preservation of probability from the state space description in the case of Markov processes.

When we consider the time instants t_0, t and $t + \Delta t$, Equation 6.31a becomes

$$p_{\mathbf{Y}}(\mathbf{y}, t+\Delta t|\mathbf{y}_0, t_0) = \int p_{\mathbf{Y}}(\mathbf{y}, t+\Delta t|\mathbf{z}, t) \cdot p_{\mathbf{Y}}(\mathbf{z}, t|\mathbf{y}_0, t_0)\, d\mathbf{z} \tag{6.31b}$$

When introducing the increment $\boldsymbol{\kappa} = \mathbf{y} - \mathbf{z}$, we have

$$p_{\mathbf{Y}}(\mathbf{y}, t+\Delta t|\mathbf{y}_0, t_0) = \int p_{\mathbf{Y}}(\mathbf{y}, t+\Delta t|\mathbf{y}-\boldsymbol{\kappa}, t) \cdot p_{\mathbf{Y}}(\mathbf{y}-\boldsymbol{\kappa}, t|\mathbf{y}_0, t_0)\, d\boldsymbol{\kappa} \tag{6.31c}$$

In the time interval $[t, t + \Delta t]$, from Equation 6.25 it is known that the incremental change of the state $\mathbf{Y}$ is

$$\Delta\mathbf{Y} = \mathbf{Y}(t+\Delta t) - \mathbf{Y}(t) = \mathbf{A}(\mathbf{Y}, t)\Delta t + o(\Delta t) \tag{6.32}$$

which is of course a random vector because $\mathbf{Y}(t)$ is random. For notational convenience, denote $\Delta\mathbf{Y}$ by $\boldsymbol{\eta} = (\eta_1, \eta_2, \ldots, \eta_m)^{\mathrm{T}}$ and the conditional probability density of $\boldsymbol{\eta}$ given $\mathbf{Y}(t) = \mathbf{y}$ by $\phi_{\boldsymbol{\eta}|\mathbf{Y}}(\boldsymbol{\eta}; \mathbf{y}, t, \Delta t)$. From Equation 6.32 it is seen that

$$\boldsymbol{\eta}|[\mathbf{Y}(t) = \mathbf{y}] = \mathbf{A}(\mathbf{y}, t)\Delta t + o(\Delta t) \tag{6.33}$$

This means that the conditional random vector $\boldsymbol{\eta}\,|\,[\mathbf{Y}(t)=\mathbf{y}]$ is essentially a deterministic vector; thus, it follows that

$$\int \phi_{\boldsymbol{\eta}|\mathbf{Y}}(\boldsymbol{\eta};\mathbf{y},t,\Delta t)\,\mathrm{d}\boldsymbol{\eta}=1 \tag{6.34a}$$

$$\int \eta_\ell \phi_{\boldsymbol{\eta}|\mathbf{Y}}(\boldsymbol{\eta};\mathbf{y},t,\Delta t)\,\mathrm{d}\boldsymbol{\eta}=A_\ell(\mathbf{y},t)\Delta t+o(\Delta t) \qquad \ell=1,2,\ldots,m \tag{6.34b}$$

$$\int \eta_\ell \eta_k \phi_{\boldsymbol{\eta}|\mathbf{Y}}(\boldsymbol{\eta};\mathbf{y},t,\Delta t)\,\mathrm{d}\boldsymbol{\eta}=o(\Delta t) \qquad \ell,k=1,2,\ldots,m \tag{6.34c}$$

Clearly, from Equations 6.32 and 6.33 it is known that

$$p_{\mathbf{Y}}(\mathbf{y}+\boldsymbol{\eta},t+\Delta t|\mathbf{y},t)=\phi_{\boldsymbol{\eta}|\mathbf{Y}}(\boldsymbol{\eta};\mathbf{y},t,\Delta t) \tag{6.35}$$

Introducing this transition probability density into Equation 6.31c and changing the notation accordingly, we have

$$p_{\mathbf{Y}}(\mathbf{y},t+\Delta t|\mathbf{y}_0,t_0)=\int \phi_{\boldsymbol{\eta}|\mathbf{Y}}(\boldsymbol{\eta};\mathbf{y}-\boldsymbol{\eta},t,\Delta t)\cdot p_{\mathbf{Y}}(\mathbf{y}-\boldsymbol{\eta},t|\mathbf{y}_0,t_0)\,\mathrm{d}\boldsymbol{\eta} \tag{6.36}$$

Expanding the integrand up to second order by the Taylor series:

$$\begin{aligned}\phi_{\boldsymbol{\eta}|\mathbf{Y}}(\boldsymbol{\eta};\mathbf{y}-\boldsymbol{\eta},t,\Delta t)\cdot p_{\mathbf{Y}}(\mathbf{y}-\boldsymbol{\eta},t|\mathbf{y}_0,t_0)&=\phi_{\boldsymbol{\eta}|\mathbf{Y}}(\boldsymbol{\eta};\mathbf{y},t,\Delta t)\cdot p_{\mathbf{Y}}(\mathbf{y},t|\mathbf{y}_0,t_0)\\ -\sum_{\ell=1}^{m}\frac{\partial \phi_{\boldsymbol{\eta}|\mathbf{Y}}p_{\mathbf{Y}}}{\partial y_\ell}\eta_\ell+\frac{1}{2}\sum_{\ell=1}^{m}\sum_{k=1}^{m}\frac{\partial^2\phi_{\eta|\mathbf{Y}}p_{\mathbf{Y}}}{\partial y_\ell\partial y_k}\eta_\ell\eta_k+\cdots\end{aligned} \tag{6.37}$$

where $\phi_{\boldsymbol{\eta}|\mathbf{Y}}p_{\mathbf{Y}}$ in the last two terms represents $\phi_{\boldsymbol{\eta}|\mathbf{Y}}(\boldsymbol{\eta};\mathbf{y},t,\Delta t)\cdot p_{\mathbf{Y}}(\mathbf{y},t|\mathbf{y}_0,t_0)$ for simplicity of notation. Substituting it in Equation 6.36 and noting Equations 6.34a–6.34c, we have

$$p_{\mathbf{Y}}(\mathbf{y},t+\Delta t|\mathbf{y}_0,t_0)=p_{\mathbf{Y}}(\mathbf{y},t|\mathbf{y}_0,t_0)-\sum_{\ell=1}^{m}\frac{\partial}{\partial y_\ell}[A_\ell(\mathbf{y},t)p_{\mathbf{Y}}(\mathbf{y},t|\mathbf{y}_0,t_0)]\Delta t+o(\Delta t) \tag{6.38}$$

Subtracting $p_{\mathbf{Y}}(\mathbf{y},t|\mathbf{y}_0,t_0)$ from both sides of Equation 6.38, dividing by Δt and then letting $\Delta t\to 0$ yields

$$\frac{\partial p_{\mathbf{Y}}(\mathbf{y},t|\mathbf{y}_0,t_0)}{\partial t}+\sum_{\ell=1}^{m}\frac{\partial}{\partial y_\ell}[A_\ell(\mathbf{y},t)p_{\mathbf{Y}}(\mathbf{y},t|\mathbf{y}_0,t_0)]=0 \tag{6.39a}$$

Multiplying this by $p_{\mathbf{Y}}(\mathbf{y}_0,t_0)$ and integrating with respect to $\mathbf{y}_0$ on both sides of Equation 6.39a, we have

$$\frac{\partial p_{\mathbf{Y}}(\mathbf{y},t)}{\partial t}+\sum_{\ell=1}^{m}\frac{\partial}{\partial y_\ell}[A_\ell(\mathbf{y},t)p_{\mathbf{Y}}(\mathbf{y},t)]=0 \tag{6.39b}$$

This is of course the Liouville equation, identical to Equation 6.29.

In the preceding derivations, particularly from Equations 6.33, 6.34b, and 6.34c, it is seen that the probability density evolution equation, here the Liouville equation, is associated tightly with the dynamical system. That is, the evolution of probability must be associated with a physical mechanism.

It should be noted that the Liouville equation governs the evolution of the probability density of a system when randomness is only involved in the initial conditions. If randomness is also involved in, say, the system parameters, then the Liouville equation does not hold. But some technique can be used so that the random system parameters can be mathematically treated as part of the random initial conditions and thus a modified Liouville equation can still be derived (Soong, 1973).

6.3.1.2 Solution of the Liouville Equation

The Liouville equation is a first-order quasi-linear partial differential equation. For such an equation, the method of characteristics works (Soong, 1973; Sarra, 2003), as has once been employed in Section 5.6.3.1. The method will be used here to yield the closed-form solution, whereas the theroretical basis and physical sense of the method will be discussed in detail later in Section 6.6.1.

First we rewrite the Liouville equation, Equation 6.39b, as

$$\frac{\partial p_{\mathbf{Y}}(\mathbf{y},t)}{\partial t}+\sum_{\ell=1}^{m}A_{\ell}(\mathbf{y},t)\frac{\partial p_{\mathbf{Y}}(\mathbf{y},t)}{\partial y_{\ell}}+\sum_{\ell=1}^{m}p_{\mathbf{Y}}(\mathbf{y},t)\frac{\partial A_{\ell}(\mathbf{y},t)}{\partial y_{\ell}}=0 \tag{6.40}$$

The subsidiary equation is then

$$\frac{\mathrm{d}t}{1}=-\frac{\mathrm{d}p_{\mathbf{Y}}(\mathbf{y},t)}{p_{\mathbf{Y}}(\mathbf{y},t)\sum_{\ell=1}^{m}\frac{\partial A_{\ell}(\mathbf{y},t)}{\partial y_{\ell}}}=\frac{\mathrm{d}y_{1}}{A_{1}(\mathbf{y},t)}=\cdots=\frac{\mathrm{d}y_{m}}{A_{m}(\mathbf{y},t)} \tag{6.41}$$

The last m equalities are essentially the state equation. The first equality will give the solution

$$p_{\mathbf{Y}}(\mathbf{y},t)=\left[p_{\mathbf{Y}_0}(\mathbf{y}_0)\exp\left\{-\int_{t_0}^{t}\sum_{\ell=1}^{m}\frac{\partial A_{\ell}[\mathbf{y}=\mathbf{H}(\mathbf{y}_0,\tau),\tau]}{\partial y_{\ell}}\,\mathrm{d}\tau\right\}\right]_{\mathbf{y}_0=\mathbf{H}^{-1}(\mathbf{y},t)} \tag{6.42}$$

where $\mathbf{y}=\mathbf{H}(\mathbf{y}_0,t)$ is the closed-form solution of the state equation from the last m equalities in Equation 6.41 and $\mathbf{H}^{-1}(\cdot)$ is the inverse function of $\mathbf{H}$.

The solution establishes the relationship between the density of $\mathbf{Y}(t)$ and that of $\mathbf{Y}(t_0)$, where the closed-form solution of the associated state equation is involved. By the way, we point out at present that the characteristic curves of the Liouville equation are in essence the trajectories on which the probability measure is invariant.

On the other hand, according to Equation 6.18a, where the map $\mathcal{G}_t$ is essentially the function $\mathbf{H}$ here, we have

$$p_{\mathbf{Y}}(\mathbf{y},t)\,\mathrm{d}\mathbf{y}=p_{\mathbf{Y}_0}(\mathbf{y}_0)\,\mathrm{d}\mathbf{y}_0 \tag{6.43}$$

The PDF is thus given in an alternative form by

$$p_{\mathbf{Y}}(\mathbf{y},t)=\left|\frac{\partial \mathbf{y}_0}{\partial \mathbf{y}}\right|[p_{\mathbf{Y}_0}(\mathbf{y}_0)]_{\mathbf{y}_0=\mathbf{H}^{-1}(\mathbf{y},t)}=|J|p_{\mathbf{Y}_0}\left[\mathbf{y}_0=\mathbf{H}^{-1}(\mathbf{y},t)\right] \tag{6.44}$$

where $J=|\partial\mathbf{y}_0/\partial\mathbf{y}|$ is the Jacobian.

Comparing Equation 6.44 with Equation 6.42, we reach the equality

$$|J| = \left[\exp\left\{-\int_{t_0}^{t}\sum_{\ell=1}^{m}\frac{\partial A_\ell[\mathbf{y}=\mathbf{H}(\mathbf{y}_0,\tau),\tau]}{\partial y_\ell}\,\mathrm{d}\tau\right\}\right]_{\mathbf{y}_0=\mathbf{H}^{-1}(\mathbf{y},t)} \tag{6.45}$$

which can sometimes be used to compute the Jacobian.

Note that, for a given $\mathbf{y}_0$, the state is determined by $\mathbf{Y}=\mathbf{H}(\mathbf{y}_0,\ t)$. Therefore, it is a deterministic vector; namely, the transition probability density is given as

$$p_{\mathbf{Y}}(\mathbf{y},t|\mathbf{y}_0,t_0) = \delta[\mathbf{y}-\mathbf{H}(\mathbf{y}_0,t)] \tag{6.46}$$

where $\delta(\cdot)$ is the Dirac delta function (see Appendix A). From the total probability formula, the density of $\mathbf{Y}(t)$ is thus

$$\begin{aligned} p_{\mathbf{Y}}(\mathbf{y},t) &= \int p_{\mathbf{Y}}(\mathbf{y},t|\mathbf{y}_0,t_0)p_{\mathbf{Y}_0}(\mathbf{y}_0)\,\mathrm{d}\mathbf{y}_0 \\ &= \int \delta[\mathbf{y}-\mathbf{H}(\mathbf{y}_0,t)]p_{\mathbf{Y}_0}(\mathbf{y}_0)\,\mathrm{d}\mathbf{y}_0 \end{aligned} \tag{6.47}$$

It is easy to verify that the transition probability density given by Equation 6.46 satisfies Equation 6.39a and the probability density given by Equation 6.47 satisfies Equation 6.39b. Actually, a further step of integrating the Dirac delta function in Equation 6.47 by changing the variables will immediately yield Equation 6.44.

The above discussion shows that Equations 6.42, 6.44 and 6.47 are all equivalent, although Equation 6.42 is obtained from the solution of the Liouville equation, Equation 6.44 is from the preservation of probability when an arbitrary ensemble domain in the state space is examined, and Equation 6.47 is in essence directly from the point of view of the sample.

The asymptotic attributes of the solution of the Liouville equation are now discussed.

For a global asymptotic stable system (for example, a damped linear system subject to deterministic external excitations), as the time elapses, the effect of the initial condition will vanish; namely:

$$\lim_{t\to\infty}\mathbf{Y}(t) = \lim_{t\to\infty}\mathbf{H}(\mathbf{Y}_0,t) = \mathbf{H}_\infty(t) \tag{6.48}$$

where $\mathbf{H}_\infty(t)$ is the asymptotic response of the system. From Equation 6.47 we have

$$\lim_{t\to\infty}p_{\mathbf{Y}}(\mathbf{y},t) = \lim_{t\to\infty}\int\delta[\mathbf{y}-\mathbf{H}(\mathbf{y}_0,t)]p_{\mathbf{Y}_0}(\mathbf{y}_0)\,\mathrm{d}\mathbf{y}_0 = \int\delta[\mathbf{y}-\mathbf{H}_\infty(t)]p_{\mathbf{Y}_0}(\mathbf{y}_0)\,\mathrm{d}\mathbf{y}_0 = \delta[\mathbf{y}-\mathbf{H}_\infty(t)] \tag{6.49}$$

This indicates that as time passes, the stochastic response of the system tends to be a deterministic process.

On the other hand, if there are some attractors with the attracting domain Ω_ℓ, $\ell=1,2,\ldots,n_{\text{attractor}}$, where $n_{\text{attractor}}$ is the number of attractors, then there are $n_{\text{attractor}}$ possible asymptotic responses $\mathbf{H}_{\infty,\ell}(t)$, $\ell=1,2,\ldots,n_{\text{attractor}}$ (Strogatz, 1994), and we have

$$\lim_{t\to\infty}p_{\mathbf{Y}}(\mathbf{y},t) = \lim_{t\to\infty}\int\delta[\mathbf{y}-\mathbf{H}(\mathbf{y}_0,t)]p_{\mathbf{Y}_0}(\mathbf{y}_0)\,\mathrm{d}\mathbf{y}_0 = \sum_{\ell=1}^{n_{\text{attractor}}}P_\ell\delta[\mathbf{y}-\mathbf{H}_{\infty,\ell}(t)] \tag{6.50}$$

where

$$P_\ell = \int_{\Omega_\ell} p_{\mathbf{Y}_0}(\mathbf{y}_0)\, \mathrm{d}\mathbf{y}_0$$

Example 6.1. Solution of a Liouvillian System To provide a visual impression, we consider the linear system[3]

$$\dot{X} = -\left[\zeta\omega + \sqrt{1-\zeta^2}\tan\left(\sqrt{1-\zeta^2}\omega t\right)\right]X \qquad X(0) = X_0 \tag{6.51a}$$

where X_0 is a random variable with known density $p_{X0}(x_0)$. The formal solution of the state is given by

$$X = H(X_0, t) = X_0\, \mathrm{e}^{-\zeta\omega t}\cos\left(\sqrt{1-\zeta^2}\omega t\right) \tag{6.51b}$$

Therefore, from Equation 6.42, the density of $X(t)$ yields

$$p_X(x,t) = \frac{\mathrm{e}^{\zeta\omega t}}{\cos(\sqrt{1-\zeta^2}\omega t)} p_{X_0}(x_0)\Bigg|_{x_0 = x\mathrm{e}^{\zeta\omega t}/\cos\left(\sqrt{1-\zeta^2}\omega t\right)} \tag{6.51c}$$

This equation also holds if we denote $p_X(x,t) = \delta(x)$ for $t = (k+\frac{1}{2})\pi/(\sqrt{1-\zeta^2}\omega)$, $k = 0, 1, 2, \ldots$.

The contour of the PDF in Equation 6.51c is shown in Figure 6.4 when $\omega = 1$ and X_0 is a normally distributed random variable with a mean of 3 mm and a standard deviation of 1 mm. It

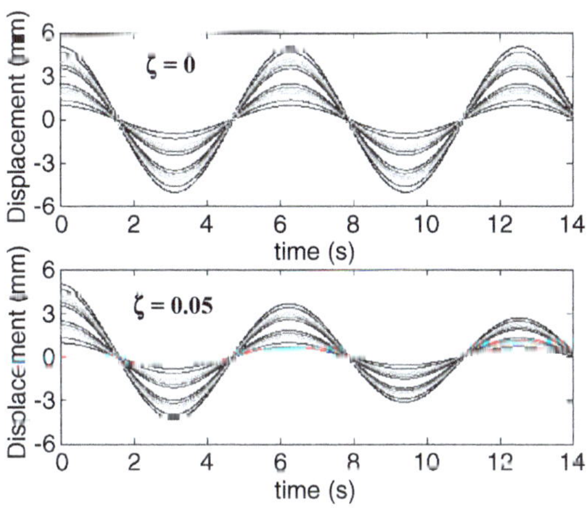

Figure 6.4 Contour of PDF.

[3] The physical sense of this system is as follows. Consider an SDOF system $\ddot{X} + 2\zeta\omega\dot{X} + \omega_2 X = 0$ with the initial condition $X(0) = X_0$. Then, according to the analytical solutions of the displacement and the velocity, we can get the state equation in Equation 6.51a except some singular points at $t_k = (k+\frac{1}{2})\pi/[(1-\zeta^2)^{1/2}\omega]$, $k = 0, 1, 2, \ldots$.

is seen that at the time instants $t = (k + \frac{1}{2})\pi/(\sqrt{1 - \zeta^2}\omega)$, $k = 0, 1, 2, \ldots$, singular points occur where the probability density is infinity because all samples are now concentrated at the same point. At these time instants, note here that Equation 6.51a is undetermined.

This also provides an example that the map $\mathcal{G}_t$ in Equation 6.18a does not have an inverse and demonstrates that $\mathcal{G}_t$ having an inverse is not a necessary condition; simultaneously, the Jacobian in Equation 6.44 does not exist. However, the PDF still makes sense if we introduce the distribution function of Dirac's delta. This is in contrast to conservation of mass in continuum mechanics. □

6.3.2 Fokker–Planck–Kolmogorov Equation Revisited

When randomness is involved in the nonhomogeneous terms of a dynamical system, especially as the excitations are white-noise processes, the state equation can be understood as the Itô stochastic differential equation, as discussed in Section 5.6.1. That is

$$\mathrm{d}\mathbf{Y}(t) = \mathbf{A}(\mathbf{Y}, t)\,\mathrm{d}t + \mathbf{B}(\mathbf{Y}, t)\,\mathrm{d}\mathbf{W}(t) \tag{6.52}$$

where $\mathbf{Y}$ and $\mathbf{A}$ are the same as defined in Equation 6.25, $\mathbf{B}(\mathbf{y}, t) = [B_{\ell k}(\mathbf{y}, t)]_{m\times r}$ is the input force influence matrix, the $B_{\ell k}(\mathbf{y}, t)$ are nonanticipating functions and $\mathbf{W}(t) = (W_1(t), W_2(t), \ldots, W_r(t))^{\mathrm{T}}$ is an r-dimensional Wiener process vector with the mean and covariance matrix of increment

$$\mathcal{E}[\mathrm{d}\mathbf{W}(t)] = \mathbf{0} \qquad \mathcal{E}[\mathrm{d}\mathbf{W}(t)\,\mathrm{d}\mathbf{W}^{\mathrm{T}}(t)] = \mathbf{D}\,\mathrm{d}t \tag{6.53}$$

in which $\mathbf{D} = [D_{ij}]_{r\times r}$ is the same as in Equation 5.179.

Equation 6.52 can be rewritten in an incremental form:

$$\Delta\mathbf{Y}(t) = \mathbf{A}(\mathbf{Y}, t)\Delta t + \mathbf{B}(\mathbf{Y}, t)\Delta\mathbf{W}(t) + o(\Delta t) \tag{6.54}$$

As elaborated in Section 5.6, the transition probability density of $\mathbf{Y}(t)$ satisfies the FPK equation. We now examine the physical sense of this equation in detail. To this end, the probability flow in the state space is studied herein. For schematic convenience, we use the case $m = 2$ in Figure 6.5. Let us consider a domain $\mathrm{d}\mathbf{y} = \mathrm{d}y_1\,\mathrm{d}y_2 \cdots \mathrm{d}y_m$. The increment of probability in this domain is

$$\Delta P = \frac{\partial p_{\mathbf{Y}}}{\partial t}\mathrm{d}y_1\mathrm{d}y_2 \cdots \mathrm{d}y_m\Delta t + o(\Delta t) \tag{6.55}$$

During the time interval $[t, t + \Delta t]$, due to the drift effect (or effect of differentiation of the first order), in the direction of y_1, the probability in the domain on the left-hand side $\Delta y_{1,\mathrm{L}}\mathrm{d}y_2\,\mathrm{d}y_3 \cdots \mathrm{d}y_m$ will enter the domain $\mathrm{d}y_1\mathrm{d}y_2 \cdots \mathrm{d}y_m$, while on the right-hand side the probability in the domain $\Delta y_{1,\mathrm{R}}\,\mathrm{d}y_2\,\mathrm{d}y_3 \cdots \mathrm{d}y_m$ will leave the domain; thus, the net imported probability in the direction of y_1 is given as (see Figure 6.5a)

$$\Delta P_{y_1} = p_{\mathbf{Y}}(y_1, y_2, t)\Delta y_{1,\mathrm{L}}\,\mathrm{d}y_2\,\mathrm{d}y_3 \cdots \mathrm{d}y_m - p_{\mathbf{Y}}(y_1 + \mathrm{d}y_1, y_2, t)\Delta y_{1,\mathrm{R}}\,\mathrm{d}y_2\,\mathrm{d}y_3 \cdots \mathrm{d}y_m + o(\Delta t) \tag{6.56a}$$

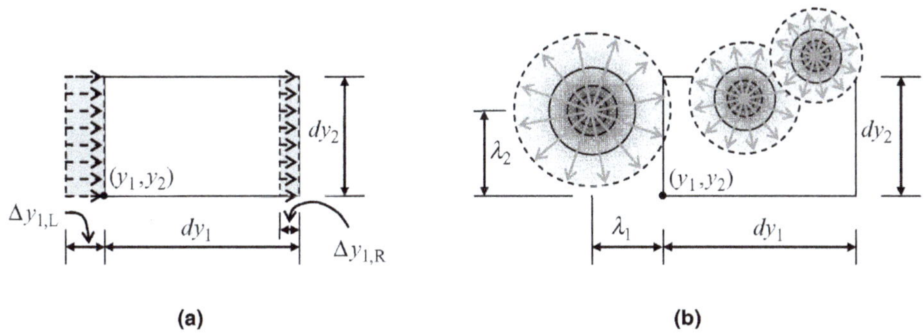

Figure 6.5 Schematic diagram of probability flow in the state space: (a) the probability flow due to drift; (b) the probability flow due to diffusion. Here, two points are illustrated; the arrow represents the diffusion and the different lengths of the arrows represent that the diffusion coefficients depend on the position. The probability of the point in the vicinity of the position $(y_1+\lambda_1, y_2+\lambda_2)$ at time t is $p_{\mathbf{Y}}(y_1+\lambda_1, y_2+\lambda_2, t)\mathrm{d}\lambda_1\mathrm{d}\lambda_2$, while the probability of this point transit to the domain $\mathrm{d}y_1\,\mathrm{d}y_2$ is $p_{\mathbf{Y}}(y_1+\lambda_1, y_2+\lambda_2, t)$ $\phi_{\boldsymbol{\lambda}}(\lambda_1, \lambda_2; y_1+\lambda_1, y_2+\lambda_2, t, \Delta t)\mathrm{d}\lambda_1\mathrm{d}\lambda_2\mathrm{d}y_1\mathrm{d}y_2$.

Noting $\Delta y_{1,\mathrm{L}} = A_1(\mathbf{y}, t)\Delta t + o(\Delta t)$ and $\Delta y_{1,\mathrm{R}} = A_1(y_1+\mathrm{d}y_1, y_2, y_3, \cdots, y_m, t)\Delta t + o(\Delta t)$, we have

$$\Delta P_{y_1} = -\frac{\partial}{\partial y_1}[A_1(\mathbf{y}, t)p_{\mathbf{Y}}(\mathbf{y}, t|\mathbf{y}_0, t_0)]\mathrm{d}\mathbf{y}\Delta t + o(\Delta t) \tag{6.56b}$$

Simultaneously, the analogous thing happens in the direction of $y_j (j = 2, 3, \ldots, m)$; therefore, the net imported probability in the domain $\mathrm{d}y_1\mathrm{d}y_2\cdots\mathrm{d}y_m$ due to the effect of drift is given as

$$\Delta P_1 = \sum_{\ell=1}^{m}\Delta P_{y_\ell} = -\sum_{\ell=1}^{m}\frac{\partial[A_\ell(\mathbf{y}, t)p_{\mathbf{Y}}]}{\partial y_\ell}\mathrm{d}\mathbf{y}\Delta t + o(\Delta t) \tag{6.56c}$$

Now we consider the effect of a diffusion field (arising from the terms of Brownian motion). Denote the instantaneous probability density of $\boldsymbol{\lambda} = \mathbf{B}(\mathbf{y}, t)\Delta\mathbf{W}(t) = \boldsymbol{\eta}(\mathbf{y}, t)(\Delta t)^{1/2}$ by $\phi_{\boldsymbol{\lambda}}(\boldsymbol{\lambda}; \mathbf{y}, t, \Delta t)$, where $\boldsymbol{\eta}(\mathbf{y}, t)$ is an m-dimensional zero-mean stochastic process vector with covariance matrix $\mathcal{E}[\boldsymbol{\eta}(\mathbf{y}, t)\boldsymbol{\eta}^{\mathrm{T}}(\mathbf{y}, t)] = \mathbf{B}(\mathbf{y}, t)\mathbf{D}\mathbf{B}^{\mathrm{T}}(\mathbf{y}, t)$.[4] Certainly, we have

$$\int \phi_{\boldsymbol{\lambda}}(\boldsymbol{\lambda}; \mathbf{y}, t, \Delta t)\,\mathrm{d}\boldsymbol{\lambda} = 1 \tag{6.57a}$$

$$\int \lambda_\ell\phi_{\boldsymbol{\lambda}}(\boldsymbol{\lambda}; \mathbf{y}, t, \Delta t)\,\mathrm{d}\boldsymbol{\lambda} = 0 \qquad \ell = 1, 2, \ldots, m \tag{6.57b}$$

$$\int \lambda_\ell\lambda_k\phi_{\boldsymbol{\lambda}}(\boldsymbol{\lambda}; \mathbf{y}, t, \Delta t)\,\mathrm{d}\boldsymbol{\lambda} = [\mathbf{B}(\mathbf{y}, t)\mathbf{D}\mathbf{B}^{\mathrm{T}}(\mathbf{y}, t)]_{\ell k}\Delta t \qquad \ell, k = 1, 2, \ldots, m \tag{6.57c}$$

The probability of the point in the vicinity of the position $(\mathbf{y}+\boldsymbol{\lambda})$ at time t is $p_{\mathbf{Y}}(\mathbf{y}+\boldsymbol{\lambda}, t|\mathbf{y}_0, t_0)\mathrm{d}\boldsymbol{\lambda}$, while the probability of this point transit to the domain $\mathrm{d}\mathbf{y}$ during the time interval $[t, t+\Delta t]$ is $p_{\mathbf{Y}}(\mathbf{y}, t+\Delta t|\mathbf{y}+\boldsymbol{\lambda}, t)p_{\mathbf{Y}}(\mathbf{y}+\boldsymbol{\lambda}, t|\mathbf{y}_0, t_0)\mathrm{d}\boldsymbol{\lambda}$, where $p_{\mathbf{Y}}(\mathbf{y}, t+\Delta t|\mathbf{y}+\boldsymbol{\lambda}, t)$ is

[4] For details, refer to Section 5.6.3.1, Equation 5.244.

the transition probability density from the time instant t to $t+\Delta t$. The probability in the domain $d\mathbf{y}$ at the time instant $t+\Delta t$ is thus (see Figure 6.5b)

$$p_{\mathbf{Y}}(\mathbf{y},t+\Delta t|\mathbf{y}_0,t_0)\,d\mathbf{y} = d\mathbf{y}\int p_{\mathbf{Y}}(\mathbf{y},t+\Delta t|\mathbf{y}+\boldsymbol{\lambda},t)p_{\mathbf{Y}}(\mathbf{y}+\boldsymbol{\lambda},t|\mathbf{y}_0,t_0)\,d\boldsymbol{\lambda} \tag{6.58a}$$

Noting that the transition PDF from $t+\Delta t$ to t is just

$$p_{\mathbf{Y}}(\mathbf{y},t+\Delta t|\mathbf{y}+\boldsymbol{\lambda},t) = \phi_{\boldsymbol{\lambda}}(\boldsymbol{\lambda};\mathbf{y}+\boldsymbol{\lambda},t,\Delta t) \tag{6.58b}$$

we have

$$p_{\mathbf{Y}}(\mathbf{y},t+\Delta t|\mathbf{y}_0,t_0)\,d\mathbf{y} = d\mathbf{y}\int p_{\mathbf{Y}}(\mathbf{y}+\boldsymbol{\lambda},t|\mathbf{y}_0,t_0)\phi_{\boldsymbol{\lambda}}(\boldsymbol{\lambda};\mathbf{y}+\boldsymbol{\lambda},t,\Delta t)\,d\boldsymbol{\lambda} \tag{6.58c}$$

The increment of probability due to diffusion is then

$$\Delta P_2 = p_{\mathbf{Y}}(\mathbf{y},t+\Delta t|\mathbf{y}_0,t_0)\,d\mathbf{y} - p_{\mathbf{Y}}(\mathbf{y},t|\mathbf{y}_0,t_0)\,d\mathbf{y} \tag{6.59}$$

Expanding the integrand in Equations 6.58a and 6.58b to the second order in the vicinity of $\mathbf{y}$, regarding $\boldsymbol{\lambda}$ as a vector of increment, we have

$$\begin{aligned} p_{\mathbf{Y}}(\mathbf{y}+\boldsymbol{\lambda},t|\mathbf{y}_0,t_0)\phi_{\boldsymbol{\lambda}}(\boldsymbol{\lambda};\mathbf{y}+\boldsymbol{\lambda},t,\Delta t) &= p_{\mathbf{Y}}(\mathbf{y},t|\mathbf{y}_0,t_0)\phi_{\boldsymbol{\lambda}}(\boldsymbol{\lambda};\mathbf{y},t,\Delta t) \\ &+ \sum_{\ell=1}^{m}\frac{\partial[p_{\mathbf{Y}}\phi_{\boldsymbol{\lambda}}]}{\partial y_\ell}\lambda_\ell + \frac{1}{2}\sum_{\ell=1}^{m}\sum_{k=1}^{m}\frac{\partial^2[p_{\mathbf{Y}}\phi_{\boldsymbol{\lambda}}]}{\partial y_\ell \partial y_k}\lambda_\ell\lambda_k + \cdots \end{aligned} \tag{6.60}$$

Introducing this equation into Equations 6.58a and 6.58b, and noting Equations 6.57a–6.57b, we therefore get

$$\Delta P_2 = \frac{1}{2}\sum_{\ell=1}^{m}\sum_{k=1}^{m}\frac{\partial^2}{\partial y_\ell \partial y_k}\{[\mathbf{B}(\mathbf{y},t)\mathbf{D}\mathbf{B}^{\mathrm{T}}(\mathbf{y},t)]_{\ell k}\,p_{\mathbf{Y}}\}\Delta t + o(\Delta t) \tag{6.61}$$

Owing to the preservation of probability, the velocity field and the diffusion field both contributing to the increment of probability, it follows that

$$\Delta P = \Delta P_1 + \Delta P_2 \tag{6.62}$$

Employing Equations 6.55, 6.56c and 6.61, dividing both sides by Δt and taking the limit $\Delta t \to 0$, we get

$$\frac{\partial p_{\mathbf{Y}}}{\partial t} = -\sum_{\ell=1}^{m}\frac{\partial[A_\ell(\mathbf{y},t)p_{\mathbf{Y}}]}{\partial y_\ell} + \frac{1}{2}\sum_{\ell=1}^{m}\sum_{k=1}^{m}\frac{\partial^2}{\partial y_\ell \partial y_k}\{[\mathbf{B}(\mathbf{y},t)\mathbf{D}\mathbf{B}^{\mathrm{T}}(\mathbf{y},t)]_{\ell k}p_{\mathbf{Y}}\} \tag{6.63}$$

This is nothing but the FPK equation identical to Equation 5.212.

The sense of Equation 6.62 is that the increment of probability in a domain during a time interval equals the net probability imported through the boundary into this domain. This is, of course, the principle of preservation of probability viewed from the state space description. The above analysis demonstrates that the FPK equation is the result of this principle. Meanwhile, because the coefficients of the FPK equation are related to the coefficients of the associated stochastic differential equation, which is the embedment of the physical law, keeping the preceding analysis in mind, we find that the transition process of probability, or the probability flow, must result from a certain physical system, which is usually a stochastic differential equation physically representing the sampling path. In other words, a transition of probability must have a physical mechanism.

The treatment in Equation 6.62 is in an intuitive way in which the effects of drift and diffusion are superposed by a summation. Employing a more rigorous treatment starting with the Chapman–Kolmogorov equation, we can derive the associated FPK equation in a different way and provide some new insights into the physical sense.

Denote

$$\boldsymbol{\eta} = \mathbf{A}(\mathbf{Y}, t)\Delta t + o(\Delta t) \qquad \boldsymbol{\lambda} = \mathbf{B}(\mathbf{Y}, t)\Delta \mathbf{W}(t) \qquad \boldsymbol{\kappa} = \Delta \mathbf{Y}(t) = \boldsymbol{\eta} + \boldsymbol{\lambda} \tag{6.64}$$

It is seen that $\boldsymbol{\eta}$ represents the contribution from the effect of drift and $\boldsymbol{\lambda}$ represents the contribution from the effect of diffusion. Because **A** and **B** are both nonanticipating functions,[5] the conditional expectations of $\boldsymbol{\eta}$ and $\boldsymbol{\lambda}$ given $\{\mathbf{Y}(t)=\mathbf{y}\}$ are given by

$$\mathcal{E}[\boldsymbol{\eta}|\mathbf{Y}(t) = \mathbf{y}] = \mathbf{A}(\mathbf{y}, t)\Delta t + o(\Delta t) \tag{6.65a}$$

$$\mathcal{E}[\boldsymbol{\eta}\boldsymbol{\eta}^{\mathrm{T}}|\mathbf{Y}(t) = \mathbf{y}] = o(\Delta t) \tag{6.65b}$$

$$\mathcal{E}[\boldsymbol{\lambda}|\mathbf{Y}(t) = \mathbf{y}] = 0 \tag{6.65c}$$

$$\mathcal{E}[\boldsymbol{\lambda}\boldsymbol{\lambda}^{\mathrm{T}}|\mathbf{Y}(t) = \mathbf{y}] = \mathbf{BDB}^{\mathrm{T}}\Delta t + o(\Delta t) = \boldsymbol{\sigma}\Delta t + o(\Delta t) \tag{6.65d}$$

where $\boldsymbol{\sigma} = [\sigma_{\ell k}]_{m\times m} \triangleq \mathbf{BDB}^{\mathrm{T}}$.

If we denote the conditional density of $\boldsymbol{\eta}$ given $\{\mathbf{Y}(t)=\mathbf{y}\}$ by $\phi_{\boldsymbol{\eta}|\mathbf{Y}}(\boldsymbol{\eta}; \mathbf{y}, t, \Delta t)$, likewise denoting that of $\boldsymbol{\lambda}$ by $\phi_{\boldsymbol{\lambda}|\mathbf{Y}}(\boldsymbol{\lambda}; \mathbf{y}, t, \Delta t)$, then Equations 6.65a and 6.65b will lead to Equations (6.34b) and (6.34c) respectively, while Equations 6.65c and 6.65d will lead to

$$\int \lambda_\ell \phi_{\boldsymbol{\lambda}|\mathbf{Y}}(\boldsymbol{\lambda}; \mathbf{y}, t, \Delta t)\, \mathrm{d}\boldsymbol{\lambda} = 0 \qquad \ell = 1, 2, \ldots, m \tag{6.66a}$$

and

$$\int \lambda_\ell \lambda_k \phi_{\boldsymbol{\lambda}|\mathbf{Y}}(\boldsymbol{\lambda}; \mathbf{y}, t, \Delta t)\, \mathrm{d}\boldsymbol{\lambda} = \sigma_{\ell k}\Delta t + o(\Delta t) \qquad \ell, k = 1, 2, \ldots, m \tag{6.66b}$$

In addition, the consistency condition requires that

$$\int \phi_{\boldsymbol{\lambda}|\mathbf{Y}}(\boldsymbol{\lambda}; \mathbf{y}, t, \Delta t)\, \mathrm{d}\boldsymbol{\lambda} = 1 \tag{6.66c}$$

Note from Equation 6.64 that $\boldsymbol{\kappa} = \boldsymbol{\eta} + \boldsymbol{\lambda}$, where $\boldsymbol{\eta}$ and $\boldsymbol{\lambda}$ are conditionally independent given $\{\mathbf{Y}(t)=\mathbf{y}\}$ because **A** and **B** are both nonanticipating functions. Therefore, using Equations (6.65a–6.65d), we have the conditional expectations of $\boldsymbol{\kappa}$:

$$\mathcal{E}[\boldsymbol{\kappa}|\mathbf{Y}(t) = \mathbf{y}] = \mathcal{E}[\boldsymbol{\eta}|\mathbf{Y}(t) = \mathbf{y}] + \mathcal{E}[\boldsymbol{\lambda}|\mathbf{Y}(t) = \mathbf{y}] = \mathbf{A}(\mathbf{y}, t)\Delta t + o(\Delta t) \tag{6.67a}$$

$$\mathcal{E}[\boldsymbol{\kappa}\boldsymbol{\kappa}^{\mathrm{T}}|\mathbf{Y}(t) = \mathbf{y}] = \mathcal{E}[\boldsymbol{\eta}\boldsymbol{\eta}^{\mathrm{T}}|\mathbf{Y}(t) = \mathbf{y}] = \boldsymbol{\sigma}(\mathbf{y}, t)\Delta t + o(\Delta t) \tag{6.67b}$$

where $\boldsymbol{\sigma}$ is the same as that in Equation (6.65d).

[5] The significance of nonanticipating functions was elaborated in Section 5.6.1.1.

Therefore, if the conditional density of $\boldsymbol{\kappa}$ given $\{\mathbf{Y}(t)=\mathbf{y}\}$ is denoted by $\phi_{\boldsymbol{\kappa}|\mathbf{Y}}(\boldsymbol{\kappa};\mathbf{y},t,\Delta t)$, then we have

$$\int \kappa_\ell \phi_{\boldsymbol{\kappa}|\mathbf{Y}}(\boldsymbol{\kappa};\mathbf{y},t,\Delta t)\,\mathrm{d}\boldsymbol{\kappa} = A_\ell(\mathbf{y},t)\Delta t + o(\Delta t) \qquad \ell = 1,2,\ldots,m \tag{6.68a}$$

$$\int \kappa_\ell \kappa_k \phi_{\boldsymbol{\kappa}|\mathbf{Y}}(\boldsymbol{\kappa};\mathbf{y},t,\Delta t)\,\mathrm{d}\boldsymbol{\kappa} = \sigma_{\ell k}(\mathbf{y},t)\Delta t + o(\Delta t) \qquad \ell,k = 1,2,\ldots,m \tag{6.68b}$$

Certainly, the consistency condition requires that

$$\int \phi_{\boldsymbol{\kappa}|\mathbf{Y}}(\boldsymbol{\kappa};\mathbf{y},t,\Delta t)\,\mathrm{d}\boldsymbol{\kappa} = 1 \tag{6.68c}$$

Because of the independence of $\mathrm{d}\mathbf{W}(t)$ and nonanticipation of $\mathbf{A}$ and $\mathbf{B}$, $\mathbf{Y}(t)$ determined by Equation 6.52 is a Markov process vector. From the Chapman–Kolmogorov equation (Equation (6.31c)):

$$p_{\mathbf{Y}}(\mathbf{y},t+\Delta t|\mathbf{y}_0,t_0) = \int p_{\mathbf{Y}}(\mathbf{y},t+\Delta t|\mathbf{y}-\boldsymbol{\kappa},t)\cdot p_{\mathbf{Y}}(\mathbf{y}-\boldsymbol{\kappa},t|\mathbf{y}_0,t_0)\,\mathrm{d}\boldsymbol{\kappa} \tag{6.69a}$$

where the transition PDF from time t to $t+\Delta t$, instead of being given by Equation 6.35, is now given by

$$p_{\mathbf{Y}}(\mathbf{y}+\boldsymbol{\kappa},t+\Delta t|\mathbf{y},t) = \phi_{\boldsymbol{\kappa}|\mathbf{Y}}(\boldsymbol{\kappa};\mathbf{y},t,\Delta t) \tag{6.69b}$$

We then have

$$p_{\mathbf{Y}}(\mathbf{y},t+\Delta t|\mathbf{y}_0,t_0) = \int \phi_{\boldsymbol{\kappa}|\mathbf{Y}}(\boldsymbol{\kappa};\mathbf{y}-\boldsymbol{\kappa},t,\Delta t)\cdot p_{\mathbf{Y}}(\mathbf{y}-\boldsymbol{\kappa},t|\mathbf{y}_0,t_0)\,\mathrm{d}\boldsymbol{\kappa} \tag{6.70}$$

Expanding the integrand up to second order by the Taylor series:

$$\begin{aligned}\phi_{\boldsymbol{\kappa}|\mathbf{Y}}(\boldsymbol{\kappa};\mathbf{y}-\boldsymbol{\kappa},t,\Delta t)\cdot p_{\mathbf{Y}}(\mathbf{y}-\boldsymbol{\kappa},t|\mathbf{y}_0,t_0) &= \phi_{\boldsymbol{\kappa}|\mathbf{Y}}(\boldsymbol{\kappa};\mathbf{y},t,\Delta t)\cdot p_{\mathbf{Y}}(\mathbf{y},t|\mathbf{y}_0,t_0)\\ &- \sum_{\ell=1}^{m}\frac{\partial \phi_{\boldsymbol{\kappa}|\mathbf{Y}}p_{\mathbf{Y}}}{\partial y_\ell}\kappa_\ell + \frac{1}{2}\sum_{\ell=1}^{m}\sum_{k=1}^{m}\frac{\partial^2 \phi_{\boldsymbol{\kappa}|\mathbf{Y}}p_{\mathbf{Y}}}{\partial y_\ell \partial y_k}\kappa_\ell\kappa_k + \cdots\end{aligned} \tag{6.71}$$

where the arguments in $\phi_{\boldsymbol{\eta}|\mathbf{Y}}$ and $p_{\mathbf{Y}}(\cdot)$ have been omitted in the last two terms for simplicity of notation. Substituting this in Equation 6.70 and noting Equations 6.68a–6.68c, we have

$$p_{\mathbf{Y}}(\mathbf{y},t+\Delta t|\mathbf{y}_0,t_0) = p_{\mathbf{Y}}(\mathbf{y},t|\mathbf{y}_0,t_0) - \sum_{\ell=1}^{m}\frac{\partial A_\ell p_{\mathbf{Y}}}{\partial y_\ell}\Delta t + \frac{1}{2}\sum_{\ell=1}^{m}\sum_{k=1}^{m}\frac{\partial^2 \sigma_{\ell k} p_{\mathbf{Y}}}{\partial y_\ell \partial y_k}\Delta t + o(\Delta t) \tag{6.72}$$

Further, subtracting $p_{\mathbf{Y}}(\mathbf{y},t|\mathbf{y}_0,t_0)$ from both sides of Equation 6.72, dividing by Δt and letting $\Delta t \to 0$ yields

$$\frac{\partial p_{\mathbf{Y}}}{\partial t} = -\sum_{\ell=1}^{m}\frac{\partial A_\ell p_{\mathbf{Y}}}{\partial y_\ell} + \frac{1}{2}\sum_{\ell=1}^{m}\sum_{k=1}^{m}\frac{\partial^2 \sigma_{\ell k} p_{\mathbf{Y}}}{\partial y_\ell \partial y_k} \tag{6.73}$$

where $p_{\mathbf{Y}}$ can either be understood as the transition probability density $p_{\mathbf{Y}}(\mathbf{y},t|\mathbf{y}_0,t_0)$ or be taken as the instantaneous PDF $p_{\mathbf{Y}}(\mathbf{y},t)$.

Equation 6.73 is nothing but the FPK equation identical to that derived in Section 5.6.2. Again, stressed here is the tight relationship between the evolution of probability density and the physical mechanism of the dynamical system. Undoubtedly, noting that the Chapman–Kolmogorov equation (Equations 6.31a, 6.31b and 6.31c) is in essence the embedment of the preservation of probability from the point of view of the state space description for the Markovian systems, the previous derivations demonstrate clearly that the FPK equation is the natural result of the preservation of probability in the state space description.

6.4 Dostupov–Pugachev Equation

6.4.1 From Equation of Motion to Random State Equation

Without loss of generality, the equation of motion of a general n_d-degree-of-freedom structural or mechanical system can be written as

$$\mathbf{M}\ddot{\mathbf{X}} + \mathbf{f}(\dot{\mathbf{X}}, \mathbf{X}) = \underset{\sim}{\mathbf{B}}(\mathbf{X}, t)\boldsymbol{\xi}(t) \qquad \dot{\mathbf{X}}(t_0) = \dot{\mathbf{X}}_0, \mathbf{X}(t_0) = \mathbf{X}_0 \tag{6.74}$$

where $\mathbf{X}$, $\dot{\mathbf{X}}$ and $\ddot{\mathbf{X}}$ are the n_d-dimensional displacement, velocity and acceleration vectors respectively, $\mathbf{M} = [M_{\ell k}]_{n_d \times n_d}$ is the mass matrix, $\mathbf{f}(\cdot)$ is the n_d-dimensional internal force vector including the damping and restoring forces, $\underset{\sim}{\mathbf{B}}(\mathbf{X}, t) = [\underset{\sim}{B}_{\ell k}(\mathbf{X}, t)]_{n_d \times r}$ is the input force influence matrix, $\boldsymbol{\xi}(t)$ is the r-dimensional external excitation vector, and $\dot{\mathbf{X}}_0$ and $\mathbf{X}_0$ are respectively the initial velocity and displacement vectors.

When introducing the state vector $\mathbf{Y} = (\dot{\mathbf{X}}^{\mathrm{T}}, \mathbf{X}^{\mathrm{T}})^{\mathrm{T}}$, Equation 6.74 can be rewritten as

$$\dot{\mathbf{Y}} = \mathbf{A}(\mathbf{Y}, t) + \mathbf{B}(\mathbf{Y}, t)\boldsymbol{\xi}(t) \qquad \mathbf{Y}(t_0) = \mathbf{Y}_0 \tag{6.75a}$$

where

$$\mathbf{A}(\mathbf{Y}, t) = \begin{Bmatrix} -\mathbf{M}^{-1}\mathbf{f}(\mathbf{Y}) \\ \dot{\mathbf{X}} \end{Bmatrix} \qquad \mathbf{B}(\mathbf{Y}, t) = \begin{bmatrix} \mathbf{M}^{-1}\underset{\sim}{\mathbf{B}}(\mathbf{X}, t) \\ \mathbf{0} \end{bmatrix} \tag{6.75b}$$

If randomness is involved in the excitations, then Equation (6.75a) can be remodeled by

$$\dot{\mathbf{Y}} = \mathbf{A}(\mathbf{Y}, t) + \mathbf{B}(\mathbf{Y}, t)\boldsymbol{\zeta}(\omega, t) \qquad \mathbf{Y}(t_0) = \mathbf{Y}_0 \tag{6.76}$$

One approach to tackling the problem is to model the random excitations as Wiener processes, thus leading to the Itô stochastic differential equation and the FPK equation, as elaborated in Sections 5.6 and 6.3.2. An alternative approach is to decompose the excitation using, for example, the Karhunen–Loève decomposition (see Equation 2.120 in Section 2.2.5):

$$\xi_j(\varpi, t) = \xi_{j0}(t) + \sum_{n=1}^{N_j} \zeta_{j,n}(\varpi)\sqrt{\lambda_{j,n}}f_{j,n}(t) \tag{6.77a}$$

where $\xi_j(\varpi, t)$ is the jth component of $\boldsymbol{\xi}(\omega, t)$, $\xi_{j0}(t)$ is the mean, $\lambda_{j,n}$ and $f_{j,n}(t)$ are the eigenvalues and eigenfunctions, $\boldsymbol{\xi}_{j,n}(\varpi)$ are the uncorrelated standard random variables and N_j is the number of truncated terms. This treatment is particularly useful for the nonstationary process encountered in practice.

If we denote $\boldsymbol{\Theta} = [\zeta_{1,1}(\varpi), \zeta_{1,2}(\varpi), \ldots, \zeta_{1,N_1}(\varpi), \zeta_{2,1}(\varpi), \ldots, \zeta_{2,N_2}(\varpi), \ldots, \zeta_{r,N_r}(\omega)]$, then according to Equation 6.77a the excitation vector can be represented explicitly by

$$\boldsymbol{\xi}(\varpi, t) = \mathbf{F}(\boldsymbol{\Theta}, t) \tag{6.77b}$$

Substituting this, Equation 6.76 can then be rewritten as

$$\dot{\mathbf{Y}} = \mathbf{G}(\boldsymbol{\Theta}, \mathbf{Y}, t) \qquad \mathbf{Y}(t_0) = \mathbf{Y}_0 \tag{6.78}$$

where $\mathbf{G} = (G_1, G_2, \ldots, G_m)^{\mathrm{T}}$ is given by

$$\mathbf{G}(\boldsymbol{\Theta}, \mathbf{Y}, t) = \mathbf{A}(\mathbf{Y}, t) + \mathbf{B}(\mathbf{Y}, t)\mathbf{F}(\boldsymbol{\Theta}, t)$$

Equation 6.78 is a random state equation with random variables made explicit.

6.4.2 The Dostupov–Pugachev Equation

To capture the probability density evolution of the system in Equation 6.78, we first start as was suggested by Dostupov and Pugachev (1957). Then we will tackle it from the unified point of view of the preservation of probability.

If we consider a given $\boldsymbol{\theta}$, the random variable $\mathbf{Y}(t + \Delta t)$ can be regarded as a linear transformation of $\mathbf{Y}(t)$; namely:

$$\mathbf{Y}(t+\Delta t) = \mathbf{Y}(t) + \mathbf{G}(\mathbf{Y}, \boldsymbol{\theta}, t)\Delta t + o(\Delta t) \tag{6.79}$$

Denoting the density of $\mathbf{Y}(t + \Delta t)$ with the parameter $\boldsymbol{\theta}$ by $p_{\mathbf{Y}\boldsymbol{\Theta}}(\mathbf{y}, \boldsymbol{\theta}, t)$, then from Equation 6.12 we have

$$p_{\mathbf{Y}\boldsymbol{\Theta}}(\tilde{\mathbf{y}}, \boldsymbol{\theta}, t+\Delta t)\, \mathrm{d}\tilde{\mathbf{y}} = p_{\mathbf{Y}\boldsymbol{\Theta}}(\mathbf{y}, \boldsymbol{\theta}, t)\, \mathrm{d}\mathbf{y} \tag{6.80}$$

where, according to Equation 6.79:

$$\tilde{\mathbf{y}} = \mathbf{y} + \mathbf{G}(\mathbf{y}, \boldsymbol{\theta}, t)\Delta t + o(\Delta t) \tag{6.81}$$

Differentiating Equation 6.81, we have[6]

$$\mathrm{d}\tilde{\mathbf{y}} = \left[1 + \sum_{\ell=1}^{m} \frac{\partial G_\ell(\mathbf{y}, \boldsymbol{\theta}, t)}{\partial y_\ell}\right] \mathrm{d}\mathbf{y} = |J|\mathrm{d}\mathbf{y} \tag{6.82}$$

where J is the Jacobian.

Introducing Equations 6.81 and 6.82 into the left-hand side of Equation 6.80 yields

$$\begin{aligned}
& p_{\mathbf{Y}\boldsymbol{\Theta}}[\mathbf{y}+\mathbf{G}(\mathbf{y},\boldsymbol{\theta},t)\Delta t+o(\Delta t),\boldsymbol{\theta},t+\Delta t]|J|\mathrm{d}\mathbf{y} \\
&= \left[p_{\mathbf{Y}\boldsymbol{\Theta}}(\mathbf{y},\boldsymbol{\theta},t+\Delta t)+\sum_{\ell=1}^{m}\frac{\partial p_{\mathbf{Y}\boldsymbol{\Theta}}}{\partial y_\ell}G_\ell(\mathbf{y},\boldsymbol{\theta},t)\Delta t\right]\left[1+\sum_{\ell=1}^{m}\frac{\partial G_\ell(\mathbf{y},\boldsymbol{\theta},t)}{\partial y_\ell}\Delta t\right]\mathrm{d}\mathbf{y}+o(\Delta t) \\
&= \left\{p_{\mathbf{Y}\boldsymbol{\Theta}}(\mathbf{y},\boldsymbol{\theta},t+\Delta t)+\left[\sum_{\ell=1}^{m}\frac{\partial p_{\mathbf{Y}\boldsymbol{\Theta}}}{\partial y_\ell}G_\ell(\mathbf{y},\boldsymbol{\theta},t)+p_{\mathbf{Y}\boldsymbol{\Theta}}(\mathbf{y},\boldsymbol{\theta},t+\Delta t)\sum_{\ell=1}^{m}\frac{\partial G_\ell(\mathbf{y},\boldsymbol{\theta},t)}{\partial y_\ell}\right]\Delta t\right\}\mathrm{d}\mathbf{y}+o(\Delta t)
\end{aligned} \tag{6.83}$$

in which the first-order terms are retained in the Taylor expansion.

---

[6] This can be obtained as a product of the differentiation of the components. From Equation 6.81, we have

$$\mathrm{d}\tilde{y}_k = \mathrm{d}y_k + \sum_{\ell=1}^{m} \frac{\partial G_k(\mathbf{y}, \boldsymbol{\theta}, t)}{\partial y_\ell}\mathrm{d}y_\ell\, \Delta t = \left\{1 + \frac{\partial G_k(\mathbf{y}, \boldsymbol{\theta}, t)}{\partial y_k}\Delta t\right\}\mathrm{d}y_k + \sum_{\ell=1,\ell\neq k}^{m} \frac{\partial G_k(\mathbf{y}, \boldsymbol{\theta}, t)}{\partial y_\ell}\mathrm{d}y_\ell\, \Delta t$$

$k = 1, 2, \ldots, m$

Multiplying $d\tilde{y}_k$ and ignoring the terms of higher order of Δt yields $\mathrm{d}\tilde{\mathbf{y}} = \mathrm{d}\tilde{y}_1\, \mathrm{d}\tilde{y}_2 \cdots \mathrm{d}\tilde{y}_m = \left[1 + \sum_{k=1}^{m} \frac{\partial G_k(\mathbf{y}, \boldsymbol{\theta}, t)}{\partial y_k}\Delta t\right] \mathrm{d}y_1\, \mathrm{d}y_2 \cdots \mathrm{d}y_m = |J|\, \mathrm{d}\mathbf{y}$

Replacing the left-hand side term of Equation 6.80 by Equation 6.83, subtracting $p_{\mathbf{Y\Theta}}(\mathbf{y}, \boldsymbol{\theta}, t)\mathrm{d}\mathbf{y}$ on both sides, then dividing it by Δt and letting $\Delta t \to 0$, we have

$$\frac{\partial p_{\mathbf{Y\Theta}}(\mathbf{y}, \boldsymbol{\theta}, t)}{\partial t} + \sum_{\ell=1}^{m} \frac{\partial [p_{\mathbf{Y\Theta}}(\mathbf{y}, \boldsymbol{\theta}, t) G_\ell(\mathbf{y}, \boldsymbol{\theta}, t)]}{\partial y_\ell} = 0 \tag{6.84}$$

where $\mathrm{d}\mathbf{y}$ on both sides is eliminated. We call this equation the Dostupov–Pugachev equation, which was first derived by Dostupov and Pugachev (1957). Comparing Equation 6.84 with Equation 6.39b, we find that Equation 6.84 may be regarded as a parametric Liouville equation where $\boldsymbol{\theta}$ occurs explicitly. This change is essential, because the Liouville equation does not hold for the system in Equation 6.78.

It is noted that the derivation of Equation 6.84 holds for every possible value of $\boldsymbol{\theta}$; that is, for any prescribed random event. In this sense, Equation 6.84 is the result from the random event description of the *preservation of probability*. Thus, in contrast to the Liouville equation, the methodology of the Dostupov–Pugachev equation is changed from incorporation of the state space description and coupling physical equations to incorporation of the random event description and coupling physical equations. To understand this point further, we will derive the Dostupov–Pugachev equation on the basis of the preservation of probability (Chen and Li, 2009).

Compared with Equation 6.25, in the visual form the difference of Equation 6.78 is that $\boldsymbol{\theta}$ is involved in the operator. This difference leads to a distinct feature that the evolution process of $\mathbf{Y}(t)$ itself may not be probability preserved because of the effect of $\boldsymbol{\theta}$. In other words, in order to form a probability preserved system, we should consider $\boldsymbol{\Theta}$ is time invariant; namely:

$$\dot{\boldsymbol{\Theta}} = \mathbf{0} \tag{6.85}$$

Actually, it is the time invariance of $\boldsymbol{\Theta}$, which is the embedment of the nondisappearance of a random event, that ensures the preservation of probability (see Section 6.2.2.1).

According to Equation 6.20, the probability being preserved leads to

$$\frac{\mathrm{D}}{\mathrm{D}t}\int_{D_t \times D_{\boldsymbol{\Theta}}} p_{\mathbf{Y\Theta}}(\mathbf{y}, \boldsymbol{\theta}, t)\,\mathrm{d}\mathbf{y}\,\mathrm{d}\boldsymbol{\theta} = 0 \tag{6.86}$$

Noting the map from time 0 to time t, Equation 6.86 can be rearranged to

$$\begin{aligned}
&\frac{\mathrm{D}}{\mathrm{D}t}\int_{D_t \times D_{\boldsymbol{\Theta}}} p_{\mathbf{Y\Theta}}(\mathbf{y}, \boldsymbol{\theta}, t) d\mathbf{y}\, d\boldsymbol{\theta} \\
&= \frac{\mathrm{D}}{\mathrm{D}t}\int_{D_{t_0} \times D_{\boldsymbol{\Theta}}} p_{\mathbf{Y\Theta}}(\mathbf{y}, \boldsymbol{\theta}, t)|J| d\mathbf{y}\, d\boldsymbol{\theta} \\
&= \int_{D_{t_0} \times D_{\boldsymbol{\Theta}}} \left(|J| \frac{\mathrm{D}p_{\mathbf{Y\Theta}}(\mathbf{y}, \boldsymbol{\theta}, t)}{\mathrm{D}t} + p_{\mathbf{Y\Theta}}(\mathbf{y}, \boldsymbol{\theta}, t) \frac{\mathrm{D}|J|}{\mathrm{D}t} \right) d\mathbf{y}\, d\boldsymbol{\theta}
\end{aligned}$$

$$= \int_{D_{t_0}\times D_{\Theta}} \left\{ |J| \left(\frac{\partial p_{\mathbf{Y\Theta}}(\mathbf{y},\boldsymbol{\theta},t)}{\partial t} + \sum_{\ell=1}^{m} G_\ell(\mathbf{y},\boldsymbol{\theta},t) \frac{\partial p_{\mathbf{Y\Theta}}(\mathbf{y},\boldsymbol{\theta},t)}{\partial y_\ell} \right) \right.$$

$$\left. + |J| p_{\mathbf{Y\Theta}}(\mathbf{y},\boldsymbol{\theta},t) \sum_{\ell=1}^{m} \frac{\partial G_\ell(\mathbf{y},\boldsymbol{\theta},t)}{\partial y_\ell} \right\} d\mathbf{y}\, d\boldsymbol{\theta}$$

$$= \int_{D_{t_0}\times D_{\Theta}} \left(\frac{\partial p_{\mathbf{Y\Theta}}(\mathbf{y},\boldsymbol{\theta},t)}{\partial t} + \sum_{\ell=1}^{m} \frac{\partial [p_{\mathbf{Y\Theta}}(\mathbf{y},\boldsymbol{\theta},t) G_l(y,\boldsymbol{\theta},t)]}{\partial y_\ell} \right) |J| d\mathbf{y}\, d\boldsymbol{\theta}$$

$$= \int_{D_t\times D_{\Theta}} \left(\frac{\partial p_{\mathbf{Y\Theta}}(\mathbf{y},\boldsymbol{\theta},t)}{\partial t} + \sum_{\ell=1}^{m} \frac{\partial [p_{\mathbf{Y\Theta}}(\mathbf{y},\boldsymbol{\theta},t) G_\ell(\mathbf{y},\boldsymbol{\theta},t)]}{\partial y_\ell} \right) d\mathbf{y}\, d\boldsymbol{\theta} \tag{6.87}$$

where the total derivative is given as

$$\frac{\mathrm{D}p_{\mathbf{Y\Theta}}(\mathbf{y},\boldsymbol{\theta},t)}{\mathrm{D}t} = \frac{\partial p_{\mathbf{Y\Theta}}(\mathbf{y},\boldsymbol{\theta},t)}{\partial t} + \sum_{\ell=1}^{m} G_\ell(\mathbf{y},\boldsymbol{\theta},t) \frac{\partial p_{\mathbf{Y\Theta}}(\mathbf{y},\boldsymbol{\theta},t)}{\partial y_\ell} \tag{6.88}$$

in which Equation 6.85 has been considered, and the total derivative of the Jacobian is (Belytschko, 2000)

$$\frac{\mathrm{D}|J|}{\mathrm{D}t} = |J| \sum_{\ell=1}^{m} \frac{\partial \dot{y}_\ell}{\partial y_\ell} = |J| \sum_{\ell=1}^{m} \frac{\partial G_\ell(\mathbf{y},\boldsymbol{\theta},t)}{\partial y_\ell} \tag{6.89}$$

In addition, we should note that when the integral domain is $D_t \times D_{\Theta}$, the corresponding arguments $\mathbf{y}$ and $\boldsymbol{\theta}$ in the integrand are Eulerian coordinates, while the integral domain is $D_{t_0} \times D_{\Theta}$, the arguments $\mathbf{y}$ and $\boldsymbol{\theta}$ in the integrand should be understood as Lagriangian coordinates $\mathbf{y}_{\mathrm{L}}, \boldsymbol{\theta}_{\mathrm{L}}$, where $(\mathbf{y},\boldsymbol{\theta}) = \mathbf{H}(\mathbf{y}_{\mathrm{L}},\boldsymbol{\theta}_{\mathrm{L}},t)$ is the solution of the system in Equations 6.78 and 6.85 and the Jacobian is given by

$$|J| = \left| \frac{\partial(\mathbf{y},\boldsymbol{\theta})}{\partial(\mathbf{y}_{\mathrm{L}},\boldsymbol{\theta}_{\mathrm{L}})} \right| = \left| \frac{\partial \mathbf{H}(\mathbf{y}_{\mathrm{L}},\boldsymbol{\theta}_{\mathrm{L}},t)}{\partial(\mathbf{y}_{\mathrm{L}},\boldsymbol{\theta}_{\mathrm{L}})} \right| \tag{6.90}$$

However, for notational simplicity, we use the same symbols for the Eulerian and Lagrangian coordinates in Equation 6.87 without inducing confusion.

Combining Equations 6.86 and 6.87 and noting the arbitrariness of $D_t \times D_{\Theta}$, it follows that

$$\frac{\partial p_{\mathbf{Y\Theta}}(\mathbf{y},\boldsymbol{\theta},t)}{\partial t} + \sum_{\ell=1}^{m} \frac{\partial [p_{\mathbf{Y\Theta}}(\mathbf{y},\boldsymbol{\theta},t) G_\ell(\mathbf{y},\boldsymbol{\theta},t)]}{\partial y_\ell} = 0 \tag{6.91}$$

This is nothing but the Dostupov–Pugachev equation identical to Equation 6.84.

It is interesting that although the random event description of the principle of preservation of probability, as discussed in Sections 6.2.1 and 6.2.2.1, seems logically more straightforward than the state space description, the latter is preferred in the history of developing the probability density evolution equations such as the Liouville equation and FPK equation. In the above derivation of Equation 6.91, however, the random event description of the principle is employed in conjunction with the coupling physical equation. In other words, the Dostupov– Pugachev equation can be regarded as the probability density evolution equation that turns the point of view from the complete state space description to partly considering the random event description.

6.5 The Generalized Density Evolution Equation

6.5.1 Derivation of the Generalized Density Evolution Equation

6.5.1.1 From Random Event Description to Generalized Density Evolution Equation

As discussed, the Dostupov-Pugachev equation is the result of cooperating the random event description of the principle of preservation of probability and the coupling physical equations in state space description. Actually, a further step will open the way to a generalized density evolution equation (GDEE). In fact, when viewed from the random event description of the principle of preservation of probability and introducing the physical solution of the system, a completely uncoupled, any arbitrary-dimensional density evolution equation can be reached (Li and Chen, 2006c, 2008).

We now consider the generic stochastic dynamical system

$$\dot{\mathbf{Y}} = \mathbf{G}(\boldsymbol{\Theta}, \mathbf{Y}, t) \qquad \mathbf{Y}(t_0) = \mathbf{Y}_0 \tag{6.92}$$

where $\mathbf{Y} = (Y_1, Y_2, \ldots, Y_m)$ is the state vector, $\mathbf{Y}_0$ is the initial value vector, m is the dimension of the system and $\boldsymbol{\Theta} = (\Theta_1, \Theta_2, \ldots, \Theta_s)$ is an s-dimensional random vector characterizing the randomness involved with known joint PDF $p_{\boldsymbol{\Theta}}(\boldsymbol{\theta})$. The randomness might come not only from the excitations, but also from the system properties. Generally, the random excitations are modeled as stochastic processes, which could be further represented by some types of random functions of some standard basic random variables, say through decompositions or physical stochastic modeling as elaborated in Sections 2.2.5 and 6.4.1 and Chapter 3. The randomness involved in the system properties might originally occur as random fields or directly as some random parameters. Again, the random fields can be discretized or decomposed to a set of standard random variables, by employing the methodologies in Section 2.3 for example. The random vector $\boldsymbol{\Theta}$ consists of these two sets of standard basic random variables coming respectively from random excitations and system properties. This is different from Equation 6.78, where the random parameters come only from random excitations.

As a random state equation, Equation 6.92 can be understood as an Eulerian description of a dynamical system, where a velocity field is specified and thus a Dostupov–Pugachev equation will be led to. However, the system can also be specified by a Lagrangian description. Without loss of generality, suppose the Lagrangian description is given by

$$\mathbf{Y} = \mathbf{H}(\boldsymbol{\Theta}, \mathbf{Y}_0, t) \quad \text{or} \quad Y_\ell = H_\ell(\boldsymbol{\Theta}, \mathbf{Y}_0, t) \qquad \ell = 1, 2, \ldots, m \tag{6.93}$$

which, of course, is the physical solution of Equation 6.92 and satisfies $\mathbf{Y}_0 = \mathbf{H}(\boldsymbol{\Theta}, \mathbf{Y}_0, t_0)$. Correspondingly, the velocity can be assumed to take the form

$$\dot{\mathbf{Y}} = \mathbf{h}(\boldsymbol{\Theta}, \mathbf{Y}_0, t) \quad \text{or} \quad \dot{Y}_\ell = h_\ell(\boldsymbol{\Theta}, \mathbf{Y}_0, t) \qquad \ell = 1, 2, \ldots, m \tag{6.94}$$

where $\mathbf{h} = \partial\mathbf{H}/\partial t$. The expressions of $H_\ell(\cdot)$ and $h_\ell(\cdot)$ need not be explicitly figured out in the present instance; it is sufficient to know that they exist.

In a general sense, if there are a set of physical quantities $\mathbf{Z}(t) = (Z_1(t), Z_2(t), \ldots, Z_{n_Z}(t))^{\mathrm{T}}$ associated with the system in Equation 6.92, then $\mathbf{Z}$ can usually be determined by its connection with the state vectors, for instance by

$$\dot{\mathbf{Z}}(t) = \psi[\dot{\mathbf{Y}}(t)] \qquad \mathbf{Z}(t_0) = \mathbf{z}_0 \tag{6.95}$$

where $\psi[\cdot]$ is a transfer operator. For example, for a structural system in which geometric nonlinearity is ignored, if $Z_\ell(t)$ are the strains at some points, then $\psi[\cdot]$ will be a linear operator bridging the displacement and the strains (Fung and Tong,2001). While if $\mathbf{Z}$ represents $\mathbf{Y}$ itself, then $\psi[\cdot]$ is an identity operator.

Introducing Equation 6.94 into Equation 6.95 we have

$$\dot{\mathbf{Z}}(t) = \psi[\dot{\mathbf{Y}}(t)] = \psi[\mathbf{h}(\mathbf{\Theta}, \mathbf{Y}_0, t)] = \mathbf{h_Z}(\mathbf{\Theta}, t) \tag{6.96a}$$

or in a component form as

$$\dot{Z}_\ell(t) = h_{\mathbf{Z},\ell}(\mathbf{\Theta}, t) \qquad \ell = 1, 2, \ldots, n_\mathbf{Z} \tag{6.96b}$$

Here, $\mathbf{h_Z} = (h_{\mathbf{Z},1}, h_{\mathbf{Z},2}, \ldots, h_{\mathbf{Z},n_\mathbf{Z}})^\mathrm{T}$, where $n_\mathbf{z}$ is the number of the physical quantities considered in $\mathbf{Z}$. For brevity, we now consider the case of deterministic $\mathbf{Y}_0$ and omit it in the equation.

From Equations (6.96a) and (6.96b), it is noted that the randomness involved in $\mathbf{Z}(t)$ results completely from $\mathbf{\Theta}$; therefore, the system $(\mathbf{Z}(t),\mathbf{\Theta})$ is a probability preserved system. From Equation 6.20, if the joint density of $(\mathbf{Z}(t),\mathbf{\Theta})$ is denoted by $p_{\mathbf{Z\Theta}}(\mathbf{z}, \boldsymbol{\theta}, t)$, where $\mathbf{z} = (z_1, z_2, \ldots, z_{n_\mathbf{Z}})$, then it follows that

$$\frac{\mathrm{D}}{\mathrm{D}t}\int_{D_t \times D_\mathbf{\Theta}} p_{\mathbf{Z\Theta}}(\mathbf{z}, \boldsymbol{\theta}, t)\, \mathrm{d}\mathbf{z}\, \mathrm{d}\boldsymbol{\theta} = 0 \tag{6.97}$$

Following a process analogous to Equation 6.87, we have

$$\begin{aligned}
\frac{\mathrm{D}}{\mathrm{D}t}\int_{D_t \times D_\mathbf{\Theta}} p_{\mathbf{Z\Theta}}(\mathbf{z}, \boldsymbol{\theta}, t)\, \mathrm{d}\mathbf{z}\, \mathrm{d}\boldsymbol{\theta} &= \frac{\mathrm{D}}{\mathrm{D}t}\int_{D_{t_0} \times D_\mathbf{\Theta}} p_{\mathbf{Z\Theta}}(\mathbf{z}, \boldsymbol{\theta}, t)|J|\, \mathrm{d}\mathbf{z}\, \mathrm{d}\boldsymbol{\theta} \\
&= \int_{D_{t_0} \times D_\mathbf{\Theta}} \left(|J|\frac{\mathrm{D}p_{\mathbf{Z\Theta}}}{\mathrm{D}t} + p_{\mathbf{Z\Theta}}\frac{\mathrm{D}|J|}{\mathrm{D}t}\right) \mathrm{d}\mathbf{z}\, \mathrm{d}\boldsymbol{\theta} \\
&= \int_{D_{t_0} \times D_\mathbf{\Theta}} \left[|J|\left(\frac{\partial p_{\mathbf{Z\Theta}}}{\partial t} + \sum_{\ell=1}^{n_\mathbf{Z}} h_{\mathbf{Z},\ell}\frac{\partial p_{\mathbf{Z\Theta}}}{\partial z_\ell}\right) + |J| p_{\mathbf{Z\Theta}}\frac{\partial h_{\mathbf{Z},\ell}}{\partial z_\ell}\right] \mathrm{d}\mathbf{z}\, \mathrm{d}\boldsymbol{\theta} \\
&= \int_{D_{t_0} \times D_\mathbf{\Theta}} \left(\frac{\partial p_{\mathbf{Z\Theta}}}{\partial t} + \sum_{\ell=1}^{n_\mathbf{Z}} h_{\mathbf{Z},\ell}\frac{\partial p_{\mathbf{Z\Theta}}}{\partial z_\ell}\right)|J|\, \mathrm{d}\mathbf{z}\, \mathrm{d}\boldsymbol{\theta} \\
&= \int_{D_t \times D_\mathbf{\Theta}} \left(\frac{\partial p_{\mathbf{Z\Theta}}}{\partial t} + \sum_{\ell=1}^{n_\mathbf{Z}} h_{\mathbf{Z},\ell}\frac{\partial p_{\mathbf{Z\Theta}}}{\partial z_\ell}\right) \mathrm{d}\mathbf{z}\, \mathrm{d}\boldsymbol{\theta}
\end{aligned} \tag{6.98}$$

where the arguments of the functions $p_{\mathbf{Z\Theta}}(\cdot)$ and $h_{\mathbf{Z},\ell}(\cdot)$ are omitted for notational simplicity. Again, as pointed out in Equation 6.87, in each step of Equation 6.98 the arguments should be carefully understood as sometimes Eulerian but sometimes as Lagrangian coordinates, according to the integral domain.

Introducing Equation 6.98 into Equation 6.97 and noting the arbitrariness of $D_t \times D_\mathbf{\Theta}$, we have

$$\frac{\partial p_{\mathbf{Z\Theta}}(\mathbf{z}, \boldsymbol{\theta}, t)}{\partial t} + \sum_{\ell=1}^{n_\mathbf{Z}} h_{\mathbf{Z},\ell}(\boldsymbol{\theta}, t)\frac{\partial p_{\mathbf{Z\Theta}}(\mathbf{z}, \boldsymbol{\theta}, t)}{\partial z_\ell} = 0 \tag{6.99a}$$

or in an alternative form, when considering Equation 6.96b, as

$$\frac{\partial p_{\mathbf{Z\Theta}}(\mathbf{z},\boldsymbol{\theta},t)}{\partial t}+\sum_{\ell=1}^{n_{\mathbf{Z}}}\dot{Z}_\ell(\boldsymbol{\theta},t)\frac{\partial p_{\mathbf{Z\Theta}}(\mathbf{z},\boldsymbol{\theta},t)}{\partial z_\ell}=0 \tag{6.99b}$$

The joint density of $\mathbf{Z}(t)$ can then be given by

$$p_{\mathbf{Z}}(\mathbf{z},t)=\int_{\Omega_{\boldsymbol{\Theta}}}p_{\mathbf{Z\Theta}}(\mathbf{z},\boldsymbol{\theta},t)\,\mathrm{d}\boldsymbol{\theta} \tag{6.100}$$

where $\Omega_{\boldsymbol{\Theta}}$ is the distribution domain of $\boldsymbol{\Theta}$.

It is noted that the dimension $n_{\mathbf{Z}}$ of Equation 6.95 depends only on the research requirement and is independent of the dimension m of the system in Equation 6.92. In this regard, we might call Equations 6.96a and 6.96b the result of an arbitrary-dimensional Lagrangian description. It is the flexibility here that makes the dimension of Equations 6.99a and 6.99b flexible rather than fixed to be the same as the dimension of the state vector of the system as in the Liouville, FPK and Dostupov–Pugachev equations.

6.5.1.2 From Multidimensions to One Dimension: A Formal Treatment

When employing a formal expression of the density of the responses and handling it directly, we can also reach the one-dimensional uncoupled density evolution equation (Chen and Li, 2005a; Li and Chen, 2005a, 2006a).

Clearly, owing to Equation 6.93, the density of $\mathbf{Y}(t)$ is given by (see Appendix A)

$$p_{\mathbf{Y}}(\mathbf{y},t)=\int\delta[\mathbf{y}-\mathbf{H}(\boldsymbol{\theta},\mathbf{y}_0,t)]p_{\boldsymbol{\Theta}}(\boldsymbol{\theta})\,\mathrm{d}\boldsymbol{\theta} \tag{6.101}$$

where $\delta(\cdot)$ is the Dirac delta function and

$$p_{\mathbf{Y\Theta}}(\mathbf{y},\boldsymbol{\theta},t)=\delta[\mathbf{y}-\mathbf{H}(\boldsymbol{\theta},\mathbf{y}_0,t)]p_{\boldsymbol{\Theta}}(\boldsymbol{\theta})=\prod_{\ell=1}^{m}\delta[y_\ell-H_\ell(\boldsymbol{\theta},\mathbf{y}_0,t)]p_{\boldsymbol{\Theta}}(\boldsymbol{\theta}) \tag{6.102}$$

is the joint density of $(\mathbf{Y}(t),\boldsymbol{\Theta})$. Here, without loss of generality and for simplicity, we consider $\mathbf{Y}_0=\mathbf{y}_0$ as a deterministic vector.

Differentiating with respect to t on both sides of Equation 6.102 yields

$$\begin{aligned}\frac{\partial p_{\mathbf{Y\Theta}}(\mathbf{y},\boldsymbol{\theta},t)}{\partial t}&=\frac{\partial}{\partial t}\prod_{\ell=1}^{m}\delta[y_\ell-H_\ell(\boldsymbol{\theta},\mathbf{y}_0,t)]p_{\boldsymbol{\Theta}}(\boldsymbol{\theta})\\&=\sum_{\ell=1}^{m}\left\{\prod_{k=1,k\neq\ell}^{m}\delta[y_k-H_k(\boldsymbol{\theta},\mathbf{y}_0,t)]\frac{\partial\delta[y_\ell-H_\ell(\boldsymbol{\theta},\mathbf{y}_0,t)]}{\partial t}\right\}p_{\boldsymbol{\Theta}}(\boldsymbol{\theta})\\&=\sum_{\ell=1}^{m}\left(\left[-\frac{\partial H_\ell(\boldsymbol{\theta},\mathbf{y}_0,t)}{\partial t}\right]\prod_{k=1,k\neq\ell}^{m}\delta[y_k-H_k(\boldsymbol{\theta},\mathbf{y}_0,t)]\frac{\partial\delta[y_\ell-H_\ell(\boldsymbol{\theta},\mathbf{y}_0,t)]}{\partial y_\ell}\right)p_{\boldsymbol{\Theta}}(\boldsymbol{\theta})\\&=\sum_{\ell=1}^{m}\left(\left[-\frac{\partial H_\ell(\boldsymbol{\theta},\mathbf{y}_0,t)}{\partial t}\right]\frac{\partial\left\{\prod_{k=1}^{m}\delta[y_k-H_k(\boldsymbol{\theta},\mathbf{y}_0,t)]p_{\boldsymbol{\Theta}}(\boldsymbol{\theta})\right\}}{\partial y_\ell}\right)\\&=-\sum_{\ell=1}^{m}h_\ell(\boldsymbol{\theta},\mathbf{y}_0,t)\frac{\partial p_{\mathbf{Y\Theta}}(\mathbf{y},\boldsymbol{\theta},t)}{\partial y_\ell}\end{aligned} \tag{6.103}$$

or in an alternative form as

$$\frac{\partial p_{\mathbf{Y}\mathbf{\Theta}}(\mathbf{y},\boldsymbol{\theta},t)}{\partial t}+\sum_{\ell=1}^{m} h_\ell(\boldsymbol{\theta},\mathbf{y}_0,t)\frac{\partial p_{\mathbf{Y}\mathbf{\Theta}}(\mathbf{y},\boldsymbol{\theta},t)}{\partial y_\ell}=0 \tag{6.104}$$

Comparing this equation with the Dostupov–Pugachev Equation 6.91 shows that the coefficients here are no longer coupled with the density in the partial differentiation operator. In mathematical form, this is because Equation 6.92 is replaced by Equation 6.94 and, therefore, $G_\ell(\cdot)$ in the coefficient is replaced by $h_\ell(\cdot)$. Here, it is worth pointing out that we have turned from the Eulerian system to the Lagrangian system; that is, from the coupling physical equations to the uncoupled physical equations (physical solutions).

Taking multiple integrals with respect to $y_1,\ldots,y_{\ell-1},y_{\ell+1},\ldots,y_m$ and denoting the marginal density by

$$p_{Y_\ell\mathbf{\Theta}}(y_\ell,\boldsymbol{\theta},t)=\int p_{\mathbf{Y}\mathbf{\Theta}}(\mathbf{y},\boldsymbol{\theta},t)\mathrm{d}y_1\cdots\mathrm{d}y_{\ell-1}\mathrm{d}y_{\ell+1}\cdots\mathrm{d}y_m \tag{6.105}$$

we immediately have

$$\frac{\partial p_{Y_\ell\mathbf{\Theta}}(y_\ell,\boldsymbol{\theta},t)}{\partial t}+h_\ell(\boldsymbol{\theta},\mathbf{y}_0,t)\frac{\partial p_{Y_\ell\mathbf{\Theta}}(y_\ell,\boldsymbol{\theta},t)}{\partial y_\ell}=0 \tag{6.106a}$$

or in an alternative form, when Equation 6.94 is considered, as

$$\frac{\partial p_{Y_\ell\mathbf{\Theta}}(y_\ell,\boldsymbol{\theta},t)}{\partial t}+\dot{Y}_\ell(\boldsymbol{\theta},t)\frac{\partial p_{Y_\ell\mathbf{\Theta}}(y_\ell,\boldsymbol{\theta},t)}{\partial y_\ell}=0 \tag{6.106b}$$

where use has been made of

$$p_{Y_\ell\mathbf{\Theta}}(y_\ell,\boldsymbol{\theta},t)\big|_{y_\ell\to\pm\infty}=0 \quad \text{and} \quad y_\ell p_{Y_\ell\mathbf{\Theta}}(y_\ell,\boldsymbol{\theta},t)\big|_{y_\ell\to\pm\infty}=0 \tag{6.107}$$

Clearly, Equations 6.106a and 6.10b are identical to Equations 6.99a and 6.99b respectively if $n_{\mathbf{Z}}=1$ and $\mathbf{Z}$ is replaced by Y_ℓ.

These equations can also be obtained in a more straightforward way. From Equation 6.93 it is known that the density of $Y_\ell(t)$ is given by (see Appendix A)

$$p_{Y_\ell}(y_\ell,t)=\int \delta[y_\ell-H_\ell(\boldsymbol{\theta},\mathbf{y}_0,t)]p_{\mathbf{\Theta}}(\boldsymbol{\theta})\,\mathrm{d}\boldsymbol{\theta} \tag{6.108}$$

Therefore, the joint density of $(Y_\ell(t),\mathbf{\Theta})$ is

$$p_{Y_\ell\mathbf{\Theta}}(y_\ell,\boldsymbol{\theta},t)=\delta[y_\ell-H_\ell(\boldsymbol{\theta},\mathbf{y}_0,t)]p_{\mathbf{\Theta}}(\boldsymbol{\theta}) \tag{6.109}$$

Differentiating this equation on both sides with respect to t yields

$$\begin{aligned}\frac{\partial p_{Y_\ell\mathbf{\Theta}}(y_\ell,\boldsymbol{\theta},t)}{\partial t}&=\frac{\partial}{\partial t}\delta[y_\ell-H_\ell(\boldsymbol{\theta},\mathbf{y}_0,t)]p_{\mathbf{\Theta}}(\boldsymbol{\theta})\\&=\frac{\partial\delta[y_\ell-H_\ell(\boldsymbol{\theta},\mathbf{y}_0,t)]p_{\mathbf{\Theta}}(\boldsymbol{\theta})}{\partial y_\ell}\frac{\partial[y_\ell-H_\ell(\boldsymbol{\theta},\mathbf{y}_0,t)]}{\partial t}\\&=-\frac{\partial H_\ell(\boldsymbol{\theta},\mathbf{y}_0,t)}{\partial t}\frac{\partial p_{Y_\ell\mathbf{\Theta}}(y_\ell,\boldsymbol{\theta},t)}{\partial y_\ell}\\&=-h_\ell(\boldsymbol{\theta},\mathbf{y}_0,t)\frac{\partial p_{Y_\ell\mathbf{\Theta}}(y_\ell,\boldsymbol{\theta},t)}{\partial y_\ell}\end{aligned} \tag{6.110}$$

which is of course the same equation as Equations 6.106a and 6.106b.

Clearly, if we replace $\mathbf{Y}$ and $\mathbf{h}$ by $\mathbf{Z}$ and $\mathbf{h}_\mathbf{Z}$ respectively, then following the above derivations we can also reach Equations 6.99a and 6.99b.

In the above manipulations, we note that when the Dirac delta function is employed, we are tackling the problem starting from the perspective of the sample. In other words, we can reach the density evolution equation by establishing a relationship between the sample and the density.

6.5.2 Linear Systems: Uncoupling of the Dostupov–Pugachev Equation

For linear systems, the GDEE can also be obtained by uncoupling the Dostupov–Pugachev equation, which is a high-dimensional partial differential equation. Actually, this is the first way that we found to reach the completely uncoupled GDEE for multidimensional problems (Li and Chen, 2004a).

Suppose Equation 6.78 takes the form

$$\dot{\mathbf{Y}} = \mathbf{a}(\mathbf{\Theta})\mathbf{Y} + \mathbf{F}(\mathbf{\Theta}, t);\ \mathbf{Y}(t_0) = \mathbf{Y}_0 \tag{6.111}$$

where $\mathbf{a} = [a_{\ell k}]_{m\times m}$ and $\mathbf{F} = (F_1, F_2, \ldots, F_m)^{\mathrm{T}}$, which can be determined from the equation of motion, say, through transformation used in Equations 6.74–6.76.

In this case, the component G_ℓ of Equation 6.78 is given by

$$\dot{Y}_\ell(\mathbf{\Theta}, t) = G_\ell(\mathbf{\Theta}, \mathbf{Y}, t) = \sum_{k=1}^{m} a_{\ell k}(\mathbf{\Theta})Y_k + F_\ell(\mathbf{\Theta}, t) \tag{6.112}$$

Substituting this in the Dostupov–Pugachev equation, Equation 6.91, yields

$$\frac{\partial p_{\mathbf{Y\Theta}}}{\partial t} + \sum_{\ell=1}^{m} \frac{\partial}{\partial y_\ell}\left[\left(\sum_{k=1}^{m} a_{\ell k} y_k + F_\ell\right)\right] p_{\mathbf{Y\Theta}} = 0 \tag{6.113}$$

Here, for notational brevity, the arguments in $a_{\ell k}(\cdot)$, $F_\ell(\cdot)$ and $p_{\mathbf{Y\Theta}}(\cdot)$ have been omitted. Equation 6.113 can further be rearranged to

$$\frac{\partial p_{\mathbf{Y\Theta}}}{\partial t} + \sum_{\ell=1}^{m}\left[\left(\sum_{k=1}^{m} a_{\ell k} y_k + F_\ell\right)\frac{\partial p_{\mathbf{Y\Theta}}}{\partial y_\ell}\right] + p_{\mathbf{Y\Theta}}\mathrm{Tr}(\mathbf{a}) = 0 \tag{6.114}$$

where $\mathrm{Tr}(\mathbf{a}) = \sum_{j=1}^{m} a_{jj}$ is the trace of matrix $\mathbf{a}$.

Integrating both sides of Equation 6.114 with respect to $y_1, \ldots, y_{\ell-1}, y_{\ell+1}, \ldots, y_m$, for the first and the third terms on the left-hand side of Equation 6.114 we need only replace $p_{\mathbf{Y\Theta}}$ by $p_{Y_\ell\mathbf{\Theta}}$ (see Equation 6.105), while for the second term we are led to

$$\begin{aligned}
&\int \sum_{j=1}^{m}\left[\left(\sum_{k=1}^{m} a_{jk} y_k + F_j\right)\frac{\partial p_{\mathbf{Y\Theta}}}{\partial y_j}\right]\mathrm{d}y_1\cdots\mathrm{d}y_{\ell-1}\mathrm{d}y_{\ell+1}\cdots\mathrm{d}y_m \\
&= \int \sum_{j=1}^{m}\left(\sum_{k=1}^{m} a_{jk} y_k \frac{\partial p_{\mathbf{Y\Theta}}}{\partial y_j}\right)\mathrm{d}y_1\cdots\mathrm{d}y_{\ell-1}\mathrm{d}y_{\ell+1}\cdots\mathrm{d}y_m + F_\ell\frac{\partial p_{Y_\ell\mathbf{\Theta}}}{\partial y_\ell}
\end{aligned}$$

$$
\begin{aligned}
&= \int \left(\sum_{k=1}^{m} a_{\ell k} y_k \frac{\partial p_{\mathbf{Y\Theta}}}{\partial y_\ell} \right) \mathrm{d}y_1 \cdots \mathrm{d}y_{\ell-1} \mathrm{d}y_{\ell+1} \cdots \mathrm{d}y_m - \sum_{j=1,\, j\neq\ell}^{m} a_{jj} p_{Y_\ell \mathbf{\Theta}} + F_\ell \frac{\partial p_{Y_\ell \mathbf{\Theta}}}{\partial y_\ell} \\
&= \sum_{k=1,k\neq\ell}^{m} \int \left(a_{\ell k} y_k \frac{\partial p_{Y_k Y_\ell \mathbf{\Theta}}}{\partial y_\ell} \mathrm{d}y_k \right) + a_{\ell\ell} y_\ell \frac{\partial p_{Y_\ell \mathbf{\Theta}}}{\partial y_\ell} - \sum_{j=1,j\neq\ell}^{m} a_{jj} p_{Y_\ell \mathbf{\Theta}} + F_\ell \frac{\partial p_{Y_\ell \mathbf{\Theta}}}{\partial y_\ell}
\end{aligned} \tag{6.115}
$$

where

$$
p_{Y_\ell Y_k \mathbf{\Theta}}(y_\ell, y_k, \mathbf{\theta}, t) = \int p_{\mathbf{Y\Theta}}(\mathbf{y}, \mathbf{\theta}, t) \mathrm{d}y_1 \cdots \mathrm{d}y_{k-1} \mathrm{d}y_{k+1} \cdots \mathrm{d}y_{\ell-1} \mathrm{d}y_{\ell+1} \cdots \mathrm{d}y_m \tag{6.116}
$$

Because, for a given $\mathbf{\theta}$, y_k can essentially take only a single value $Y_k(\mathbf{\theta})$, Equation 6.115 then becomes

$$
\begin{aligned}
&\int \sum_{j=1}^{m} \left[\left(\sum_{k=1}^{m} a_{jk} y_k + F_j \right) \frac{\partial p_{\mathbf{Y\Theta}}}{\partial y_j} \right] \mathrm{d}y_1 \cdots \mathrm{d}y_{\ell-1} \mathrm{d}y_{\ell+1} \cdots \mathrm{d}y_m \\
&= \left[\sum_{k=1}^{m} a_{\ell k} Y_k(\mathbf{\theta}) + F_\ell \right] \frac{\partial p_{Y_\ell \mathbf{\Theta}}}{\partial y_\ell} - \sum_{j=1,j\neq\ell}^{m} a_{jj} p_{Y_\ell \mathbf{\Theta}}
\end{aligned} \tag{6.117}
$$

Substituting this in Equation 6.114, we have

$$
\frac{\partial p_{Y_\ell \mathbf{\Theta}}}{\partial t} + \left[\sum_{k=1}^{m} a_{\ell k} Y_k(\mathbf{\theta}) + F_\ell \right] \frac{\partial p_{Y_\ell \mathbf{\Theta}}}{\partial y_\ell} + a_{\ell\ell} p_{Y_\ell \mathbf{\Theta}} = 0 \tag{6.118}
$$

According to the definition of the state vector, when Y_ℓ here is the displacement, namely $\ell = n_{\mathrm{d}} + 1, n_{\mathrm{d}} + 2, \ldots, m$, it is seen that $a_{\ell\ell} = 0$, then the above equation becomes

$$
\frac{\partial p_{Y_\ell \mathbf{\Theta}}}{\partial t} + \left[\sum_{k=1}^{m} a_{\ell k} Y_k(\mathbf{\theta}) + F_\ell \right] \frac{\partial p_{Y_\ell \mathbf{\Theta}}}{\partial y_\ell} = 0 \tag{6.119a}
$$

In contrast to the original Dostupov–Pugachev equation, which is an m-dimensional partial differential equation, Equation 6.119a is a one-dimensional equation. Further, noting Equation 6.112, we find that the coefficient is essentially the velocity; therefore, the equation can also be rewritten as

$$
\frac{\partial p_{Y_\ell \mathbf{\Theta}}}{\partial t} + \dot{Y}_\ell(\mathbf{\theta}, t) \frac{\partial p_{Y_\ell \mathbf{\Theta}}}{\partial y_\ell} = 0 \tag{6.119b}
$$

This is nothing but Equation 6.106b.

The above manipulations uncouple the high-dimensional density evolution equation to a one-dimensional equation and simplify the problem greatly in the case of linear systems. Although it does not work in nonlinear systems, the idea that we have to consider the evolution of probability from the point of view of random event description has essentially been used, particularly in the derivation of Equation 6.117.

6.5.3 Initial and Boundary Conditions

If the initial value vector in Equation 6.95

$$\mathbf{Z}(t_0) = \mathbf{z}_0 \tag{6.120}$$

is a deterministic vector, where $\mathbf{z}_0 = (z_{0,1}, z_{0,2}, \ldots, z_{0,n_\mathrm{Z}})^\mathrm{T}$, then the initial condition of Equation 6.99 will be given by

$$p_{\mathbf{Z}\boldsymbol{\Theta}}(\mathbf{z}, \boldsymbol{\theta}, t_0) = \delta(\mathbf{z} - \mathbf{z}_0) p_{\boldsymbol{\Theta}}(\boldsymbol{\theta}) = \prod_{\ell=1}^{n_\mathrm{Z}} \delta(z - z_{0,\ell}) p_{\boldsymbol{\Theta}}(\boldsymbol{\theta}) \tag{6.121a}$$

In the case that randomness is also involved in the initial conditions, the corresponding random variables can be absorbed into and then become part of $\boldsymbol{\Theta}$. For such cases, there are

$$p_{\mathbf{Z}\boldsymbol{\Theta}}(\mathbf{z}, \boldsymbol{\theta}, t_0) = p_{\mathbf{Z}_0}(\mathbf{z}) p_{\boldsymbol{\Theta}}(\boldsymbol{\theta}) \tag{6.121b}$$

where $p_{\mathbf{Z}_0}(\mathbf{z}_0)$ is the joint density of $\mathbf{Z}_0$.

For the system without external constraint on $\mathbf{Z}(t)$, the boundary condition for Equations 6.99a and 6.99b can take

$$p_{\mathbf{Z}\boldsymbol{\Theta}}(\mathbf{z}, \boldsymbol{\theta}, t)\big|_{z_\ell \to \pm\infty} = 0 \qquad \ell = 1, 2, \ldots, n_\mathrm{Z} \tag{6.122}$$

while for some special cases (for instance, in the first-passage reliability assessment), some other conditions (such as absorbing boundary conditions) might be imposed on the equation (Li and Chen, 2005a; Chen and Li, 2005a). This will be discussed in Chapter 8.

6.5.4 Physical Sense of the Generalized Density Evolution Equation

Equation 6.99 holds for arbitrary physical quantities involved in a physical system. For stochastic structural systems, $\mathbf{Z}$ as determined in Equations 6.99a and 6.99b can be a vector of, say the stress, strain, internal force, displacement, velocity and acceleration, and so on. In particular, if $\mathbf{Z}$ represents $\mathbf{Y}$, then Equations 6.99a and 6.99b essentially become Equation 6.104, while if $\mathbf{Z}$ represents one component of $\mathbf{Y}$, say $Y_\ell(t)$, then Equations 6.99a and 6.99b are then identical to Equations 6.106a and 6.106b. In the case $n_\mathrm{Z} = 1$, Equations 6.99a and 6.99b reduce to a one-dimensional partial differential equation which, for clarity, can be rewritten as

$$\frac{\partial p_{Z\boldsymbol{\Theta}}(z, \boldsymbol{\theta}, t)}{\partial t} + h_Z(\boldsymbol{\theta}, t) \frac{\partial p_{Z\boldsymbol{\Theta}}(z, \boldsymbol{\theta}, t)}{\partial z} = 0 \tag{6.123a}$$

or in an alternative form as

$$\frac{\partial p_{Z\boldsymbol{\Theta}}(z, \boldsymbol{\theta}, t)}{\partial t} + \dot{Z}(\boldsymbol{\theta}, t) \frac{\partial p_{Z\boldsymbol{\Theta}}(z, \boldsymbol{\theta}, t)}{\partial z} = 0 \tag{6.123b}$$

where $\dot{Z}(\boldsymbol{\theta}, t)$ is the velocity given $\boldsymbol{\Theta} = \boldsymbol{\theta}$.

If two physical quantities are involved, then Equations 6.99a and 6.99b become

$$\frac{\partial p_{Z_1Z_2\boldsymbol{\Theta}}(z_1, z_2, \boldsymbol{\theta}, t)}{\partial t} + h_{\mathbf{Z},1}(\boldsymbol{\theta}, t)\frac{\partial p_{Z_1Z_2\boldsymbol{\Theta}}(z_1, z_2, \boldsymbol{\theta}, t)}{\partial z_1} + h_{\mathbf{Z},2}(\boldsymbol{\theta}, t)\frac{\partial p_{Z_1Z_2\boldsymbol{\Theta}}(z_1, z_2, \boldsymbol{\theta}, t)}{\partial z_2} = 0 \tag{6.124a}$$

or in an alternative form

$$\frac{\partial p_{Z_1Z_2\boldsymbol{\Theta}}(z_1, z_2, \boldsymbol{\theta}, t)}{\partial t} + \dot{Z}_1(\boldsymbol{\theta}, t)\frac{\partial p_{Z_1Z_2\boldsymbol{\Theta}}(z_1, z_2, \boldsymbol{\theta}, t)}{\partial z_1} + \dot{Z}_2(\boldsymbol{\theta}, t)\frac{\partial p_{Z_1Z_2\boldsymbol{\Theta}}(z_1, z_2, \boldsymbol{\theta}, t)}{\partial z_2} = 0 \tag{6.124b}$$

where $\dot{Z}_1(\boldsymbol{\theta}, t)$ and $\dot{Z}_2(\boldsymbol{\theta}, t)$ are the corresponding velocities given $\boldsymbol{\Theta} = \boldsymbol{\theta}$.

From Equation 6.123b, the physical sense of the GDEE is clear that the change of the probability density is due to change of the position; therefore, the time rate of change of the probability density is associated with the time rate of change of the position. In this equation, the inseparable ties between the physical system and the evolution of probability are clearly exhibited. It is quite interesting to revisit Section 6.5.2 and find that the same result is reached in Equation 6.119b, which is obtained in a quite different way from that in this section.

Further, the most important point distinct from the classical Liouville equation and FPK equation is that the probability density of each physical quantity is due to the change of the state itself, not by other components, no matter whether the physical quantities are coupled or not. Because of this, the GDEE can be of any arbitrary dimension without constraints on $n_{\mathbf{Z}}$. Then it is possible to extract the probabilistic information of any single or any two or more physical quantities, which can be at least numerically feasible, as will be elaborated in Chapter 7. The crux of the uncoupling is to view the principle of preservation of probability from the random event description incorporating with the physical solution instead of the state space description together with the coupling physical equation.

The embedded reason of the above difference of the two different descriptions essentially lies in the different methodologies. In the state space description, as elaborated in Section 6.3, the investigator is focused on the transition of probability in a fixed domain in different ways according to their different phenomenological origins; for example, the effect of drift and diffusion. This requires globally taking into account the state vector, because each component of the vector is an indispensable dimension in the state space where the fixed domain examined is located. On the other hand, in the random event description, it is stressed that the transition of probability is adherent to random events and their probability measure; namely, the transition of probability results, in essence, not from the superfluous phenomenological exhibition of change of state, but from its embedded random events. In other words, the transition of probability density is adherent to a physical evolution process. Thus, the transition of probability is treated in a unified way by its association with the random event. In fact, such a principle is suitable for all physical stochastic systems.

In the above sense, Equations 6.99a and 6.99b are called the generalized density evolution equation (GDEE). The methodology for tackling stochastic dynamical problems through solving the GDEE is called the *probability density evolution theory* (Li and Chen, 2006c, 2009).

6.6 Solution of the Generalized Density Evolution Equation

6.6.1 Analytical Solution

6.6.1.1 The Method of Characteristics

We have encountered the method of characteristics several times, in Sections 5.6.3 and 6.3.1 for instance. Here, we will discuss its basic idea and the embedded physical sense. For simplicity of concept, we first consider the one-dimensional first-order partial differential equation

$$\frac{\partial p(x,t)}{\partial t}+a(x,t)\frac{\partial p(x,t)}{\partial x}+b(x,t)p(x,t)=0 \tag{6.125}$$

where $p(x,t)$ is the unknown function of x and t and $a(x,t)$ and $b(x,t)$ are known functions of x and t. Clearly, we note that if $b(x,t)=\partial a(x,t)/\partial x$, then Equation 6.125 becomes

$$\frac{\partial p(x,t)}{\partial t}+\frac{\partial}{\partial x}[p(x,t)a(x,t)]=0 \tag{6.126}$$

which is the Liouville equation elaborated in Section 6.3.1.

If we introduce a parameter τ and use the parametric equation

$$\begin{cases} x=x(\tau) \\ t=t(\tau) \end{cases} \tag{6.127}$$

then we have

$$\frac{\mathrm{d}p[x(\tau),t(\tau)]}{\mathrm{d}\tau}=\frac{\partial p(x,t)}{\partial t}\frac{\mathrm{d}t}{\mathrm{d}\tau}+\frac{\partial p(x,t)}{\partial x}\frac{\mathrm{d}x}{\mathrm{d}\tau} \tag{6.128}$$

Comparing this equation with Equation 6.125, when setting

$$\frac{\mathrm{d}t}{\mathrm{d}\tau}=1 \tag{6.129a}$$

$$\frac{\mathrm{d}x}{\mathrm{d}\tau}=a(x,t) \tag{6.129b}$$

it is found that Equation 6.125 becomes

$$\frac{\mathrm{d}p[x(\tau),t(\tau)]}{\mathrm{d}\tau}+b[x(\tau),t(\tau)]p[x(\tau),t(\tau)]=0 \tag{6.130}$$

Here, we see that Equations 6.129a and 6.129b determine a family of curves represented by a parametric equation, Equation 6.127, in the x–t plane. Along this family of curves, the partial differential equation Equation 6.125 becomes an ordinary differential equation Equation 6.130. These curves are called the *characteristic curves* or *characteristics*.

Note that the initial condition of Equations 6.129a and 6.129b can take

$$t(0)=0 \qquad x(0)=\chi \tag{6.131}$$

The coordinate system (x, t), therefore, is transformed to the coordinate system (χ, τ) by

$$\begin{cases} x = x(\chi, \tau) \\ t = t(\chi, \tau) = \tau \end{cases} \tag{6.132}$$

This implies that, for a given initial value χ, Equation 6.132 (through integration of Equations 6.129a and 6.129b) determines a curve in the x–t plane, which is one specified characteristic curve. Figure 6.6 shows two typical characteristic curves for different initial values χ_1 and χ_2. The transformation in Equation 6.132 is nonsingular, so that the inverse transformation exists:

$$\begin{cases} \chi = \chi(x, t) \\ \tau = \tau(x, t) = t \end{cases} \tag{6.133}$$

It is well known that the solution of Equation 6.130 is given by

$$f(\chi, \tau) \triangleq p[x(\chi, \tau), t(\chi, \tau)] = p[x(\chi, \tau_0), t(\chi, \tau_0)]\exp\left\{-\int_{\tau_0}^{\tau} b[x(\chi, \tau), t(\chi, \tau)]\,\mathrm{d}\tau\right\} \tag{6.134a}$$

Here, $x(\tau)$ and $t(\tau)$ used in Equation 6.130 are replaced by $x(\chi,\tau)$ and $t(\chi,\tau)$ respectively.

Noting Equations 6.131 and 6.132, Equation 6.134a becomes

$$f(\chi, \tau) = p[x(\chi, \tau), t(\chi, \tau)] = p_0(\chi)\exp\left\{-\int_0^{\tau} b[x(\chi, \tau), t(\chi, \tau)]\,\mathrm{d}\tau\right\} \tag{6.134b}$$

where $p_0(x)$ is the initial function of $p(x,t)$; that is, $p(x,0) = p_0(x)$.

Introducing the inverse transformation in Equation 6.133 will then give the solution of the original Equation 6.125 by

$$\begin{aligned} p(x, t) &= f(\chi, \tau)\big|_{\chi=\chi(x,t),\tau=\tau(x,t)} \\ &= p_0[\chi(x, t)]\exp\left\{-\int_0^{\tau} b[x(\chi, \tau), t(\chi, \tau)]\,\mathrm{d}\tau\right\}\Bigg|_{\chi=\chi(x,t),\tau=\tau(x,t)} \end{aligned} \tag{6.135}$$

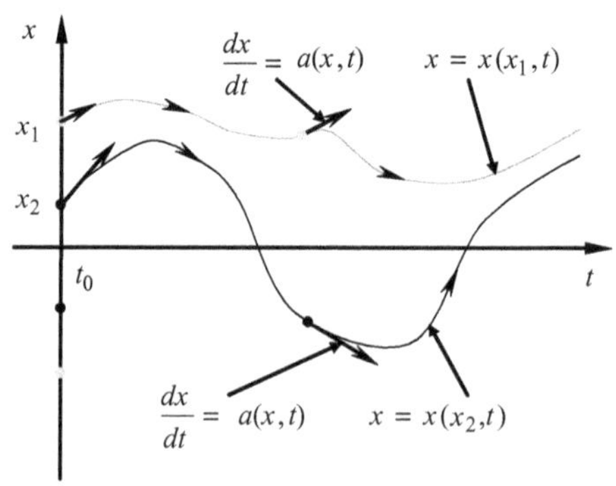

Figure 6.6 Characteristics of the first-order partial differential equation.

The above method for solution of the first-order partial differential equation is called the *method of characteristics* (Hodge,1950; Petrovsky,1954; Farlow,1993; Sarra,2003), which has been employed in Sections 5.6.2 and 6.3.1.2.

The physical sense of the method of characteristics is of particular interest. From Equations 6.129a, 6.129b and 6.131 it is seen that the characteristic curves are determined by

$$\frac{dx}{dt} = a(x,t) \qquad x(0) = \chi \tag{6.136}$$

If $b(x,t) = 0$, then from Equation 6.130 we have

$$\frac{dp[x(\chi,\tau),t(\chi,\tau)]}{d\tau} = 0 \tag{6.137}$$

Here, we note that χ is invariant against τ. This indicates that along the characteristic curve for a specified χ, which is determined by Equation 6.132, $p(x,t)$ is invariant and equal to $p(\chi,0)$. In the case $p(x,t)$ is a PDF, this underlies that the probability density will be preserved along the characteristic curves, which is nothing but what is embedded in the random event description of the principle of preservation of probability. Referring to Section 6.2.2.1 will provide more insight. Actually, if randomness is only involved in the initial condition, then it is seen clearly that the characteristic curves have a Lagrangian coordinate description of the position of the particle being studied. This is particularly clear when we view Equation 6.132 and rewrite it as

$$x = x(\chi,t) \tag{6.138}$$

where $\tau = t$ is used.

When Equation 6.136 is viewed as a state space equation in an Eulerian system, Equation 6.138 as its solution is nothing but the corresponding Lagrangian description. Here, once again, we find the essential relation between the Lagrangian description and the preservation of probability.

6.6.1.2 Analytical Solution of Generalized Density Evolution Equation

We now consider the special case when Equation 6.125 becomes

$$\frac{\partial p(x,t)}{\partial t} + a(t)\frac{\partial p(x,t)}{\partial x} = 0 \tag{6.139}$$

This is essentially in the same form as the one-dimensional generalized density equation (Equation 6.123a) for a specified $\boldsymbol{\theta}$. In this case, from Equations 6.129a and 6.129b and the initial condition in Equation 6.131, the characteristic curves are given by

$$x = \chi + \int_0^t a(t)\,dt = \chi + \psi(t) \tag{6.140}$$

where $\psi(t) = \int_0^t a(t)\,dt$.

From Equation 6.135 it is known that the solution of Equation 6.139 reads

$$p(x,t) = p_0[x - \psi(t)] \tag{6.141}$$

where $p_0(x)$ is the initial function of $p(x,t)$. This is actually a one-way wave propagating at a velocity of $a(t)$ (Graff, 1975).

Now we consider Equation 6.123a again, repeated here for convenience:

$$\frac{\partial p_{Z\boldsymbol{\Theta}}(z,\boldsymbol{\theta},t)}{\partial t} + h_Z(\boldsymbol{\theta},t)\frac{\partial p_{Z\boldsymbol{\Theta}}(z,\boldsymbol{\theta},t)}{\partial z} = 0 \tag{6.123a}$$

According to Equation 6.141, the solution is

$$p_{Z\boldsymbol{\Theta}}(z,\boldsymbol{\theta},t) = p_0[z - H(\boldsymbol{\theta},t)] \tag{6.142}$$

where

$$H(\boldsymbol{\theta},t) = \int_0^t h_Z(\boldsymbol{\theta},t)\,\mathrm{d}t \tag{6.143}$$

Note that

$$p_0(z) = \delta(z - z_0)p_{\boldsymbol{\Theta}}(\boldsymbol{\theta}) \tag{6.144}$$

Combining Equations 6.142 and 6.144, we have

$$p_Z(z,t) = \int_{\Omega_{\boldsymbol{\Theta}}} p_{Z\boldsymbol{\Theta}}(z,\boldsymbol{\theta},t)\,\mathrm{d}\boldsymbol{\theta} = \int_{\Omega_{\boldsymbol{\Theta}}} \delta[z - H(\boldsymbol{\theta},t)]p_{\boldsymbol{\Theta}}(\boldsymbol{\theta})\,\mathrm{d}\boldsymbol{\theta} \tag{6.145}$$

If only one random parameter Θ is involved, then we have

$$p_Z(z,t) = \int_{\Omega_\Theta} \delta[z - H(\theta,t)]p_\Theta(\theta)\,\mathrm{d}\theta \tag{6.146}$$

Considering the integration rules of the Dirac delta function (see Appendix A), this can be changed further to

$$p_Z(z,t) = |J|p_\Theta(\theta)|_{\theta=H^{-1}(z,t)} \tag{6.147}$$

where $|J| = |\partial H^{-1}/\partial z|$.

The closed-form solution is consistent with what was employed in Section 6.5.1.2, where the solution of Equations 6.106a and 6.106b will be Equation 6.109 according to Equation 6.145.

Example 6.2. Response of an Uncertain SDOF System Consider the SDOF system

$$\ddot{X} + \omega^2 X = 0 \qquad \dot{X}(0) = 0 \qquad X(0) = x_0 \tag{6.148}$$

where ω is a random variable uniformly distributed over $[\omega_1, \omega_2]$.

Because ω is a random variable, the response $X(t)$ is a stochastic process. The formal solution of the displacement and velocity of the system in Equation 6.148 is given by

$$X = x_0 \cos(\omega t) \tag{6.149a}$$

$$\dot{X} = -x_0 \omega \sin(\omega t) \tag{6.149b}$$

For clarity, denote ω by Θ. The GDEE of the process (X,Θ) is given by

$$\frac{\partial p_{X\Theta}(x,\theta,t)}{\partial t} - x_0 \theta \sin(\theta t) \frac{\partial p_{X\Theta}(x,\theta,t)}{\partial x} = 0 \tag{6.150}$$

According to Equation 6.142, under the initial condition

$$p_{X\Theta}(x,\theta,t_0) = \delta(x - x_0) p_\Theta(\theta) \tag{6.151}$$

the solution of Equation 6.150 is

$$p_{X\Theta}(x,\theta,t) = \delta(x - x_0 \cos\theta t) p_\Theta(\theta) \tag{6.152}$$

Thus, according to Equation 6.146, we have

$$p_X(x,t) = \int \delta(x - x_0 \cos\theta t) p_\Theta(\theta)\, \mathrm{d}\theta \tag{6.153}$$

Or further (see. Equation 6.147):

$$p_X(x,t) = \begin{cases} \dfrac{1}{\sqrt{x_0^2 - x^2}} \displaystyle\sum_{\ell=0}^{\infty} \left\{ p_\eta \left[2\ell\pi + 2\pi - \cos^{-1}\left(\frac{x}{x_0}\right), t \right] + p_\eta \left[2\ell\pi + \cos^{-1}\left(\frac{x}{x_0}\right), t \right] \right\} & |x| \le |x_0| \\ 0 & \text{otherwise} \end{cases} \tag{6.154}$$

where $\eta = \Theta t$ and

$$p_\eta(x) = \frac{1}{t} p_\Theta(x/t) \tag{6.155a}$$

$$p_\Theta(\theta) = \begin{cases} \dfrac{1}{\omega_2 - \omega_1} & \text{for } \omega_1 \le \theta \le \omega_2 \\ 0 & \text{otherwise} \end{cases} \tag{6.155b}$$

Figure 6.7 shows the typical PDF at different time instants given by Equation 6.154 (Li and Chen, 2004a).

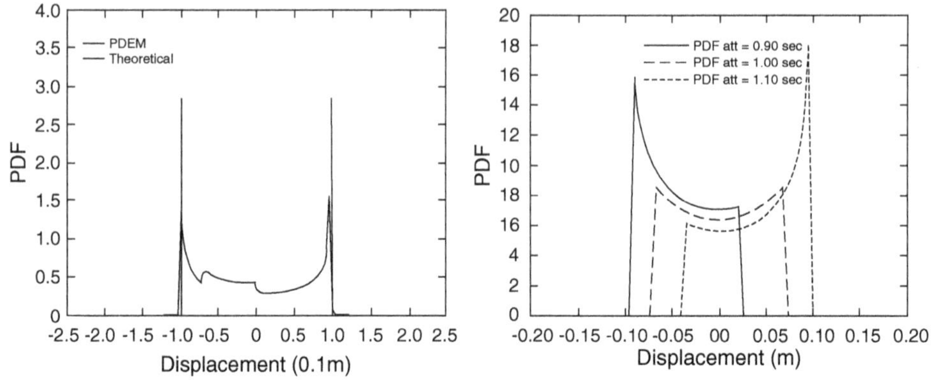

Figure 6.7 PDF at typical different time instants.

□

6.6.2 Numerical Solving Flow of the Generalized Density Evolution Equation

6.6.2.1 Numerical Solving Flow

To resolve a stochastic dynamical system by the GDEE, some special methods must be developed. Here, we only give a brief description. A more detailed discussion is given in Chapter 7.

It is seen that Equations 6.99a and 6.99b are linear partial differential equations where there are no terms with respect to partial differentiation in terms of $\boldsymbol{\theta}$ involved. That is, for a given $\boldsymbol{\theta}$, Equations 6.99a and 6.99b become partial differential equations where the arguments of the unknown function are only $\mathbf{z}$ and t. This equation can of course be solved through a numerical algorithm once the time-variant coefficients $\dot{Z}_\ell(\boldsymbol{\theta}, t)$ are available.

Therefore, to solve the GDEE, we should first select a set of representative points in the random parameter space $\Omega_{\boldsymbol{\Theta}}$. Then, for each representative point chosen, carry out a deterministic dynamic analysis to obtain $\dot{Z}_\ell(\boldsymbol{\theta}, t)$. These results are then introduced into the GDEE and solved by, say, some type of numerical method. Finally, the results associated with all the representative points are synthesized to obtain the instantaneous probability density of the responses of interest. For simplicity of illustration, we will consider the solution of the one-dimensional GDEE (Equation 6.123b). The same idea can be used for Equations 6.99a and 6.99b.

Explicitly, the above procedures usually involve the following four steps:

Step 1. Select representative points in the random parameter space $\Omega_{\boldsymbol{\Theta}}$.
Denote this point set by $\mathcal{P} = \{\boldsymbol{\theta}_q = (\boldsymbol{\theta}_{1,q}, \boldsymbol{\theta}_{2,q}, \ldots, \boldsymbol{\theta}_{s,q}); q = 1, 2, \ldots, n_{\text{sel}}\}$, where s is the total number of random variables involved, as discussed in Section 6.5.1.1, and n_{sel} is the cardinal number of the point set. For each representative point $\boldsymbol{\theta}_q$, a representative volume (domain), say the Voronoi cell V_q, exists (Conway and Sloane, 1999). These cells form a partition of $\Omega_{\boldsymbol{\Theta}}$. The probability measure over this domain is assigned to this point and

denoted by P_q; namely:[7]

$$P_q = \int_{V_q} p_{\boldsymbol{\Theta}}(\boldsymbol{\theta})\, d\boldsymbol{\theta} \tag{6.156}$$

Clearly, $\sum_{q=1}^{n_{\text{sel}}} P_q = 1$. The initial conditions (Equations 6.121a and 6.121b) are partially discretized correspondingly to

$$p_{Z\boldsymbol{\Theta}}(z, \boldsymbol{\theta}_q, t_0) = \delta(z - z_0) P_q \qquad q = 1, 2, \ldots, n_{\text{sel}} \tag{6.157a}$$

$$p_{Z\boldsymbol{\Theta}}(z, \boldsymbol{\theta}_q, t_0) = p_{Z_0}(z) P_q \qquad q = 1, 2, \ldots, n_{\text{sel}} \tag{6.157b}$$

Step 2. For each representative point $\boldsymbol{\theta}_q$, carry out deterministic analysis on the dynamical system (Equation 6.92) when setting $\boldsymbol{\Theta} = \boldsymbol{\theta}_q$ and then obtain the velocity of $\dot{Z}(\boldsymbol{\theta}_q, t)$ from Equation 6.95.
Step 3. For each representative point $\boldsymbol{\theta}_q$, introduce $\dot{Z}_\ell(\boldsymbol{\theta}_q, t)$ obtained in Step 2 into the discretized version of Equation 6.123b:

$$\frac{\partial p_{Z\boldsymbol{\Theta}}(z, \boldsymbol{\theta}_q, t)}{\partial t} + \dot{Z}(\boldsymbol{\theta}_q, t) \frac{\partial p_{Z\boldsymbol{\Theta}}(z, \boldsymbol{\theta}_q, t)}{\partial z} = 0 \qquad q = 1, 2, \ldots, n_{\text{sel}} \tag{6.158}$$

Then solve this equation under the initial conditions (Equations 6.121a and 6.121b) with, say, the finite difference method. In this step, the space (z,t) should be meshed. Denote the nodes of the mesh by (z_i, t_k), $i = 0; \pm1, \pm2, \ldots$, $k = 0, 1, 2, \ldots$, where $z_i = i\Delta z$, $t_k = k\Delta t$, $k = 0, 1, 2, \ldots, \Delta z$ is the space step in the direction of z and Δt is the time step. Equation 6.158 is then transformed to an algebraic equation set and can be solved to give the values of the density at the nodes, denoted by $p_{Z\boldsymbol{\Theta}}(z_i, \boldsymbol{\theta}_q, t_k)$.
Step 4. Synthesize the results in Step 3 to obtain the instantaneous density through the discretized version of Equation 6.100:

$$p_Z(z_i, t_k) = \sum_{q=1}^{n_{\text{sel}}} p_{Z\boldsymbol{\Theta}}(z_i, \boldsymbol{\theta}_q, t_k) \tag{6.159}$$

6.6.2.2 Schematic Solution Process of a One-Dimensional Liouvillian System

To illustrate further the meanings of the GDEE and its difference from the Liouville equation, we consider the solving flow of a one-dimensional system with randomness involved only in the initial condition:

$$\dot{X} = A(X, t) \qquad X(t_0) = X_0 \tag{6.160}$$

where X_0 is a random variable with the density $p_{X_0}(x_0)$.

Denote the density of $X(t)$ by $p_X(x,t)$. According to Equation 6.29, the Liouville equation and its initial conditions are respectively

$$\frac{\partial p_X(x, t)}{\partial t} + \frac{\partial}{\partial x}[A(x, t) p_X(x, t)] = 0 \tag{6.161a}$$

$$p_X(x, t_0) = p_{X_0}(x) \tag{6.161b}$$

[7] This will be elaborated in Section 7.2.2.

On the other hand, the Lagrangian description of the system in Equation 6.160 and its velocity are assumed to be given respectively by (see Equations 6.93 and 6.94)

$$X = H(X_0, t) \quad \text{and} \quad \dot{X} = h(X_0, t) \tag{6.162}$$

When $(X(t), X_0)$ is examined, it is a probability preserved system and the GDEE is given by

$$\frac{\partial p_{XX_0}(x, x_0, t)}{\partial t} + h(x_0, t)\frac{\partial p_{XX_0}(x, x_0, t)}{\partial x} = 0 \tag{6.163a}$$

with the initial condition

$$p_{XX_0}(x, x_0, t_0) = \delta(x - x_0)p_{X_0}(x_0) \tag{6.163b}$$

where $p_{XX_0}(x, x_0, t)$ is the joint density of $(X(t), X_0)$.

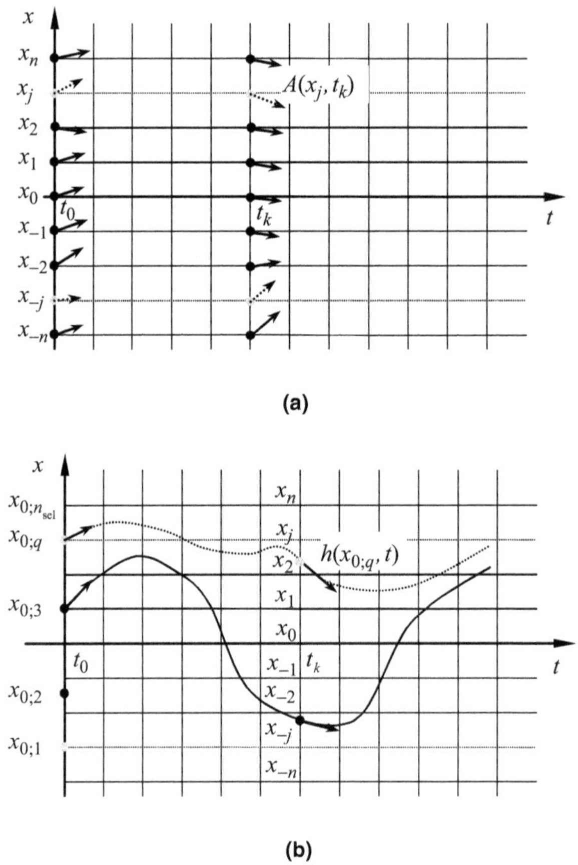

Figure 6.8 Schematic solution process through (a) the Liouville equation and (b) the GDEE.

The density of $X(t)$ can then be obtained by

$$p_X(x, t) = \int p_{XX_0}(x, x_0, t)\, \mathrm{d}x_0 \tag{6.163c}$$

If we solve the above sets of equations numerically by the difference method, for the system in Equations 6.161a and 6.161b the solving process is implemented on a meshed x–t plane, the initial condition (Equation 6.161b) will be first discretized and then Equation 6.161a is solved through a difference scheme, where the velocity field $A(x, t)$ can be computed instantaneously at the discretized mesh node points (x_j, t_k) (Figure 6.8a). For the system in Equations 6.163a –6.163c, however, what is done first is to select some representative point of the random parameter X_0, which can be denoted by $x_{0;1}, x_{0;2}, \ldots, x_{0;n_{\mathrm{sel}}}$, then take time integration on Equation 6.160 for each given $x_{0;q}$ to obtain the time history of the velocity in Equation 6.162. After that the time histories of the velocity are employed in the finite difference method to solve Equations 6.163a and 6.163b and finally synthesize all the results to obtain the joint density, as shown in Equation 6.163c (Figure 6.8b).

Here, it is seen clearly that in the solving process of the Liouville equation a velocity field in the state space is used and computed instantaneously, whereas in the solving process of the GDEE the tracing of some representative trajectories is needed instead of computing the velocity field. This is just the difference between the state space description and the random event description.

It seems, in this example, that the implementation process of the probability density evolution method is more complicated than the Liouville equation. However, in the analysis of a large system, the Liouville equation might be impossible because a multidimensional partial differential equation must be handled, which is usually impossible for large dimensions. In contrast, there are no essential difficulties arising compared with the one-dimensional problems in the probability density evolution method.

7

Probability Density Evolution Analysis: Numerical Methods

7.1 Numerical Solution of First-Order Partial Differential Equation

7.1.1 The Finite-Difference Method

Despite the endeavors devoted to analytical solutions of first-order partial differential equations, it is more feasible to seek numerical solutions for most problems of practical interest. Because the generalized density evolution equation, as a first-order partial differential equation, is in a form similar to conservative equations in fluid dynamics, some of the approaches developed there can be applied in the probability density evolution method. Actually, the numerical methods for first-order partial differential equation, such as the finite-difference method, the finite-volume method and the cell-mapping method, and so on, have been well developed, particularly some special approaches or schemes stimulated by demands in computational fluid dynamics (Anderson, 1995; Wesseling, 2001). The finite-difference method, which has been well dealt with in general textbooks (Mitchell and Griffiths, 1980; Smith, 1985; Strikwerda, 1989) and especially in tackling problems of conservative laws in physical systems (Godlewski and Raviart, 1996), exhibits satisfactory performances in probability density evolution analysis when the appropriate difference schemes are employed (Chen and Li, 2004a).

Without loss of generality, we first deal with the equation in the form of Equation 6.139, which is given again here for convenience:

$$\frac{\partial p(x,t)}{\partial t} + a(t)\frac{\partial p(x,t)}{\partial x} = 0 \tag{7.1}$$

This is a hyperbolic partial differential equation. The basic idea of the finite-difference method is to discretize the partial differential Equation 7.1 into an algebraic equation, referred to as the difference equation.

To use the finite-difference method, the x–t plane will be meshed by the two families of lines

$$x = x_j \quad t = t_k \qquad j = 0, \pm 1, \pm 2, \ldots, \; k = 0, 1, 2, \ldots \tag{7.2}$$

such that a uniform grid with time step Δt and spatial mesh size Δx is determined. For notational convenience, denote the value $p(x_j, t_k)$ at the point $(x_j = j\Delta x, t_k = k\Delta t)$ by $p_j^{(k)}$. Representing the partial differentiation by the difference between the values on the nodes will give an algebraic equation, then solving the algebraic equation will give the approximate values of $p(x_j, t_k)$. Obviously, different approximate representations of the partial differentiations will result in different difference schemes.

7.1.1.1 One-Sided Difference Schemes

Using the first-order Taylor expansion in terms of t

$$p_j^{(k+1)} = p_j^{(k)} + \left[\frac{\partial p}{\partial t}\right]_j^{(k)} \Delta t + o(\Delta t) \tag{7.3}$$

we can approximate the partial differentiation in terms of t by

$$\left[\frac{\partial p}{\partial t}\right]_j^{(k)} \doteq \frac{p_j^{(k+1)} - p_j^{(k)}}{\Delta t} \tag{7.4}$$

Likewise, in terms of x, we have the first-order Taylor expansion

$$p_j^{(k)} = p_{j-1}^{(k)} + \left[\frac{\partial p}{\partial x}\right]_{j-1}^{(k)} \Delta x + o(\Delta x) \tag{7.5}$$

Therefore, partial differentiation in terms of x can be approximated by

$$\left[\frac{\partial p}{\partial x}\right]_{j-1}^{(k)} \doteq \frac{p_j^{(k)} - p_{j-1}^{(k)}}{\Delta x} \tag{7.6}$$

Substituting Equations 7.4 and 7.6 in Equation 7.1, replacing $a(t)$ by $a^{(k)}$ and rearranging the equation yields

$$p_j^{(k+1)} = p_j^{(k)} - \lambda a^{(k)} [p_j^{(k)} - p_{j-1}^{(k)}] \tag{7.7a}$$

or in the alternative form

$$p_j^{(k+1)} = (1 - \lambda a^{(k)}) p_j^{(k)} + \lambda a^{(k)} p_{j-1}^{(k)} \tag{7.7b}$$

where $\lambda = \Delta t / \Delta x$ is the ratio of the time step to the spacial mesh size.

A schematic illustration of this difference scheme is shown in Figure 7.1.

The most important attributes of a difference scheme for a hyperbolic partial differential equation are the consistency, convergence and stability. A difference scheme is *consistent* if the difference equation tends to the original partial differential equation as $\Delta x \to 0$ and $\Delta t \to 0$. On the other hand, a difference scheme is *convergent* if the solution of the difference equation at the grid tends to the solution of the original partial differential equation as $\Delta x \to 0$ and $\Delta t \to 0$. Generally, a consistent scheme cannot guarantee convergence. The *stability*, here the numerical stability, requires that the increase of the computed value of the solution

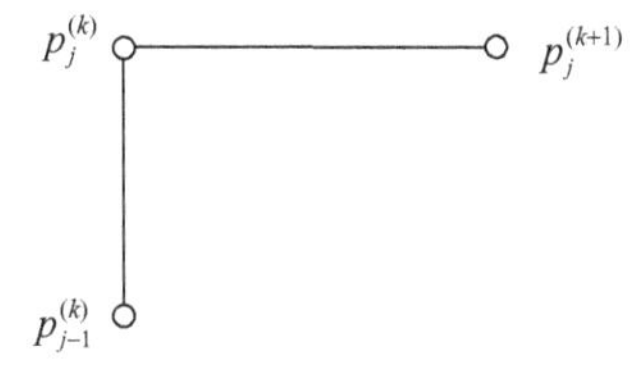

Figure 7.1 One-sided scheme (Equation 7.7).

of the difference equation should be bounded. The *Lax–Richtmyer equivalent theorem* asserts that *a consistent finite-difference scheme for a partial differential equation for which the initial value problem is well posed is convergent if and only if it is stable* (Stricwerda, 1989). Because of this, the stability of a difference scheme should be stressed. Otherwise it will be of no practical use.

The scheme in Equation 7.7 is, of course, consistent because Equations 7.4 and 7.6 are consistent.

To understand the other features of the scheme in Equation 7.7, we consider the special case $a^{(k)} \equiv a$. In this case, the characteristic line will be a family of parallel lines with slope a, of which the one from the origin is shown in Figures 7.2a and 7.2b respectively for $a > 0$ and $a < 0$.

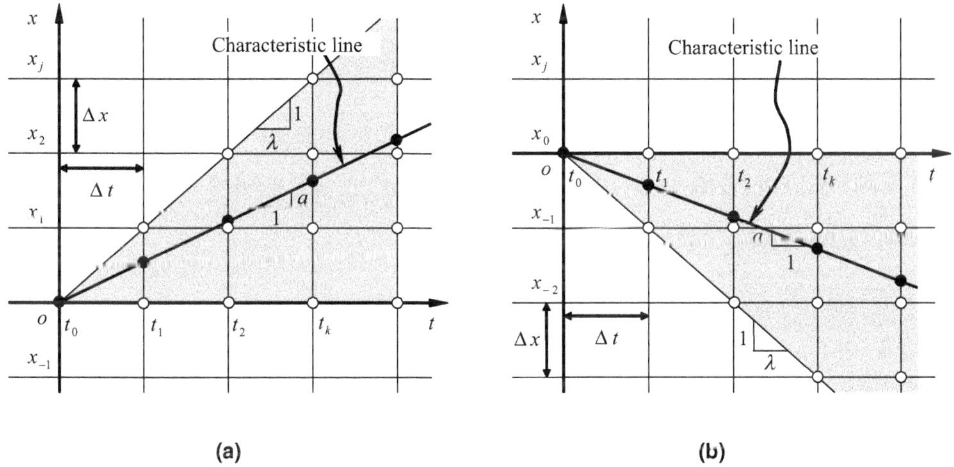

Figure 7.2 Characteristics and propagation of probability.

For simplicity, we consider the initial value condition[1]

$$p_j^{(0)} = \delta_{0j} = \begin{cases} 1 & \text{for } j = 0 \\ 0 & \text{otherwise} \end{cases} \tag{7.8}$$

[1]It should be noted that using this initial condition is also without loss of generality, because any discretized initial condition $p_j^{(0)} = p_{j,0}$, $j = 0, \pm 1, \pm 2, \ldots$, can be represented by a linear combination. In addition, Equation 7.8 is also the discretized initial condition for many practical problems, say in discretization of Equation 6.121a, except from a difference in a constant factor.

which means that $p_j^{(0)}$ is nonzero only at the origin. Here, δ is the Kronecker delta. Using Equation 7.7b, it is noticed that the nonzero points at the time instants $t_1, t_2, \ldots$, are limited in the shaded area in Figure 7.2a, denoted by the small hollow circles. In the case $a > 0$, the actual propagation of the probability is along the characteristic line from the origin on which the points at $t_1, t_2, \ldots$ are denoted by the black points. In addition, we notice that

$$x_0 p_0^{(1)} + x_1 p_1^{(1)} = a\Delta t \tag{7.9a}$$

and it is easy to verify that

$$\sum_j x_j p_j^{(k)} = ka\Delta t \tag{7.9b}$$

This means that, at a specified time instant, the actual propagation point (on the characteristic line) is the mean point of the nonzero points in the numerical solution. In particular, noting that we require the probability[2] $1 \geq p_0^{(1)} \geq 0$; $1 \geq p_1^{(1)} \geq 0$ and $x_0 = 0$, $x_1 = \Delta x$, from Equation 7.9a we have

$$0 \leq p_1^{(1)} = a\Delta t/\Delta x \leq 1 \quad \text{or} \quad 0 \leq \lambda a \leq 1 \tag{7.10}$$

What will happen if Equation 7.10 is not satisfied? From Equation 7.7b, it is seen that if $\lambda a > 1$, then

$$p_0^{(1)} = (1-\lambda a)p_0^{(0)} < 0 \qquad p_1^{(1)} = \lambda a p_0^{(0)} > 1 \tag{7.11a}$$

and further

$$p_0^{(2)} = (1-\lambda a)^2 p_0^{(0)} \qquad p_1^{(2)} = 2(\lambda a - \lambda^2 a^2)p_0^{(0)} \qquad p_2^{(2)} = \lambda^2 a^2 p_0^{(0)} \tag{7.11b}$$

Actually, we have in general

$$p_j^{(k)} = \binom{k}{j}(1-\lambda a)^{k-j}\lambda^j a^j p_0^{(0)} \tag{7.11c}$$

where

$$\binom{k}{j} = \frac{k!}{j!(k-j)!}$$

is the combinatorial number. This leads to

$$p_k^{(k)} = \lambda a p_{k-1}^{(k-1)} = \lambda^k a^k p_0^{(0)} = \lambda^k a^k \tag{7.11d}$$

This is an unbounded quantity and increasing very rapidly against k if $\lambda a > 1$. Therefore, if Equation 7.10 is violated, the scheme in Equation 7.7 is unstable.

[2] The quantity $p(x, t)$ is the PDF. However, it is usually more convenient to understand the discretized value as a value of probability. This can be achieved when we replace the discretized initial condition $p_j^{(0)} = \delta_{j0}/\Delta x$ by $p_j^{(0)} = \delta_{j0}$, as is done in Equation 7.8. Then, after we get the numerical solution, we in turn replace $p(x, t)$ by $p(x, t)/\Delta x$.

On the other hand, if $a<0$, then the practical propagation of probability is along the characteristic line shown in Figure 7.2b; however, the numerical probability in the scheme in Equation 7.7 propagates in the shaded area in Figure 7.2a. In this case, from Equation 7.7b or Equation 7.11c:

$$p_0^{(k)} = (1-\lambda a)p_0^{(k-1)} = (1-\lambda a)^k p_0^{(0)} = (1-\lambda a)^k \tag{7.12}$$

Because now $a<0$, and thus $1-\lambda a > 1$ and then $(1-\lambda a)^k$ is an unbounded quantity and increasing very rapidly against k, this means that the scheme in Equation 7.7 is now unstable. Therefore, in this case, the scheme should be modified such that the propagation direction should be the shaded area in Figure 7.2b.

Modify Equation 7.6 to

$$\frac{\partial p(x,t)}{\partial x} \doteq \frac{p_{j+1}^{(k)} - p_j^{(k)}}{\Delta x} \tag{7.13}$$

Substituting Equations 7.13 and 7.4 in Equation 7.1 yields

$$p_j^{(k+1)} = p_j^{(k)} - \lambda a^{(k)}[p_{j+1}^{(k)} - p_j^{(k)}] \tag{7.14a}$$

or in an alternative form as

$$p_j^{(k+1)} = [1+\lambda a^{(k)}]p_j^{(k)} - \lambda a^{(k)} p_{j+1}^{(k)} \tag{7.14b}$$

Likewise, to guarantee the stability of the scheme in Equation 7.14 requires that

$$[1+\lambda a^{(k)}] \geq 0 \quad \text{or} \quad -1 \leq \lambda a^{(k)} \leq 0 \tag{7.15}$$

The schematic illustration of the scheme in Equation 7.14 is shown in Figure 7.3.

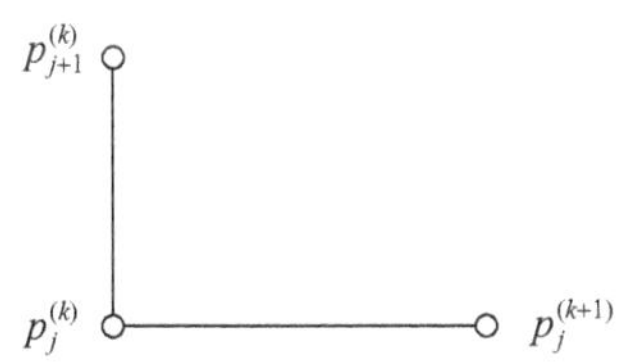

Figure 7.3 One-sided scheme (Equation 7.14).

Equation 7.7 is sometimes called a forward-time backward-space scheme and Equation 7.14 is called a forward-time forward-space scheme. It should be stressed that, according to the above analysis, the appropriate scheme should be chosen according to the sign of $a(t)$ such that the propagation direction of the numerical solution coincides with the propagation of the real solution, which is determined by the characteristic curves.

The schemes for $a > 0$ (Equation 7.7) and $a<0$ (Equation 7.14) can also be written in a unified way by

$$p_j^{(k+1)} = (1-|\lambda a|)p_j^{(k)} + \frac{1}{2}(|\lambda a| - \lambda a)p_{j+1}^{(k)} + \frac{1}{2}(|\lambda a| + \lambda a)p_{j-1}^{(k)} \tag{7.16a}$$

or

$$p_j^{(k+1)} = -\lambda a p_{j+1}^{(k)} u(-a) + (1 - |\lambda a|) p_j^{(k)} + \lambda a p_{j-1}^{(k)} u(a) \tag{7.16b}$$

where $u(\cdot)$ is the Heaviside unit step function, of which the value is unity when the argument is nonnegative and otherwise zero (see Appendix A), and a here represents $a^{(k)}$ for simplicity.

The conditions in Equations 7.10 and 7.15 now become

$$|\lambda a^{(k)}| \leq 1 \tag{7.17}$$

Actually, this is just the celebrated Courant–Friedrichs–Lewy (CFL) condition (Courant *et al.*,1928).

One of the advantages of the one-sided scheme in Equation 7.16 is that the nonnegativeness of probability can be preserved. In addition, the total probability is conversative; namely:

$$\sum_j p_j^{(k)} = \sum_j p_j^{(0)} = 1 \tag{7.18}$$

which can be verified by introducing the unified scheme in Equation 7.16. However, it is of first-order accuracy because only the first-order expansion is employed in the Taylor series approximations in Equations 7.3, 7.5 and 7.13. The schemes of higher accuracy should retain higher-order terms, which will be elaborated in the following sections.

Before leaving this section, we spend a little more time on the unified scheme in Equation 7.16. It is seen that Equation 7.16 is a one-step linear scheme; namely:

$$p_j^{(k+1)} = \sum_{\ell=-\nu}^{\nu} c_\ell p_{j+\ell}^{(k)} \tag{7.19}$$

For the present case, $\nu = 1$, $c_{-1} = \frac{1}{2}(|\lambda a| + \lambda a)$, $c_0 = 1 - |\lambda a|$ and $c_1 = \frac{1}{2}(|\lambda a| - \lambda a)$. This shows that the value at the grid at time t_{k+1} is the linear combination of the values at the grid at time t_k.

The physical sense of Equation 7.19, however, is not so clear. In particular, preservation of probability cannot be seen directly. If we let $F(p) = ap(x, t)$, which is the flux of probability, Equation 7.1 now becomes

$$\frac{\partial p}{\partial t} + \frac{\partial F(p)}{\partial x} = 0 \tag{7.20}$$

This equation is usually called the convertive type partial differential equation. When it is discretized using Equations 7.4 and 7.6, where p is replaced by F, we get

$$p_j^{(k+1)} = p_j^{(k)} - \lambda(F_j^{(k)} - F_{j-1}^{(k)}) = p_j^{(k)} - \lambda \Delta F_{j-\frac{1}{2}}^{(k)} \tag{7.21a}$$

where $F_j^{(k)}$ and $F_{j-1}^{(k)}$ are the numerical flux and $\Delta F_{j-\frac{1}{2}}^{(k)} = F_j^{(k)} - F_{j-1}^{(k)}$ is the difference in the numerical flux.[3]

[3]Such convention for notation is widely used in numerical analysis and computational fluid dynamics and will be widely used in the following sections. For any quantity p_j, we denote $\Delta p_{j+[(2m-1)/2]} = p_{j+m} - p_{j+m-1}$ for $m = 0, \pm 1, \pm 2, \ldots$. For instance, $\Delta p_{j+\frac{3}{2}} = p_{j+2} - p_{j+1}$, while $\Delta p_{j-\frac{3}{2}} = p_{j-1} - p_{j-2}$.

The physical sense of Equation 7.21a is, of course, clearer than Equation 7.19. We can now rewrite Equation 7.16 in the form of Equation 7.21a by employing the numerical flux:

$$F_j^{(k),\text{One-sided}} \triangleq F_j^{(k)} = \frac{1}{2}(a-|a|)p_{j+1}^{(k)} + \frac{1}{2}(a+|a|)p_j^{(k)} \tag{7.22}$$

Using similar notation, Equation 7.21a can also be rewritten as

$$p_j^{(k+1)} = p_j^{(k)} - \frac{1}{2}(\lambda a - |\lambda a|)\Delta p_{j+\frac{1}{2}}^{(k)} - \frac{1}{2}(\lambda a + |\lambda a|)\Delta p_{j-\frac{1}{2}}^{(k)} \tag{7.21b}$$

where $\Delta p_{j+\frac{1}{2}}^{(k)} = p_{j+1}^{(k)} - p_j^{(k)}$ and $\Delta p_{j-\frac{1}{2}}^{(k)} = p_j^{(k)} - p_{j-1}^{(k)}$.

7.1.1.2 Two-Sided Difference Schemes

We now construct the difference schemes with second-order accuracy. If we retain the terms up to second order in the Taylor expansion in terms of t, then Equation 7.3 becomes

$$p_j^{(k+1)} = p_j^{(k)} + \left[\frac{\partial p}{\partial t}\right]_j^{(k)} \Delta t + \frac{1}{2}\left[\frac{\partial^2 p}{\partial t^2}\right]_j^{(k)} \Delta t^2 + o(\Delta t^2) \tag{7.23}$$

Differentiating Equation 7.1 with respect to t on both sides yields

$$\begin{aligned}\frac{\partial^2 p}{\partial t^2} &= \frac{\partial}{\partial t}\left(\frac{\partial p}{\partial t}\right)\\ &= \frac{\partial}{\partial t}\left(-a(t)\frac{\partial p}{\partial x}\right)\\ &= -\dot{a}(t)\frac{\partial p}{\partial x} - a(t)\frac{\partial}{\partial t}\left(\frac{\partial p}{\partial x}\right)\\ &= -\dot{a}(t)\frac{\partial p}{\partial x} - a(t)\frac{\partial}{\partial x}\left(\frac{\partial p}{\partial t}\right)\\ &= -\dot{a}(t)\frac{\partial p}{\partial x} + a^2(t)\frac{\partial^2 p}{\partial x^2}\end{aligned} \tag{7.24a}$$

If $a(t)$ is slowly varying in the time interval $[t, t+\Delta t]$, then $\dot{a}(t) \approx 0$ and we have

$$\frac{\partial^2 p}{\partial t^2} = a^2(t)\frac{\partial^2 p}{\partial x^2} \tag{7.24b}$$

Substituting Equations 7.1 and 7.24b in Equation 7.23 yields

$$p_j^{(k+1)} = p_j^{(k)} - a(t)\left[\frac{\partial p}{\partial x}\right]_j^{(k)} \Delta t + \frac{a^2(t)}{2}\left[\frac{\partial^2 p}{\partial x^2}\right]_j^{(k)} \Delta t^2 + o(\Delta t^2) \tag{7.25}$$

Using the difference to approximate the partial differentiation with accuracy up to second order, $[\partial p/\partial x]_j^{(k)}$ should be represented by a central difference; namely:

$$\left[\frac{\partial p}{\partial x}\right]_j^{(k)} = \frac{p_{j+1}^{(k)} - p_{j-1}^{(k)}}{2\Delta x} + o(\Delta x^2) \tag{7.26a}$$

Simultaneously, the second-order partial differentiation can be approximated by a second-order difference:

$$\left[\frac{\partial^2 p}{\partial x^2}\right]_j^{(k)} = \frac{p_{j+1}^{(k)} + p_{j-1}^{(k)} - 2p_j^{(k)}}{\Delta x^2} + o(\Delta x^2) \tag{7.26b}$$

Substituting Equations 7.26a and 7.26b in Equation 7.25 and ignoring the effect of $o(\Delta t^2)$ and $o(\Delta x^2)$, we have

$$p_j^{(k+1)} = p_j^{(k)} - \frac{\lambda a}{2}\left[p_{j+1}^{(k)} - p_{j-1}^{(k)}\right] + \frac{\lambda^2 a^2}{2}\left[p_{j+1}^{(k)} + p_{j-1}^{(k)} - 2p_j^{(k)}\right] \tag{7.27a}$$

where $a^{(k)}$ is simplified to a for notational convenience.[4] The scheme can also be rewritten as

$$p_j^{(k+1)} = (1 - \lambda^2 a^2)p_j^{(k)} + \frac{1}{2}(\lambda^2 a^2 - \lambda a)p_{j+1}^{(k)} + \frac{1}{2}(\lambda^2 a^2 + \lambda a)p_{j-1}^{(k)} \tag{7.27b}$$

This is the widely used Lax–Wendroff scheme (LeVeque, 1992), a schematic illustration of which is shown in Figure 7.4.

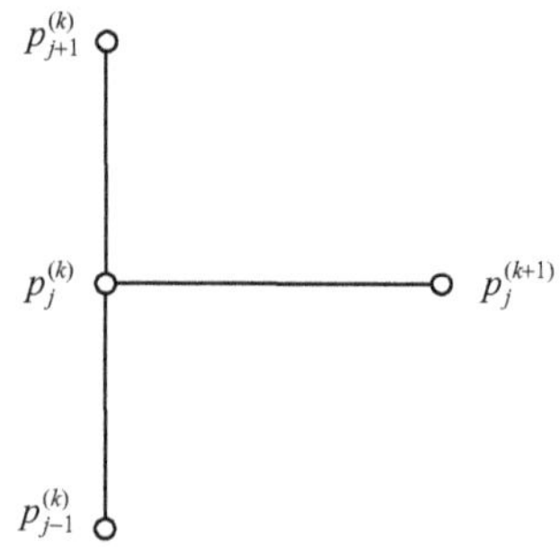

Figure 7.4 Two-sided scheme (Equation 7.27).

Rewriting Equation 7.27b similar to the numerical flux form in Equation 7.21, the numerical flux is now

$$F_j^{(k),\mathrm{LW}} \triangleq F_j^{(k)} = \frac{1}{2}(a - \lambda a^2)p_{j+1}^{(k)} + \frac{1}{2}(a + \lambda a^2)p_j^{(k)} \tag{7.27c}$$

Comparing this with Equation 7.22, we find that

$$F_j^{(k),\mathrm{LW}} = F_j^{(k),\text{One-sided}} + \frac{1}{2}(|a| - \lambda a^2)\Delta p_{j+\frac{1}{2}}^{(k)} \tag{7.27d}$$

This indicates that the numerical flux of the Lax–Wendroff scheme can be regarded as the numerical flux of the one-sided scheme plus a correction of a second-order term.

[4]We point out here that if we consider the effect of the term $-\dot{a}(t)(\partial p/\partial x)$ in Equation 7.24a, the value of a in the Lax–Wendroff scheme should take $a^{(k+\frac{1}{2})} = \frac{1}{2}\left[a^{(k)} + a^{(k+1)}\right]$. The proof is left to the reader.

To understand the properties of the Lax–Wendroff scheme (Equation 7.27b), again we consider the case when the initial condition is given by

$$p_j^{(0)} = \delta_{0j} \tag{7.8}$$

in which only the origin is the nonzero point at the initial time.

The propagation of the nonzero points at time instants $t_1, t_2, \ldots$ is shown in the shaded area in Figure 7.5, where the characteristic line when a is a constant is also plotted. Clearly, in contrast to the one-sided schemes, it is seen that the scheme of Equations 7.27a and 7.27b works both for $a > 0$ and $a < 0$.

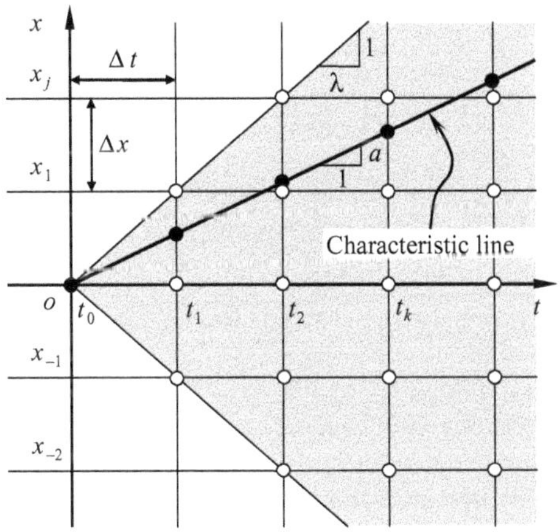

Figure 7.5 The Lax-Wendroff scheme.

Combining Equations 7.8 and 7.27b, it follows that

$$p_{-1}^{(1)} = \frac{1}{2}(\lambda^2 a^2 - \lambda a)p_0^{(0)} \qquad p_0^{(1)} = (1 - \lambda^2 a^2)p_0^{(0)} \qquad p_1^{(1)} = \frac{1}{2}(\lambda^2 a^2 + \lambda a)p_0^{(0)} \tag{7.28a}$$

and further:

$$p_{-k}^{(k)} = \left(\frac{\lambda^2 a^2 - \lambda a}{2}\right)^k p_0^{(0)} \qquad p_k^{(k)} = \left(\frac{\lambda^2 a^2 + \lambda a}{2}\right)^k p_0^{(0)} \tag{7.28b}$$

To meet the requirement that $p_{-k}^{(k)}$ and $p_k^{(k)}$ be bounded against increasing k, it must be satisfied that

$$\left|\frac{\lambda^2 a^2 - \lambda a}{2}\right| \le 1 \quad \text{and} \quad \left|\frac{\lambda^2 a^2 + \lambda a}{2}\right| \le 1 \tag{7.28c}$$

Consequently:

$$|\lambda a| \le 1, \quad \text{or} \quad |\lambda a^{(k)}| \le 1 \tag{7.29}$$

This is the CFL condition for the Lax–Wendroff scheme.

Re-examining Figures 7.2a and 7.2b and 7.5 and the CFL conditions for the one-sided scheme and the Lax–Wendroff scheme, we can see that no matter whether for the one-sided or the two-sided schemes, the CFL conditions require that the angle of the diagonal line of the grid should cover the angle of the characteristic line.

The Lax–Wendroff scheme is of second-order accuracy because in Equations 7.23, 7.26a and 7.26b the second-order terms in terms of both Δt and Δx are retained. Using the Lax–Wendroff scheme, the total probability is still conservative; namely:

$$\sum_j p_j^{(k)} = \sum_j p_j^{(0)} = 1 \tag{7.30a}$$

and the mean points still coincide with the point determined by the characteristic line; for example, at the first time step, this can be rewritten as

$$x_{-1}p_{-1}^{(1)} + x_0 p_0^{(1)} + x_1 p_1^{(1)} = a\Delta t \tag{7.30b}$$

which can be verified by Equation 7.28a. However, unlike the one-sided scheme, the nonnegativeness of the probability cannot be retained.

7.1.2 Dissipation, Dispersion and Total Variation Diminishing Schemes

The features of the difference schemes play one of the central roles in judging if the numerical solution is a physically reasonable solution. To understand this point, we now first examine a numerical example where the exact solution is a piece-wise continuous function, as shown in Figure 7.6. It is seen that, if the one-sided difference scheme is employed, the numerical solution is greatly smoothed in the vicinity of the left-hand side discontinuity point while a tiny high-frequency oscillation occurs in the vicinity of the right-hand side discontinuity point. If the Lax–Wendroff scheme is employed, on the other hand, the numerical solution is closer to the exact solution in the vicinity of the left-hand side discontinuity point. However, severe high-frequency oscillation occurs to the left of the two discontinuity points.

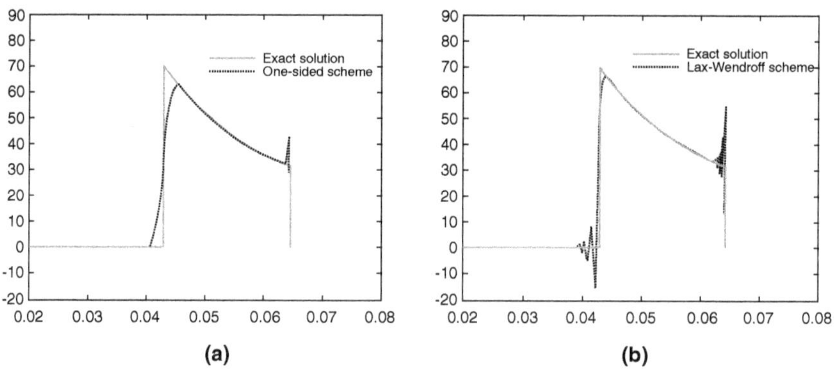

Figure 7.6 Numerical solutions computed by different schemes.

The effect of smoothing is related to *dissipation*, while the high-frequency oscillation is due to *dispersion*. Intuitively, when we gave an analysis on stability of the schemes, from

Equations 7.11b, 7.11d and 7.12 it is seen that the amplitude of the numerical solution is nonincreasing when the CFL condition is satisfied. This means that the scheme is dissipating and the results will be smoother than the real solution. A more rigorous basis is provided by the modified equation (Warming and Hyett, 1974; Hedstrom, 1975; LeVeque, 1992) or the von Neumann analysis (Stricwerda, 1989).

7.1.2.1 Modified Partial Differential Equations for Difference Schemes

In the preceding sections, the original partial differential equation is discretized on a uniform grid through truncation of a Taylor series. The $p_j^{(k)}$ values obtained by the difference schemes are approximations of the solution at the node (x_j, t_k). We now investigate what happens if we regard the $p_j^{(k)}$ values as the exact $p(x_j, t_k)$ values and replace them in the difference equation. Certainly, because the difference equation is an approximation of the original differential equation, not an exact replacement, we expect that the $p(x_j, t_k)$ values will satisfy some kind of partial differential equation as an approximation of the original partial differential equation.

We first consider the one-sided schemes. For convenience we use Equation 7.7a and replace $a^{(k)}$ by a ($a > 0$) here:

$$p_j^{(k+1)} = p_j^{(k)} - \lambda a[p_j^{(k)} - p_{j-1}^{(k)}] \tag{7.31}$$

Replacing $p_j^{(k)}$ by the exact value $p(x_j, t_k)$, Equation 7.31 becomes

$$p(x_j, t_k + \Delta t) = p(x_j, t_k) - \lambda a[p(x_j, t_k) - p(x_j - \Delta x, t_k)] \tag{7.32a}$$

Performing the Taylor expansion in the vicinity of (x_j, t_k):

$$p(x_j, t_k + \Delta t) = p(x_j, t_k) + \frac{\partial p(x_j, t_k)}{\partial t}\Delta t + \frac{1}{2}\frac{\partial^2 p(x_j, t_k)}{\partial t^2}\Delta t^2 + \frac{1}{6}\frac{\partial^3 p(x_j, t_k)}{\partial t^3}\Delta t^3 + \cdots \tag{7.32b}$$

$$p(x_j - \Delta x, t_k) = p(x_j, t_k) - \frac{\partial p(x_j, t_k)}{\partial x}\Delta x + \frac{1}{2}\frac{\partial^2 p(x_j, t_k)}{\partial x^2}\Delta x^2 - \frac{1}{6}\frac{\partial^3 p(x_j, t_k)}{\partial x^3}\Delta x^3 + \cdots \tag{7.32c}$$

Using the similar manipulation in Equation 7.24a, we have

$$\frac{\partial^2 p}{\partial t^2} = a^2\frac{\partial^2 p}{\partial x^2} \quad \frac{\partial^3 p}{\partial t^3} = -a^3\frac{\partial^3 p}{\partial x^3} \tag{7.32d}$$

Substituting Equations 7.32b–7.32d in Equation 7.32a yields[5]

$$\frac{\partial p}{\partial t} + a\frac{\partial p}{\partial x} = \frac{a(1-\lambda a)\Delta x}{2}\frac{\partial^2 p}{\partial x^2} + \frac{a(\lambda^2 a^2 - 1)\Delta x^2}{6}\frac{\partial^3 p}{\partial x^3} \tag{7.33}$$

[5] This equation is different from that in LeVeque (1992), where the second term on the right-hand side of the equation is ignored. The treatment here is more complete.

We see that, as $\Delta x \to 0$, this partial differential equation tends to the original partial differential equation, Equation 7.1. Equation 7.33 is the *modified equation* of the difference scheme (Equation 7.31). The right-hand side of the equation is the *local truncation error* of the difference scheme.

Physically speaking, the first term on the right-hand side of Equation 7.33 is the diffusive term and the second one is the dispersive term. This means that, in the one-sided difference scheme, artificial diffusion and dispersion are introduced. Of course, the term of diffusion is a type of artificial viscosity which results in dissipation. In particular, because the coefficient of the dispersive term is in the higher order of Δx, its effect is usually much less than the dissipation term. When it is ignored, Equation 7.33 becomes

$$\frac{\partial p}{\partial t} + a\frac{\partial p}{\partial x} = \frac{a(1-\lambda a)\Delta x}{2}\frac{\partial^2 p}{\partial x^2} \tag{7.34a}$$

In thermodynamics, this is an advection–diffusion equation. Actually, it is also a one-dimensional FPK equation with the diffusion coefficient $a(1-\lambda a)\Delta x/2$. As we have understood in Section 5.6.3.1, this equation will make the solution smoother and smoother as time passes. For example, an initial delta function will become a normal distribution with increasing standard deviation. By the way, we know that for a physical meaning system it is required that

$$\frac{a(1-\lambda a)\Delta x}{2} \geq 0 \tag{7.35}$$

This essentially coincides with the CFL condition given by Equation 7.10.

More generally, for both $a > 0$ and $a < 0$, Equation 7.34a becomes

$$\frac{\partial p}{\partial t} + a\frac{\partial p}{\partial x} = \frac{|a|(1-\lambda|a|)\Delta x}{2}\frac{\partial^2 p}{\partial x^2} \tag{7.34b}$$

If we note Equation 7.27d, we see that the corrections in the numerical flux of the Lax–Wendroff scheme are of course consistent with the right-hand side of Equation 7.34b; this is how the second-order accuracy is achieved.

Similar treatment works for the Lax–Wendroff scheme. In this case, we replace Equation 7.32a by

$$\begin{aligned} p(x_j, t_k+\Delta t) = p(x_j, t_k) - \frac{\lambda a}{2}\left[p(x_j+\Delta x, t_k) - p(x_j-\Delta x, t_k)\right] \\ + \frac{\lambda^2 a^2}{2}\left[p(x_j+\Delta x, t_k) + p(x_j-\Delta x, t_k) - 2p(x_j, t_k)\right] \end{aligned} \tag{7.36}$$

Substituting the Taylor expansion in the vicinity of (x_j, t_k) and rearranging the equation yields

$$\frac{\partial p}{\partial t} + a\frac{\partial p}{\partial x} = \frac{a(\lambda^2 a^2 - 1)\Delta x^2}{6}\frac{\partial^3 p}{\partial x^3} \tag{7.37}$$

Here, we see that the local truncation error is the dispersive term and the diffusive term disappears. This is because the second term in the Taylor expansion has been considered in Equation 7.25 and, therefore, the dissipation is not shown here. Actually, the dissipation of the

Lax–Wendroff scheme is very small, as will be seen in the next section. In this case, however, the effect of dispersion dominates the error. This accounts for the nonphysical spurious phenomena near the discontinuity points in Figure 7.6.

7.1.2.2 Amplification Factors of Difference Schemes: von Neumann Analysis

To capture the features of the difference schemes for all possible initial function $p_0(x)$ is not easy. However, we have successfully adopted the methodology that some types of special function can be used to understand the properties of the system, as discussed in Section 5.2.1. Similar to what is done in dynamical systems, we can also understand the features of the difference scheme through tracing the propagation of a harmonic wave in the difference equation. As a matter of fact, in the case where a Kronecker delta initial condition is used (see Equation 7.8), we are trying to understand the features of the difference scheme by examining the propagation of an impulse through the difference equation.

From Section 5.2.1 we know that any physical practical function can be represented by a Fourier transform pair:

$$p_0(x) = \frac{1}{2\pi}\int_{-\infty}^{\infty} \tilde{p}_0(\kappa)e^{i\kappa x}\,d\kappa \qquad \tilde{p}_0(\kappa) = \int_{-\infty}^{\infty} p_0(x)e^{-i\kappa x}\,dx \tag{7.38}$$

For this reason, we can just consider the initial function as a unit harmonic function $p_0(x) = e^{i\kappa x}$, where κ is the wave number.

Note that both the one-sided and the Lax–Wendroff schemes are one-step linear schemes; namely, they can be written in a unified scheme as in Equation 7.19, for convenience of reference, repeated here as Equation 7.39:

$$p_j^{(k+1)} = \sum_{\ell=-\nu}^{\nu} c_\ell p_{j+\ell}^{(k)} \tag{7.39}$$

It is easy to see that for the one-sided and the Lax–Wendroff schemes $\nu = 1$ and the coefficients c_ℓ can be determined by comparing this equation with Equations 7.16 and 7.27b respectively.

The discretized initial condition of $p_0(x) = e^{i\kappa x}$ reads

$$p_j^{(0)} = e^{i\kappa x_j} = e^{i\kappa\Delta x j} \tag{7.40}$$

Here, use has been made of $x_j = j\Delta x$. We now consider the first step by Equation 7.39

$$\begin{aligned} p_j^{(1)} &= \sum_{\ell=-\nu}^{\nu} c_\ell p_{j+\ell}^{(0)} \\ &= \sum_{\ell=-\nu}^{\nu} c_\ell e^{i\kappa\Delta x(j+\ell)} \\ &= \sum_{\ell=-\nu}^{\nu} c_\ell e^{i\kappa\Delta x\ell} e^{i\kappa\Delta x j} \\ &= \left(\sum_{\ell=-\nu}^{\nu} c_\ell e^{i\kappa\Delta x\ell}\right) p_j^{(0)} = g(\vartheta) p_j^{(0)} \end{aligned} \tag{7.41}$$

where $\vartheta = \kappa\Delta x$ and

$$g(\vartheta) = \sum_{\ell=-\nu}^{\nu} c_\ell e^{i\vartheta\ell} \tag{7.42}$$

is an amplification factor.

Because Equation 7.39 is a linear operator, it is clear that

$$p_j^{(k+1)} = g(\vartheta)p_j^{(k)} \tag{7.43a}$$

$$p_j^{(k)} = g^k(\vartheta)p_j^{(0)} = g^k(\vartheta)e^{i\vartheta j} \tag{7.43b}$$

Intuitively, for a stable scheme we must require that

$$|g(\vartheta)| \leq 1 \tag{7.44}$$

Otherwise, from Equation 7.43, $p_j^{(k)}$ will increase rapidly and be unbounded as $k \to \infty$. This is the case for a uniform grid with constant $\lambda = \Delta t/\Delta x$.

For the case Δt and Δx are not constant, Equation 7.44 can be relaxed to

$$|g(\vartheta, \Delta t, \Delta x)| \leq 1 + K\Delta t \tag{7.45}$$

for all ϑ, $0 \leq \Delta t \leq \Delta t_b$ and $0 \leq \Delta x \leq \Delta x_b$, where K is a constant (independent of ϑ, Δt and Δx) and Δt_b and Δx_b are some positive grid spacings (Stricwerda, 1989).

According to Equation 7.42, it is easy to obtain the amplification factor and then give the condition of stability by Equation 7.44 or 7.45. A more direct approach is to introduce Equation 7.43b into the difference scheme. For example, examining the one-sided scheme (Equation 7.7b) for a contant $a > 0$:

$$p_j^{(k+1)} = (1 - \lambda a)p_j^{(k)} + \lambda a p_{j-1}^{(k)} \tag{7.46a}$$

when replacing $p_j^{(k)}$ by $g^k\, e^{i\vartheta j}$, we have

$$g^{k+1}e^{i\vartheta j} = (1 - \lambda a)g^k e^{i\vartheta j} + \lambda a g^k e^{i\vartheta(j-1)} \tag{7.46b}$$

Eliminating $g^k\, e^{i\vartheta j}$ on both sides yields

$$g(\vartheta) = 1 - \lambda a(1 - e^{-i\vartheta}) \quad \text{and} \quad |g(\vartheta)|^2 = 1 - 4\lambda a(1 - \lambda a)\sin^2\frac{\vartheta}{2} \tag{7.46c}$$

For constant λ, using Equations 7.44 and 7.46c, we obtain

$$\lambda a \leq 1 \tag{7.46d}$$

This, again, gives the CFL condition for the scheme in Equation 7.46a.

Similar manipulation can be performed for the Lax–Wendroff scheme. Substituting Equation 7.43b in the scheme in Equation 7.27b gives

$$g^{k+1}e^{i\vartheta j} = (1 - \lambda^2 a^2)g^k e^{i\vartheta j} + \frac{1}{2}(\lambda^2 a^2 - \lambda a)g^k e^{i\vartheta(j+1)} + \frac{1}{2}(\lambda^2 a^2 + \lambda a)g^k e^{i\vartheta(j-1)} \tag{7.47a}$$

and thus

$$g(\vartheta) = (1 - \lambda^2 a^2) + \lambda^2 a^2 \cos\vartheta - \mathrm{i}\lambda a \sin\vartheta \tag{7.47b}$$

and

$$|g(\vartheta)|^2 = 1 - 4\lambda^2 a^2 (1 - \lambda^2 a^2) \sin^4 \frac{\vartheta}{2} \tag{7.47c}$$

From Equation 7.44, it follows that

$$|\lambda a| \leq 1 \tag{7.47d}$$

which is the CFL condition for the Lax–Wendroff scheme given by Equation 7.29.

On the other hand, from Equation 7.43, for a constant λ, in the case $|g(\vartheta)| = 1$, there will be no dissipation because the amplitude does not decay. However, dissipation exists when $|g(\vartheta)| < 1$. It is seen from Equations 7.46c and 7.47c that both the one-sided and Lax–Wendroff schemes have dissipation; but, for the same λ and a, the amplification factor of the Lax–Wendroff is closer to unity than the one-sided scheme is, which is why stronger dissipation is seen in the latter scheme.

7.1.2.3 Dispersion

As pointed out in Section 6.6.1, the solution of Equation 7.1 is a wave. Thus, to understand the properties of the numerical methods, it is useful to consider the propagation of a real wave in the original system and the propagation of the corresponding numerical wave in the discretized system. As has been done in the preceding section, we consider the propagation of an initial wave with wave number κ, $\mathrm{e}^{\mathrm{i}\kappa x}$. The analytical solution, according to Equation 6.141, is given by

$$p(x,t) = \mathrm{e}^{\mathrm{i}\kappa(x - at)} \tag{7.48}$$

In the numerical solution, there might be some distortion such that the velocity might not exactly be a. Denoting it by α, the numerical solution can then be written as

$$p(x,t) = \mathrm{e}^{\mathrm{i}\kappa(x-\alpha t)} = \mathrm{e}^{\mathrm{i}(\kappa x - \omega t)} \quad \text{and} \quad p_j^{(k)} = p(x_j, t_k) = \mathrm{e}^{\mathrm{i}\kappa(x_j - \alpha t_k)} = \mathrm{e}^{\mathrm{i}(\kappa x_j - \omega t_k)} \tag{7.49}$$

where the frequency $\omega = \kappa\alpha$.

Substituting Equation 7.49 in the Lax–Wendroff scheme in Equation 7.27b we have

$$\begin{aligned} \mathrm{e}^{\mathrm{i}(\kappa x_j - \omega t_k - \omega\Delta t)} = {} & (1 - \lambda^2 a^2)\mathrm{e}^{\mathrm{i}(\kappa x_j - \omega t_k)} + \frac{1}{2}(\lambda^2 a^2 - \lambda a)\mathrm{e}^{\mathrm{i}(\kappa x_j + \kappa\Delta x - \omega t_k)} \\ & + \frac{1}{2}(\lambda^2 a^2 + \lambda a)\mathrm{e}^{\mathrm{i}(\kappa x_j - \kappa\Delta x - \omega t_k)} \end{aligned} \tag{7.50a}$$

Eliminating $\mathrm{e}^{\mathrm{i}(\kappa x_j - \omega t_k)}$ on both sides yields

$$\mathrm{e}^{-\mathrm{i}\omega\Delta t} = (1 - \lambda^2 a^2) + \frac{1}{2}(\lambda^2 a^2 - \lambda a)\mathrm{e}^{\mathrm{i}\kappa\Delta x} + \frac{1}{2}(\lambda^2 a^2 + \lambda a)\mathrm{e}^{-\mathrm{i}\kappa\Delta x} \tag{7.50b}$$

or in an alternative form as

$$\tan(\omega\Delta t) = \frac{\lambda a \sin(\kappa\Delta x)}{(1-\lambda^2 a^2) + \lambda^2 a^2 \cos(\kappa\Delta x)} = \frac{\lambda a \sin(\kappa\Delta x)}{1 - 2\lambda^2 a^2 \sin^2(\kappa\Delta x/2)} \tag{7.50c}$$

This demonstrates that the frequency ω is a nonlinear function of the wave number κ. Therefore, the wave velocity $\alpha = \omega/\kappa$, which is now called the *phase velocity*[6] for a specified κ, is dependent on the wave number κ. Further, if we expand the above terms in the vicinity of $\kappa = 0$ by using the Taylor expansion series $\sin(x) = x\left[1 - \frac{1}{6}x^2 + \mathcal{O}(x^4)\right]$ and $\tan^{-1}(x) = x\left[1 - \frac{1}{3}x^2 + \mathcal{O}(x^4)\right]$, from Equation 7.50c we have

$$\omega \doteq a\kappa\left[1 - \frac{1}{6}\kappa^2\Delta x^2(1 - \lambda^2 a^2)\right] \tag{7.51}$$

Thus, the phase velocity is given by

$$\alpha(\kappa\Delta x) = \frac{\omega}{\kappa} = a\left[1 - \frac{1}{6}\kappa^2\Delta x^2(1 - \lambda^2 a^2)\right] \tag{7.52}$$

Equation 7.52 can also be reached by directly introducing Equation 7.49 into the dispersive equation (Equation 7.37) and then eliminating the common terms on both sides.

For practical situations, the wave is composed of many, or infinite, waves of different wave numbers κ_j; they form a wave packet or wave group. For the jth component, the change of the phase $\varphi_j = (\kappa_j x - \omega_j t)$ in dt is

$$\mathrm{d}\varphi_j = \mathrm{d}(\kappa_j x - \omega_j t) = \kappa_j \mathrm{d}x - \omega_j \mathrm{d}t \tag{7.53}$$

The same thing happens for the ℓth component with the wave number κ_ℓ and frequency ω_ℓ. In order for the wave group to be maintained, the change in the phase of different components should be the same; that is, $\mathrm{d}\varphi_j = \mathrm{d}\varphi_\ell$. This leads to

$$(\kappa_j - \kappa_\ell)\mathrm{d}x - (\omega_j - \omega_\ell)\mathrm{d}t = 0 \tag{7.54}$$

Since κ_j and κ_ℓ and ω_j and ω_ℓ differ only slightly, we have

$$\frac{\mathrm{d}x}{\mathrm{d}t} = \frac{\omega_j - \omega_\ell}{\kappa_j - \kappa_\ell} = \frac{\mathrm{d}\omega}{\mathrm{d}\kappa} \tag{7.55}$$

Thus, the *group velocity*, the velocity of the wave group, is defined (Graff, 1975) as

$$\alpha_g = \frac{\mathrm{d}\omega}{\mathrm{d}\kappa} \tag{7.56}$$

For the Lax–Wendroff scheme, from Equation 7.51 it follows that

$$\alpha_g = \frac{\mathrm{d}\omega}{\mathrm{d}\kappa} = a\left[1 - \frac{1}{2}\kappa^2\Delta x^2(1 - \lambda^2 a^2)\right] \tag{7.57}$$

Because $|\lambda a| \le 1$, it is found from Equations 7.52 and 7.57 that

[6] This term comes from the fact that it is the velocity that occurs in the phase angle $\kappa(x - \alpha t)$ of the wave $e^{i\kappa(x-\alpha t)}$, in contrast to the group velocity introduced later.

$$\alpha_g(\kappa) \leq \alpha(\kappa) \leq a \tag{7.58}$$

That is, the wave obtained by the difference equation lags behind the real wave. Moreover, this degree of lag is different for the components with different wave numbers. Thus, the components with different wave numbers will separate as time elapses. This leads to dispersion (Trefethen, 1982). The phenomenon is particularly severe when the real wave possesses discontinuity, because a wide range of (particularly higher) wave number is needed in this case. For the Lax–Wendroff scheme, Equation 7.58 also accounts for why the high-frequency oscillation is always to the left of the discontinuity points (the propagation direction is from left to right). However, for different schemes, the relationship among α, α_g and a might be different from Equation 7.58; therefore, the properties of the high-frequency oscillation will also be different.

Generally, there is competition between the dissipation and the dispersion. This can be seen from the modified Equations 7.33 and 7.37. Usually, one of them dominates the local error. The schemes with less dispersion usually exhibit stronger dissipation, and vice versa.

7.1.2.4 Total Variation Diminishing Schemes

The one-sided difference scheme is nonnegativeness preserving and the numerical results are usually smooth, but the dissipation is too large. On the other hand, the two-sided scheme (for example, the Lax–Wendroff scheme) is much less dissipative, though much more dispersive, especially in the vicinity of discontinuity. Can we have a balance between them by some type of hybrid scheme? This is possible by constructing a total variation diminishing (TVD) scheme (Harten, 1983; Shu, 1988).

The nonnegativeness preserving of the one-sided scheme, as noted in Section 7.1.1.1 and Figure 7.6a, is more rigorously called *monotonicity preserving*, which means that if the initial data $p_j^{(0)}$ is monotone as a function of j, then the solution $p_j^{(k)}$ should have the same property for all k. Actually, *a linear, monotonicity-preserving scheme is at most first-order accurate* (van Leer, 1974; LeVeque, 1992).

Intuitively, from Figure 7.6b we see that, compared with the real solution, the numerical solution by the Lax–Wendroff scheme is more irregular because of the high-frequency oscillation. This can be measured by a quantity called *total variation* of a function defined by

$$\mathrm{TV}[p(\cdot, t)] = \int_{-\infty}^{\infty} \left| \frac{\partial p(x,t)}{\partial x} \right| \mathrm{d}x \tag{7.59}$$

of which the discretized form could be written as

$$\mathrm{TV}(p_{\cdot}^{(k)}) = \sum_{j=-\infty}^{\infty} \left| p_{j+1}^{(k)} - p_j^{(k)} \right| \tag{7.60}$$

It can be proved that the real solution of Equation 7.1 satisfies (LeFloch, 2002)

$$\mathrm{TV}[p(\cdot, t_2)] \leq \mathrm{TV}[p(\cdot, t_1)] \leq \mathrm{TV}[p(\cdot, t_0)] \quad \text{for} \quad t_2 > t_1 > t_0 \tag{7.61}$$

This attribute of the solution function is TVD. Any TVD scheme is monotonicity preserving.

Clearly, if a difference scheme is TVD, then spurious nonphysical phenomena will be expected to reduce or even disappear. It can be verified that the one-sided scheme is TVD but that the Lax–Wendroff scheme is not. Actually, it is readily seen that the high-frequency oscillation makes the total variation of the numerical solution larger than the exact solution.

To construct a scheme having a trade-off between the one-sided and the Lax–Wendroff schemes, we first examine the relationship between them further. Comparing the modified Equations 7.33 and 7.37, we judge that the Lax–Wendroff scheme must be some type of modified version of the one-sided scheme where additional terms are imposed to suppress the dissipation. This is actually the case, as we have seen in Equation 7.27d, which indicates that the numerical flux of the Lax–Wendroff scheme $F_j^{(k),\mathrm{LW}}$ is a modified version of that of the one-sided scheme $F_j^{(k),\text{One-sided}}$, where a second-order correction term is added. We can then rewrite the Lax–Wendroff scheme Equation 7.27b to a flux-difference form:

$$\begin{aligned} p_j^{(k+1)} &= p_j^{(k)} - \lambda(F_j^{(k),\mathrm{LW}} - F_{j-1}^{(k),\mathrm{LW}}) \\ &= p_j^{(k)} - \lambda(F_j^{(k),\text{One-sided}} - F_{j-1}^{(k),\text{One-sided}}) - \frac{1}{2}(|\lambda a| - \lambda^2 a^2)(\Delta p_{j+\frac{1}{2}}^{(k)} - \Delta p_{j-\frac{1}{2}}^{(k)}) \end{aligned} \tag{7.62}$$

where $\Delta p_{j+\frac{1}{2}}^{(k)} = p_{j+1}^{(k)} - p_j^{(k)}$, $\Delta p_{j-\frac{1}{2}}^{(k)} = p_j^{(k)} - p_{j-1}^{(k)}$ can be regarded as the difference of numerical flux (divided by a). Here, it is clearly seen that a second-order correction term is imposed on the one-sided scheme to construct the Lax–Wendroff scheme. It is this correction term that greatly reduces the dissipation but simultaneously makes the dispersion obvious. We need to modify this term such that in the vicinity of discontinuity this term almost does not work while in the smooth part this term works well. This means that the modification should be based on the data of the solution.

The most intuitive approach to balance between the one-sided and the Lax–Wendroff scheme is to construct a hybrid scheme with the numerical flux as combination of numerical flux of the Lax–Wendroff scheme $F_j^{(k),\mathrm{LW}}$ and that of the one-sided scheme $F_j^{(k),\text{One-sided}}$; namely, for $0 \le 1 \le \beta$:

$$F_j^{(k),\mathrm{Hybrid}} = (1-\beta)F_j^{(k),\text{One-sided}} + \beta F_j^{(k),\mathrm{LW}} = F_j^{(k),\text{One-sided}} + \beta\frac{1}{2}(|\lambda a| - \lambda^2 a^2)\Delta p_{j+\frac{1}{2}}^{(k)} \tag{7.63a}$$

where $F_j^{(k),\mathrm{LW}}$ is defined in Equation 7.27d.

If β is a constant (that is, not dependent on the data of the numerical solution), then Equation 7.63a is a one-step linear scheme in the form of Equations 7.19 and 7.39. As pointed out above, a linear, monotonicity-preserving scheme is at most first-order accurate; the scheme constructed above cannot be a second-order accurate scheme.

To retain second-order accuracy, β must be a nonlinear factor dependent on the data, denoted by $\psi_{j+\frac{1}{2}}$; thus Equation 7.63a is rewritten as

$$F_j^{(k),\mathrm{Hybrid}} = F_j^{(k),\text{One-sided}} + \frac{1}{2}(|\lambda a| - \lambda^2 a^2)\psi_{j+\frac{1}{2}}\Delta p_{j+\frac{1}{2}}^{(k)} \tag{7.63b}$$

Because $\psi_{j+\frac{1}{2}}$ is a factor less than unity, imposed on and modifying the numerical flux according to the data of the numerical solution, it is called the *flux limiter*. Equation 7.62 is now modified to

$$p_j^{(k+1)} = p_j^{(k)} - \lambda(F_j^{(k),\text{Hybrid}} - F_{j-1}^{(k),\text{Hybrid}})$$
$$= p_j^{(k)} - \lambda(F_j^{(k),\text{One-sided}} - F_{j-1}^{(k),\text{One-sided}}) - \frac{1}{2}(|\lambda a| - \lambda^2 a^2)(\psi_{j+\frac{1}{2}}\Delta p_{j+\frac{1}{2}}^{(k)} - \psi_{j-\frac{1}{2}}\Delta p_{j-\frac{1}{2}}^{(k)}) \tag{7.64}$$

or in a complete form when introducing $F_j^{(k),\text{One-sided}}$ from Equation 7.22:

$$p_j^{(k+1)} = p_j^{(k)} - \frac{1}{2}(\lambda a - |\lambda a|)\Delta p_{j+\frac{1}{2}}^{(k)} - \frac{1}{2}(\lambda a + |\lambda a|)\Delta p_{j-\frac{1}{2}}^{(k)}$$
$$- \frac{1}{2}(|\lambda a| - \lambda^2 a^2)(\psi_{j+\frac{1}{2}}\Delta p_{j+\frac{1}{2}}^{(k)} - \psi_{j-\frac{1}{2}}\Delta p_{j-\frac{1}{2}}^{(k)}) \tag{7.65}$$

Note that in the case where the flux limiters $\psi_{j+\frac{1}{2}} \equiv \psi_{j-\frac{1}{2}} \equiv 0$, Equations 7.64 and 7.65 reduce to the one-sided scheme, whereas in the case where $\psi_{j+\frac{1}{2}} \equiv \psi_{j-\frac{1}{2}} \equiv 1$ they become the Lax–Wendroff scheme. Therefore, we require

$$0 \le \psi_{j+\frac{1}{2}} \le 1 \qquad 0 \le \psi_{j-\frac{1}{2}} \le 1 \tag{7.66}$$

As analyzed, the modification should adapt to the data by judging if the change of the curve is abrupt. This can be measured by the ratios of a sequent difference; that is:

$$r_{j+\frac{1}{2}}^{+} = \frac{\Delta p_{j+\frac{3}{2}}^{(k)}}{\Delta p_{j+\frac{1}{2}}^{(k)}} = \frac{p_{j+2}^{(k)} - p_{j+1}^{(k)}}{p_{j+1}^{(k)} - p_j^{(k)}} \qquad r_{j+\frac{1}{2}}^{-} = \frac{\Delta p_{j-\frac{1}{2}}^{(k)}}{\Delta p_{j+\frac{1}{2}}^{(k)}} = \frac{p_j^{(k)} - p_{j-1}^{(k)}}{p_{j+1}^{(k)} - p_j^{(k)}} \tag{7.67a, b}$$

For instance, if $r_{j+\frac{1}{2}}^{+} = 1$, then the $(j+2)$th, the $(j+1)$th and the jth points are on a straight line and the curve is smooth; that is, the change of the curve is not abrupt. The same thing happens for $r_{j+\frac{1}{2}}^{-} = 1$. But, if $\left|r_{j+\frac{1}{2}}^{+}\right|$ is very large, then the $(j+2)$th point is very far from the line determined by the $(j+1)$th and the jth points and, thus, there is an abrupt change in the curve.

If we assume $a > 0$, then imposing the TVD condition on the scheme in Equations 7.64 and 7.65 will give the conditions that the flux limiter ψ should satisfy (Sweby, 1984; Roe, 1986). According to computational experiences, the following flux limiter is recommended:

$$\psi_0(r) = \max(0, \min(2r, 1), \min(r, 2)) \tag{7.68}$$

Further, in a unified way for both $a > 0$ and $a < 0$, we use

$$\psi_{j+\frac{1}{2}}(r_{j+\frac{1}{2}}^{+}, r_{j+\frac{1}{2}}^{-}) = u(-a)\psi_0(r_{j+\frac{1}{2}}^{+}) + u(a)\psi_0(r_{j+\frac{1}{2}}^{-}) \tag{7.69}$$

where $u(\,)$ is the Heaviside's unit step function (see Appendix A). Replacing the subscript by $j-\frac{1}{2}$ gives $\psi_{j-\frac{1}{2}}$.

Investigations have proved that away from the extreme value the scheme in Equations 7.64 and 7.65 is second-order accurate, whereas near the extreme value it is first-order accurate (LeVeque, 1992).

The TVD scheme does work well in the probability density evolution analysis of most problems.

Example 7.1. Comparison between Lax–Wendroff and TVD Schemes Again, we study the SDOF system with uncertain natural frequency in Example 6.2. The PDF of the displacement is solved both by the Lax–Wendroff scheme and the TVD scheme. Figures 7.7a and 7.7b shows the PDFs at 1.00 s computed by the two schemes and Figures 7.7c and 7.7d shows the PDFs evolving against time in the period 0.9–1.1 s. These figures show that the Lax–Wendroff scheme can capture the exact results in most places but that it does not work well in the vicinity of discontinuity because of dispersion, whereas in the TVD scheme the accuracy is high even in the vicinity of discontinuity and the high-frequency oscillation disappears.

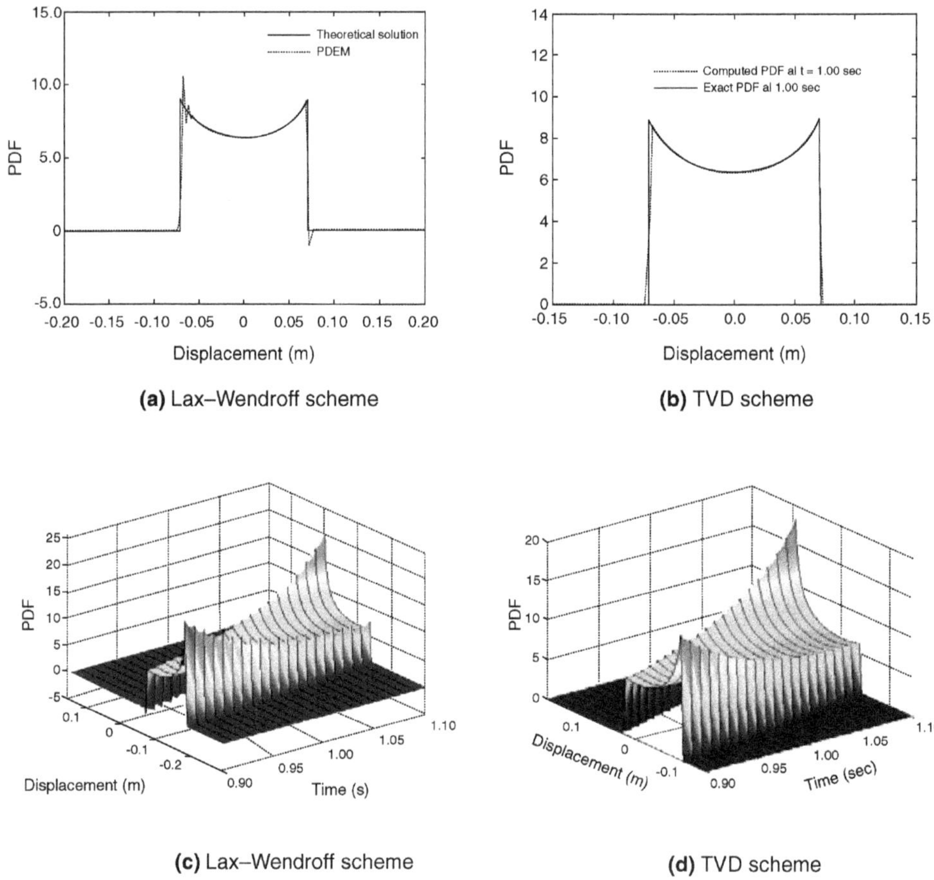

Figure 7.7 Numerical solution computed by Lax–Wendroff scheme and TVD scheme.

Computational experiences demonstrate that the accuracy of the TVD scheme in Equation 7.65 sometimes deteriorates severely compared with the Lax–Wendroff scheme, particularly in the time interval near the initial time. This might require reducing the time step to increase the accuracy. The selection of the time step usually depends on the frequency of the time history of the velocity $a(t)$, and should be carefully calibrated, say, through comparing the mean and the standard deviation with those obtained by the Lax–Wendroff scheme.

When the smaller time steps are used, an interpolation between two sample time instants of $a(t)$ is usually needed (Chen and Li, 2005a). □

7.2 Representative Point Sets and Assigned Probabilities

As we have discussed in the solution flow of the probability density evolution method in Section 6.6.2, to solve the generalized density evolution equation numerically, a set of values of the parameters $\boldsymbol{\theta} = (\theta_1, \theta_2, \ldots, \theta_s)$ should first be specified. In other words, a set of points scattered in the s-dimensional region $\Omega_{\boldsymbol{\Theta}} \subset \mathbb{R}^s$ needs to be chosen. Here $\mathbb{R}^s$ is the s-dimensonal real Euclidian space. To select these points in a smart way, it is required to understand the configuration of s-dimensional space. To this end, we first revisit the sphere packings and covering problems and then come to the strategies of determining the representative points.

7.2.1 Sphere Packings, Covering and Partition of Space

7.2.1.1 Sphere Packings

The celebrated *Kepler conjecture* asserts that the highest dense sphere packing in three-dimensional space is $\pi/\sqrt{18} = 0.740\,480\cdots$, which is closely related to the problem of largest kissing number (coordination number or contact number) of equal spheres. Essentially, this problem deals with how to pack a given space with equal spheres in an efficient way. More significantly, research on this problem in the past hundreds of years provides deep insight into the understanding of multidimensional spaces (Conway and Sloane, 1999; Zong, 1999; Martinet, 2003).

We consider the sphere packing problem in an s-dimensional space; that is, to pack the space by a set of nonoverlapping, equal spheres. The case for $s = 2$ is shown in Figure 7.8, where the nonoverlapping equal circles are placed in different ways. Clearly, there is always some room not being occupied by the circles. Visually, the packing in pattern (b) is more efficient than that in pattern (a); that is, the unoccupied room in pattern (b) is less than that in pattern (a). More rigorously, the efficiency of the packing can be measured by the density of packing defined as the proportion of the space that is occupied by the spheres. Note that a particular polyhedron is related to and covering each sphere; for example, in Figure 7.8a it is a square contact to the circle and in Figure 7.8b it is a regular hexagon contact to the circle. We call the volume of this polyhedron the *fundamental region* (or *representative region*) of the sphere.

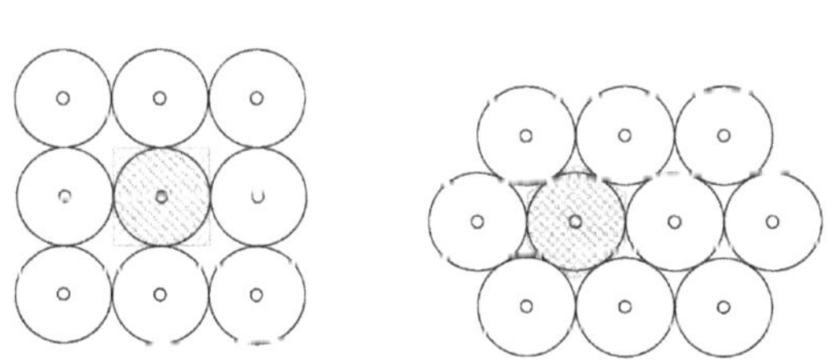

(a) (b)

Figure 7.8 Patterns of sphere packings.

Therefore, in an infinite space, the density of packing is equivalent to

$$\rho = \frac{\text{Volume of all the spheres}}{\text{Volume of the space}} = \frac{\mathcal{V}(\text{one sphere})}{\mathcal{V}(\text{fundamental region})} \tag{7.70}$$

where $\mathcal{V}(\cdot)$ is the volume. Certainly, there is always $\rho < 1$ and the packing with higher density is more efficient.

It is easy to obtain the density of packing for the patterns in Figure 7.8a and b as

$$\rho = \frac{\pi r^2}{(2r)^2} = \frac{\pi}{4} = 0.785\,398\cdots \quad \text{and} \quad \rho = \frac{\pi r^2}{2\sqrt{3}r^2} = \frac{\pi}{\sqrt{12}} = 0.906\,899\cdots$$

respectively. They are obviously distinct. Actually, the pattern in Figure 7.8b is the densest packing in two-dimensional space.

The definition in Equation 7.70 also applies in higher dimension. The volume of an s-dimensional sphere $\mathcal{B}(r,s) = \{\boldsymbol{x} = (x_1, x_2, \ldots, x_s) : \|\mathbf{x}\|^2 = x_1^2 + x_2^2 + \cdots + x_s^2 \le r^2\}$ is given by

$$\mathcal{V}(\mathcal{B}(r,s)) = \int_{\boldsymbol{x}\in\mathcal{B}(r,s)} \mathrm{d}x_1\,\mathrm{d}x_2\,\cdots\,\mathrm{d}x_s = \frac{\pi^{s/2}r^s}{\Gamma(1+\frac{s}{2})} = \begin{cases} \dfrac{\pi^m r^s}{m!} & \text{for} \quad s = 2m \\ \dfrac{2(2\pi)^m r^s}{\prod_{j=0}^{m}(2j+1)} & \text{for} \quad s = 2m+1 \end{cases} \tag{7.71}$$

where $\Gamma(\cdot)$ is the Gamma function and

$$\mathcal{V}(\text{fundamental region}) = \int_{\boldsymbol{x}\in\text{fundamental region}} \mathrm{d}x_1\,\mathrm{d}x_2\,\cdots\,\mathrm{d}x_s \tag{7.72}$$

In dimension $s = 3$, it has been proved recently that the Kepler conjecture holds; that is, the highest density is $\pi/\sqrt{18} = 0.740\,480\cdots$ (Hsiang, 2002; Hales, 2006; Hales and Ferguson, 2006). For dimensions $s > 3$, the densest packings have not been found, except for the *lattice* packings.[7] However, the densest possible lattice packings are available now in dimensions $s \le 8$. For example, face-centered cubic (fcc) packing is one of only two structures that maximize a local density in dimension 3 (Figure 7.9).

By the way, a problem closely related to the sphere packings is the *kissing number* problem, which asks how many balls can be arranged so that they all just touch, or 'kiss,' another ball of the same size. The kissing number is sometimes also called the *Newton number*, *contact number*, *coordination number* or *ligancy number* (Conway and Sloane, 1999). It is well known that the maximum kissing numbers λ in dimensions 2 and 3 are respectively $\lambda_2 = 6$ and $\lambda_3 = 12$. We see that the fcc packings in Figures 7.8b and 7.9 reach the maximum kissing number, but this is not the case in Figure 7.8a. Again, the maximum kissing number in high dimensions is far from easy to obtain. We only know that the maximum kissing number in dimension 4 might be 24 or 25, and in dimension 8 it is 240. No further information is available in other dimensions (Conway and Sloane, 1999).

[7] The lattice is defined in Section 7.3.2.

Figure 7.9 Patterns of sphere packings.

The centers of these nonoverlapping spheres form a point set, of course uniformly scattered over the space in a sense, with the attributes measured by the density of packing.

7.2.1.2 Covering

We now come to the dual problem of sphere packings. If the spheres overlap, then we encounter the problem of covering a space Ω by the overlapping equivalent spheres. For example, Figure 7.10 shows two patterns of sphere covering. Visually, the covering in pattern (b) is more efficient than that in the pattern (a) because the overlapping area is less. In other words, to cover a given space the number of spheres in pattern (b) is less than that in pattern (a). It is seen that, because the spheres are overlapping, in contrast to the sphere packings, the fundamental region related to each sphere is smaller than the sphere itself. We now call the ratio of the total volume of the spheres to the volume of the space the *thickness* of the covering, which is equivalent to

$$\mho = \frac{\mathcal{V}(\text{one sphere})}{\mathcal{V}(\text{fundamental region})} \tag{7.73}$$

Certainly, there is always $\mho > 1$ and the covering by the pattern with a smaller thickness is more efficient.

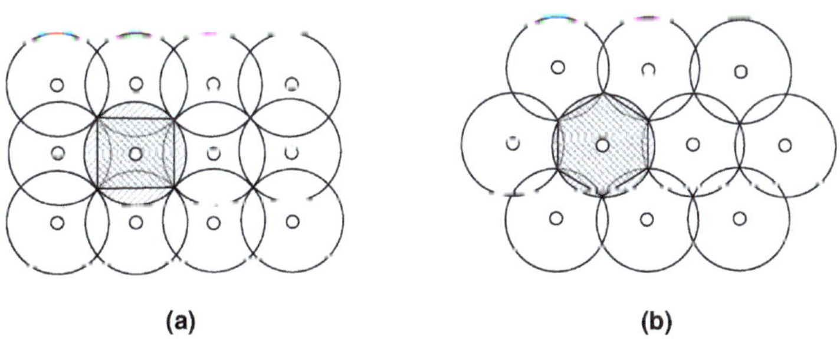

Figure 7.10 Equal spheres covering space.

For patterns (a) and (b) in Figure 7.10, the thicknesses are

$$\mho = \frac{\pi r^2}{(\sqrt{2}r)^2} = \frac{\pi}{2} = 1.570\,796\cdots \quad \text{and} \quad \mho = \frac{\pi r^2}{\frac{3\sqrt{3}}{2}r^2} = \frac{2\pi}{3\sqrt{3}} = 1.209\,199\cdots. \tag{7.74}$$

respectively. This means that the covering by pattern (b), whose thickness is smaller, is more efficient. In fact, in the two-dimensional space (plane), pattern (b) is the thinnest (Conway and Sloane, 1999).

Analogous to the packing problem, the thinnest covering is only known in dimensions 1 and 2. For lattice covering, the thinnest patterns are known in dimensions 1–5.

Like the packing problem, the centers of the covering spheres form a point set, possibly different from the packing point, but also uniformly scattered over the space in a sense.

7.2.1.3 Partition of Space

We now come to something like the inverse problem of the packing and covering. If there is a point set $\mathcal{P} = \{\boldsymbol{\theta}_q = (\theta_{1,q}, \theta_{2,q}, \ldots, \theta_{s,q});\ q = 1, 2, \ldots, n_{\text{pt}}\}$ in an s-dimensional space Ω, where n_{pt} is the number of points (cardinal number of the set), we consider the problem of using equal spheres (balls) with radius r located at these points, denoted by $\mathcal{B}_q(r, s)$, to pack or cover the space. Here:

$$\mathcal{B}_q(r, s) = \left\{ \boldsymbol{x} = (x_1, x_2, \ldots, x_s) \in \mathbb{R}^s : \|\boldsymbol{x} - \boldsymbol{\theta}_q\| = \sqrt{\sum_{j=1}^{s} (x_j - \theta_{j,q})^2} \le r \right\} \tag{7.75}$$

where $\|\cdot\|$ is the 2-norm and $\mathbb{R}^s$ is the s-dimensional real Euclidean space.

There exists a maximum value such that the packing of space by the nonoverlapping, equal spheres with this value as radius centered at the prescribed points reaches a highest density. In other words, if the radius is larger than this value, then the spheres cannot be nonoverlapping. This value is called the *packing radius*, denoted by r_{pk}; clearly:

$$r_{\text{pk}} = \frac{1}{2} \inf_{\boldsymbol{\theta}_i, \boldsymbol{\theta}_j \in P} (\|\boldsymbol{\theta}_i - \boldsymbol{\theta}_j\|) \tag{7.76}$$

Figure 7.11a schematically shows the packing radius for a given point set in a plane. It is seen that, for a given point set, equivalent spheres with packing radius located at the given points are usually not mutually tangent. Actually, it can only be ensured that at least two spheres are tangent. It is the distance between these two points that determines the packing radius via Equation 7.76.

On the other hand, there exists a value such that the covering of space by spheres with this value of radius centered at the prescribed points reaches a minimum thickness; namely, if the radius is less than this value, then the space cannot be completely covered (occupied) by the spheres centered at the prescribed points. This value is called the *covering radius*, denoted by r_{cv}, and is given by

$$r_{\text{cv}} = \sup_{\mathbf{x} \in \mathbb{R}^s} \inf_{\boldsymbol{\theta}_q \in P} (\|\boldsymbol{x} - \boldsymbol{\theta}_q\|) \tag{7.77}$$

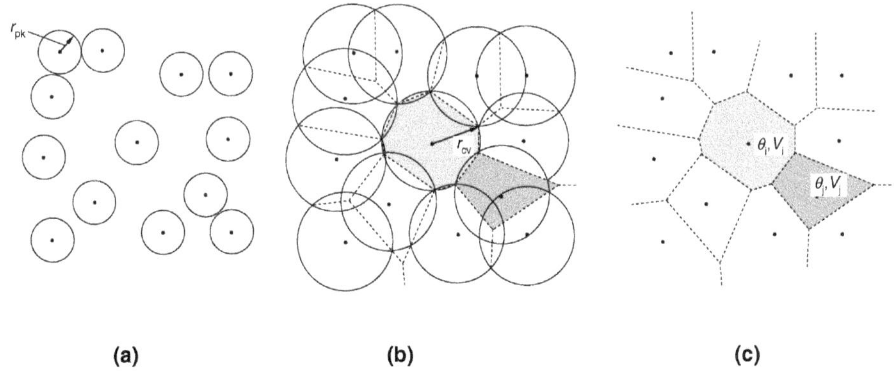

Figure 7.11 The point set and the corresponding Voronoi cells in a plane.

Figure 7.11b shows schematically the covering radius for the same set of points as in Figure 7.11a.

For example, if we regard the centers of the circles in Figures 7.8a and 7.8b and 7.10a and 7.10b as the specified points, then we have for the point sets (a) and (b)

$$r_{\mathrm{cv}} = \sqrt{2}r_{\mathrm{pk}} = 1.4142r_{\mathrm{pk}} \quad \text{and} \quad r_{\mathrm{cv}} = \frac{2}{\sqrt{3}}r_{\mathrm{pk}} = 1.1547r_{\mathrm{pk}} \tag{7.78}$$

respectively. Again, the problems in dimensions $s \geq 3$ are much more complicated.

The covering related to point set $\mathcal{P}$, as visualized in Figure 7.11b in the case $s = 2$, determines a representative region for each point to which belong all the points in the space with minimum distance to this point. The representative region so determined is the *Voronoi cell* of the point, denoted by $V(\boldsymbol{\theta}_q)$ or simply V_q, and is given by

$$V(\boldsymbol{\theta}_q) = V_q \triangleq \{\mathbf{x} \in \mathbb{R}^s : \|\mathbf{x} - \boldsymbol{\theta}_q\| \leq \|\mathbf{x} - \boldsymbol{\theta}_j\| \quad \text{for all } j, j \neq q\} \tag{7.79}$$

Other terms are *nearest-neighbor region*, *Dirichlet region*, *Brillouuin zone* and *Wigner–Seitz cell* (Barndorff-Nielsen *et al.*, 1999; Conway and Sloane, 1999; Zong, 1999).

Schematically shown in Figure 7.11c are the Voronoi cells of the points. Comparing Figure 7.11b and Figure 7.11c will immediately find the relationship between the sphere covering and the Voronoi cells. It is not difficult to understand that the packing radius and covering radius are essentially the minimum *inradius* and maximum *circumradius* of the Voronoi cells respectively.

Because all the Voronoi cells are mutually exclusive except in a zero-measure set related to the boundary surfaces, they form a complete but nonoverlapping *partition* of the space Ω; namely:

$$\bigcup_{q=1}^{n_{\mathrm{pt}}} V_q = \Omega \quad \text{and} \quad \mathcal{V}\left(V_i \bigcap V_j\right) = 0 \quad \text{for any different } i,j \tag{7.80}$$

where $\mathcal{V}(\cdot)$ is the volume measure in the s-dimensional space. If $\mathcal{V}(\Omega)$ is finite, then from Equation 7.80 we have

$$\mathcal{V}\left(\bigcup_{q=1}^{n_{\mathrm{pt}}} V_q\right) = \sum_{q=1}^{n_{\mathrm{pt}}} \mathcal{V}(V_q) = \mathcal{V}(\Omega) \quad \text{and} \quad \mathcal{V}\left(V_i \bigcap V_j\right) = 0 \quad \text{for any different } i,j \tag{7.81}$$

7.2.2 Representative Point Sets and Assigned Probabilities

7.2.2.1 Representative Points and Their Assigned Probabilities

We now come back to the solution flow of the probability density evolution method as discussed in Section 6.6.2. The point sets

$$\mathcal{P}_{\text{sel}} = \{\boldsymbol{\theta}_q = (\theta_{1,q}, \theta_{2,q}, \ldots, \theta_{s,q});\ q = 1, 2, \ldots, n_{\text{sel}}\} \tag{7.82}$$

should be first determined in the space $\Omega_{\boldsymbol{\Theta}}$. The chosen point set is termed the *representative point set*, of which each point is called a *representative point*. Here, n_{sel} is the cardinal number of the selected representative point set.

Because the representative points are scattered in a space where a probability measure is assigned, to each representative point the probability over its Voronoi cell should be assigned

$$P_q = \Pr\{\boldsymbol{\Theta} \in V_q\} = \int_{V_q} p_{\boldsymbol{\Theta}}(\boldsymbol{\theta})\mathrm{d}\boldsymbol{\theta} \qquad q = 1, 2, \ldots, n_{\text{sel}} \tag{7.83}$$

which is called the *assigned probability* of $\boldsymbol{\theta}_q$. Here, $p_{\boldsymbol{\Theta}}(\boldsymbol{\theta})$ is the joint density of the random parameters $\boldsymbol{\Theta} = (\Theta_1, \Theta_2, \ldots, \Theta_s)$. By doing this, the joint density $p_{\boldsymbol{\Theta}}(\boldsymbol{\theta})$ has in fact been discretized through replacing it by (see Appendix A)

$$\tilde{p}_{\boldsymbol{\Theta}}(\boldsymbol{\theta}) = \sum_{q=1}^{n_{\text{sel}}} [P_q \delta(\boldsymbol{\theta} - \boldsymbol{\theta}_q)] = \sum_{q=1}^{n_{\text{sel}}} \left[P_q \prod_{j=1}^{s} \delta(\boldsymbol{\theta} - \boldsymbol{\theta}_{j,q}) \right] \tag{7.84}$$

Clearly, we have

$$\lim_{r_{\text{cv}} \to 0} \tilde{p}_{\boldsymbol{\Theta}}(\boldsymbol{\theta}) = p_{\boldsymbol{\Theta}}(\boldsymbol{\theta}) \tag{7.85}$$

where r_{cv} is the covering radius of the point set $\mathcal{P}_{\text{sel}}$ and is also the maximum circumradius of the Voronoi cells. In addition, in consideration of Equation 7.81, it follows that

$$\int_{\Omega_{\boldsymbol{\Theta}}} p_{\boldsymbol{\Theta}}(\boldsymbol{\theta})\mathrm{d}\boldsymbol{\theta} = \int_{\Omega_{\boldsymbol{\Theta}}} \tilde{p}_{\boldsymbol{\Theta}}(\boldsymbol{\theta})\mathrm{d}\boldsymbol{\theta} = \sum_{q=1}^{n_{\text{sel}}} P_q = \sum_{q=1}^{n_{\text{sel}}} \int_{V_q} p_{\boldsymbol{\Theta}}(\boldsymbol{\theta})\mathrm{d}\boldsymbol{\theta} = \int_{\bigcup_{q=1}^{n_{\text{sel}}} V_q} p_{\boldsymbol{\Theta}}(\boldsymbol{\theta})\mathrm{d}\boldsymbol{\theta} = 1 \tag{7.86}$$

For visual convenience, we illustrate the assigned probabilities when only one random parameter is involved with the PDF $p_\Theta(\theta)$ shown in Figure 7.12a. Denote the representative point set $\mathcal{P}_{\text{sel}} = \{\theta_1, \theta_1, \ldots, \theta_{n_{\text{sel}}}\}$. If the Voronoi cell of the point θ_q is the interval $V_q = [\underline{\theta}_q, \bar{\theta}_q]$ (noting that the interval may be different for different q), then the assigned probability of the point θ_q is given by

$$P_q = \int_{\underline{\theta}_q}^{\bar{\theta}_q} p_\Theta(\theta)\mathrm{d}\theta \tag{7.87a}$$

Thus, the original PDF $p_\Theta(\theta)$ is discretized to

$$\tilde{p}_\Theta(\theta) = \sum_{q=1}^{n_{\text{sel}}} P_q \delta(\theta - \theta_q) \tag{7.87b}$$

as shown in Figure 7.12b.

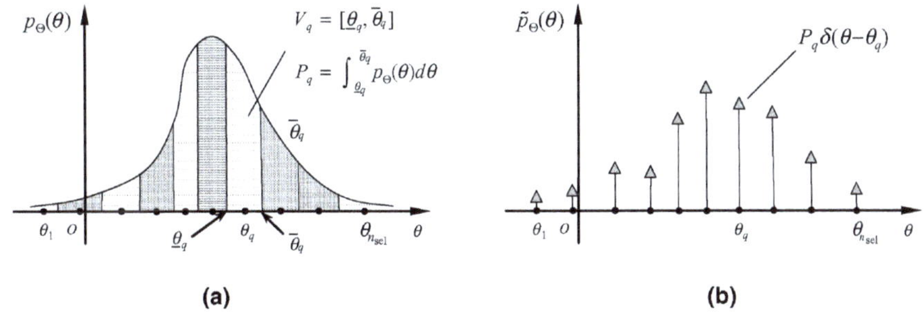

Figure 7.12 Assigned probabilities.

7.2.2.2 Discrepancy and F-Discrepancy of Point Sets

In the solution procedure of the probability density evolution equation, determination of the representative point set $\mathcal{P}_{\text{sel}}$ is undoubtedly of paramount importance. Recalling Equation 6.100, we see that in the mathematical form the right-hand side is a multidimensional integral with respect to $\boldsymbol{\Theta}$ in which the joint PDF $p_{\mathbf{Z}\boldsymbol{\Theta}}(\mathbf{z}, \boldsymbol{\theta}, t)$ is the integrand. If the information of the integrand is well or partly known, then a variety of approaches have been developed to reduce the dimensionality (He, 2001) or to select reasonable point sets, for example, in the numerical multiple integral (Engels, 1980; Genz, 1986; Sobolev and Vaskevich, 1997; Xu, 1998). However, in many cases little is known about the information of the integrand, which is the case in most problems of practical interest, because the closed form or even the qualitative features of the integrand may depend on closed-form solutions of the complex nonlinear system which are usually unfeasible. In this case, intuitively, the representative point is better scattered uniformly in a sense. For instance, choose the point set with a fixed cardinal number n_{sel} that makes

(a) the packing radius maximized; or
(b) the covering radius minimized; or
(c) some other indices minimized.

Investigations show that these criteria are usually not equivalent and will result in different point sets. Some of the point sets will be generated by these criteria in the following sections.

In order to do so, we introduce an additional family of indices here, named discrepancies, which are also usually employed in measuring the uniformity of a point set.

Without loss of generality, we consider the point sets over a unit hypercube

$$C^s = [0, 1]^s = \{\mathbf{x} = (x_1, x_2, \ldots, x_s) : x_j \in [0, 1] \text{ for all } j = 1, 2, \ldots, s\}$$

in an s-dimensional space. Denote a point set $\mathcal{P} = \{\boldsymbol{x}_k = (x_{1,k}, x_{2,k}, \ldots, x_{s,k}) \; k = 1, 2, \ldots, n\}$. If $\mathcal{P} \subset C^s$, then the *discrepancy* of $\mathcal{P}$ is defined by (Hua and Wang, 1981)

$$\mathcal{D}(n, \mathcal{P}) = \sup_{\boldsymbol{\nu} \in C^s} \left| \frac{N(\boldsymbol{\nu}, \mathcal{P})}{n} - \mathcal{V}([\mathbf{0}, \boldsymbol{\nu}]) \right| \tag{7.88}$$

where $\boldsymbol{\nu} = (\nu_1, \nu_2, \ldots, \nu_s) \in C^s, 0 \le \nu_i \le 1, i = 1, 2, \ldots, s, N(\boldsymbol{\nu}, \mathcal{P})$ is the number of the points satisfying $\mathbf{x}_k \le \boldsymbol{\nu}$ and $\mathcal{V}([0, \boldsymbol{\nu}])$ is the volume of the hyper-rectangle $[0, \boldsymbol{\nu}] = \prod_{j=1}^{s} [0, \nu_j]$ and given by $\mathcal{V}([0, \boldsymbol{\nu}]) = \nu_1 \nu_2 \cdots \nu_s$.

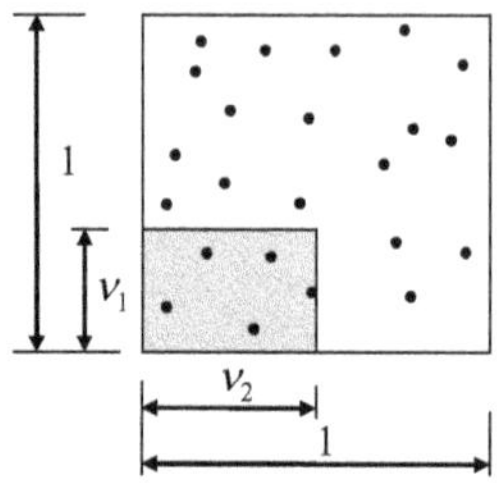

Figure 7.13 Discrepancy of a point set.

It is seen from Figure 7.13 in dimension $s = 2$ that the discrepancy defined above is the maximum error when replacing the ratio of the areas by the ratio of the number of points contained in the areas. Visually, if the points are scattered uniformly, the discrepancy will be small. An important theorem related to this discrepancy makes it quite valuable (Hua and Wang, 1981): *if $f(\boldsymbol{x})$ is a function of bounded variation in the sense of Hardy and Krause,*[8] *then*

$$\left| \int_{C^s} f(\boldsymbol{x}) \mathrm{d}\boldsymbol{x} - \frac{1}{n} \sum_{k=1}^{n} f(\boldsymbol{x}_k) \right| \le \mathrm{TV}(f) \mathcal{D}(n, \mathcal{P}) \tag{7.89}$$

where TV(f) *is the total variation of the function* f. TV(f) in dimension 1 is the same as defined in Equation 7.59. This means that the discrepancy $\mathcal{D}(n, \mathcal{P})$ bounds the error of the multi-integral.

According to the definition in Equation 7.88, the discrepancy of the uniform grid point (UGP) set (Figure 7.14a)

$$\mathcal{P}_{\mathrm{UGP}} = \left\{ \left(\frac{2\ell_1 - 1}{m}, \frac{2\ell_2 - 1}{m}, \ldots, \frac{2\ell_s - 1}{m} \right), \quad 1 \le \ell_j \le m, j = 1, 2, \ldots, s \right\} \tag{7.90}$$

satisfies

$$c_1(s) n^{-1/s} \le \mathcal{D}(n, \mathcal{P}_{\mathrm{UGP}}) \le c_2(s) n^{-1/s} \quad \text{or} \quad \mathcal{D}(n, \mathcal{P}_{\mathrm{UGP}}) = \mathcal{O}(n^{-1/s}) \tag{7.91}$$

Here, c_1 and c_2 are two constants dependent on s but not on n.

While for the Monte Carlo-sampled (MCS) points $\mathcal{P}_{\mathrm{MCS}}$ (Figure 7.14b), the discrepancy

$$\mathcal{D}(n, \mathcal{P}_{\mathrm{MCS}}) = \mathcal{O}(n^{-1/2} (\mathrm{loglog}n)^{1/2}) \tag{7.92}$$

with unity probability.

[8] The variation of a function in the sense of Hardy and Krause is to measure the irregularity and smoothness of a function. If the function is too irregular, then the variation in the sense of Hardy and Krause is usually large. For its exact definition, refer to Hua and Wang (1981) and Niederreiter (1992) for example.

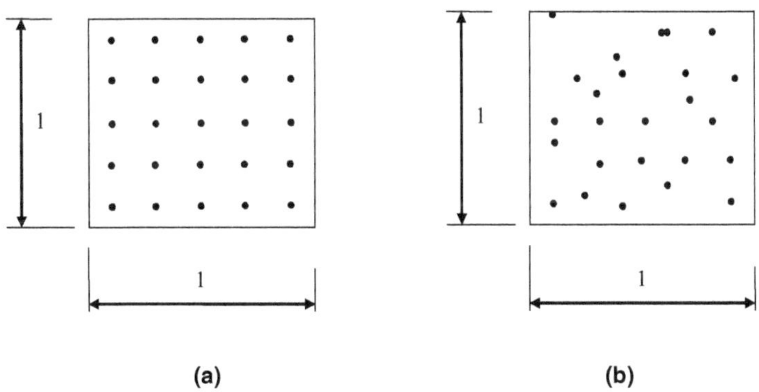

Figure 7.14 Two typical point sets.

It is seen that $\mathcal{D}(n, \mathcal{P}_{\text{MCS}}) < \mathcal{D}(n, \mathcal{P}_{\text{UGP}})$ in dimensions $s \geq 3$ with unity probability. This accounts for why the MCS set is more efficient than the UGP set for numerical multi-integrals.

However, in $\mathcal{D}(n, \mathcal{P})$, the probability density assigned to the space over which the point set $\mathcal{P}$ scatters is not considered. To take into account this information, an *F-discrepancy* can be defined by (Fang and Wang, 1994)

$$\mathcal{D}_{\mathcal{F}}(n, \mathcal{P}) = \sup_{\boldsymbol{x} \in \mathbb{R}^s} |\mathcal{F}_n(\boldsymbol{x}) - \mathcal{F}(\boldsymbol{x})| \tag{7.93}$$

where $\mathcal{F}(\mathbf{x})$ is the CDF and $\mathcal{F}_n(\mathbf{x})$ is the empirical distribution function given by

$$\mathcal{F}_n(\mathbf{x}) = \frac{1}{n} \sum_{q-1}^{n} I\{\mathbf{x}_q \leq \mathbf{x}\} \tag{7.94}$$

where $I\{\cdot\}$ is the indicator function whose value is one if the event is true and zero otherwise. The F-discrepancy is essentially the Kolmogorov–Smirnov statistic for the goodness-of-fit (Robinstein, 1981); see Figure 7.15 for $s = 1$.

It is easy to see that $\mathcal{D}_{\mathcal{F}}(n, \mathcal{P})$ defined by Equation 7.93 becomes $\mathcal{D}(n, \mathcal{P})$ defined by Equation 7.88 if the probability distribution is a uniform distribution over $\mathcal{C}^s$.

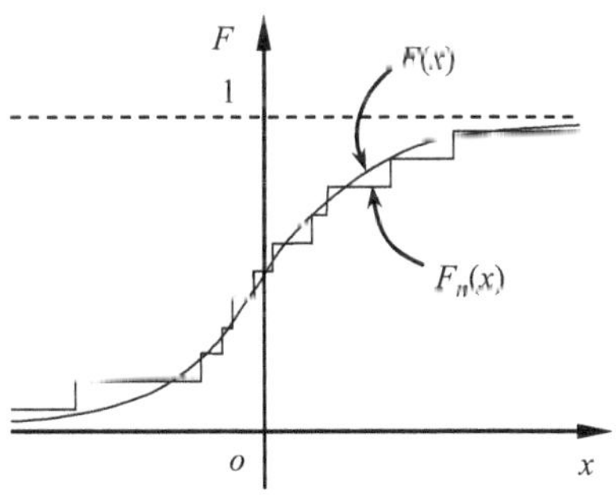

Figure 7.15 Schematic picture of F-discrepancy.

Equation 7.94 essentially possesses an acquiescence that all of the sampled points have the same weights. This is not the case, because each point $\mathbf{x}_q$ is associated with an assigned probability P_q as given by Equation 7.83; therefore, it is reasonable to modify Equation 7.94 to

$$\mathcal{F}_n(\mathbf{x}) = \sum_{q=1}^{n} P_q \cdot I\{\mathbf{x}_q \le \mathbf{x}\} \tag{7.95}$$

The F-discrepancy so obtained is called the *modified F-discrepancy* or *true F-discrepancy.*

Clearly, the discrepancy $\mathcal{D}(n, \mathcal{P})$, F-discrepancy $\mathcal{D}_{\mathcal{F}}(n, \mathcal{P})$ and the modified F-discrepancy can also be used as the indices to be minimized in the above-mentioned criterion (c).

7.2.3 First- and Second-Order Discrepancies of Point Sets

From Section 6.6.2, the PDF $p_{\mathbf{Z}}(\mathbf{z}, t)$ obtained by Equation 6.100 is approximated by Equation 6.159 (note that the latter is the case in dimension 1). The error in approximating the PDF, ignoring that in the finite-difference method, reads

$$e(\mathbf{z}, t) = \left| \int_{\Omega} p_{\mathbf{Z\Theta}}(\mathbf{z}, \boldsymbol{\theta}, t) p_{\boldsymbol{\Theta}}(\boldsymbol{\theta}) \mathrm{d}\boldsymbol{\theta} - \sum_{q=1}^{n_{\mathrm{sel}}} [p_{\mathbf{Z\Theta}}(\mathbf{z}, \boldsymbol{\theta}_q, t) P_q] \right| \tag{7.96}$$

It should be noted here that, to make Equations 6.100 and 6.159 consistent, $p_{\mathbf{Z\Theta}}(\mathbf{z}, \boldsymbol{\theta}, t)$ here is not the same as that in Equation 6.159 but equivalent to $p_{\mathbf{Z\Theta}}(\mathbf{z}, \boldsymbol{\theta}, t)/p_{\boldsymbol{\Theta}}(\boldsymbol{\theta})$.[9]

It follows from Equations 7.83 and 7.86 that

$$\begin{aligned} e(\mathbf{z}, t) &= \left| \sum_{q=1}^{n_{\mathrm{sel}}} \int_{V_q} p_{\mathbf{Z\Theta}}(\mathbf{z}, \boldsymbol{\theta}, t) p_{\boldsymbol{\Theta}}(\boldsymbol{\theta}) \mathrm{d}\boldsymbol{\theta} - \sum_{q=1}^{n_{\mathrm{sel}}} \left[p_{\mathbf{Z\Theta}}(\mathbf{z}, \boldsymbol{\theta}_q, t) \int_{V_q} p_{\boldsymbol{\Theta}}(\boldsymbol{\theta}) \mathrm{d}\boldsymbol{\theta} \right] \right| \\ &= \left| \sum_{q=1}^{n_{\mathrm{sel}}} \int_{V_q} [p_{\mathbf{Z\Theta}}(\mathbf{z}, \boldsymbol{\theta}, t) - p_{\mathbf{Z\Theta}}(\mathbf{z}, \boldsymbol{\theta}_q, t)] p_{\boldsymbol{\Theta}}(\boldsymbol{\theta}) \mathrm{d}\boldsymbol{\theta} \right| \end{aligned} \tag{7.97}$$

Using the Taylor expansion and retaining the second-order terms, we have

$$p_{\mathbf{Z\Theta}}(\mathbf{z}, \boldsymbol{\theta}, t) - p_{\mathbf{Z\Theta}}(\mathbf{z}, \boldsymbol{\theta}_q, t) \cong \sum_{i=1}^{s} \psi_{i,q}(p_{\mathbf{Z\Theta}})(\theta_i - \hat{\theta}_{i,q}) + \frac{1}{2} \sum_{j=1}^{s} \sum_{i=1}^{s} \Psi_{ij,q}(p_{\mathbf{Z\Theta}})(\theta_i - \hat{\theta}_{i,q})(\theta_j - \hat{\theta}_{j,q}) \tag{7.98}$$

where

$$\psi_{i,q}(p_{\mathbf{Z\Theta}}) = \left. \frac{\partial p_{\mathbf{Z\Theta}}(\mathbf{z}, \boldsymbol{\theta}, t)}{\partial \theta_i} \right|_{\boldsymbol{\theta} = \boldsymbol{\theta}_q} \qquad \Psi_{ij,q}(p_{\mathbf{Z\Theta}}) = \left. \frac{\partial^2 p_{\mathbf{Z}\Theta}(\mathbf{z}, \boldsymbol{\theta}, t)}{\partial \theta_i \partial \theta_j} \right|_{\boldsymbol{\theta} = \boldsymbol{\theta}_q} \tag{7.99}$$

[9] It is noted that the generalized density evolution Equation 6.109 is invariant if the initial condition (Equation 6.121a) is changed to $p_{\mathbf{Z\Theta}}(\mathbf{z}, \boldsymbol{\theta}, \mathbf{t}_0) = \delta(\mathbf{z} - \mathbf{z}_0)$ and simultaneously Equation 6.100 is changed to $p_{\mathbf{Z}}(\mathbf{z}, t) = \int_{\Omega_{\boldsymbol{\Theta}}} p_{\mathbf{Z\Theta}}(z, \boldsymbol{\theta}, t) p_{\boldsymbol{\Theta}}(\boldsymbol{\theta}) \mathrm{d}\boldsymbol{\theta}$. The expression in Equation 7.96 adopts the present treatment.

are respectively the first- and second-order sensitivities of the function $p_{\mathbf{Z}\Theta}(\mathbf{z}, \boldsymbol{\theta}, t)$ in terms of $\boldsymbol{\theta}$ and $\hat{\theta}_{i,q}$ are the coordinates of the representative point set, where the hat is added to avoid notational confusion.

Taking Equation 7.98 into Equation 7.97 yields

$$e(\mathbf{z},t) \cong \left| \sum_{q=1}^{n_{\text{sel}}} \int_{V_q} \left[\sum_{i=1}^{s} \psi_{i,q}(p_{\mathbf{Z}\boldsymbol{\Theta}})(\theta_i - \hat{\theta}_{i,q}) + \frac{1}{2} \sum_{j=1}^{s} \sum_{i=1}^{s} \Psi_{ij,q}(p_{\mathbf{Z}\boldsymbol{\Theta}})(\theta_i - \hat{\theta}_{i,q})(\theta_j - \hat{\theta}_{j,q}) \right] p_{\boldsymbol{\Theta}}(\boldsymbol{\theta}) \mathrm{d}\boldsymbol{\theta} \right| \tag{7.100}$$

Further:

$$\begin{aligned} e(\mathbf{Z}, t) &\leq \left| (\max_{i,q} |\psi_{i,q}(p_{\mathbf{Z}\boldsymbol{\Theta}})|) \sum_{q=1}^{n_{\text{sel}}} \int_{V_q} \sum_{i=1}^{s} |\theta_i - \hat{\theta}_{i,q}| p_{\boldsymbol{\Theta}}(\boldsymbol{\theta}) \mathrm{d}\boldsymbol{\theta} \right| \\ &\quad + \frac{1}{2} (\max_{i,j,q} |\Psi_{ij,q}(p_{\mathbf{Z}\boldsymbol{\Theta}})|) \sum_{q=1}^{n_{\text{sel}}} \int_{V_q} \sum_{j=1}^{s} \sum_{i=1}^{s} |(\theta_i - \hat{\theta}_{i,q})(\theta_j - \hat{\theta}_{j,q})| p_{\boldsymbol{\Theta}}(\boldsymbol{\theta}) \mathrm{d}\boldsymbol{\theta} \\ &= \phi_1(p_{\mathbf{Z}\boldsymbol{\Theta}}) \mathcal{D}_1(\mathcal{P}_{\text{sel}}) + \phi_2(p_{\mathbf{Z}\boldsymbol{\Theta}}) \mathcal{D}_2(\mathcal{P}_{\text{sel}}) \end{aligned} \tag{7.101}$$

where $\phi_1(p_{\mathbf{Z}\Theta})$ and $\phi_2(p_{\mathbf{Z}\Theta})$ are the functionals as the maxima of absolute value of the first- and second-order sensitivities of the function $p_{\mathbf{Z}\boldsymbol{\Theta}}(\cdot)$:

$$\phi_1(p_{\mathbf{Z}\boldsymbol{\Theta}}) = \max_{i,q} |\psi_{i,q}(p_{\mathbf{Z}\boldsymbol{\Theta}})| \qquad \phi_2(p_{\mathbf{Z}\boldsymbol{\Theta}}) = \frac{1}{2} \max_{i,j,q} |\Psi_{ij,q}(p_{\mathbf{Z}\boldsymbol{\Theta}})| \tag{7.102}$$

and $\mathcal{D}_1(\mathcal{P}_{\text{sel}})$ and $\mathcal{D}_2(\mathcal{P}_{\text{sel}})$ are the measures of discrepancy of the point set $\mathcal{P}_{\text{sel}}$ defined by

$$\mathcal{D}_1(\mathcal{P}_{\text{sel}}) = \sum_{q=1}^{n_{\text{sel}}} \int_{V_q} \sum_{i=1}^{s} |\theta_i - \hat{\theta}_{i,q}| p_{\boldsymbol{\Theta}}(\boldsymbol{\theta}) \mathrm{d}\boldsymbol{\theta} \tag{7.103}$$

and

$$\mathcal{D}_2(\mathcal{P}_{\text{sel}}) = \sum_{q=1}^{n_{\text{sel}}} \int_{V_q} \sum_{j=1}^{s} \sum_{i=1}^{s} |(\theta_i - \hat{\theta}_{i,q})(\theta_j - \hat{\theta}_{j,q})| p_{\boldsymbol{\Theta}}(\boldsymbol{\theta}) \mathrm{d}\boldsymbol{\theta} \tag{7.104}$$

which might as well be called the *first*- and the *second-order discrepancy* respectively.

Equation 7.101 means that the error of the numerical algorithm depends on the configuration of the point set (measured by $\mathcal{D}_1(\mathcal{P}_{\text{sel}})$ and $\mathcal{D}_2(\mathcal{P}_{\text{sel}})$) and the sensitivities with respect to the parameters (measured by $\phi_1(p_{\mathbf{Z}\boldsymbol{\Theta}})$ and $\phi_2(p_{\mathbf{Z}\boldsymbol{\Theta}})$). Therefore, a good algorithm should, in principle, consider these two factors.

7.2.4 Two-Step Procedure of Constructing Representative Points

According to the above analysis, to improve the accuracy in approximating the PDF, a point set $\mathcal{P}_{\text{sel}}$ should make the modified F-discrepancy $\mathcal{D}_{\mathcal{F}}(n, \mathcal{P})$ and the first- and second-order discrepancies $\mathcal{D}_1$ and $\mathcal{D}_2$ as small as possible. Simultaneously, the sensitivities $\phi_1(p_{\mathbf{Z}\boldsymbol{\Theta}})$ and $\phi_2(p_{\mathbf{Z}\boldsymbol{\Theta}})$ of $p_{\mathbf{Z}\boldsymbol{\Theta}}(\mathbf{z}, \boldsymbol{\theta}, t)$ should be taken into account.

Generally, in the region where the sensitivities of $p_{\mathbf{Z}\boldsymbol{\Theta}}(\mathbf{z}, \boldsymbol{\theta}, t)$ in terms of $\boldsymbol{\theta}$ are larger, the points should be denser. Unfortunately, because $p_{\mathbf{Z}\boldsymbol{\Theta}}(\mathbf{z}, \boldsymbol{\theta}, t)$ is an unknown function to be obtained through the generalized density evolution equation, it is usually hard to get information on its sensitivities in terms of $\boldsymbol{\theta}$, particularly for nonlinear systems. This makes it reasonable to have the points scattered uniformly; namely, to make the discrepancy $\mathcal{D}(n, \mathcal{P})$ as small as possible for the given n. Point sets so obtained, however, cannot guarantee the minimization of the modified F-discrepancy $\mathcal{D}_{\mathcal{F}}(n, \mathcal{P})$ and the first- and second-order discrepancies $\mathcal{D}_1(\mathcal{P})$ and $\mathcal{D}_2(\mathcal{P})$. Actually, the point set with small $\mathcal{D}(n, \mathcal{P})$ may have a very large modified F-discrepancy $\mathcal{D}_{\mathcal{F}}(n, \mathcal{P})$ if the joint probability density of the parameters is not uniform, as will be seen in Section 7.4. To make $\mathcal{D}_{\mathcal{F}}(n, \mathcal{P})$, $\mathcal{D}_1(\mathcal{P})$ and $\mathcal{D}_2(\mathcal{P})$ as small as possible, a density-related transformation on the uniformly scattered point sets obtained can be imposed, to adjust the density of points partly according to the density of the parameters.

According to these considerations, we may employ a two-step procedure of constructing representative point sets (Chen *et al.*, 2009):

i. construct a uniformly scattered point set as a basic point set, denoted by $\mathcal{P}_{\text{Basic}}$;
ii. perform the density-related transformation on $\mathcal{P}_{\text{Basic}}$ to yield the representative point set $\mathcal{P}_{\text{sel}}$.

In the next two sections we will elaborate the approaches in these two steps.

7.3 Strategy for Generating Basic Point Sets

To determine point sets uniformly scattered over a given space has long received the attention of mathematicians, physicists and chemists. As mentioned, the sphere packings and covering will lead to uniform point sets in a space (Conway and Sloane, 1999). In addition, the number theoretical method can also generate uniform point sets (Hua and Wang, 1978, 1981). All these are deterministic point sets. The MCS points (including their improvement, the Latin hypercube sampled points), on the other hand, will generate random point sets uniformly, usually with higher discrepancy (Robinstein, 1981; Fang and Wang, 1994). This section will discuss deterministic point sets by tangent spheres, lattices and the number theoretical method.

7.3.1 From Sphere Packings: Tangent Sphere Method

7.3.1.1 Construction of Point Sets by Tangent Sphere Method

The sphere packings problem provides a possible way to construct uniformly scattered point sets as the basic or representative points. Considering first the case for $s = 2$. In a plane, it is well known that the kissing number of a circle is six. In this case, the centers of the kissing circles form the vertices of a regular hexagon (shown in Figure 7.16a) and a total of seven circles forms a fixed-shape substructure. Therefore, the pattern of packing generated by using this substructure as the basis structures will have a tight packing with the highest density. This is essentially the pattern in Figure 7.8b. However, using the generating process as shown in Figure 7.16a will make it easy to locate and number the centers in computer programs. Actually, we see that the circles form different loops (layers) outward, and in each loop we numerate the circles anticlockwise. By doing this, we can get the polar coordinates (r_i, φ_i) of the

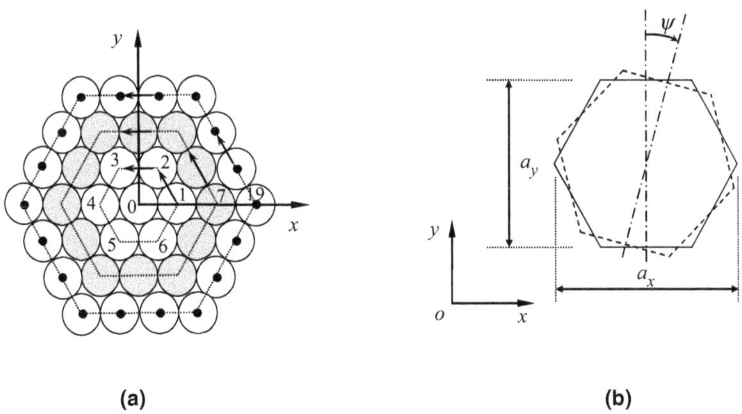

Figure 7.16 Tangent circles in a plane.

center of the ith circle:

$$i = (3\ell^2 - 3\ell + 1) + \ell j + k \qquad j = 0, 1, \ldots, 5; \quad k = 0, 1, \ldots, \ell - 1 \tag{7.105a}$$

$$r_i = r\sqrt{(2\ell - k)^2 + (\sqrt{3}k)^2} \qquad j = 0, 1, \ldots, 5; \quad k = 0, 1, \ldots, \ell - 1 \tag{7.105b}$$

$$\varphi_i = j\frac{\pi}{3} + \tan^{-1}\left(\frac{\sqrt{3}k}{2\ell - k}\right) \qquad j = 0, 1, \ldots, 5; \quad k = 0, 1, \ldots, \ell - 1, \tag{7.105c}$$

where $\ell = 0, 1, \ldots, L$ is the order number of the loops with the origin being labeled 0 and r is the radius of the tangent spheres.

It follows from Equation 7.105a that the total number of circles is

$$n_{\text{total}} = 3(L+1)^2 - 3(L+1) + 1 = 3L^2 + 3L + 1 \tag{7.106}$$

It is noted, however, that the numbers of projections of centers to the x and y coordinate axes are different; thus, the information reflected by the points is not equivalent in different coordinates. This is because the projections of the regular hexagon, on which the centers of the circles in the same loop are located, to the x and y coordinate axes, $a_x = 2a$ and $a_y = \sqrt{3}a$ (where a is the edge length), are not equal. Rotating the hexagon by an appropriate angle ψ to make the modified projections a'_x and a'_y equal gives $\psi = \pi/12$ (Figure 7.16b). Equation 7.105c, therefore, should be modified to

$$\varphi_i = \frac{\pi}{3}\left(j - \frac{1}{4}\right) + \tan^{-1}\left(\frac{\sqrt{3}k}{2\ell - k}\right) \qquad j = 0, 1, \ldots, 5; \; k = 0, 1, \ldots, \ell - 1. \tag{7.107}$$

No modification is needed in Equation 7.105b.

The Cartesian coordinates of the points are then given by

$$\begin{cases} x_i = r_i \cos\varphi_i \\ y_i = r_i \sin\varphi_i \end{cases} \qquad i = 0, 1, 2, \ldots \tag{7.108}$$

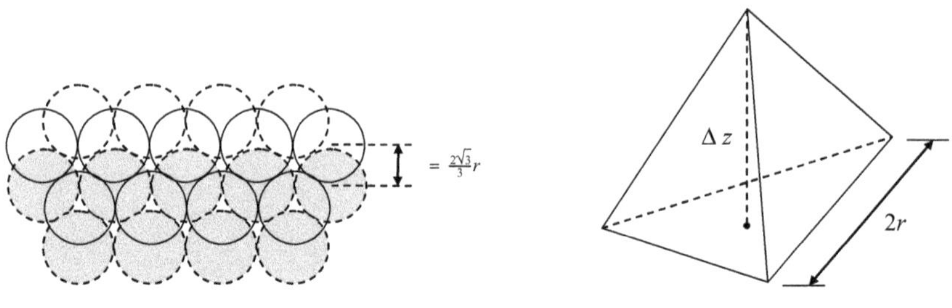

(a) (b)

Figure 7.17 Pattern of tangent spheres in three-dimensional space.

As mentioned above, in the three-dimensional space ($s = 3$), the highest kissing number of the sphere is 12. The pattern of packing with kissing number 12 can be constructed using the pattern in the two-dimensional space discussed above as the pattern in one layer (Figure 7.17a; also see Figure 7.9). It is seen that the spheres in one layer (see the white solid circles as projections of the spheres in the same layer to the plane) can be regarded as a translation of the spheres (see the dashed, shaded circles) in the adjacent layers, the translation being $\Delta x = 0$, $\Delta y = 2\sqrt{3}r/3$ before rotation in Equation 7.107. After rotation, we have

$$\Delta x = \frac{2\sqrt{3}r}{3}\sin\psi \qquad \Delta y = \frac{2\sqrt{3}r}{3}\cos\psi \tag{7.109}$$

Simultaneously, the centers of one sphere in the upper layer and the three kissing spheres in the lower layer will form the vertices of a regular tetrahedron with edge length $2r$ (Figure 7.17b); thus, the distance between the two layers is the height Δz of the regular tetrahedron and given by

$$\Delta z = \frac{2\sqrt{6}r}{3} \tag{7.110}$$

For convenience, we get the coordinate in the z-direction of different layers in a symmetric way

$$\tilde{z}_k = k\Delta z \qquad k = 0, \pm 1, \pm 2, \ldots \tag{7.111}$$

Denoting the coordinates of the centers located in the kth layer $z = \tilde{z}_k$ as $(\tilde{x}_{i,k}, \tilde{y}_{i,k}, \tilde{z}_k)$, then we can get

$$\begin{cases} \tilde{x}_{i,k} = x_{i,0} + \dfrac{1}{2}\left[1 - (-1)^k\right]\Delta x \\ \tilde{y}_{i,k} = y_{i,0} + \dfrac{1}{2}\left[1 - (-1)^k\right]\Delta y \end{cases} \qquad (k = 0, \pm 1, \ldots, \pm N_z) \tag{7.112}$$

where $x_{i,0}$, $y_{i,0}$ are the Cartesian coordinates of the centers located in the plane $z = 0$ and given by Equation 7.108 and N_z is the total number of layers.

Renumber the points and change the subscript indices (i, k) to a single index j and denote the coordinates of the points by $(\bar{x}_j, \bar{y}_j, \bar{z}_j)$. Similar to the situation for $s=2$, it is necessary to perform a rotation here with x as the rotational axis and finally the points are given by

$$\begin{cases} x_j = \bar{x}_j \\ y_j = \bar{y}_j \cos\psi_x + \bar{z}_j \sin\psi_x \qquad (j = 1, 2, \ldots) \\ z_j = \bar{z}_j \cos\psi_x - \bar{y}_j \sin\psi_x \end{cases} \tag{7.113}$$

where ψ_x are the rotational angles circling the x-axis. Considering the symmetry of x, y and z in the three-dimensional space, one can reasonably take $\psi_x = \psi = \pi/12$.

Likewise, the construction for $s=4$ can also be made. But it is more direct now to generate the points in the four-dimensional space by a generator matrix. For details, refer to Chen and Li (2008).

7.3.1.2 Discrepancy and Projection Ratio of Tangent Sphere Points

Denote the point sets obtained through the tangent sphere method by $\mathcal{P}_{\mathrm{TaS}}$.

Investigations show that for $s=2$ the discrepancy of $\mathcal{P}_{\mathrm{TaS}}$ defined by Equation 7.88 is (Chen and Li, 2008)

$$\mathcal{D}(n, \mathcal{P}_{\mathrm{TaS}}) = c\mathcal{O}(n^{-1+\varepsilon}) \tag{7.114a}$$

where $c = (16\sqrt{3} - 3)/12 = 2.06$ is a constant and $\varepsilon = o(1)$. In dimensions $s=3$ and 4, fitting the computational results gives respectively

$$\mathcal{D}(n, \mathcal{P}_{\mathrm{TaS}}) = c\mathcal{O}(n^{-(3/4)+\varepsilon}) \quad \text{and} \quad \mathcal{D}(n, \mathcal{P}_{\mathrm{TaS}}) = c\mathcal{O}(n^{-(2/3)+\varepsilon}) \tag{7.114b}$$

The constant c is different for different s. Uniformly, Equations 7.114a and 7.114b can be rewritten as

$$\mathcal{D}(n, \mathcal{P}_{\mathrm{TaS}}) = \mathcal{O}(n^{-(1/2)-\{1/[2(s-1)]\}+\varepsilon}) \tag{7.115}$$

We will compare them later with the number theoretical method.

Another index attached to the point sets which is related to the symmetry in different coordinate directions is the projection ratio. For a point $\mathcal{P} = \{\mathbf{x}_k = (x_{1,k}, x_{2,k}, \ldots, x_{s,k}),\ k = 1, 2, \ldots, n$, we define the *projection ratio* by

$$\eta_j = \frac{N(\mathrm{Proj}(\mathcal{P}, j))}{n} \qquad j = 1, 2, \ldots, s \tag{7.116}$$

Here, $N(\mathrm{Proj}(\mathcal{P}, j))$ stands for the number of the projections of the point set to the coordinate x_j (Figure 7.18). Clearly, for any arbitrary point set there exists

$$n^{-1} \leq \eta_j \leq 1 \tag{7.117a}$$

The projection ratio reflects the marginal information that is contained in a point set. For example, for the uniform grid point sets $\mathcal{P}_{\mathrm{UGP}}$ in Equation 7.90

$$\eta_j = n^{-1/s} \qquad j = 1, 2, \ldots, s \tag{7.117b}$$

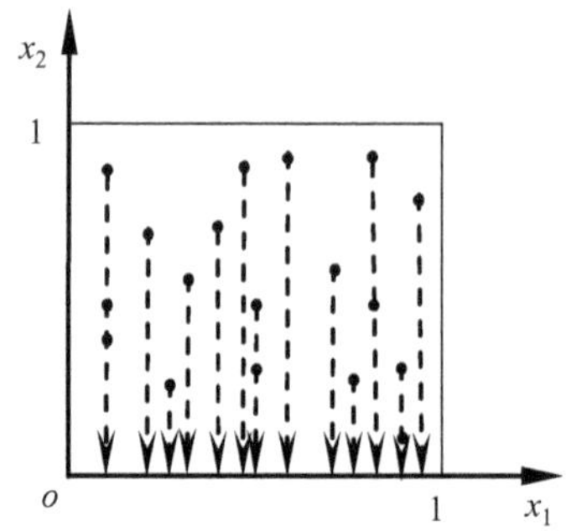

Figure 7.18 Schematic projection ratio. A total of 18 points are scattered over the square [0, 1]^2. The number of points projected to the x_1-axis is 13 because some points have identical x_1-coordinate values. Thus, in this figure, $n = 18$, $N(\text{Proj}(\mathcal{P}, 1)) = 13$ and $\eta_1 = 13/18$.

while for the MCS point sets $\mathcal{P}_{\text{MCS}}$

$$\eta_j = 1 \qquad j = 1, 2, \ldots, s \tag{7.117c}$$

with unity probability.

For the tangent sphere points before rotation in dimension 2 (see Figure 7.16a), computations result in (Chen and Li, 2008)

$$\eta_x = \sqrt[4]{12}\mathcal{O}(n^{-(1/2)+\varepsilon}) = 1.86\mathcal{O}(n^{-(1/2)+\varepsilon}) \quad \eta_y = \sqrt[4]{\frac{4}{3}}\mathcal{O}(n^{-(1/2)+\varepsilon}) = 1.07\mathcal{O}(n^{-(1/2)+\varepsilon}) \tag{7.118}$$

so that $\eta_x = \sqrt{3}\eta_y$. The projection ratios of the points after rotation are given by

$$\eta_x = \eta_y \geq 4.65\mathcal{O}(n^{-(1/2)+\varepsilon}) \tag{7.119}$$

This indicates that the rotation can change and improve the projection ratio greatly. Actually, in dimensions 3 and 4, the projection ratios can be improved from the magnitude of order $\mathcal{O}(n^{-1/s})$ to the magnitude of order $\mathcal{O}(1/2)$. That is why the rotation in Section 7.3.1.1 can improve the features of the points.

7.3.2 From Thinnest Covering: Lattices Approach

A lattice $\mathcal{P}_{\text{Lattice}}$ is an infinite set of points in $\mathbb{R}^s$ with the following three properties (Sloan, 1985):

(a) if $\mathbf{x}, \mathbf{x}' \in \mathcal{P}_{\text{Lattice}}$, then $\mathbf{x} \pm \mathbf{x}' \in \mathcal{P}_{\text{Lattice}}$;
(b) $\mathcal{P}_{\text{Lattice}}$ contains s linear independent points; and
(c) there exists a sphere center at the origin O that contains no points of $\mathcal{P}_{\text{Lattice}}$ other than O itself.

These properties mean that $\mathcal{P}_{\text{Lattice}}$ is invariant to certain translation groups (properties (a) and (c)) and could be generated by a generator matrix composed of the coordinates of the s

linear independent points (property (b)). Obviously, the above-constructed point set $\mathcal{P}_{\text{TaS}}$ belongs to the family of lattices. The simplest lattice is the rectangular lattice $\left\{\left(\frac{j_1}{n},\frac{j_2}{n},\ldots,\frac{j_s}{n}\right),\ j_i \in \mathbb{Z},\ 1 \leq i \leq s\right\}$. For example the uniform grid point set shown in Figure 7.14a is the subset of a rectangular lattice.

The dual of a $\mathcal{P}_{\text{Lattice}}$ is defined by

$$\mathcal{P}_{\text{Lattice}}^{\perp} = \{y \in \mathbb{R}^s : y \cdot x \in \mathbb{Z},\ \forall x \in \mathcal{P}_{\text{Lattice}}\} \tag{7.120}$$

where $\mathbb{Z}$ is the integer set.

Applications of lattices in multiple integrals have received attentions for years (Haber, 1970; Sloan, 1985; Sloan and Kachoyan, 1987). As mentioned above, although little is known about the densest packings and thinnest covering problem in dimensions $s \geq 3$, the information on the densest lattice packings and the thinnest lattice covering, in contrast, is much richer. For instance, the pattern of tangent spheres elaborated in the preceding section is one of the densest lattice packings in dimension 3. In this section, on the other hand, we will introduce the point sets via a possible thinnest lattice covering.

It can be verified that the set

$$A_s = \{(x_0, x_1, \ldots, x_s) \in \mathbb{Z}^{s+1} : x_0 + x_1 + \cdots + x_s = 0\} \tag{7.121}$$

which uses $s+1$ coordinates to define an s-dimensional point set, is a lattice lying on the hyperplane $\sum_{i=1}^{s+1} x_i = 0$ in $\mathbb{R}^{s+1}$. The pattern of tangent spheres in the preceding section is essentially equivalent to this lattice in dimensions 2 and 3. According to the investigations (Conway and Sloane, 1999), the dual lattices A_s^* of A_s have the known best efficient covering for all $s \leq 23$; namely, they have the known smallest thickness for lattices. From Equations 7.120 and 7.121, the generator matrix of the dual lattice A_s^* in the s dimension is

$$\mathbf{S} = \begin{bmatrix} \mathbf{S}_1 \\ \mathbf{S}_2 \\ \mathbf{S}_3 \\ \vdots \\ \mathbf{S}_{s-1} \\ \mathbf{S}_s \end{bmatrix} = \begin{bmatrix} 1 & -1 & 0 & \cdots & 0 & 0 \\ 1 & 0 & -1 & \cdots & 0 & 0 \\ \cdots & \cdots & \cdots & \cdots & \cdots & \cdots \\ 1 & 0 & 0 & \cdots & -1 & 0 \\ \dfrac{-s}{s+1} & \dfrac{1}{s+1} & \dfrac{1}{s+1} & \cdots & \dfrac{1}{s+1} & \dfrac{1}{s+1} \end{bmatrix} \tag{7.122}$$

which is an $s \times (s+1)$ matrix, $\mathbf{S}_j$ ($j = 1, 2, \ldots, s$) is the jth $(s+1)$-dimensional row vector. If $\mathbf{z} = (z_1, z_2, \ldots, z_s)^{\mathrm{T}} \in \mathbb{Z}^s$ is an arbitrary s-dimensional integer vector, then the point $\bar{\mathbf{x}} = (x_1, \ldots, x_{s+1})$ determined by

$$\bar{\mathbf{x}} = \sum_{j=1}^{s} z_j \mathbf{S}_j = \mathbf{z}^{\mathrm{T}}\mathbf{S} \tag{7.123}$$

is a point of the lattice A_s^*.

For example, consider the case $s = 2$. Equation 7.122 becomes

$$\mathbf{S} = \begin{bmatrix} \mathbf{S}_1 \\ \mathbf{S}_2 \end{bmatrix} = \begin{bmatrix} 1 & -1 & 0 \\ -\dfrac{2}{3} & \dfrac{1}{3} & \dfrac{1}{3} \end{bmatrix} \tag{7.124}$$

Thus, letting $\mathbf{z}_1 = (1,1)^T$, according to Equation 7.123 we get $\bar{\mathbf{x}}_1 = (\frac{1}{3}, -\frac{2}{3}, \frac{1}{3})$. Likewise, we have the pairs

$$\mathbf{z}_2 = (1, 0)^T \qquad \bar{x}_2 = (1, -1, 0)$$

$$\mathbf{z}_3 = (0, 1)^T \qquad \bar{x}_3 = \left(-\frac{2}{3}, \frac{1}{3}, \frac{1}{3}\right)$$

$$\cdots$$

If we denote $\bar{\mathbf{x}}_0 = (0, 0, 0)$, then the distances between arbitrary two points are given by

$$d(\bar{\mathbf{x}}_0, \bar{\mathbf{x}}_1) = \sqrt{\frac{2}{3}} \quad d(\bar{\mathbf{x}}_0, \bar{\mathbf{x}}_2) = \sqrt{2} \quad d(\bar{\mathbf{x}}_0, \bar{\mathbf{x}}_3) = \sqrt{\frac{2}{3}}$$

$$d(\bar{\mathbf{x}}_1, \bar{\mathbf{x}}_2) = \sqrt{\frac{2}{3}} \quad d(\bar{\mathbf{x}}_1, \bar{\mathbf{x}}_3) = \sqrt{2} \quad d(\bar{\mathbf{x}}_1, \bar{\mathbf{x}}_3) = \sqrt{\frac{14}{3}}$$

$$\cdots$$

Actually, it can be found that the minimum distance is $\sqrt{2/3}$, and thus the packing radius is $\sqrt{2/3}/2$.

In addition, it is clear that the points $\bar{\mathbf{x}}_0, \bar{\mathbf{x}}_1, \bar{\mathbf{x}}_2, \bar{\mathbf{x}}_3, \ldots,$ are all located in the plane $\bar{\mathbf{x}}_1 + \bar{\mathbf{x}}_2 + \bar{\mathbf{x}}_3 = 0$. Thus, we can transform the coordinate system to a new coordinate system such that in the new system $ox_1x_2x_3$ the points are located in the plane $x_3 = 0$. This can be achieved when we introduce the base vectors $\mathbf{e}_1$, $\mathbf{e}_2$ and $\mathbf{e}_3$ for the new coordinate system such that

$$\mathbf{e}_i \cdot \mathbf{e}_j = \delta_{ij} \qquad i,j = 1, 2, 3 \tag{7.125}$$

and

$$\mathbf{e}_3 = \frac{1}{\sqrt{3}}(1, 1, 1)^T \tag{7.126}$$

that is, $\mathbf{e}_3$ is normal to the plane $\bar{x}_1 + \bar{x}_2 + \bar{x}_3 = 0$ and, therefore, $\mathbf{e}_1$ and $\mathbf{e}_1$ are in the plane $\bar{x}_1 + \bar{x}_2 + \bar{x}_3 = 0$. Satisfying these conditions, we can choose

$$\mathbf{e}_1 = \frac{1}{\sqrt{2}}(1, -1, 0)^T \qquad \mathbf{e}_2 = \frac{1}{\sqrt{6}}(1, 1, -2)^T \tag{7.127}$$

Thus, the new coordinates of the point (x_1, x_2, x_3) are given by

$$x_j = \bar{\mathbf{x}} \cdot \mathbf{e}_j \qquad j = 1, 2, 3 \tag{7.128}$$

Using this transformation, the points $\bar{\mathbf{x}}_1, \bar{\mathbf{x}}_2, \bar{\mathbf{x}}_3, \ldots,$ can be given in the new coordinate system by

$$\mathbf{x}_1 = \left(\frac{1}{\sqrt{2}}, -\frac{1}{\sqrt{6}}, 0\right) \qquad \mathbf{x}_2 = (\sqrt{2}, 0, 0) \qquad \mathbf{x}_3 = \left(-\frac{1}{\sqrt{2}}, -\frac{1}{\sqrt{6}}, 0\right) \qquad \cdots$$

respectively. It is seen clearly that these points are located in the plane $x_3 = 0$. Therefore, it is more convenient just to use the former two coordinates such that the points are essentially in a two-dimensional space

$$\mathbf{x}_1 = \left(\frac{1}{\sqrt{2}}, -\frac{1}{\sqrt{6}}\right) \qquad \mathbf{x}_2 = (\sqrt{2}, 0) \qquad \mathbf{x}_3 = \left(-\frac{1}{\sqrt{2}}, -\frac{1}{\sqrt{6}}\right) \qquad \cdots$$

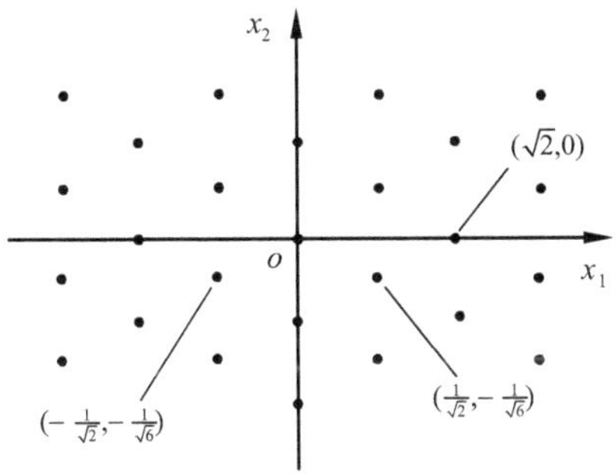

Figure 7.19 The dual point set in two-dimensional space.

The points so generated are shown in Figure 7.19. The configuration of the points is in essence the same as that in Figure 7.8b. They can be further transformed by a scale factor λ such that the configuration is the same but the scale of the points can be varied. By this, Equation 7.128 can be changed to

$$x_j = \lambda \bar{\mathbf{x}} \cdot \mathbf{e}_j \qquad j = 1, 2, 3 \tag{7.129}$$

Or we can achieve the same thing by changing Equation 7.123 to

$$\tilde{\mathbf{x}} = \lambda \sum_{j=1}^{s} z_j \mathbf{S}_j = \lambda \mathbf{z}^{\mathrm{T}} \mathbf{S} \tag{7.130}$$

The above treatment can be extended to higher dimensions. Actually, the packing radius and covering radius of the sphere are given respectively by

$$r_{\mathrm{pk}} = \frac{1}{2}\sqrt{\frac{s}{s+1}} \quad \text{and} \quad r_{\mathrm{cv}} = r_{\mathrm{pk}}\sqrt{\frac{s+2}{3}} = \frac{1}{2}\sqrt{\frac{s(s+2)}{12(s+1)}} \tag{7.131}$$

while the kissing number is $\kappa = 2s + 2$. Here we find the kissing number of A_s^* is much less than the known highest kissing number in packings. For example, in dimension 8, the highest kissing number is 240 while the kissing number of A_s^* is only 18. In turn, this implies that, by the known thinnest covering, the number of points can be minimized. This is one of the advantages of using the lattice covering.

Letting $\tilde{\mathbf{x}}_k = (\tilde{x}_{1,k}, \ldots, \tilde{x}_{s+1,k})$, the coordinates of the lattice could then be given by Equation 7.130. Obviously, the points generated by Equation 7.130 satisfy

$$\tilde{x}_{1,k} + \cdots + \tilde{x}_{s+1,k} = 0 \tag{7.132}$$

That is, these points are located on the $(s + 1)$-dimensional hyperplane

$$x_1 + \cdots + x_{s+1} = 0 \tag{7.133}$$

Thus, Equation 7.130 gives the lattices on a hyperplane in dimension $(s + 1)$.

We now proceed to extract the s independent coordinates and give the explicit expressions. In order to do so, we introduce a new coordinate system with the base vectors $\mathbf{e}_1, \mathbf{e}_2, \ldots, \mathbf{e}_{s+1}$ such that

$$\mathbf{e}_i \cdot \mathbf{e}_j = \delta_{ij} \tag{7.134}$$

where δ_{ij} is the Kronecker delta. For convenience, let

$$\mathbf{e}_{s+1} = \frac{1}{\sqrt{s+1}} \underbrace{(1, 1, \ldots, 1)}_{\text{totally } (s+1) \text{ of '1'}} \tag{7.135}$$

Clearly, $\|\mathbf{e}_{s+1}\| = \mathbf{e}_{s+1} \cdot \mathbf{e}_{s+1} = 1$. Then the other base vectors can be chosen according to Equation 7.134. The typical base vectors for $s = 4, 5, \ldots, 23$ are shown in Appendix D.

Using the new base vectors, the new coordinates of the points can then be computed by

$$x_{j,k} = \tilde{\mathbf{x}}_k \cdot \mathbf{e}_j \qquad j = 1, 2, \ldots, s \tag{7.136}$$

From Equations 7.133 and 7.130 it follows that $x_{s+1,k} = 0$. Therefore, the coordinate x_{s+1} is trivial and the nontrivial coordinates are the former s components.

Likewise, a rotational transformation can then be performed on the points by Equation 7.136 such that the features of the projection ratio (discussed in Section 7.3.1.2) can be improved.

The volume of the Voronoi cells of the point set A_s^* reads

$$\mathcal{V}(V) = \frac{2^s r_{\text{pk}}^s (s+1)^{(s-1)/2}}{s^{s/2}} \tag{7.137}$$

and the thickness of the lattice is

$$\mho = \frac{\mathcal{V}(\mathcal{B}(r_{\text{cv}}, s))}{\mathcal{V}(V)} = \mathcal{V}(\mathcal{B}(1, s))\sqrt{s+1}\left[\frac{s(s+2)}{12(s+1)}\right]^{s/2} \tag{7.138}$$

While the covering ratios of the Voronoi cell to the contact covering hypercube is

$$\gamma = \frac{\mathcal{V}(V)}{(2r_{\text{cv}})^s} = \frac{2^s(s+1)^{(s-1)/2}}{2^s s^{s/2} r_{\text{cv}}^s} r_{\text{pk}}^s = \frac{(s+1)^{(s-1)/2}}{s^{s/2}} \left(\frac{r_{\text{pk}}}{r_{\text{cv}}}\right)^s = \frac{(s+1)^{(s-1)/2}}{s^{s/2}} \left(\frac{3}{s+2}\right)^{s/2} \tag{7.139}$$

The quantities in Equations 7.138 and 7.139 can be used to check the accuracy when employing lattices to compute the assigned probabilities.

7.3.3 Number Theoretical Method

Another family of uniform point sets is the number theoretical net (NT-net) (Hua and Wang, 1981). Since the 1950s, NT-nets with low discrepancy have been studied extensively and applied widely in science and engineering (Niederreiter, 1992; Fang and Wang, 1994; Sobol' 1998; Nie and Ellingwood, 2004; Li and Chen, 2007a).

The basic idea of the method is to employ an integer generator vector $(n, Q_1, Q_2, \ldots, Q_s)$ to generate a point set $\mathcal{P}_{\text{NTM}} = \{x_k = (x_{1,k}, x_{2,k}, \ldots, x_{s,k}) : k = 1, 2, \ldots, n\}$ by

$$\begin{aligned} \hat{x}_{j,k} &= (2kQ_j - 1)\bmod(2n) \qquad j = 1, 2, \ldots, s;\ k = 1, 2, \ldots, n \\ x_{j,k} &= \frac{\hat{x}_{j,k}}{2n} \end{aligned} \tag{7.140a}$$

where the meaning of the modulo is identical to Equations 4.70a–4.70c. Equation 7.140a is equivalent to

$$x_{j,k} = \frac{2kQ_j - 1}{2n} - \text{int}\left(\frac{2kQ_j - 1}{2n}\right) \qquad j = 1, 2, \ldots, s; \quad k = 1, 2, \ldots, n \tag{7.140b}$$

Here, int $(\cdot)$ takes the integer part of the value and n is the cardinal number of the point set $\mathcal{P}_{\text{NTM}}$.

Equations 7.140a and 7.140b mean that the coordinates of the point set take decimal fractions of the value $(2kQ_j - 1)/2n$; thus

$$0 < x_{j,k} < 1 \qquad j = 1, 2, \ldots, s; \quad k = 1, 2, \ldots, n \tag{7.141}$$

This is of course related closely to congruence and the Diophantine equations and is one of the central topics in algebraic number theory (Manin and Panchishkin, 2005). Through units of the cyclotomic field, we can get different sets of integer vector $(n, Q_1, Q_2, \ldots, Q_s)$ so that the point set generated by Equations 7.140a and 7.140b is a point set uniformly scattered over the unit hypercube $\mathcal{C}^s$. Hua and Wang (1981) proved that the discrepancy of such a constructed point set, referred to as a Hua–Wang point set and denoted by $\mathcal{P}_{\text{H-W}}$, is

$$\mathcal{D}(n, \mathcal{P}_{\text{H-W}}) = \mathcal{O}(n^{-(1/2)-\{1/[2(s-1)]\}+\varepsilon}) \tag{7.142}$$

Generally, we can set $Q_1 = 1$. The generator vectors with low discrepancy for $s = 2$–18 for different n can be found in, say, Hua and Wang (1981) and Fang and Wang (1994) and are given in Appendix G. In dimension $s = 1$, from Equations 7.140a and 7.140b it is seen that $\mathcal{P}_{\text{H-W}}$ is identical to the uniform grid point set $\mathcal{P}_{\text{UGP}}$ (given by Equation 7.90). In dimension $s = 2$, the Fibonacci sequence F_ℓ, $\ell = 0, 1, 2, \ldots$, can be used:

$$F_\ell = F_{\ell-1} + F_{\ell-2}; \qquad \ell = 2, 3, \ldots; \ F_0 = 1; \ F_1 = 1 \tag{7.143}$$

In this case, letting $n = F_\ell$, $Q_1 = 1$ and $Q_2 = F_{\ell-1}$, Equations 7.140a and 7.140b will generate a uniformly scattered point set in the square $\mathcal{C}^2 = [0, 1] \times [0, 1]$.

Comparing Equation 7.142 with Equation 7.115, it is seen that the discrepancy of the Hua–Wang point sets $\mathcal{P}_{\text{H-W}}$ is in the same order as that of the tangent sphere points $\mathcal{P}_{\text{TaS}}$, at least for $s = 2$–4. Actually, this similarity can also be seen from the scattered configuration, as shown in Figure 7.20 in dimension 2 for similar n. However, we should note that they are not always so similar for different n. In fact, the possible n for $\mathcal{P}_{\text{H-W}}$ and $\mathcal{P}_{\text{TaS}}$ might differ greatly: we cannot always generate a $\mathcal{P}_{\text{H-W}}$ with n close to a given $\mathcal{P}_{\text{TaS}}$, let alone similar configurations.

Figure 7.21 shows the discrepancies of different patterns of point sets, including the tangent spheres point set $\mathcal{P}_{\text{TaS}}$, the Hua–Wang point set $\mathcal{P}_{\text{H-W}}$, the uniform grid point set $\mathcal{P}_{\text{UGP}}$ and the point sets obtained by Latin hypercube sampling $\mathcal{P}_{\text{LHS}}$. Because $\mathcal{P}_{\text{LHS}}$ is a random point set,

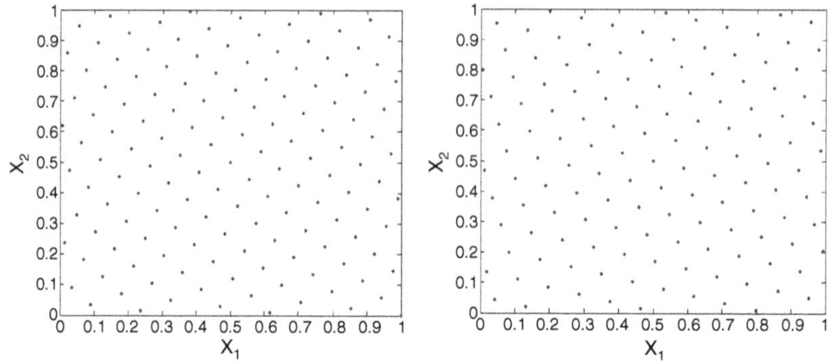

(a) H-W points $\mathcal{P}_{\text{H-W}}$ $(n = 144)$ **(b) TaS points** $\mathcal{P}_{\text{TaS}}$ $(n = 135)$

Figure 7.20 Configuration of different uniform scattered point sets.

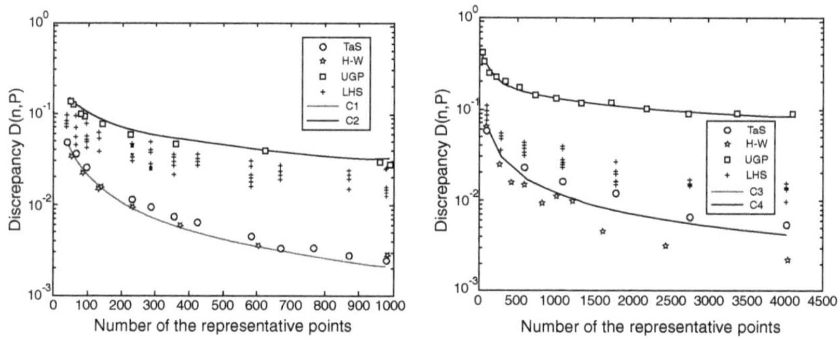

(a) $s = 2$ **(b)** $s = 3$

Figure 7.21 Discrepancies of different patterns of point sets.
(**C1**: $\tilde{\mathcal{D}}(n) = n^{-1/2}$, **C2**: $\tilde{\mathcal{D}}(n) = 2.06n^{-1}$, **C3**: $\tilde{\mathcal{D}}(n) = 1.2n^{-1/3}$, **C4**: $\tilde{\mathcal{D}}(n) = 206n^{-3/4}$)

we plotted six samples of the point set (McKay *et al.*, 1979). This indicates clearly that the discrepancy in $\mathcal{P}_{\text{TaS}}$ is of the same order of magnitude as $\mathcal{P}_{\text{H-W}}$, whereas the discrepancy in $\mathcal{P}_{\text{LHS}}$ is almost surely larger than that in $\mathcal{P}_{\text{TaS}}$ and $\mathcal{P}_{\text{H-W}}$. Thus, employing $\mathcal{P}_{\text{H-W}}$ is usually preferred.

The above-generated tangent sphere point sets $\mathcal{P}_{\text{TaS}}$, the lattices $\mathcal{P}_{\text{Lattice}}$ and the Hua–Wang point set $\mathcal{P}_{\text{H-W}}$ and the like can be applied sometimes directly as the representative points. But more reasonably, they can be used as basic point sets as dicussed in Section 7.2.4. For notational convenience, we denote all these points by $\mathcal{P}_{\text{Basic}}$.

7.4 Density-Related Transformation

7.4.1 Affine Transformation

The point sets generated in the preceding sections, including the tangent spheres points, the lattices and the Hua–Wang points, denoted uniformly by $\mathcal{P} = \{\mathbf{x}_k = (x_{1,k}, x_{2,k}, \ldots, x_{s,k}),$

$k = 1, 2, \ldots, n_{\text{pt}}\}$, are all point sets in a sense uniformly scattered over the unit hypercube $\mathcal{C}^s$. Here, n_{pt} is the cardinal number of $\mathcal{P}$. The distribution domain of the basic random variables $\Omega_{\boldsymbol{\Theta}}$, however, is usually not $\mathcal{C}^s$. We need to perform an affine transformation to map the point set $\mathcal{P}$ over $\mathcal{C}^s$ to the point set $\tilde{\mathcal{P}}$ over $\Omega_{\boldsymbol{\Theta}}$.

Without loss of generality, we assume the distribution domain

$$\Omega_{\boldsymbol{\Theta}} = \{\boldsymbol{\theta} = (\theta_1, \theta_2, \ldots, \theta_s) : b_{j,\text{L}} \leq \theta_j \leq b_{j,\text{R}}; j = 1, 2, \ldots, s\} = \prod_{j=1}^{s} [b_{j,\text{L}}, b_{j,\text{R}}] \quad (7.144)$$

where $b_{j,\text{L}}$ and $b_{j,\text{R}}$ are respectively the left-hand side and right-hand side boundaries. We can then construct the point set $\tilde{\mathcal{P}} = \{\tilde{\boldsymbol{\theta}}_k = (\tilde{\theta}_{1,k}, \tilde{\theta}_{2,k}, \ldots, \tilde{\theta}_{s,k}),\ k = 1, 2, \ldots, n_{\text{pt}}\}$ over $\Omega_{\boldsymbol{\Theta}}$ by employing an affine transformation:

$$\tilde{\theta}_{j,k} = b_{j,\text{L}} + x_{j,k}(b_{j,\text{R}} - b_{j,\text{L}}) \qquad j = 1, 2, \ldots, s;\ k = 1, 2, \ldots, n_{\text{pt}} \quad (7.145)$$

In the case the distribution domain $\Omega_{\boldsymbol{\Theta}}$ is an infinite domain (for example, when the basic random variables are normal or lognormal variables), the domain should be truncated to a bounded domain. How to select the truncation boundary is somewhat an open problem (Fang and Wang, 1994). We will come back to this issue again later.

The point set $\tilde{\mathcal{P}}$ generated by Equation 7.145 can be used as the representative point set $\mathcal{P}_{\text{sel}}$; for example, in dimensions $s \leq 4$ (Chen and Li, 2008). Otherwise, it can be used as the basic point set as discussed in Section 7.2.4 and denoted in a unified way by

$$\mathcal{P}_{\text{Basic}} = \{\tilde{\boldsymbol{\theta}}_k = (\tilde{\theta}_{1,k}, \tilde{\theta}_{2,k}, \ldots, \tilde{\theta}_{s,k}),\ k = 1, 2, \ldots, n_{\text{pt}}\} \quad (7.146)$$

7.4.2 Density-Related Transformation

The point set $\mathcal{P}_{\text{Basic}}$ generated in the preceding section is essentially based on the uniform distribution of basic random variables. If the density $p_{\boldsymbol{\Theta}}(\boldsymbol{\theta})$ is nonuniform, then the assigned probabilities of the points in $\mathcal{P}_{\text{Basic}}$ might differ so severely that the representative of different points varies too severely. In this case, further transformation imposed on the basic point set is needed; namely:

$$\mathcal{P}_{\text{sel}} = \mathcal{T}(\mathcal{P}_{\text{Basic}}) \quad (7.147)$$

This transformation, of course, should change the density of the points and simultaneously change the assigned probabilities to make them more uniform than before. In other words, Equation 7.147 should reduce the modified F-discrepancy $\mathcal{D}_{\mathcal{F}}(n, \mathcal{P})$ defined by Equations 7.93 and 7.95; that is:

$$\mathcal{D}_{\mathcal{F}}(n, \mathcal{P}_{\text{sel}}) \lll \mathcal{D}_{\mathcal{F}}(n, \mathcal{P}_{\text{Basic}}) \quad (7.148)$$

Because of this, we call this the *density-related transformation*.

Besides Equation 7.148, the density-related transformation in Equation 7.147 should also try to minimize the first- and the second-order discrepancies $\mathcal{D}_1$ and $\mathcal{D}_2$, as discussed in Section 7.2.3.

7.4.3 *Radial Decay Distribution: Spherical Sieving and Expansion–Contraction Transformation*

In many problems of practical interest, the density of the basic random variables exhibits the property of radial decay. For example, the multidimensional normal distribution exhibits a more fascinating property of spherical symmetric decay. In this case, a spherical sieving operator can greatly reduce the number of points finally selected. In addition, an isotropic expansion–contraction transformation can be adopted as the density-related transformation.

7.4.3.1 Spherical Sieving

For radial decay densities, the assigned probabilities of the points around the corners of the domain Ω_{Θ} are very small compared with those of the inner points and, thus, can be eliminated. This could be achieved by sieving the points in the point set $\mathcal{P}$, which is uniformly scattered over $\mathcal{C}^s$, through the inner contact sphere:

$$\left(x_{1,k}-\frac{1}{2}\right)^2+\left(x_{2,k}-\frac{1}{2}\right)^2+\cdots+\left(x_{s,k}-\frac{1}{2}\right)^2\leq\left(\frac{1}{2}r_0\right)^2 \tag{7.149}$$

where $r_0 \geq 1$.

Figure 7.22 schematically illustrates Equation 7.149. Clearly, for $r_0 \geq \sqrt{s}$, the condition Equation 7.149 does not work, while the number of the points satisfying Equation 7.149 will be smaller than n_{pt} in the case $1 \leq r_0 < \sqrt{s}$. In particular, in the case $r_0 = 1$, the ratio of the number of remaining points n_{sel} to the total number n_{pt} is given by

$$\gamma_{\mathrm{sv}}=\frac{n_{\mathrm{sel}}}{n_{\mathrm{pt}}}\to\frac{\mathcal{V}(\mathcal{B}(r,s))|_{r=1/2}}{\mathcal{V}(\mathcal{C}^s)}\qquad n_{\mathrm{pt}}\to\infty \tag{7.150}$$

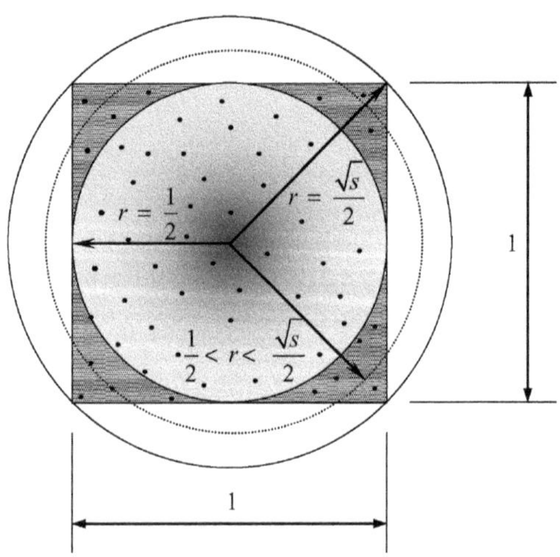

Figure 7.22 Sphere sieving ($s = 2$).

where $\mathcal{V}(\mathcal{B}(r,s))$ is the volume of an s-dimensional ball given by Equation 7.71. Therefore, the sieving ratio γ_{sv} satisfies

$$\lim_{n_{pt}\to\infty}\gamma_{sv} = \lim_{n\to\infty}\frac{n_{sel}}{n_{pt}} = \frac{\pi^{s/2}}{2^s\Gamma(1+\frac{s}{2})} = \begin{cases} \dfrac{\pi^m}{2^{2m}m!} & \text{for } s = 2m \\ \dfrac{(2\pi)^m}{2^{2m}\prod_{j=0}^{m}(2j+1)} & \text{for } s = 2m+1 \end{cases} \tag{7.151}$$

The nominal sieving ratio $\lim_{n_{pt}\to\infty}\gamma_{sv}$, shown in Figure 7.23, decreases very quickly (in fact, faster than exponentially) as s increases. For instance, $\lim_{n_{pt}\to\infty}\gamma_{sv} = 0.785$, 0.524, $\sim$1/400 and $<10^{-6}$ in dimensions 2, 3, 10 and 18 respectively.

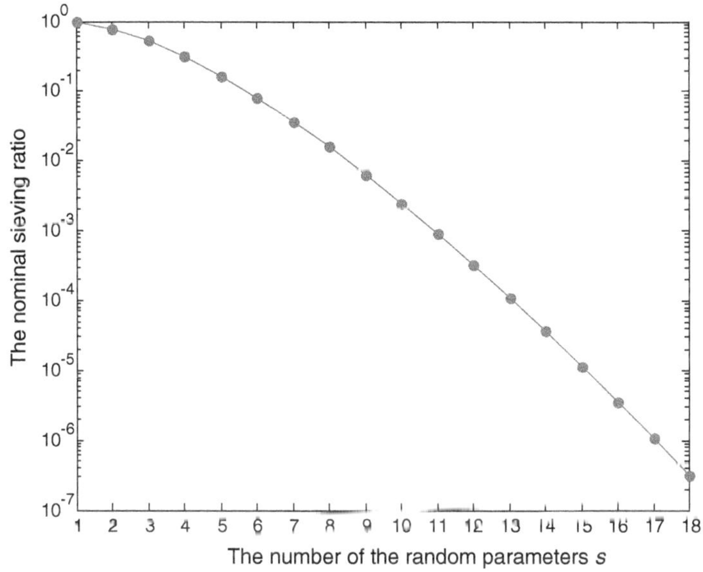

Figure 7.23 The nominal sieving ratio versus the dimension s.

The number of the uniformly scattered points over C^s usually increases in degree as the dimension s increases; spherical sieving will compensate for this increase and can make the number of points finally selected almost invariant against s for acceptable accuracy.

Figure 7.24 illustrates an example of spherical sieving in dimension 2. The original points come from the tangent spheres, as shown in Figure 7.20b. These points are sieved by a circle and then transformed by Equation 7.145 from the square $[0,1]^2$ to the square $[-4,4]^2$. Also plotted in Figure 7.24 are the contours of the joint normal distribution, each concentric circle being equi-PDF. This intuitively justifies sphere sieving.

7.4.3.2 Expansion–Contraction Transformation

For the radial decay parametric density, transformation $\mathcal{T}$ in Equation 7.147 can take the isotropic *expansion–contraction transformation* (Chen et al., 2009); namely:

$$\theta_{j,q} = g(\|\tilde{\boldsymbol{\theta}}_q\|)\tilde{\theta}_{j,q} \qquad q = 1,2,\ldots,n_{sel};\ j = 1,2,\ldots,s \tag{7.152}$$

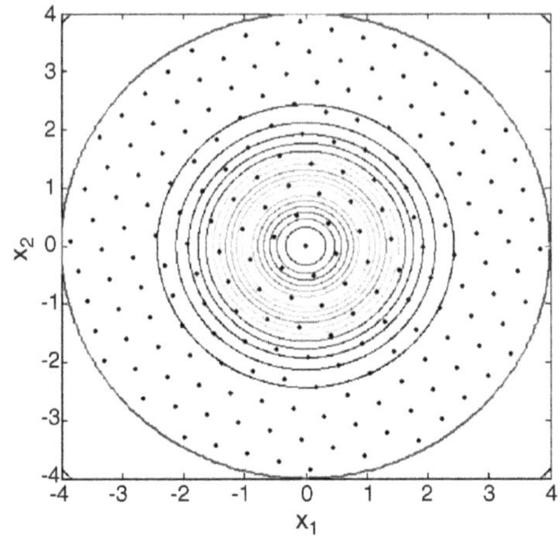

Figure 7.24 Spherical sieving.

where $\tilde{\boldsymbol{\theta}}_q = (\tilde{\theta}_{1,q}, \tilde{\theta}_{2,q}, \ldots, \tilde{\theta}_{s,q})$ and $g(\cdot)$ is an operator with respect to the norm of the point. Through the transformation in Equation 7.152, all the points will move toward or outward from the origin in an isotropic radial manner.

Actually, $g(\cdot)$ is the ratio of expansion or contraction. It is reasonable to make points on the edge of the distribution domain invariant, whereas points near the origin (peak density) should contract at an appropriate ratio β; namely:

$$g(r)|_{r \to 0} = \beta \qquad g(r)|_{r=\rho} = 1 \tag{7.153}$$

A simple one might take the form

$$g(r) = ar^m + b \qquad (m \geq 0,\ m \in \mathbb{Z}) \tag{7.154}$$

Combining Equations 7.153 and 7.154 gives

$$a = \frac{1-\beta}{\rho^m} \qquad b = \beta \tag{7.155}$$

The isotropic expansion–contraction transformation in a normal density field in dimension 2 is shown in Figure 7.25. It is seen that more points are scattered in the area of larger probability density after the isotropic expansion–contraction transformation.

In the present case, the F-discrepancy defined by Equation 7.93 may be transformed to the *radial F-discrepancy*

$$\mathcal{D}_{\mathrm{R}}(\mathcal{P}) = \max_{0 \leq r \leq r_b} |\mathcal{F}_{\mathcal{P}}(r) - \mathcal{F}(r)| \tag{7.156}$$

where r_{b} is the radius of the hyperball which covers Ω and

$$\mathcal{F}_{\mathcal{P}}(r) = \sum_{q=1}^{n_{\mathrm{sel}}} P_q \cdot I\{||\boldsymbol{\theta}_q|| \leq r\} = \sum\nolimits_{||\boldsymbol{\theta}_q|| \leq r} P_q = \sum\nolimits_{||\boldsymbol{\theta}_q|| \leq r} \int_{V_q} p_{\boldsymbol{\Theta}}(\boldsymbol{\theta}) \mathrm{d}\boldsymbol{\theta} \tag{7.157}$$

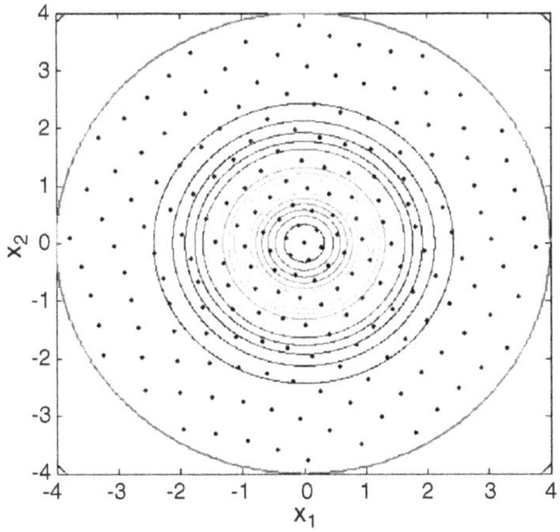

Figure 7.25 The point sets after expansion–contraction transformation ($n = 199$, $m = 1$, $\beta = 0.5$).

and

$$\mathcal{F}(r) = \int_{\Omega} (p_{\boldsymbol{\Theta}}(\boldsymbol{\theta}) \cdot I\{||\boldsymbol{\theta}|| \leq r\}) \mathrm{d}\boldsymbol{\theta} = \int_{||\boldsymbol{\theta}|| \leq r} p_{\boldsymbol{\Theta}}(\boldsymbol{\theta}) \mathrm{d}\boldsymbol{\theta} \tag{7.158}$$

are respectively the sum of the assigned probabilities and the probability contained in the hyperball $\mathcal{B}(r, s)$ of radius r in dimension s. In contrast to Equation 7.93, only one variable is involved in the distribution functions in Equation 7.156, which makes it much more tractable.

The closed-form expression of Equation 7.158 for a standard normal distribution is elaborated in Appendix E. The probability contained for different r and s is illustrated in Figure 7.26. What is particularly important and shown in the figure is that, as the dimension s increases, the probability in the area near the origin decreases greatly while the probability

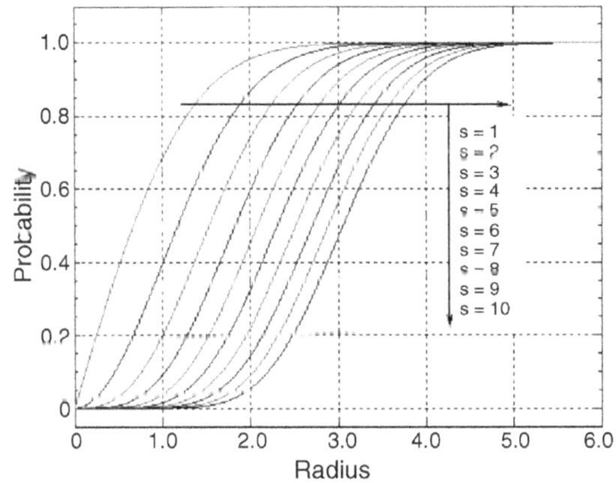

Figure 7.26 The probability contained in a hyperball, $F(r, s)$.

distributed over the outer area increases correspondingly. For instance, for $r=2$, in dimension 1 more than 0.7 is contained, whereas in dimension 10 the probability is less than 0.05. In addition, to have a probability of more than 0.99, for example, $r=3.5$ is enough in dimension 1, whereas r should be around 5.0 in dimension 10. This implies that, when we truncate an infinite distribution, we should be very careful and should have sufficient understanding of the distribution over the space.

Through appropriate isotropic expansion–contraction transformations, the radial F-discrepancy can usually be greatly reduced. For instance, in dimensions $s=2$ and $s=8$, Figure 7.27 shows the distributions given respectively by Equations 7.157 and 7.158. This demonstrates that the isotropic expansion–contraction transformation can greatly reduce the F-discrepancy. The effect is particularly obvious in larger dimension s. Correspondingly, the computational results when employing such selected point sets are usually satisfactory (Chen *et al.*, 2009).

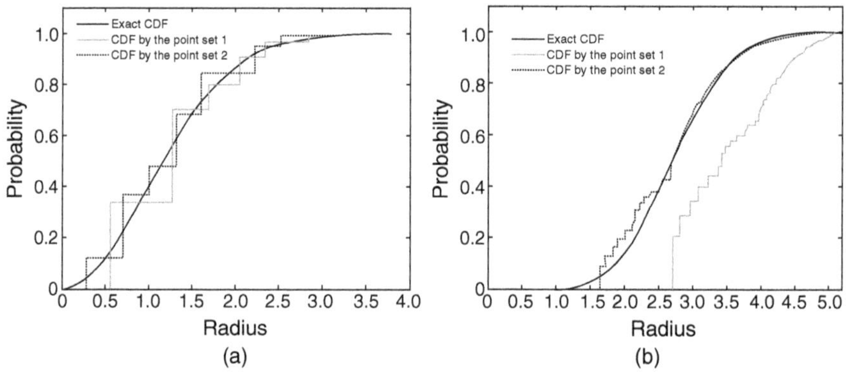

Figure 7.27 The CDF before and after expansion–contraction transformation (NTM). (a) $s=2, N_{\text{sel}}=68$; point set 1: $m=1, \beta=1$; point set 2: $m=2, \beta=0.5$. (b) $s=8, N_{\text{sel}}=360$; point set 1: $m=1, \beta=1$; point set 2: $m=6$, $\beta=0.\ 6$.

7.5 Stochastic Response Analysis of Nonlinear MDOF Structures

7.5.1 Responses of Nonlinear Stochastic Structures

Consider a 10-story shear frame as shown in Figure 7.28 subjected to earthquake. Randomness is involved in the mass parameters, stiffness parameters, the parameters of the nonlinear restoring force model and the peak parameters of the ground acceleration.

The means of the lumped masses $m_1, m_2, \ldots, m_{10}$ are listed in Table 7.1. From the bottom to the top, the first four stories are grouped into a subset that these four masses are completely correlated and the randomness can be characterized by a single random variable. The other six stories are grouped into the other subset. Thus, two random parameters are involved in the masses. In a similar way, another two random parameters are involved in the initial elastic modulus. The geometric size of the section of the columns is 500 mm × 400 mm. The Bouc–Wen model discussed in Section 5.5.2 is employed to characterize the interstory restoring force versus the interstory drift. Here, the model parameters β, γ, d_ν, d_η are taken as random parameters of which the probabilistic information is given in Table 7.2. The other model parameters take deterministic values: $\alpha=0.01, A=1, n=1, q=0, p=600$,

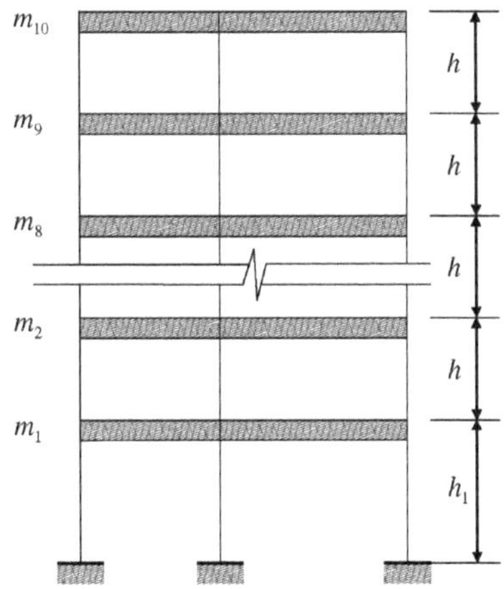

Figure 7.28 Structural model.

$d_\psi = 0$, $\lambda = 0.5$, $\zeta_s = 0.95$, $\psi = 0.2$. For simplicity, Rayleigh damping is assumed such that $\mathbf{C} = a\mathbf{M} + b\mathbf{K}$, where $\mathbf{K}$ is the initial stiffness matrix, $a = 0.01$ and $b = 0.005$. The ground acceleration is assumed to be the random combination of the El Centro accelerograms in the N–S and E–W directions with random combination coefficients $\Theta_{PGA,1}$ and $\Theta_{PGA,2}$. Regarding

Table 7.1 Probabilistic information of the mass and stiffness parameters.

Story number	Mean value		Coefficient of variation	
	Lumped mass (×10 000 kg)	Initial elastic modulus (×100 000 MPa)	Lumped mass	Initial elastic modulus
10	0.5	2.8	0.2	0.2
9	1.1	2.8		
8	1.1	3.0		
7	1.0	3.0		
6	1.1	3.0		
5	1.1	3.0		
4	1.3	3.25	0.2	0.2
3	1.2	3.25		
2	1.2	3.25		
1	1.2	3.25		

Table 7.2 Probabilistic information of parameters in the Bouc–Wen model and excitation.

Parameters	β	γ	d_ν	d_η	$\Theta_{PGA,1}$	$\Theta_{PGA,2}$
Mean value	60	10	200	200	2.0 m/s^2	2.0 m/s^2
Coefficient of variation	0.2	0.2	0.2	0.2	0.2	0.2

all the randomness, a total of 10 random variables are involved. The partition of probability-assigned space using the NT-nets as the basic point sets is employed to generate representative points and then the generalized density evolution equation is solved to obtain the probabilistic information of the stochastic response of the nonlinear structure.

Figures 7.29–7.31 illustrate some of the results. Figure 7.29 shows that both the mean and the standard deviation of the response accord well with the MCS. In the computation, only 570 representative points are selected in the PDEM, but 16 000 reanalyses are carried out in the MCS. One of the most important advantages of the PDEM is that the instantaneous PDF of the

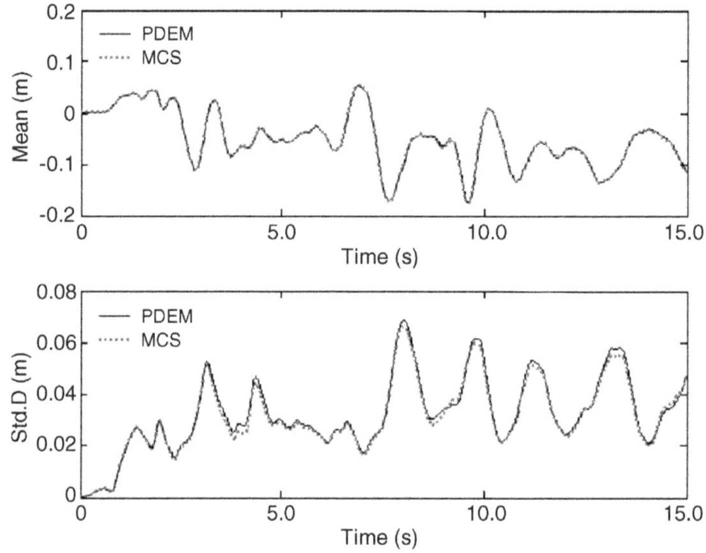

Figure 7.29 Mean and standard deviation of stochastic responses.

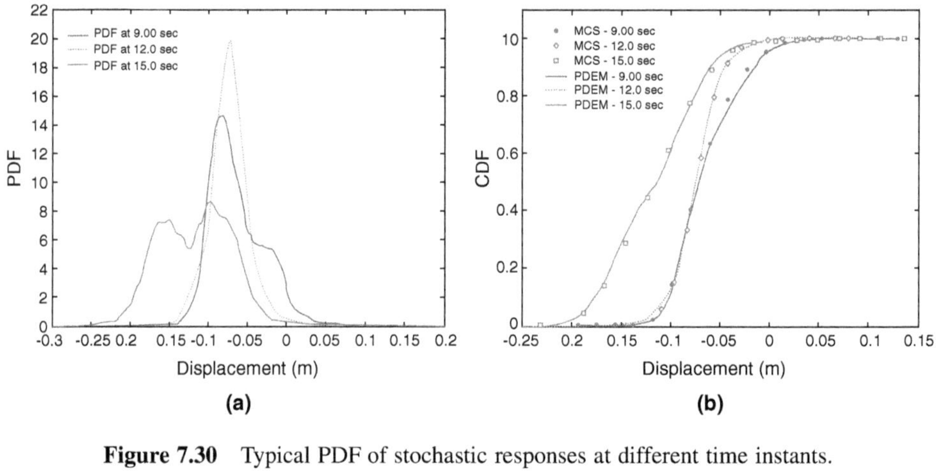

Figure 7.30 Typical PDF of stochastic responses at different time instants.

response can be captured. This can be seen in Figure 7.30a, where the PDFs at three different instants of time are illustrated. Figure 7.30b shows the comparisons between the CDF by the PDEM and the empirical CDF by the MCS. Clearly, it is seen that the PDFs at different instants of time are irregular and quite different. This can be seen further in Figure 7.31a, where the PDFs evolving against time construct a PDF surface. Simultaneously plotted in Figure 7.31b is the contour, where it is seen that the probability transits against time just like water flowing in a river with many whirls.

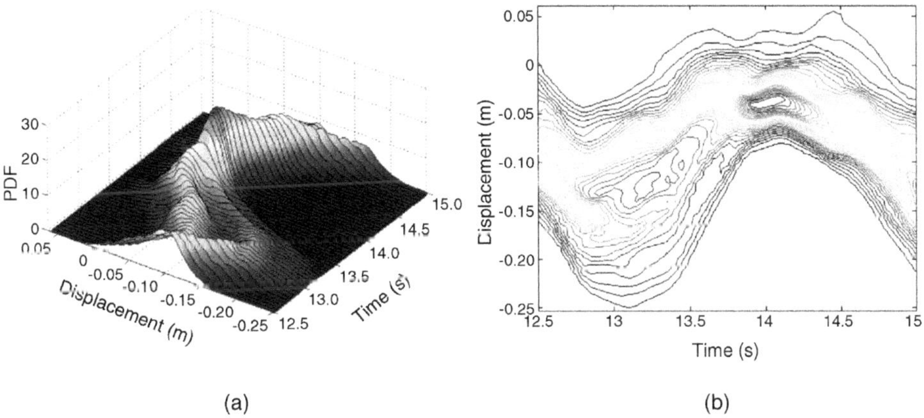

Figure 7.31 PDF evolution surface and the contour.

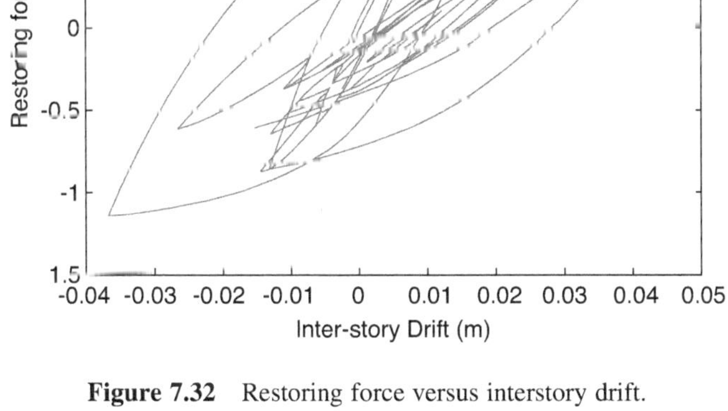

Figure 7.32 Restoring force versus interstory drift.

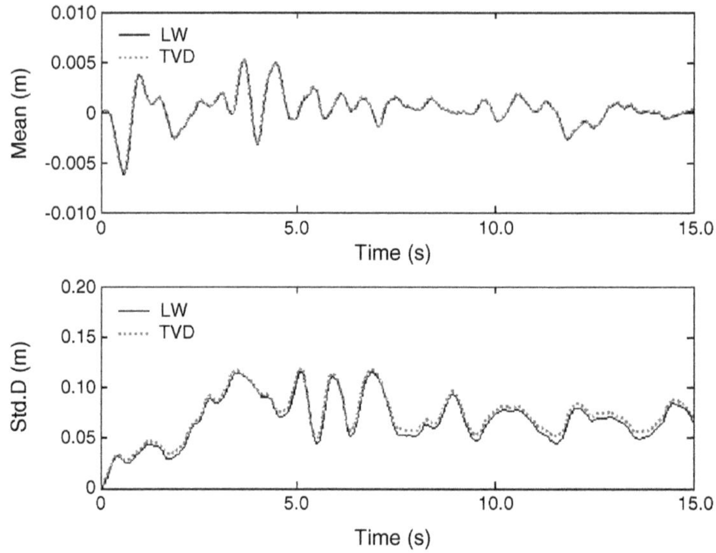

Figure 7.33 Mean and standard deviation of stochastic responses.

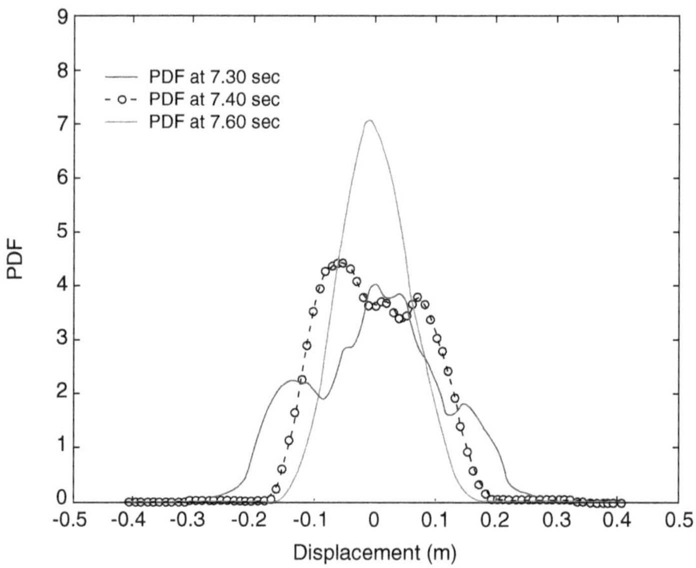

Figure 7.34 Typical PDF at different time instants.

7.5.2 Stochastic Seismic Response of Nonlinear Structures

Now a nonlinear structure subjected to stochastic ground motions is considered. In the present context, the structural properties are regarded as deterministic, taking the same values as those in Section 7.5.1, except that the geometric size of the columns is 400 mm × 400 mm and that the parameters regarded as random variables in Section 7.5.1 take values identical to the mean values. The physical stochastic model of ground motion elaborated in Section 3.2.3 is employed here to represent a stochastic ground acceleration. In total, 221 representative points generated by the strategy of tangent spheres are employed here to generate 221 representative time histories of the ground acceleration. Then the PDEM is performed to obtain the probabilistic information of the stochastic response.

Figure 7.32 shows a typical curve of the restoring force versus interstory drift in one of the representative time histories. Figure 7.33 shows the mean and standard deviation of the stochastic response. It is noted that the amplitude of the mean process has a magnitude of small order. The PDF of the response is illustrated in Figures 7.34–7.36. We can see that the shape of the PDF is quite different at different instants of time.

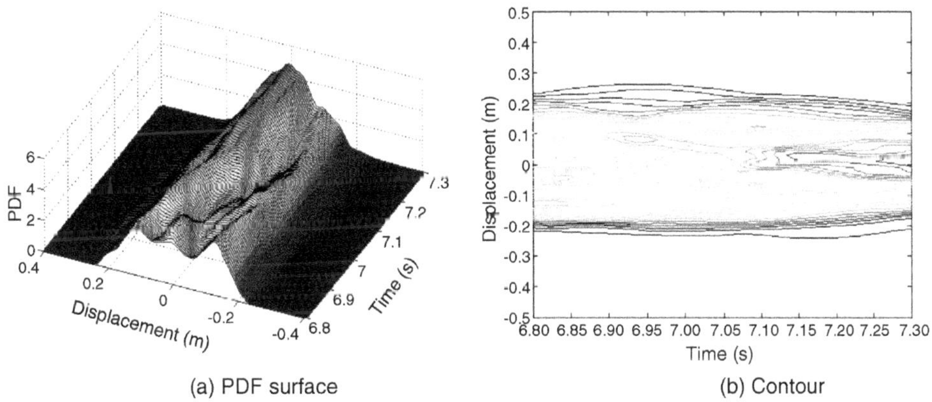

Figure 7.35 PDF surface and contours of stochastic responses.

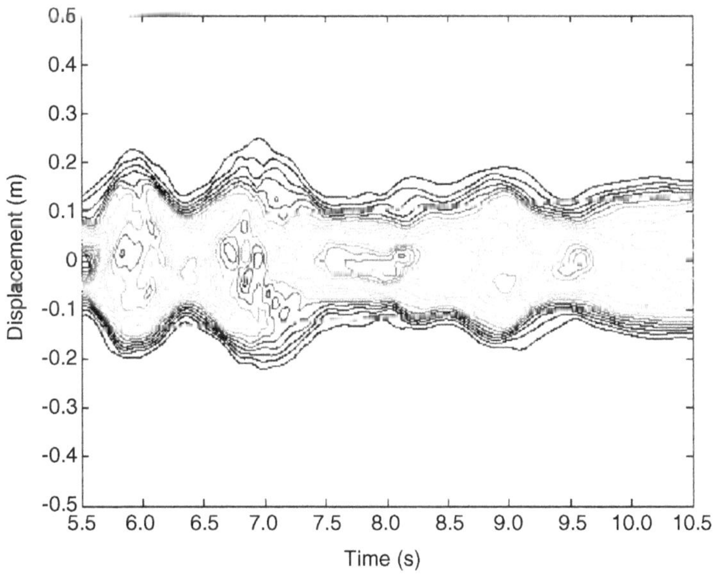

Figure 7.36 Contours of stochastic response.

8

Dynamic Reliability of Structures

8.1 Fundamentals of Structural Reliability Analysis

8.1.1 Structural Reliability

One of the important purposes of stochastic response analysis of structures is to provide a quantitative basis to enable the designed structure to satisfy the requirements for safety or serviceability in the expected service life. Structural safety and serviceability may be comprehensively called structural reliability. Herein, reliability means the probability of success. In this chapter the concern is only dynamic reliability that is related to dynamic responses of stable structures.

If a structure does not meet a certain prescribed requirement (that is, the whole structure or part of it exceeds a certain specified state), then this certain state is called the *limit state*. Accordingly, limit states are boundaries that distinguish whether the working state of the structure is reliable or not. The limit states of engineering structures can usually be classified into two basic categories: ultimate states and serviceable states. The ultimate states correspond to cases in which the structure or members of the structure reach the ultimate bearing capacity or deformation states that are not suitable to continue to carry loads. The serviceability limit states refer to those states where the structure or members of the structure reach the prescribed limit values for serviceability or durability.

According to the background, the main target of structural reliability analysis is to evaluate the probability of the structural response not exceeding the limit states. This target can be accomplished equivalently by calculating the probability of the structural response exceeding the limit states (namely, failure).

In general, the limit states of the structure can be defined by limit state functions. Assume that $\xi_1, \xi_2, \ldots, \xi_n$ are the n random variables influencing the structural response, then the random function

$$Z = g(\xi_1, \xi_2, \ldots, \xi_n) \tag{8.1}$$

Stochastic Dynamics of Structures Jie Li and Jianbing Chen

is called the *limit state function* if it satisfies

when $Z > 0$, the structure retains the prescribed function – that is, the structure is reliable;
when $Z < 0$, the structure loses the prescribed function – that is, the structure is in failure;
when $Z = 0$, the structure is in a critical state – namely, the structure reaches the limit state.

Assume the limit state function is prescribed by Equation 8.1. The reliability $P_r = \Pr\{Z > 0\}$ can be given by the integral of the joint density function of the set of basic random variables $p_{\xi_1\xi_2\cdots\xi_n}(x_1, x_2, \ldots, x_n)$; that is:

$$P_r = \int \cdots \int_{z=g(x_1,x_2,\ldots,x_n)>0} p_{\xi_1\xi_2\cdots\xi_n}(x_1, x_2, \ldots, x_n)\,\mathrm{d}x_1\,\mathrm{d}x_2 \ldots \mathrm{d}x_n \tag{8.2}$$

Similarly, the probability that the limit state function is less than zero is called the *failure probability* in terms of the certain functions. Obviously, the failure probability $P_f = \Pr\{Z < 0\}$ can be expressed by

$$P_f = \int \cdots \int_{z=g(x_1,x_2,\ldots,x_n)<0} p_{\xi_1\xi_2\cdots\xi_n}(x_1, x_2, \ldots, x_n)\,\mathrm{d}x_1\,\mathrm{d}x_2 \ldots \mathrm{d}x_n \tag{8.3}$$

Then, there exists

$$P_f = 1 - P_r \tag{8.4}$$

Therefore, computation of the structural failure probability is equivalent to computation of the structural reliability.

It should be pointed out that, in structural reliability evaluation, the probability that a catastrophic load occurs in the service life of the structures is not taken into account. If it is considered, then the aforementioned results should be properly revised. Taking an earthquake as an example, if the probability that the peak of the ground motion Y exceeds the prescribed value is $\Pr\{Y > y\}$, then the PDF $p_Y(y)$ in terms of parameter Y can be determined. The failure probability in the service life of the structures can thus be expressed by

$$P_f = \Pr(Z < 0) = \int_0^{\infty} \Pr\{Z < 0 | Y = y\} p_Y(y)\,\mathrm{d}y \tag{8.5}$$

in which the conditional probability $\Pr\{Z < 0 \mid Y = y\}$ can be computed by employing Equation 8.3 according to the prescribed y.

8.1.2 Dynamic Reliability Analysis of Structures

Structural dynamic reliability still complies with the definition for the general structural reliability problem. The new particularity only lies in the problem that dynamic action and dynamic response are time-varying processes. For a dynamic system, the probability of exceeding a limit state can be described as the first-passage probability or a fatigue failure probability (Lin, 1967). In this chapter, only the first-passage problem is of concern.

Suppose $X(t)$ is the response of a stochastic dynamical system. The general definition of the first-passage problem can be described as

$$F_s = \Pr\{X(t) \in \Omega_s, t \in (0, T]\} \tag{8.6}$$

where Ω_s is the safe domain. This means that the response will never exceed the boundary of the safety domain during the time duration (0, T]. In other words, once the response exceeds the boundary, the structure will fail.

Depending on the specific background, the boundary of Ω_s may be a one-sided boundary, a double-sided boundary, circle boundary and so on. For a simple double-sided boundary problem, the structural dynamic reliability in time T is defined as

$$F_s(-a_1, a_2, T) = \Pr\{(X(t) > -a_1) \cap (X(t) < a_2) | 0 \le t \le T\} \tag{8.7}$$

in which $-a_1$ and $a_2(a_1 > 0,\ a_2 > 0)$ are the permissive lower limit and upper limit of the stochastic structural responses respectively.

For convenience, we define the dynamic reliability function $R_a(T)$ as the probability that $Y(t)$ does not exceed the limit value in the time interval [0, T]; namely:

$$R_a(T) = \Pr\{X(t) < a | 0 \le t \le T\} \tag{8.8}$$

which is of course the reliability of a one-sided boundary problem.

8.1.3 Global Reliability of Structures

Generally speaking, the above-defined reliability is that when only one limit state function is involved; in other words, it is only defined when one element (or member) of the structure fails or when one specified failure mode is considered. For the reliability evaluation of a structure, we usually need to take into account more than one index or more than one failure mode. For instance, when we consider the serviceability of a multistory frame structure, not only might it be required that the interstory drift between the first floor and the second floor not exceed a threshold, but also require that all the other interstory drifts not exceed the corresponding threshold. In this case, a family of limit state functions should be considered:

$$\begin{aligned} Z_1 &= g_1(\xi_1, \xi_2, \ldots, \xi_n) \\ Z_2 &= g_2(\xi_1, \xi_2, \ldots, \xi_n) \\ &\ldots \\ Z_m &= g_m(\xi_1, \xi_2, \ldots, \xi_n) \end{aligned} \tag{8.9}$$

The failure event might be a combination of these limit state functions in different logical relationships; that is:

$$P_f = \Pr\{\aleph_{j=1}^{m}(Z_j < 0)\} \tag{8.10}$$

Here, we use $\aleph$ to denote different logical combination operators. For instance, when it is a series system, Equation 8.10 becomes

$$P_f = \Pr\left\{\bigcup_{j=1}^{m}(Z_j < 0)\right\} \tag{8.11}$$

whereas if it is a parallel system, Equation 8.10 becomes

$$P_{\mathrm{f}} = \Pr\left\{\bigcap_{j=1}^{m}(Z_j < 0)\right\} \tag{8.12}$$

In addition, it can also be a combined system; say:

$$P_{\mathrm{f}} = \Pr\left\{\left[\bigcap_{j=1}^{m_1}(Z_j < 0)\right] \cup \left[\bigcap_{j=m_1+1}^{m}(Z_j < 0)\right]\right\} \tag{8.13}$$

or in any other type of combination.

Of course, the reliability is given by

$$P_{\mathrm{r}} = 1 - P_{\mathrm{f}} \tag{8.14}$$

The reliability problem (Equation 8.10) is generally known as the *structural system reliability*. Because of correlation and combinatorial explosion problems, this is one of the most difficult topics in reliability theory. Actually, for most of the so-called system reliability problems encountered in engineering practice, it might be more reasonable to call it the *global reliability* of structures. We will come to this problem in detail later.

8.2 Dynamic Reliability Analysis: First-Passage Probability Based on Excursion Assumption

8.2.1 Excursion Rates

As pointed out in Equation 8.6, the first-passage failure problem requires that in the time duration of interest the response of the system will never exceed the boundary of the safety domain. One of the treatments to tackle this problem is that we can first set a virtual boundary, if we can evaluate the times of excursion in the time duration, then we can capture the probability that the times of excursion is zero. This, of course, is the reliability of the system. To this end, we first consider the excursion rate of a response exceeding the virtual boundary. This is also known as the *level-crossing* problem.

Figure 8.1 illustrates a sample function of the stochastic process and the situations that the sample time process $X(t)$ passes the level $x(t) = a$ ($a > 0$) with a positive and negative rate of slope (passage upward and downward respectively). Obviously, for the stochastic process, the time of passing the level in a time interval is a random variable. The probability distribution density of this random variable at any time is called the rate of passage (it is also sometimes called the expected rate of threshold crossing), which can be designated as $\lambda(t)$.

The condition that the stochastic process passes the threshold with positive rate of slope in the time interval $(t, t + dt)$ can be expressed by

$$\begin{cases} x(t) < a \\ x(t + \mathrm{d}t) > a \end{cases} \tag{8.15}$$

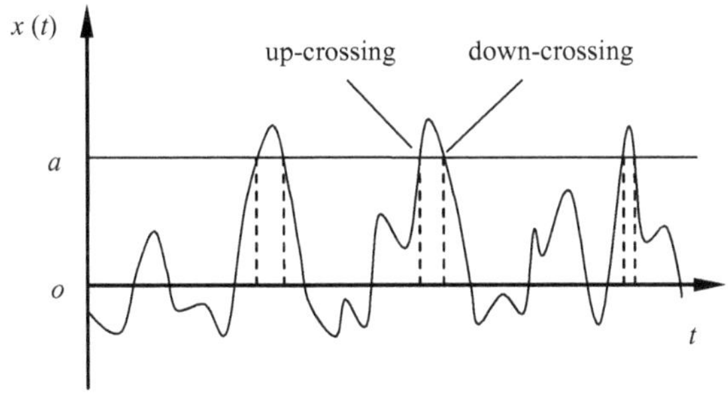

Figure 8.1 Excursion of a sample process.

Considering that $x(t+\mathrm{d}t) = x(t) + \dot{x}(t)\,\mathrm{d}t$, the condition can be rewritten as

$$\begin{cases} x(t) < a \\ \dot{x}(t)\,\mathrm{d}t > a - x(t) \end{cases} \tag{8.16}$$

which denotes the shaded area shown in Figure 8.2.

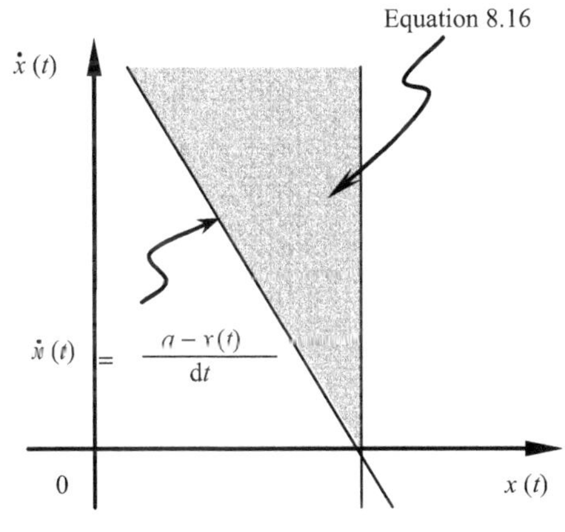

Figure 8.2 The area determined by Equation 8.16.

Assume the joint PDF of $X(t)$ and $\dot{X}(t)$ is $p_{X\dot{X}}(x, \dot{x}, t)$, then the probability that passages happen in the time interval $(t, t + \mathrm{d}t)$ is the integral of $p_{X\dot{X}}(x, \dot{x}, t)$ in the shaded area; that is:

$$\lambda_a(t)\,\mathrm{d}t = \mathrm{d}t \int_0^{\infty} \int_{a-\dot{x}\mathrm{d}t}^{a} p_{X\dot{X}}(x, \dot{x}, t)\,\mathrm{d}x\,\mathrm{d}\dot{x} \tag{8.17}$$

According to the intermediate value theorem, the above integral is equal to

$$\lambda_a(t)\,\mathrm{d}t = \mathrm{d}t \int_0^{\infty} \dot{x} p_{X\dot{X}}(a, \dot{x}, t)\,\mathrm{d}\dot{x} \tag{8.18}$$

Namely:

$$\lambda_a(t) = \int_0^{\infty} \dot{x} p_{X\dot{X}}(a, \dot{x}, t)\, \mathrm{d}\dot{x} \tag{8.19}$$

This result is known as Rice's formula (Rice, 1944), which is the rate of crossing threshold at time instant t with positive rate of slope.

Similarly, the rate of passing $x(t) = -a$ at time instant t with negative rate of slope can be deduced as

$$\lambda_{-a}(t) = \int_{-\infty}^{0} \dot{x} p_{X\dot{X}}(-a, \dot{x}, t)\, \mathrm{d}\dot{x} \tag{8.20}$$

Equations 8.19 and 8.20 demonstrate that, for any stochastic process, the rate of passage can be evaluated as long as the joint probability density of the process and its derivative process are known. In particular, for the zero-mean stationary Gaussian stochastic process, we have

$$\lambda_a = \lambda_{-a} = \frac{1}{2\pi}\frac{\sigma_{\dot{X}}}{\sigma_X}\exp\left(-\frac{a^2}{2\sigma_X^2}\right) \tag{8.21}$$

In the case $a = 0$, this becomes

$$\lambda_0 = \frac{1}{2\pi}\frac{\sigma_{\dot{X}}}{\sigma_X} \tag{8.22}$$

Usually, λ_0 is called the expected rate of crossing zero with positive (or negative) rate of slope. Here, σ_X and $\sigma_{\dot{X}}$ are the standard deviations of $X(t)$ and $\dot{X}(t)$ respectively.

For a general nonstationary Gaussian stochastic process there exists (Zhu, 1992)

$$\lambda_a(t) = \frac{\sigma_{\dot{X}}}{2\pi\sigma_X}\left[\sqrt{1-\rho^2}\exp\left(-\frac{1}{1-\rho^2}\frac{a^{*2}}{2\sigma_X^2}\right) + \sqrt{2\pi}\frac{\rho a^*}{\sigma_X}\exp\left(-\frac{a^{*2}}{2\sigma_X^2}\right)\Phi\left(\frac{\rho}{1-\rho^2}\frac{a^*}{2\sigma_X}\right)\right] \tag{8.23}$$

in which $a^* = a - \varepsilon[X(t)]$, $\rho = \rho(t)$ is the correlation coefficient of X and $\dot{X}$ and $\Phi(\cdot)$ is the standard normal distribution function.

8.2.2 Excursion Assumption and First-Passage Probability

Consider first the simplest situation of the one-sided boundary problem. In some cases, this can also be a basis for more complicated problems. Although it seems much easier than other cases, even for the simplest stationary response process the accurate solution of dynamic reliability function unfortunately remains to be found. Therefore, most studies have been limited to obtaining approximate solutions. In these solutions, those based on the Poisson crossing assumption (Coleman, 1959) and the Markovian crossing assumption (Crandall *et al.*, 1966) are two representative methods.

The Poisson crossing assumption assumes that:

(a) in a tiny time interval, the positive (negative) crossing occurs at most once;
(b) the times of crossing happening in different time intervals are independent.

The above assumption is essentially to regard the excursion process as a Poisson process. Consequently, the probability that the response process $X(t)$ passes the threshold with positive rate of slope for k times in a time interval $(t,\ t + dt)$ is

$$\begin{aligned} P_a(k, t+\mathrm{d}t) &= P_a(k,t)P_a(0,\mathrm{d}t) + P_a(k-1,t)P_a(1,\mathrm{d}t) \\ &= P_a(k,t)[1-\lambda_a(t)\,\mathrm{d}t] + P_a(k-1,t)\lambda_a(t)\,\mathrm{d}t \end{aligned} \tag{8.24}$$

Rewriting the equation as

$$\frac{P_a(k,t+\mathrm{d}t) - P_a(k,t)}{\mathrm{d}t} = -\lambda_a(t)P_a(k,t) + \lambda_a(t)P_a(k-1,t)$$

that is:

$$\frac{\mathrm{d}P_a(k,t)}{\mathrm{d}t} + \lambda_a(t)P_a(k,t) = \lambda_a(t)P_a(k-1,t) \tag{8.25}$$

The general solution of this difference-differential equation is

$$P_a(k,t)\exp\left[\int_0^t \lambda_a(\tau)\,\mathrm{d}\tau\right] = C_k + \int_0^t P_a(k-1,\tau)\lambda_a(\tau)\exp\left[\int_0^t \lambda_a(u)\,\mathrm{d}u\right]\mathrm{d}\tau \tag{8.26}$$

In the case $k = 0$, the equation becomes

$$P_a(0,t)\exp\left[\int_0^t \lambda_a(\tau)\,\mathrm{d}\tau\right] = C_0 + \int_0^t P_a(-1,\tau)\lambda_a(\tau)\exp\left[\int_0^t \lambda_a(u)\,\mathrm{d}u\right]\mathrm{d}\tau \tag{8.27}$$

Because k takes only positive numbers, $P_a(-1,\ \tau) = 0$, and at time $t = 0$ the event passing $y = a$ is impossible (namely, $P_a(0,\ 0) = 1$); accordingly, $C_0 = 1$ and consequently

$$P_a(0,t) = \exp\left[-\int_0^t \lambda_a(\tau)\,\mathrm{d}\tau\right] \tag{8.28}$$

According to the definition of the dynamic reliability function, there exists

$$R_a(T) = P_a(0,T) = \exp\left[-\int_0^T \lambda_a(t)\,\mathrm{d}t\right] \tag{8.29}$$

Note that, in the present assumption, crossing events are independent; thus, the expected excursion rate of the double-sided boundary problem is the sum of the expected excursion rate of two one-sided boundary problems. Therefore, the structural dynamic reliability in the time interval $(0,\ T)$ for the double-sided boundary problem can be obtained as follows:

$$F_s(-a_1,a_2,T) = \exp\left\{-\int_0^T [\lambda_{-a_1}(t) + \lambda_{a_2}(t)]\,\mathrm{d}t\right\} \tag{8.30}$$

When the threshold values $a_1 = a_2 = a$ (the situation with a symmetric threshold value), the above equation becomes

$$F_s(-a_1,a_2,T) = \exp\left\{-\int_0^T [\lambda_{-a}(t) + \lambda_a(t)]\,\mathrm{d}t\right\} \tag{8.31}$$

Therefore, if the stochastic structural response is a nonstationary normal process, then the structural dynamic reliability under a specified threshold can be obtained by substituting Equation 8.23 in the above two equations. If the stochastic structural response is a zero-mean stationary Gaussian process, then the structural dynamic reliability can be evaluated by substituting Equation 8.21 in the above two equations.

Related studies demonstrate that, for stationary normal structural responses, when the threshold approaches an infinitely high value, the above method may give an accurate solution of the dynamic reliability (Cramer, 1966). However, when the threshold is not so high, there are some errors in the results. For narrow-band processes the results deviate to conservation (the computed reliability deviates to a lower value). In fact, for narrow-band processes, if the threshold is not very high, then crossing events are not independent but occur in clusters (Cramer, 1966; Vanmarcke, 1972, 1975).

In the Markovian crossing assumption, it is assumed that the next crossing event is related to the present crossing event whereas it is independent of the past events. Therefore, the crossing process is a Markov process. The structural dynamic reliability (for the symmetry limit situation) under a general nonstationary normal process can be obtained by

$$F_s(-a, a, T) = \exp\left[-\int_0^T \alpha(t)\,\mathrm{d}t\right] \tag{8.32}$$

in which

$$\alpha(t) = \frac{\gamma_2(t)}{\pi}\exp\left(-\frac{r^2(t)}{2}\right)\frac{1-\exp\left[-\sqrt{(\pi/2)}q^{1+b}(t)r(t)\right]}{1-\exp[-(r^2(t))/2]} \tag{8.33}$$

where $\gamma_2(t)$ and $q(t)$ are the spectral parameters as defined by Equations F.7 and F.8 respectively in Appendix F, b is an empirical parameter, usually $b = 0.2$, and

$$r = \frac{a}{\sigma_X(t)} \tag{8.34}$$

For stationary response processes, the structural dynamic reliability is given by

$$F_s(-a, a, T) = \exp\left\{\frac{\gamma_2}{\pi}T\exp\left(-\frac{r^2}{2}\right)\frac{1-\exp[-\sqrt{(\pi/2)}q^{1+b}r]}{1-\exp(-r^2/2)}\right\} \tag{8.35}$$

in which γ_2 and q are the spectral parameters as defined by Equations F.3 and F.4 in Appendix F respectively and

$$r = \frac{a}{\sigma_X} \tag{8.36}$$

8.2.3 *First-Passage Probability Considering Random Thresholds*

In the above analysis, the values of the thresholds a_1 and a_2 are regarded as deterministic variables. However, in most practical problems the thresholds might be random variables. Therefore, structural reliability analysis with random thresholds should be discussed.

First consider the situation with asymmetric failure thresholds (namely, $a_1 \neq a_2$). Assume that the joint PDF of the two-sided thresholds is $f_{a_1 a_2}(u_1, u_2)$; thus, the reliability that a_1 takes the value varying in the range (u_1, $u_1 + \mathrm{d}u_1$) and a_2 in the range (u_2, $u_2 + \mathrm{d}u_2$) is

$$\mathrm{d}F_s'(-a_1, a_2, T) = F_s(-u_1, u_2, T) f_{a_1 a_2}(u_1, u_2)\, \mathrm{d}u_1\, \mathrm{d}u_2 \tag{8.37}$$

in which $F_s(-u_1, u_2, T)$ and $F'_s(-a_1, a_2, T)$ are the reliabilities with deterministic thresholds and thresholds limits respectively.

According to above equation, we have

$$F_s'(-a_1, a_2, T) = \int_0^\infty \int_0^\infty F_s(-u_1, u_2, T) f_{a_1 a_2}(u_1, u_2)\, \mathrm{d}u_1\, \mathrm{d}u_2 \tag{8.38}$$

Usually, the lower threshold a_1 and the upper threshold a_2 can be treated as independent random variables with probability density $f_{a_1}(u_1)$ and $f_{a_2}(u_2)$ respectively; therefore, combining Equations 8.30 and 8.38 gives

$$F_s'(-a_1,a_2,T) = \left\{ \int_0^\infty \exp\left[-\int_0^T \lambda_{-u_1}(t)\, \mathrm{d}t \right] f_{a_1}(u_1)\, \mathrm{d}u_1 \right\} \left\{ \int_0^\infty \exp\left[-\int_0^T \lambda_{u_2}(t)\, \mathrm{d}t \right] f_{a_2}(u_2)\, \mathrm{d}u_2 \right\} \tag{8.39}$$

For the situation with symmetric failure thresholds ($a_1 = a_2 = a$), this is equivalent to the situation that a_1 and a_2 have identical probability distributions and are completely correlated. Assume the PDF of the threshold a is $f_a(u)$; accordingly, the structural dynamic reliability is given by

$$\begin{aligned} F_s'(-a, a, T) &= \int_0^\infty F_s(-u, u, T) f_a(u)\, \mathrm{d}u \\ &= \int_0^\infty \exp\left\{ -\int_0^T [\lambda_{-u}(t) + \lambda_u(t)]\, \mathrm{d}t \right\} f_a(u)\, \mathrm{d}u \end{aligned} \tag{8.40}$$

The basic characteristic of dynamic reliability evaluation based on the crossing assumptions is to estimate the reliability by the crossing characteristic of the response process for specified thresholds. In the procedures, the structural analysis does not intervene in the structural reliability evaluation. Accordingly, this family of methods in essence belongs to the separated algorithms. The stochastic response analysis, which is the basis of the dynamic reliability evaluation, can employ the methods described in Chapters 4 and 5.

As noted, although solutions based on Poisson or Markovian assumptions might have acceptable accuracy in some cases, the accuracy usually cannot be guaranteed because these assumptions are mainly based on empirical intuitiveness rather than rigorous mathematical approximation. The underlying reason is that, to obtain an accurate solution of first-passage reliability, all the finite-dimensional correlation information is needed, rather than only two-dimensional information. This will be clear in Section 8.4.4.

8.2.4 Pseudo-Static Analysis Method

The essence of the pseudo-static analysis method is to convert the time process response of the structures to some indices of them and compute the structural dynamic reliability by the probability distribution or statistic moments of the indices.

Usually, structural failure in a time interval [0, T] may be simplified as

$$D \geq D_c \tag{8.41}$$

in which D is the extreme value of the response in the time interval [0, T] corresponding to the failure criteria of the strength and/or deformations and D_c is the structural failure threshold and is related to a certain criterion.

If the joint probability density of D and D_c, denoted by $f_{DD_c}(u_1, u_2)$, is known, then the structural dynamic reliability can be computed by

$$\begin{aligned} F_s(T) &= \Pr\{D < D_c \mid 0 \leq t \leq T\} \\ &= \iint\limits_{D<D_c} f_{DD_c}(u_1, u_2)\,\mathrm{d}u_1\,\mathrm{d}u_2 \end{aligned} \tag{8.42}$$

If D and D_c are mutually independent, then

$$f_{DD_c}(u_1, u_2) = f_D(u_1) f_{D_c}(u_2) \tag{8.43}$$

where $f_D(u_1)$ and $f_{D_c}(u_2)$ are the PDFs of D and D_c respectively.

Consequently, the structural dynamic reliability can be computed by

$$\begin{aligned} F_s(T) &= \iint\limits_{D<D_c} f_D(u_1) f_{D_c}(u_2)\,\mathrm{d}u_1\,\mathrm{d}u_2 \\ &= \int_0^{\infty}\left[\int_0^{u_1} f_D(u_1)\,\mathrm{d}u_1\right] f_{D_c}(u_2)\,\mathrm{d}u_2 \\ &= \int_0^{\infty} F_D(u_1) f_{D_c}(u_2)\,\mathrm{d}u_2 \end{aligned} \tag{8.44}$$

in which

$$F_D(u_1) = \int_0^{u_1} f_D(u_1)\,\mathrm{d}u_1 \tag{8.45}$$

is the CDF of D.

Obviously, for the strength and deformation failure criteria, the essence of Equation 8.44 and the first equality of Equation 8.42 are identical. In particular, in the case the threshold value D_c is a deterministic variable, the structural dynamic reliability becomes

$$F_s(T) = \int_0^{D_c} f_D(u_1)\,\mathrm{d}u_1 \tag{8.46}$$

Equations 8.42, 8.44 and 8.46 are accurate expressions for structural dynamic reliability evaluation. If the PDFs required in the equations are available, then the structural dynamic reliability can be evaluated directly by the corresponding equations with the analytic method or numerical integral method. Unfortunately, the PDFs are difficult to obtain, especially for nonlinear structural systems. This difficulty had to be approached based on the probability density evolution method.

8.3 Dynamic Reliability Analysis: Generalized Density Evolution Equation-Based Approach

From the point of view of the GDEE, the dynamic reliability assessment can be performed in two different ways. One is to view the first-passage problem from the transition and absorbing of probability, resulting in the absorbing boundary condition method (Chen and Li, 2005a); the other is to view the problem by transforming it to an issue related to the corresponding extreme value (Chen and Li, 2007a). Clearly, these two ways can be viewed as counterpart ideas, having once been employed in Sections 8.2.2 and 8.2.4 respectively.

8.3.1 Absorbing Boundary Condition Method

Let us examine the reliability of the first-passage problem defined in Equation 8.6:

$$F_s = \Pr\{X(t) \in \Omega_s,\ t \in (0, T]\} \tag{8.6}$$

As is pointed out, this means that the structural failure happens once the response exceeds the safety boundary. The generalized density evolution equation in terms of the probability-preserved system ($X(t)$, $\boldsymbol{\Theta}$) is given by (see Section 6.5 and Equations 6.123a and 6.123b)

$$\frac{\partial p_{X\boldsymbol{\Theta}}(x, \boldsymbol{\theta}, t)}{\partial t} + h_X(\boldsymbol{\theta}, t)\frac{\partial p_{X\boldsymbol{\Theta}}(x, \boldsymbol{\theta}, t)}{\partial x} = 0 \tag{8.47}$$

Here, $\boldsymbol{\Theta}$ is the basic random variable vector involved, $p_{X\boldsymbol{\Theta}}(x, \boldsymbol{\theta}, t)$ is the joint PDF and $h_X(\boldsymbol{\theta}, t)$ is the formal solution of the velocity.

We now recall the random event description of the evolution of probability as elaborated in Section 6.2. Equation 8.6 requires that all the samples must satisfy the criterion to ensure the safety of the structure, otherwise the structure will fail. Of course, if a sample (a realized event) violates the criterion, then this sample will contribute to the failure probability, but not contribute to the reliability.[1] Thus, equivalently, we can impose an absorbing boundary condition on Equation 8.47:

$$p_{X\boldsymbol{\Theta}}(x, \boldsymbol{\theta}, t) = 0 \quad \text{for } x \in \Omega_f \tag{8.48}$$

where Ω_f is the failure domain, which is the complementary set of the safety domain Ω_s in Equation 8.6.

Combining Equations 8.48 and 8.47 will give the basic equations to obtain the PDFs $\breve{p}_{X\boldsymbol{\Theta}}(x, \boldsymbol{\theta}, t)$. We see that physical meaning of the absorbing boundary condition says that, once a sample (a realized event) violates the safety criterion, the associated (adherent) probability will never return to the safety domain. In this sense, we call such an obtained probability density the *remaining probability density* and denoted by $\breve{p}_X(x, t)$ (or $\breve{p}_{X\boldsymbol{\Theta}}(x, \boldsymbol{\theta}, t)$ for the joint PDF):

$$\breve{p}_X(x, t) = \int_{\Omega_{\boldsymbol{\Theta}}} \breve{p}_{X\boldsymbol{\Theta}}(x, \boldsymbol{\theta}, t)\, d\boldsymbol{\theta} \tag{8.49}$$

[1] Rigorously speaking, the analysis here has to exclude the so-called zero-measure set. But this will not lead to any essential difference for the present problem.

The reliability is then given by

$$F_s = \int_{\Omega_s} \breve{p}_X(x,t)\,\mathrm{d}x = \int_{-\infty}^{\infty} \breve{p}_X(x,t)\,\mathrm{d}x \tag{8.50}$$

The second equality holds because of Equation 8.48.

For instance, for the symmetric double-boundary problem of which the reliability is defined by

$$F_s = \Pr\{|X(t)| \le x_b,\ t \in (0, T]\} \tag{8.51}$$

where x_b is the threshold. The absorbing boundary condition (Equation 8.48) becomes

$$p_{X\boldsymbol{\Theta}}(x, \boldsymbol{\theta}, t) = 0 \quad \text{for } |x| > x_b \tag{8.52}$$

Thus, the reliability is given by

$$F_s = \int_{-x_b}^{x_b} \breve{p}_X(x,t)\,\mathrm{d}x = \int_{-\infty}^{\infty} \breve{p}_X(x,t)\,\mathrm{d}x \tag{8.53}$$

Except for the absorbing boundary condition, all the solving procedures for Equations 8.47 and 8.48 is the same as that elaborated in Section 6.6 and Chapter 7.

Figure 8.3 illustrates the effect of the absorbing boundary condition on the remaining PDFs. They are the contours of the probability density surface against time (probability density evolution surface) of the displacement response of a nonlinear structure subjected to earthquake ground motions. From the contours it is seen that, because part of the probability (related to the failure events) is absorbed, the remaining probability density is quite different from the original probability density. Figure 8.4 illustrates the dynamic reliability of the nonlinear system evaluated through the approach elaborated in this section.

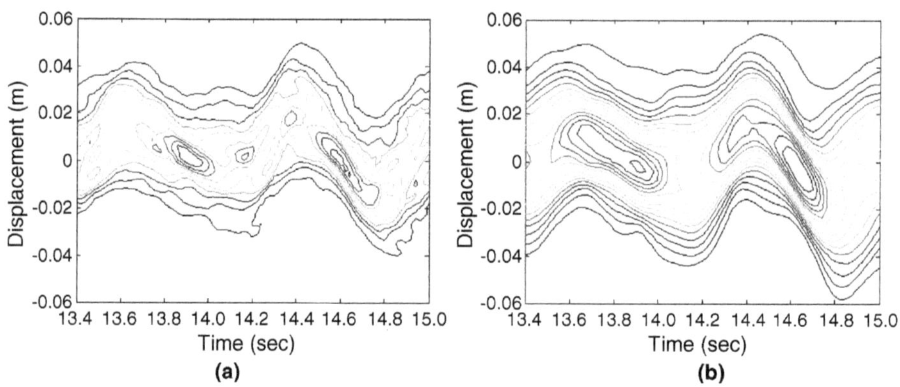

Figure 8.3 Contour of the PDF surface: (a) without absorbing boundary condition; (b) with absorbing boundary condition.

8.3.2 Extreme-Value Distribution of the Stochastic Dynamical Response

In general, the extreme value of a stochastic process is a random variable. As discussed in the section above, how to get the extreme-value distribution (EVD) of a general stochastic process

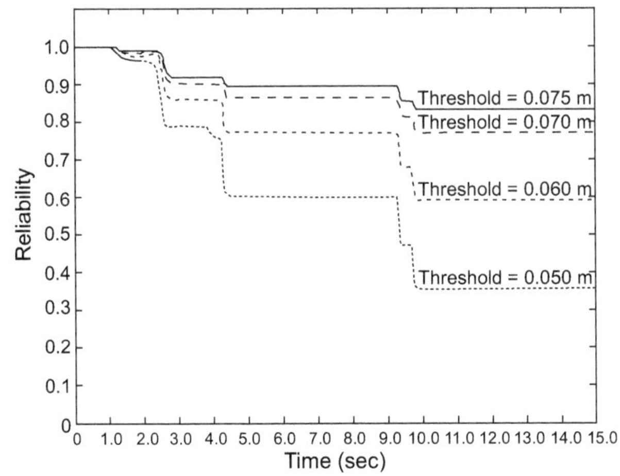

Figure 8.4 First-passage reliability.

is a difficult problem. Only some special results were achieved for some particular stochastic processes (Newland, 1993; Finkenstädt and Rootzén, 2004). In contrast, based on the GDEE described in Chapter 6, the EVD can be evaluated through constructing a virtual stochastic process.

Denote the extreme value of the response $X(t)$ of a stochastic system as

$$X_{\text{ext}} = \underset{t \in [0,T]}{\text{ext}} X(\mathbf{\Theta}, t) \tag{8.54}$$

For instance, if one considers the maximum absolute value of $X(t)$, $t \in [0, T]$, then Equation 8.54 essentially stands for

$$|X|_{\max} = \max_{\tau \in [0,T]} |X(\mathbf{\Theta}, \tau)| \tag{8.55}$$

From Equation 8.54, it is seen that the extreme value of $X(t)$, $t \in [0, T]$, depends on $\mathbf{\Theta}$. For convenience, it can be assumed to take the form

$$X_{\text{ext}} = W(\mathbf{\Theta}, T) \tag{8.56}$$

which means that the extreme value of $X(t)$, $t \in [0, T]$, is existent, unique and is a function of $\mathbf{\Theta}$ and T.

Introducing a virtual stochastic process:

$$Y(\tau) = \tilde{\varphi}[W(\mathbf{\Theta}, T), \tau] = \varphi(\mathbf{\Theta}, \tau) \tag{8.57}$$

where τ is somewhat like the time and is termed the 'virtual time.' $Y(\tau)$ is a 'virtual stochastic process' whose randomness comes from the random vector $\mathbf{\Theta}$. Usually, we require that the virtual stochastic process satisfies the conditions

$$Y(\tau)|_{\tau=0} = 0 \qquad Y(\tau)|_{\tau=\tau_c} = W(\mathbf{\Theta}, T) \tag{8.58}$$

For instance, if we let

$$Y(\tau) = W(\boldsymbol{\Theta}, T) \cdot \tau \tag{8.59}$$

and $\tau_0 = 1$, then this process satisfies the conditions in Equation 8.58.

Differentiating Equation 8.57 on both sides with regard to τ will yield

$$\dot{Y}(\tau) = \frac{\partial \varphi(\boldsymbol{\Theta}, \tau)}{\partial \tau} = \dot{\varphi}(\boldsymbol{\Theta}, \tau) \tag{8.60}$$

It is easy to find that Equation 8.60 is in a form similar to Equation 6.96b. Therefore, a generalized density evolution equation, described in Chapter 6, could be employed here to obtain the PDF of $Y(\tau)$. After similar deductions like that in Chapter 6, we can obtain the following equation (Chen and Li, 2007a):

$$\frac{\partial p_{Y\boldsymbol{\Theta}}(y, \boldsymbol{\theta}, \tau)}{\partial \tau} + \dot{\varphi}(\boldsymbol{\theta}, \tau) \frac{\partial p_{Y\boldsymbol{\Theta}}(y, \boldsymbol{\theta}, \tau)}{\partial y} = 0 \tag{8.61}$$

with the initial condition (from Equation 8.58)

$$p_{Y\boldsymbol{\Theta}}(y, \boldsymbol{\theta}, \tau)|_{\tau=0} = \delta(y) p_{\boldsymbol{\Theta}}(\boldsymbol{\theta}) \tag{8.62}$$

where $p_{Y\boldsymbol{\Theta}}(y, \boldsymbol{\theta}, \tau)$ is the joint PDF of $(Y(\tau), \boldsymbol{\Theta})$.

Once the initial-value problem (Equations 8.61 and 8.62) is solved, the PDF of $Y(\tau)$ will be given by

$$p_Y(y, \tau) = \int_{\Omega_{\boldsymbol{\Theta}}} p_{Y\boldsymbol{\Theta}}(y, \boldsymbol{\theta}, \tau) \, \mathrm{d}\boldsymbol{\theta} \tag{8.63}$$

From Equation 8.58 it can be seen that the extreme value X_{ext} equals the value of the virtual stochastic process $Y(\tau)$ at the instant of time $\tau = \tau_c$; that is:

$$X_{\text{ext}} = Y(\tau)|_{\tau=\tau_c} \tag{8.64}$$

According to Equations 8.63 and 8.64, the PDF of X_{ext} can be obtained immediately:

$$p_{X_{\text{ext}}}(x) = p_Y(y = x, \tau)|_{\tau=\tau_c} \tag{8.65}$$

Example 8.1. The EVD of a Set of Random Variables Consider a set of mutually independent random variables $(X_1, X_2, \ldots, X_r)$ with identical PDF $p_X(x)$. Let

$$X_{\max} = \max(X_1, X_2, \ldots, X_r) \tag{8.66}$$

Then the closed form of the PDF of $X_{\max}$ is available (Ang and Tang, 1984):

$$p_{X_{\max}}(x) = r[P_X(x)]^{r-1} p_X(x)$$

where

$$P_X(x) = \int_{-\infty}^{x} p_X(x)\, dx$$

is the CDF of $p_X(x)$.

If we note that Equation 8.66 can be viewed as a special case of Equation 8.56, then the approach elaborated above can be employed to obtain the PDF of X_{max}. Figure 8.5 shows the comparison between the analytical solution and the EVD obtained by the PDEM when the original distribution is uniform distribution over [1, 2] and normal distribution with mean being 4 and unity variance respectively. It is seen that, for $r = 2$ and 3, the EVDs obtained by the PDEM are almost identical to the analytical solution except in the vicinity of discontinuity.

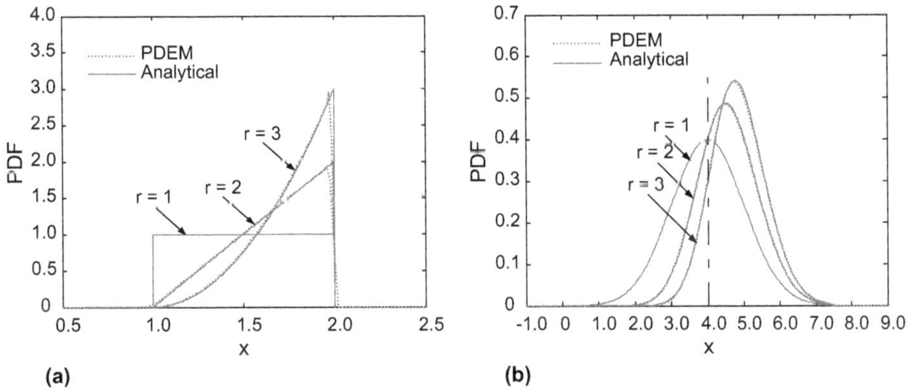

Figure 8.5 EVD: (a) $X_i \sim U[1,2]$; (b) $X_i \sim N(4,1)$.

By the way, we point out that the approach elaborated here can also be extended to obtain the PDF of an arbitrary function of some basic random variables. □

8.3.3 *Extreme-Value Distribution-based Dynamical Reliability Evaluation of Stochastic Systems*

Dynamic reliability of stochastic systems could be evaluated in a straightforward way through integration of the above EVD. For example, for the symmetric double boundary problem, the dynamic reliability of the structure in time T can be described as

$$F_s(-a, a, T) = \Pr\{|X(\tau)| \le a,\ \tau \in [0, T]\} \tag{8.67}$$

where a is the value of the symmetric boundary. Viewed from the EVD, the above equation can be rewritten as

$$F_s(-a, a, T) = \Pr\{|X(\tau)|_{max} \le a\} \tag{8.68}$$

Since the EVD $p_{X_{ext}}(x)$ can be captured in the preceding section, it is quite easy to evaluate the reliability in Equation 8.68 through a simple integration:

$$F_s(-a, a, T) = \int_{-a}^{a} p_{X_{ext}}(x)\, dx \tag{8.69}$$

If the boundary a is a random variable with the PDF $p_A(a)$, then the dynamic reliability will be

$$R = \int_{\Omega_A} \left[\int_{-a}^{a} p_{X_{\text{ext}}}(x)\,\mathrm{d}x \right] p_A(a)\,\mathrm{d}a \tag{8.70}$$

where Ω_A is the distribution domain of a.

The above analysis indicates that, viewed from the EVD, the problem of dynamic reliability evaluation is transformed to a simple integration problem. In contrast to reliability theory based on the level-crossing process, the above two methods based on the GDEE require neither the joint PDF of the response and its velocity, nor the assumptions on properties of the level-crossing events.

Figure 8.6 shows the PDF and CDF of the absolute maximum displacement at the top of a 10-story frame structures. Figure 8.6a shows that the EVD obtained is obviously different from the widely used regular distribution with the same mean and standard deviation. Figure 8.6b shows the comparison between the CDF of the EVD obtained and that obtained by the Monte Carlo simulation. Clearly, if the abscissa of Figure 8.6b is understood as the threshold, then the ordinate gives the reliability, and thus the complementary to one gives the failure probability.

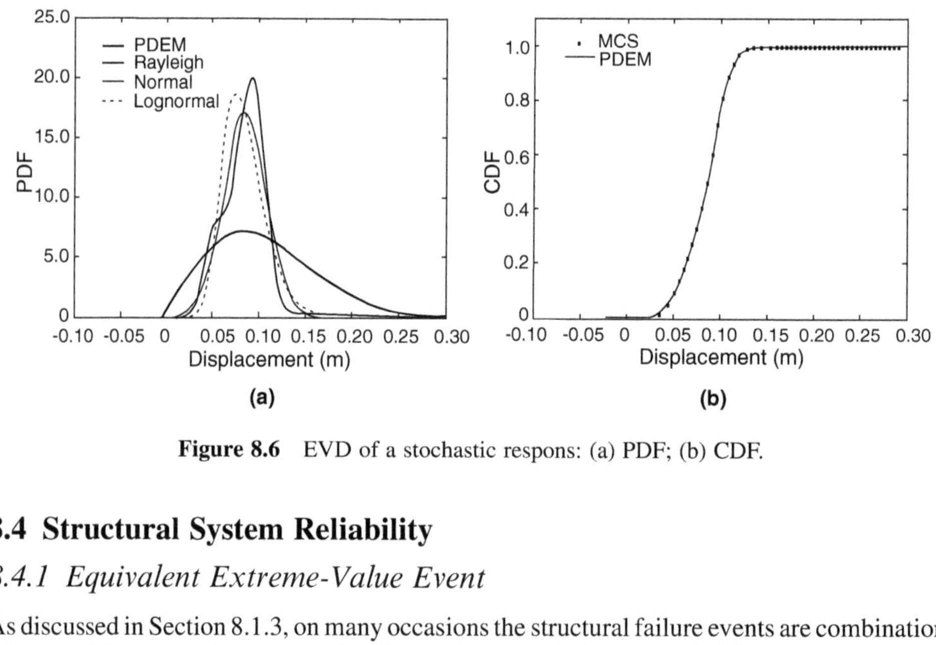

Figure 8.6 EVD of a stochastic respons: (a) PDF; (b) CDF.

8.4 Structural System Reliability

8.4.1 Equivalent Extreme-Value Event

As discussed in Section 8.1.3, on many occasions the structural failure events are combinations of some different random events, leading to the so-called system reliability; for instance, the reliability of a structure might be defined by

$$P_r = \Pr\{(G_1(\boldsymbol{\Theta}) > 0) \cap (G_2(\boldsymbol{\Theta}) > 0)\} \tag{8.71}$$

where $G_1(\cdot)$ and $G_2(\cdot)$ are two different limit state functions corresponding to different failure modes of a structure.

When the random events are combinations of more than one inequality, it is found that the probability could generally be evaluated through an equivalent extreme-value event

(Li *et al.*, 2007). In order to give insight to the idea, we will start with the simplest occasions involving two random variables.

Lemma 8.1. *Suppose X and Y are correlated random variables, W_{min} being the minimum value of X and Y and, therefore, also being a random variable, then there exists*

$$\Pr\{(X>a)\cap(Y>a)\} = \Pr\{W_{min}>a\} \tag{8.72}$$

Although Equation 8.72 evidently holds from the point of view of a logical relationship, it is worth giving a rigorous proof because this will help to understand the idea of inherent correlation, as will be discussed later.

Proof: Denote the joint PDF of (X, Y) by $p_{XY}(x, y)$, then the probability of the random event $\{(X>a)\cap(Y>a)\}$ is

$$\begin{aligned}\Pr\{(X>a)\cap(Y>a)\} &= \iint_{x>a,y>a} p_{XY}(x,y)\,\mathrm{d}x\,\mathrm{d}y \\ &= \int_a^\infty\left[\int_a^\infty p_{XY}(x,y)\,\mathrm{d}x\right]\mathrm{d}y\end{aligned} \tag{8.73}$$

Because

$$W_{\min} = \min(X, Y) = \begin{cases} X & \text{if } X \le Y \\ Y & \text{otherwise}\end{cases} \tag{8.74}$$

it follows that

$$\begin{aligned}\Pr\{W_{\min}>a\} &= \int_a^\infty p_{W_{\min}}(z)\,\mathrm{d}z \\ &= \Pr\{\min(X,Y)>a\} \\ &= \iint_{x<y,x>a} p_{XY}(x,y)\,\mathrm{d}x\,\mathrm{d}y + \iint_{y<x,y>a} p_{XY}(x,y)\,\mathrm{d}x\,\mathrm{d}y \\ &= \int_a^\infty\left[\int_x^\infty p_{XY}(x,y)\,\mathrm{d}y\right]\mathrm{d}x + \int_a^\infty\left[\int_y^\infty p_{XY}(x,y)\,\mathrm{d}x\right]\mathrm{d}y \\ &= \int_a^\infty\left[\int_a^y p_{XY}(x,y)\,\mathrm{d}x\right]\mathrm{d}y + \int_a^\infty\left[\int_y^\infty p_{XY}(x,y)\,\mathrm{d}x\right]\mathrm{d}y \\ &= \int_a^\infty\left[\int_a^\infty p_{XY}(x,y)\,\mathrm{d}x\right]\mathrm{d}y\end{aligned} \tag{8.75}$$

Comparing Equation 8.73 with Equation 8.75 will immediately yield Equation 8.72. □

Lemma 8.2. *Suppose X, Y are correlated random variables, W_{max} being the maximum value of X and Y and therefore also a random variable, then there exists*

$$\Pr\{(X>a)\cup(Y>a)\} = \Pr\{W_{max}>a\} \tag{8.76}$$

Proof: The probability of the random event $\{(X>a)\cup(Y>a)\}$ is

$$\begin{aligned}\Pr\{(X>a)\cup(Y>a)\} &= \iint_{x>a} p_{XY}(x,y)\,\mathrm{d}x\,\mathrm{d}y + \iint_{y>a,x<a} p_{XY}(x,y)\,\mathrm{d}x\,\mathrm{d}y \\ &= \int_{-\infty}^{\infty}\left[\int_a^{\infty} p_{XY}(x,y)\,\mathrm{d}x\right]\mathrm{d}y + \int_a^{\infty}\left[\int_{-\infty}^{a} p_{XY}(x,y)\,\mathrm{d}x\right]\mathrm{d}y\end{aligned} \tag{8.77}$$

Bearing in mind that

$$W_{\max} = \max(X,Y) = \begin{cases} Y & \text{if } X\le Y \\ X & \text{otherwise}\end{cases} \tag{8.78}$$

we can get

$$\begin{aligned}\Pr\{W_{\max}>a\} &= \int_a^{\infty} p_{W_{\max}}(z)\,\mathrm{d}z = \Pr\{\max(X,Y)>a\} \\ &= \iint_{x<y,y>a} p_{XY}(x,y)\,\mathrm{d}x\,\mathrm{d}y + \iint_{y<x,x>a} p_{XY}(x,y)\,\mathrm{d}x\,\mathrm{d}y \\ &= \int_a^{\infty}\left[\int_{-\infty}^{y} p_{XY}(x,y)\,\mathrm{d}x\right]\mathrm{d}y + \int_a^{\infty}\left[\int_{-\infty}^{x} p_{XY}(x,y)\,\mathrm{d}y\right]\mathrm{d}x \\ &= \left\{\int_a^{\infty}\left[\int_{-\infty}^{a} p_{XY}(x,y)\,\mathrm{d}x\right]\mathrm{d}y + \int_a^{\infty}\left[\int_a^{y} p_{XY}(x,y)\,\mathrm{d}x\right]\mathrm{d}y\right\} \\ &\quad + \left\{\int_a^{\infty}\left[\int_{-\infty}^{a} p_{XY}(x,y)\,\mathrm{d}y\right]\mathrm{d}x + \int_a^{\infty}\left[\int_a^{x} p_{XY}(x,y)\,\mathrm{d}y\right]\mathrm{d}x\right\}\end{aligned} \tag{8.79}$$

Exchanging the order of integration with respect to x and y in the last two terms will yield

$$\begin{aligned}\Pr\{W_{\max}>a\} &= \left\{\int_a^{\infty}\left[\int_{-\infty}^{a} p_{XY}(x,y)\,\mathrm{d}x\right]\mathrm{d}y + \int_a^{\infty}\left[\int_a^{y} p_{XY}(x,y)\,\mathrm{d}x\right]\mathrm{d}y\right\} \\ &\quad + \left\{\int_{-\infty}^{a}\left[\int_a^{\infty} p_{XY}(x,y)\,\mathrm{d}x\right]\mathrm{d}y + \int_a^{\infty}\left[\int_y^{\infty} p_{XY}(x,y)\,\mathrm{d}x\right]\mathrm{d}y\right\} \\ &= \int_a^{\infty}\left[\int_{-\infty}^{a} p_{XY}(x,y)\,\mathrm{d}x\right]\mathrm{d}y + \left\{\int_a^{\infty}\left[\int_a^{y} p_{XY}(x,y)\,\mathrm{d}x\right]\mathrm{d}y\right. \\ &\quad \left. + \int_{-\infty}^{a}\left[\int_a^{\infty} p_{XY}(x,y)\,\mathrm{d}x\right]\mathrm{d}y + \int_a^{\infty}\left[\int_y^{\infty} p_{XY}(x,y)\,\mathrm{d}x\right]\mathrm{d}y\right\} \\ &= \int_a^{\infty}\left[\int_{-\infty}^{a} p_{XY}(x,y)\,\mathrm{d}x\right]\mathrm{d}y + \int_{-\infty}^{\infty}\left[\int_a^{\infty} p_{XY}(x,y)\,\mathrm{d}x\right]\mathrm{d}y\end{aligned} \tag{8.80}$$

Comparing Equation 8.77 with Equation 8.80, we find that the first term on the right-hand side of Equation 8.80 is identical to the second term on the right-hand side of Equation 8.77 while the second term on the right-hand side of Equation 8.80 is identical to the first term on the right-hand side of Equation 8.77. This means that Equation 8.76 holds true. □

According to Lemmas 8.1 and 8.2, this indicates that if one wants to evaluate the probability of a compound random event as a combination of two random events represented by inequalities, one just needs to evaluate the probability of an event related to an extreme value which is defined according to the logical relationship between the original two inequalities. In this sense, say in the case of Lemma 8.1, the random event $\{W_{\min} > a\}$ could be referred to as the *equivalent extreme-value event* of $\{(X>a) \cap (Y>a)\}$ and $W_{\min}$ as the equivalent extreme-value random variable. Likewise, the random event $\{W_{\max} > a\}$ is the equivalent extreme-value event of $\{(X>a) \cup (Y>a)\}$ and $W_{\max}$ is the corresponding equivalent extreme-value random variable.

Evidently, the rules holds true when the compound random event is a combination of more than two component random events. This idea leads to the following theorems.

Theorem 8.1. *Suppose $X_1, X_2, \ldots, X_m$ are m random variables. Let $W_{min} = min_{1 \le j \le m}(X_j)$, then it goes that*

$$\Pr\left\{\bigcap_{j=1}^{m}(X_j > a)\right\} = \Pr\{W_{\min} > a\} \tag{8.81}$$

Proof: Denote the minimum value of $X_1, X_2, \ldots, X_j$ $(2 \le j \le m)$ as $W_{\min}^{(j)}$; namely:

$$W_{\min}^{(j)} = \min(X_1, X_2, \ldots, X_j) \tag{8.82}$$

Define $W_{\min}^{(1)} = X_1$ and $W_{\min}^{(m)} = W_{\min}$. There is a recursive relation that $W_{\min}^{(j)} = \min(W_{\min}^{(j-1)}, X_j)$, $2 \le j \le m$. Using Lemma 8.1 recursively we obtain

$$\begin{aligned}
\Pr\left\{\bigcap_{j=1}^{m}(X_j > a)\right\} &= \Pr\{[(X_1 > a) \cap (X_2 > a)]\bigcap_{j=3}^{m}(X_j > a)\} \\
&= \Pr\{[(W_{\min}^{(2)} > a) \cap (X_3 > a)]\bigcap_{j=4}^{m}(X_j > a)\} \\
&= \ldots \\
&= \Pr\{(W_{\min}^{(m-1)} > a) \cap (X_m > a)\} \\
&= \Pr\{(W_{\min}^{(m)} > a)\} \\
&= \Pr\{(W_{\min} > a)\}
\end{aligned} \tag{8.83}$$

□

Theorem 8.2. *Suppose $X_1, X_2, \ldots, X_m$ are m random variables. Let $W_{max} = max_{1 \le j \le m}(X_j)$; then it goes that*

$$\Pr\left\{\bigcup_{j=1}^{m}(X_j > a)\right\} = \Pr\{W_{\max} > a\} \tag{8.84}$$

The proof is similar to Theorem 8.1 and will not be detailed. Likewise, for any arbitrary types of combination of random events, we can always construct the according equivalent extreme-value events; for instance, the following theorem holds.

Theorem 8.3. *Suppose* X_{ij}, $i = 1, 2, \ldots, n$; $j = 1, 2, \ldots, m$, *are* $m \times n$ *random variables. Let* $W_{ext} = max_{1 \le i \le n}(min_{1 \le j \le m}(X_{ij}))$; *then it goes that*

$$\Pr\left\{\bigcup_{i=1}^{n}\left[\bigcap_{j=1}^{m}(X_{ij} > a)\right]\right\} = \Pr\{W_{\text{ext}} > a\} \tag{8.85}$$

For the sake of clarity, in the lemmas and theorems, the thresholds for different random variables take the same value a. It appears at first glance that this will lead to loss of generality; but this is not true, because the inequality can be equivalently transformed with a linear transformation. For instance, consider a random event $\{(\tilde{X} < b) \cap (\tilde{Y} > c)\}$ where $\tilde{X}$ and $\tilde{Y}$ are random variables with joint PDF $p_{\tilde{X}\tilde{Y}}(\tilde{x}, \tilde{y})$. If we introduce a couple of new random variables

$$X = -\tilde{X} + b + a \tag{8.86a}$$

$$Y = \tilde{Y} - c + a \tag{8.86b}$$

of which the joint PDF is

$$p_{XY}(x, y) = p_{\tilde{X}\tilde{Y}}(-x + b + a, y + c - a) \tag{8.87}$$

then the random event $\{(\tilde{X} < b) \cap (\tilde{Y} > c)\}$ is equivalently transformed to $\{(X > a) \cap (Y > a)\}$, which is the case discussed in Lemma 8.1.

8.4.2 Inherent Correlation Property of Equivalent Extreme-Value Event

In constructing the equivalent extreme-value event, the random variables can be either mutually dependent or independent. In other words, this indicates that although in the equivalent extreme-value event only one equivalent random variable is employed explicitly instead of the original multiple random variables, the correlation information in the original random variables, together with the effects on the computed probability, is retained in the equivalent extreme-value event.

In practical situations, the random variables involved in the probability computation of random events, say as shown in Equation 8.72 or 8.76, are usually not basic random variables. Instead, they may be functions of the same set of basic random variables where the randomness comes from. For instance, when different response indices of a structure involving randomness characterized by $\boldsymbol{\Theta}$ are considered, these response indices, if denoted by X and Y, are evidently functions of $\boldsymbol{\Theta}$; that is:

$$X = H_X(\boldsymbol{\Theta}) \qquad Y = H_Y(\boldsymbol{\Theta}) \tag{8.88}$$

where $\boldsymbol{\Theta} = (\Theta_1, \Theta_2, \ldots, \Theta_s)$ are the basic random variables with joint PDF $p_{\boldsymbol{\Theta}}(\boldsymbol{\theta})$. In this situation, the random variables X and Y are unlikely mutually independent in general, except in some pretty special cases.[2] The joint PDF $p_{XY}(x, y)$ could be computed in

[2] There are indeed some very special cases where two random variables as functions of the same set of random variables are independent or uncorrelated. For example, refer to Wang (1976). However, in practical engineering, the chance of encountering these cases is very rare.

principle by

$$p_{XY}(x,y) = \frac{\partial^2}{\partial x \partial y} \int_{H_X(\boldsymbol{\Theta}) < x, H_Y(\boldsymbol{\Theta}) < y} p_{\boldsymbol{\Theta}}(\boldsymbol{\theta})\, \mathrm{d}\boldsymbol{\theta} \tag{8.89}$$

Usually, the computation of Equation 8.89 is practically far from easy.

According to Equations 8.73 and 8.89, we can get

$$\begin{aligned} \Pr\{(X>a) \cap (Y>a)\} &= \iint_{x>a, y>a} p_{XY}(x,y)\, \mathrm{d}x\, \mathrm{d}y \\ &= \int_{H_X(\boldsymbol{\Theta})>a, H_Y(\boldsymbol{\Theta})>a} p_{\boldsymbol{\Theta}}(\boldsymbol{\theta})\, \mathrm{d}\boldsymbol{\theta} \end{aligned} \tag{8.90}$$

From Equation 8.90 it is seen that the correlation information between X and Y is involved in turn in the form of integration with regard to $\boldsymbol{\theta}$ in the domain determined by Equation 8.88. In other words, in Equation 8.90, the correlation information is involved.

On the other hand, according to Equations 8.74, 8.75 and 8.54, if we define

$$W_{\min} = \min(X, Y) = \min(H_X(\boldsymbol{\Theta}), H_Y(\boldsymbol{\Theta})) = H_W(\boldsymbol{\Theta}) \tag{8.91}$$

then Equation 8.75 becomes

$$\begin{aligned} \Pr\{W_{\min} > a\} &= \int_a^\infty p_{W_{\min}}(w)\, \mathrm{d}w \\ &= \Pr\{\min(X,Y) > a\} \\ &= \int_{H_W(\boldsymbol{\Theta})>a} p_{\boldsymbol{\Theta}}(\boldsymbol{\theta})\, \mathrm{d}\boldsymbol{\theta} \end{aligned} \tag{8.92}$$

Noting Equations 8.73 and 8.91, it is seen that although using the equivalent extreme-value event $\{W_{\min} > a\}$ in place of $\{(X>a) \cap (Y>a)\}$ in computation of the probability with the first equality in Equation 8.92 appears not to involve the correlation information between X and Y explicitly, the correlation information is indeed retained.

The above discussions on the inherent correlation property in the equivalent extreme-value event are obviously true in the case that more than two random variables are involved. Using the equivalent extreme-value event in place of the original random events as combinations of more than one random event makes it possible to reduce the multidimensional probability integration to a one-dimensional probability integration, provided the PDF of the equivalent extreme-value random variable is available. The correlation information is inherent in the equivalent extreme-value event; therefore, in this process, no correlation information disappears.

8.4.3 Differences between the Equivalent Extreme-Value Event and the Weakest Link Assumption

In structural reliability evaluation, the weakest link assumption is often employed. Consider a structural system whose probability of failure is

$$P_{\mathrm{f}} = \Pr\{(X>a) \cup (Y>a)\} \tag{8.93}$$

Denote $P_{f1} = \Pr\{X>a\}$, $P_{f2} = \Pr\{Y>a\}$. When the weakest link assumption is adopted, we may use

$$P_f = \max(P_{f1}, P_{f2}) \tag{8.94}$$

in place of Equation 8.93 (Madsen *et al.*, 1986; Melchers, 1999).

According to Lemma 8.2 in Section 8.4.1, the probability in Equation 8.93 is equal to

$$P_f = \Pr\{W_{\max}>a\} \tag{8.95}$$

It is easy to see that, in the weakest link assumption, the failure probability of the system is replaced by the maximum of the failure probabilities of the basic failure events; that is, the following equality is assumed:

$$P_f = \max(\Pr\{X>a\}, \Pr\{Y>a\}) = \Pr\{\max(X,Y)>a\} \tag{8.96}$$

However, because the orders of the operator $\Pr\{\cdot\}$ and $\max\{\cdot\}$ cannot be exchanged, Equation 8.96 does not hold true in general.

In fact, Equation 8.93 could be computed by

$$\begin{aligned} P_f' &= \int_{(X>a)\cup(Y>a)} p_{XY}(x,y)\,\mathrm{d}x\,\mathrm{d}y \\ &= \int_a^{\infty}\left[\int_{-\infty}^{\infty} p_{XY}(x,y)\,\mathrm{d}y\right]\mathrm{d}x + \int_{-\infty}^{a}\left[\int_a^{\infty} p_{XY}(x,y)\,\mathrm{d}y\right]\mathrm{d}x \\ &= P_{f1} + \Delta P_1 \end{aligned} \tag{8.97}$$

where

$$\Delta P_1 = \int_{-\infty}^{a}\left[\int_a^{\infty} p_{XY}(x,y)\,\mathrm{d}y\right]\mathrm{d}x = \iint_{A_1} p_{XY}(x,y)\,\mathrm{d}x\,\mathrm{d}y \tag{8.98}$$

in which A_1 is the area indicated in Figure 8.7.

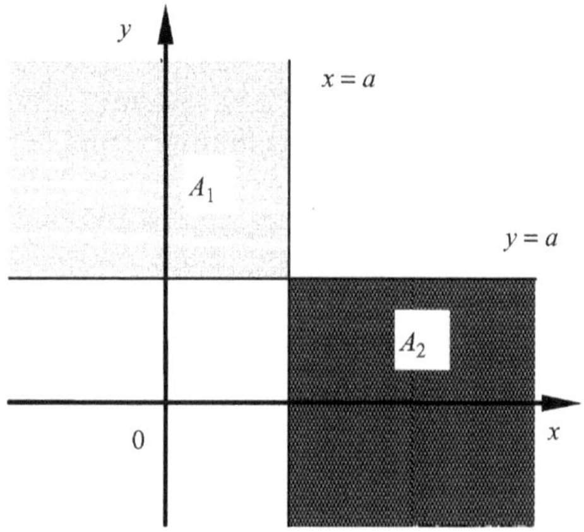

Figure 8.7 Partition of the integral domain.

Likewise, we can get

$$P_{\mathrm{f}} = P_{\mathrm{f}2} + \Delta P_2 \tag{8.99}$$

where $\Delta P_2 = \iint_{A_2} p_{XY}(x, y)\,\mathrm{d}x\,\mathrm{d}y$ and A_2 is the area shown in Figure 8.7.

Because $p_{XY}(x, y) \geq 0$, there are $\Delta P_1 \geq 0$ and $\Delta P_2 \geq 0$; consequently:

$$P_{\mathrm{f}} \geq \max(P_{\mathrm{f}1}, P_{\mathrm{f}2}) \tag{8.100}$$

This shows that Equation 8.94 does not hold in general cases.

If X and Y are completely positively correlated random variables, that is, $Y = kX + b (k > 0)$ and the joint PDF is thus

$$p_{XY}(x, y) = p_X(x)\delta(y - kx - b) \tag{8.101}$$

where $\delta(\cdot)$ is the Dirac delta function. According to Equation 8.98 there is

$$\begin{aligned} \Delta P_1 &= \int_{-\infty}^{a} \left(\int_{a}^{\infty} p_{XY}(x, y)\,\mathrm{d}y \right) \mathrm{d}x \\ &= \int_{-\infty}^{a} \left(\int_{a}^{\infty} p_X(x)\delta(y - kx - b)\,\mathrm{d}y \right) \mathrm{d}x \\ &= \int_{-\infty}^{a} p_X(x) u(x - (a - b)/k)\,\mathrm{d}x \end{aligned} \tag{8.102}$$

in which $u(\cdot)$ is the Heaviside step function (see Appendix A).

Likewise, we can obtain

$$\Delta P_2 = \int_{-\infty}^{a} p_Y(y) u(y - ka - b)\,\mathrm{d}y \tag{8.103}$$

It is easy to prove that either

$$\frac{a - b}{k} < a < ka + b \tag{8.104}$$

or

$$ka + b < a < \frac{a - b}{k} \tag{8.105}$$

holds true provided $k > 0$.

Therefore, from Equations 8.102 and 8.103 we can see that either $\Delta P_1 = 0$ or $\Delta P_2 = 0$ holds. Therefore, either $P_{\mathrm{f}} = P_{\mathrm{f}1}$ or $P_{\mathrm{f}} = P_{\mathrm{f}2}$ holds according to Equations 8.97 and 8.99. That means only in this case the weakest link assumption holds true.

In summary, the equivalent extreme-value event is different from the weakest link assumption in essence. Only in the case that the basic failure events are completely positively correlated the weakest link assumption is identical to the equivalent extreme-value event.

8.4.4 Evaluation of Structural System Reliability

For the first-passage problem, the reliability against the response index $X(t)$ can be generally described as

$$R = \Pr\{X(\boldsymbol{\Theta}, t) \in \Omega_s,\ t \in [0, T]\} \tag{8.106}$$

where Ω_s is the safe domain.

For most practical problems, Equation 8.106 can be rewritten as

$$R = \Pr\{G(\boldsymbol{\Theta}, t) > 0,\ t \in [0, T]\} \tag{8.107}$$

where $G(\cdot)$ is a time-dependent limit state function. For instance, if Equation 8.106 takes the form (as a double boundary condition)

$$R = \Pr\{|X(\boldsymbol{\Theta}, t)| < x_b,\ t \in [0, T]\} \tag{8.108}$$

where x_b is the threshold, then we can get

$$G(\boldsymbol{\Theta}, t) = x_b - |X(\boldsymbol{\Theta}, t)| \tag{8.109}$$

Equation 8.107 could also be written equivalently in a different form as

$$R = \Pr\left\{ \bigcap_{t \in [0,T]} (G(\boldsymbol{\Theta}, t) > 0) \right\} \tag{8.110}$$

According to the situation similar to Theorem 8.1 in Section 8.4.1, if we define an extreme value as

$$W_{\min} = \min_{t \in [0,T]} (G(\boldsymbol{\Theta}, t)) \tag{8.111}$$

of which the PDF can be captured according to Section 8.3, then the reliability in Equation 8.110 equals

$$R = \Pr\{W_{\min} > 0\} \tag{8.112}$$

It is worth pointing out that if we want to evaluate the reliability in Equation 8.110 directly with the probability integration analogous to Equation 8.73, then the infinite-dimensional joint PDF of the stochastic process $G(\boldsymbol{\Theta},t)$ is needed; that is, the correlation information among any different time instants is required. Noting that in the widely used out-crossing-process theory on the first-passage reliability problem, either with the Poisson assumption or with the Markovian assumption (see Section 8.2.2), usually only the correlation information between two different time instants is considered. Consequently, in general situations, the out-crossing-process theory on the first-passage reliability problem unlikely yields the exact solution. Whereas using the equivalent extreme-value event, as is discussed in the preceding sections, total information of the correlation is inherent and an exact solution can be derived easily.

For system reliability (that is, if there is more than one limit state function combined together to be considered), say, consider

$$R = \Pr\left\{\bigcap_{j=1}^{m}(G_j(\boldsymbol{\Theta}, t) > 0,\ t \in [0, T_j])\right\} \tag{8.113}$$

where T_j is the time duration corresponding to $G_j(\cdot)$. Combining Equation 8.110 and Theorem 8.1 in Section 8.4.1, we can define the equivalent extreme value as

$$W_{\text{ext}} = \min_{1 \le j \le m}\left[\min_{t \in [0, T_j]}(G_j(\boldsymbol{\Theta}, t))\right] \tag{8.114}$$

Therefore, the reliability in Equation 8.113 can be computed directly by

$$R = \Pr\{W_{\text{ext}} > 0\} \tag{8.115}$$

Example 8.2. System Reliability of Nonlinear Structures under Earthquake The reliability evaluation of a 10-story nonlinear structure subjected to random seismic ground motions is illustrated (Li *et al.*, 2007). Denote the interstory drifts from the floor to the top by $X_1(t), X_2(t), \ldots, X_{10}(t)$ and the heights of the stories by $h_1, h_2, \ldots, h_{10}$. The system reliability of the structure can be defined by

$$R = \Pr\left\{\bigcap_{j=1}^{10}\left\{\left|\frac{X_j(t)}{h_j}\right| < \varphi_{\text{b}},\ t \in [0, T]\right\}\right\} \tag{8.116}$$

where $\varphi_{\text{b}} = 1/50$ is the threshold of the interstory angle. For clarity, we define the dimensionless interstory drift as

$$\bar{X}_j(t) = \left|\frac{X_j(t)}{h_j \varphi_{\text{b}}}\right| \qquad j = 1, 2, \ldots, 10 \tag{8.117}$$

Thus, Equation 8.116 becomes

$$\begin{aligned} R &= \Pr\left\{\bigcap_{j=1}^{10}\{\bar{X}_j(t) < 1,\ t \in [0, T]\}\right\} \\ &= \Pr\left\{\bigcap_{j=1}^{10}\{\bar{X}_{j,\max} < 1\}\right\} \end{aligned} \tag{8.118}$$

where $\bar{X}_{j,\max} = \max_{t \in [0,T]}\{\bar{X}_j(t)\}$. The PDFs of the $\bar{X}_{j,\max}$ obtained by the approach in Section 8.3.2 are pictured in Figure 8.8. Further, we define an equivalent extreme value by

$$\bar{X}_{\max} = \max_{1 \le j \le 10}(\bar{X}_{j,\max}) \tag{8.119}$$

The PDF of the equivalent extreme value $\bar{X}_{\max}$ is also shown in Figure 8.8.

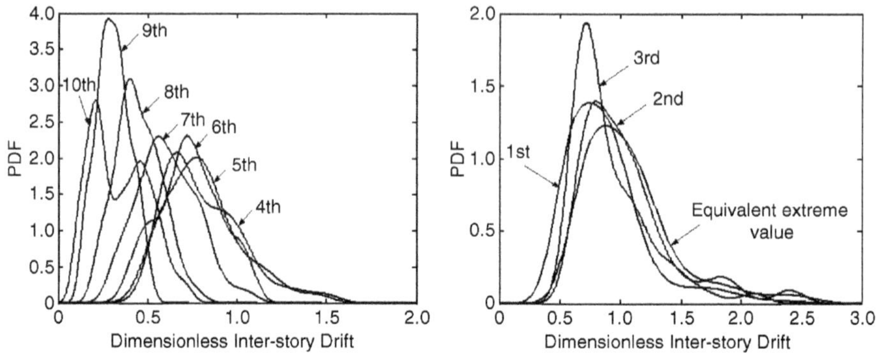

Figure 8.8 EVD and the distribution of the equivalent extreme value.

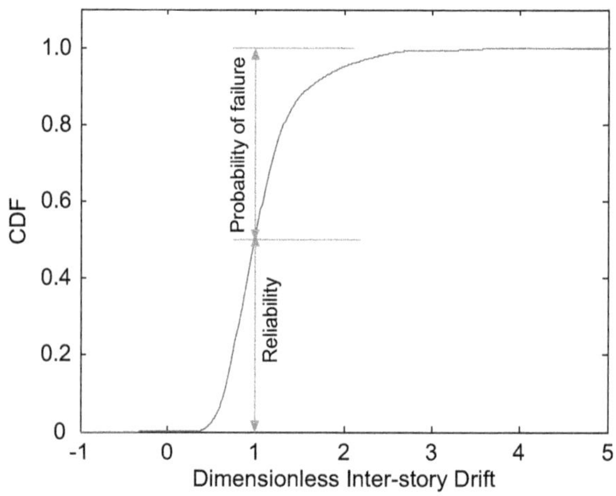

Figure 8.9 System reliability.

Table 8.1 The failure probability of the structure against inter-story drifts.

Story number	Probability of failure
10	3.360 183e-008
9	0.000 000
8	0.000 000
7	0.022 199
6	0.086 191
5	0.179 421
4	0.194 678
3	0.303 173
2	0.449 662
1	0.279 903
Failure probability of the structure	0.491 547

The integral of the PDF of this equivalent extreme value random variable will then give the system reliability and the probability of failure (Figure 8.9); that is:

$$R = \Pr\{\bar{X}_{\max} < 1\} = \int_0^1 p_{\bar{X}_{\max}}(x)\,\mathrm{d}x \qquad P_{\mathrm{f}} = 1 - R \tag{8.120}$$

The reliability and the probability of failure of each story can be defined by

$$R_j = \Pr\{\bar{X}_{j,\max} < 1\} = \int_0^1 p_{\bar{X}_{\mathrm{j,\,max}}}(x)\,\mathrm{d}x \qquad P_{\mathrm{f},j} = 1 - R_j \qquad j = 1, 2, \ldots, 10 \tag{8.121}$$

Table 8.1 lists the probability of failure of each story (Equation 8.121) and the probability of failure of the structure (Equation 8.120). It is seen that the probability of failure of the structure is larger than that of each story. In addition, in the present case, it is noted that the largest probability of failure of the stories occurs in the second floor, not in the first. □

9

Optimal Control of Stochastic Systems

9.1 Introduction

In a general sense, structural control can be understood as designing a structure which will behave in a desired way. This is actually what structural engineers have been doing from the past to modern times, although only part of the aim has been achieved so far. In a narrower and more special sense, structural control is to attach some additional substructure or device appropriately on a structure to regulate structural behavior; for instance, to mitigate vibrations of structures (Housner *et al.*, 1997).

Since Yao's pioneering work (Yao, 1972), structural controls have been developed extensively; for instance, in earthquake- and wind-induced structural vibration mitigation. When considering if and how much energy input is needed by the control device, they are in general classified into passive control, active control, hybrid control and semi-active control (Soong, 1990). When consideration is based on the control aim and the corresponding algorithm, they include common control, optimal control, intelligent control and so on.

Considering the aim of this book, this chapter will only deal with the theory of active structural control when uncertainties are involved. Stochastic optimal control and reliability-based stochastic control will be stressed in this chapter.

Without loss of generality, the equation of motion of a nonlinear deterministic MDOF structural system can be rewritten as a state-space equation (see Equation 5.178)

$$\dot{\mathbf{x}} = \mathbf{f}[\mathbf{x}(t), t] \tag{9.1}$$

where $\mathbf{x}(t) = [x_1(t), x_2(t), \ldots, x_n(t)]^{\mathrm{T}}$ is the n-dimensional state vector, usually consisting of the displacement and velocity vectors of the structural system, and $\mathbf{f}(\cdot) = [f_1(\cdot), f_2(\cdot), \ldots, f_n(\cdot)]^{\mathrm{T}}$ is an n-dimensional operator vector. We note that the deterministic excitations can be incorporated in $\mathbf{f}(\cdot)$.

If there are control devices incorporated in the structural system, then a control vector $\mathbf{u}(t) = [u_1(t), u_2(t), \ldots, u_m(t)]^{\mathrm{T}}$ will have effects on the system response and, thus, Equation 9.1 becomes a state-control equation:

$$\dot{\mathbf{x}} = \mathbf{f}[\mathbf{x}(t), \mathbf{u}(t), t] \tag{9.2}$$

where m is the dimension of the control vector.

The task of optimal control is to seek a control process vector $\mathbf{u}(t)$, $t_0 \le t \le t_f$, which minimizes a given *performance index*. This index is usually a functional of the state and control vector over a given time interval $[t_0, t_f]$, for instance, taking the form (Stengel, 1994)

$$\mathcal{J} = \phi[\mathbf{x}(t_f), t_f] + \int_{t_0}^{t_f} \mathcal{L}[\mathbf{x}(t), \mathbf{u}(t), t]\, dt \tag{9.3}$$

where t_0 is the initial time, t_f is the terminal time, $\phi[\cdot]$ is the portion with respect to the terminal constraint and $\mathcal{L}[\cdot]$ is called the *Lagrangian*. Generally, we require $\phi[\cdot] \ge 0$ and $\mathcal{L}[\cdot] \ge 0$ to hold for all possible values of the arguments. The performance index is also referred to as the *cost function*, or *cost functional*, and so on. Incidentally, without considering the effect of $\phi[\cdot]$, the cost function is in a form similar to the action integral in Lagrangian analytical dynamics, that is why there is an analogy between optimal control theory and analytical dynamics and we called $\mathcal{L}[\cdot]$ the Lagrangian.

The first- and second-order variations of the performance index are given respectively by

$$\delta\mathcal{J} = \left.\frac{\partial\phi}{\partial\mathbf{x}}\right|_{t=t_f} \delta\mathbf{x}(t_f) + \int_{t_0}^{t_f} \left[\frac{\partial\mathcal{L}}{\partial\mathbf{x}}\delta\mathbf{x} + \frac{\partial\mathcal{L}}{\partial\mathbf{u}}\delta\mathbf{u}\right] dt \tag{9.4}$$

and

$$\delta^2\mathcal{J} = \delta\mathbf{x}^{\mathrm{T}}(t_f)\left.\frac{\partial^2\phi}{\partial\mathbf{x}^2}\right|_{t=t_f} \delta\mathbf{x}(t_f) + \int_{t_0}^{t_f} \left[\delta\mathbf{x}^{\mathrm{T}}\frac{\partial^2\mathcal{L}}{\partial\mathbf{x}^2}\delta\mathbf{x} + \delta\mathbf{u}^{\mathrm{T}}\frac{\partial^2\mathcal{L}}{\partial\mathbf{u}^2}\delta\mathbf{u} + 2\delta\mathbf{x}^{\mathrm{T}}\frac{\partial^2\mathcal{L}}{\partial\mathbf{x}\partial\mathbf{u}}\delta\mathbf{u}\right] dt \tag{9.5}$$

For notational simplicity, denote the Hessian matrices by

$$\mathbf{S}_f = \frac{\partial^2\phi[\mathbf{x}(t_f), t_f]}{\partial\mathbf{x}^2} \quad \mathbf{Q} = \frac{\partial^2\mathcal{L}}{\partial\mathbf{x}^2} \quad \mathbf{R} = \frac{\partial^2\mathcal{L}}{\partial\mathbf{u}^2} \quad \text{and} \quad \mathbf{M}' = \frac{\partial^2\mathcal{L}}{\partial\mathbf{u}\partial\mathbf{x}} \tag{9.6}$$

Clearly, in the case $\mathbf{S}_f$ and $\mathbf{Q}$ are symmetric, positive semi-definite, $\mathbf{R}$ is symmetric, positive definite and $\mathbf{M}' = \mathbf{0}$, it follows from Equation 9.5 that the second-order variation of the performance index

$$\delta^2\mathcal{J} > 0 \tag{9.7}$$

In applications, one of the simplest cases is that $\mathbf{S}_f$, $\mathbf{Q}$ and $\mathbf{R}$ are independent of $\mathbf{x}(t_f)$, $\mathbf{x}(\cdot)$ and $\mathbf{u}(\cdot)$, but $\mathbf{Q}$ and $\mathbf{R}$ can be time varying, then the performance index (Equation 9.3) is in a quadratic form:

$$\mathcal{J} = \frac{1}{2}\mathbf{x}^{\mathrm{T}}(t_f)\mathbf{S}_f\mathbf{x}(t_f) + \frac{1}{2}\int_{t_0}^{t_f} [\mathbf{x}^{\mathrm{T}}(t)\mathbf{Q}(t)\mathbf{x}(t) + \mathbf{u}^{\mathrm{T}}(t)\mathbf{R}(t)\mathbf{u}(t)]\, dt \tag{9.8}$$

If randomness is involved in the system parameters and excitations, usually, the system randomness can be modeled as a random function $\boldsymbol{\zeta}(\varpi)$, while the stochastic excitations can be modeled by a stochastic process vector $\boldsymbol{\xi}(\varpi, t)$. The system in Equation 9.2 can then be extended to

$$\dot{\mathbf{X}} = \mathbf{f}[\mathbf{X}(t), \mathbf{U}(t), \boldsymbol{\zeta}(\varpi), \boldsymbol{\xi}(\varpi, t), t] \tag{9.9}$$

As discussed in Chapters 2 and 3, through discretization or decomposition of the random field and stochastic process, a set of basic random variables $\boldsymbol{\Theta} = (\Theta_1, \Theta_2, \ldots, \Theta_s)$ can be used to represent the randomness involved in the system in Equation 9.9, where s is the number of the random variables. By doing this, the system in Equation 9.9 becomes a random state-control equation:

$$\dot{\mathbf{X}} = \mathbf{f}[\mathbf{X}(t), \mathbf{U}(t), t, \boldsymbol{\Theta}(\varpi)] \tag{9.10}$$

Here, we note that, because randomness is involved, the control vector $\mathbf{U}(\cdot)$ and the state vector $\mathbf{X}(\cdot)$ are both stochastic processes. Meanwhile, when $\mathbf{x}$ and $\mathbf{u}$ are replaced respectively by $\mathbf{X}$ and $\mathbf{U}$, the performance index in Equation 9.3 is usually a random variable rather than a deterministic value.

9.2 Optimal Control of Deterministic Systems

9.2.1 Optimal Control of Structural Systems

Before we come to the control of stochastic systems, let us first explore the optimal control of deterministic structural systems. For simplicity, we will consider the sytems with perfect observation and complete control. Without loss of generality, we consider the system in Equation 9.2, which is repeated here as Equation 9.11 for convenience:

$$\dot{\mathbf{x}} = \mathbf{f}[\mathbf{x}(t), \mathbf{u}(t), t] \qquad \mathbf{x}(t_0) = \mathbf{x}_0 \tag{9.11}$$

The goal of optimal control is to find a control process $\mathbf{u}(\cdot)$ that minimizes the performance index in Equation 9.3:

$$\mathcal{J}[\mathbf{u}(\cdot)] = \phi[\mathbf{x}(t_\mathrm{f}), t_\mathrm{f}] + \int_{t_0}^{t_\mathrm{f}} \mathcal{L}[\mathbf{x}(t), \mathbf{u}(t), t]\,\mathrm{d}t \tag{9.12}$$

Here, we note that although the state process $\mathbf{x}(t)$ is involved in the Lagrangian $\mathcal{L}[\cdot]$, the performance index $\mathcal{J}$ is essentially only a functional of the control process $\mathbf{u}(\cdot)$, because once the process $\mathbf{u}(\cdot)$ is specified, $\mathbf{x}(t)$ can be determined by solving Equation 9.11; in other words, $\mathbf{x}(t)$ is a functional, not independent, of $\mathbf{u}(\cdot)$.

Thus, the problem we encounter here is an optimal problem of minimizing the performance index $\mathcal{J}$ in Equation 9.12 with the state-control Equation 9.11 serving as a dynamic constraint. This can be resolved by the variational approach (Lanczos, 1970; Yong and Zhou, 1999; Naidu, 2003).

When introducing a Lagrange multiplier vector $\boldsymbol{\lambda}(t) = [\lambda_1(t), \lambda_2(t), \ldots, \lambda_n(t)]^\mathrm{T}$, the necessary condition of minimizing $\mathcal{J}$ subject to the dynamic constraint in Equation 9.11 is that the control $\mathbf{u}(\cdot)$ makes the augmented performance index

$$\mathcal{J}_\mathrm{A} \triangleq \phi[\mathbf{x}(t_\mathrm{f}), t_\mathrm{f}] + \int_{t_0}^{t_\mathrm{f}} (\mathcal{L}[\mathbf{x}(t), \mathbf{u}(t), t] + \boldsymbol{\lambda}^\mathrm{T}(t)\{\mathbf{f}[\mathbf{x}(t), \mathbf{u}(t), t] - \dot{\mathbf{x}}(t)\})\,\mathrm{d}t \tag{9.13}$$

stationary; that is:

$$\delta \mathcal{J}_{\mathrm{A}} = 0 \tag{9.14}$$

In fact, it is seen that if the state-control Equation 9.11 as a constraint is strictly satisfied, then the augmented performance index $\mathcal{J}_{\mathrm{A}}$ in Equation 9.13 is equal to the original performance index $\mathcal{J}$ in Equation 9.12.

Before computing the variation $\delta\mathcal{J}_{\mathrm{A}}$, integrating by parts with regard to $\dot{\mathbf{x}}(t)$ in Equation 9.13 yields

$$\begin{aligned}\mathcal{J}_{\mathrm{A}} = {} & \phi[\mathbf{x}(t_{\mathrm{f}}), t_{\mathrm{f}}] + \boldsymbol{\lambda}^{\mathrm{T}}(t_0)\mathbf{x}(t_0) - \boldsymbol{\lambda}^{\mathrm{T}}(t_{\mathrm{f}})\mathbf{x}(t_{\mathrm{f}}) \\ & + \int_{t_0}^{t_{\mathrm{f}}} \{\mathcal{L}[\mathbf{x}(t), \mathbf{u}(t), t] + \boldsymbol{\lambda}^{\mathrm{T}}(t)\mathbf{f}[\mathbf{x}(t), \mathbf{u}(t), t] + \dot{\boldsymbol{\lambda}}^{\mathrm{T}}(t)\mathbf{x}(t)\}\,\mathrm{d}t\end{aligned} \tag{9.15}$$

For notational convenience, denote

$$\mathcal{H}\,[\mathbf{x}(t), \mathbf{u}(t), \boldsymbol{\lambda}(t), t] = \mathcal{L}[\mathbf{x}(t), \mathbf{u}(t), t] + \boldsymbol{\lambda}^{\mathrm{T}}(t)\mathbf{f}[\mathbf{x}(t), \mathbf{u}(t), t] \tag{9.16}$$

Then, Equation 9.15 becomes

$$\mathcal{J}_{\mathrm{A}} = \phi[\mathbf{x}(t_{\mathrm{f}}), t_{\mathrm{f}}] + \boldsymbol{\lambda}^{\mathrm{T}}(t_0)\mathbf{x}(t_0) - \boldsymbol{\lambda}^{\mathrm{T}}(t_{\mathrm{f}})\mathbf{x}(t_{\mathrm{f}}) + \int_{t_0}^{t_{\mathrm{f}}} \{\mathcal{H}\,[\mathbf{x}(t), \mathbf{u}(t), \boldsymbol{\lambda}(t), t] + \dot{\boldsymbol{\lambda}}^{\mathrm{T}}(t)\mathbf{x}(t)\}\,\mathrm{d}t \tag{9.17}$$

Later, we will find that $\mathcal{H}\,[\cdot]$ in Equation 9.16 plays a role similar to that of the Hamiltonian function in analytical dynamics and is thus called the *Hamiltonian*.

Now we consider the first order variation of $\mathcal{J}_{\mathrm{A}}$. Note that $\delta\mathbf{x}$ and $\delta\mathbf{u}$ are the variations of $\mathbf{x}$ and $\mathbf{u}$ themselves, and $\delta\mathbf{x}(t_0) = 0$ because $\mathbf{x}(t_0)$ is prescribed in Equation 9.11. The variation of the first three terms in Equation 9.17 gives

$$\delta\{\phi[\mathbf{x}(t_{\mathrm{f}}), t_{\mathrm{f}}] + \boldsymbol{\lambda}^{\mathrm{T}}(t_0)\mathbf{x}(t_0) - \boldsymbol{\lambda}^{\mathrm{T}}(t_{\mathrm{f}})\mathbf{x}(t_{\mathrm{f}})\} = \left.\frac{\partial\phi}{\partial\mathbf{x}}\right|_{t=t_{\mathrm{f}}} \delta\mathbf{x}(t_{\mathrm{f}}) - \boldsymbol{\lambda}^{\mathrm{T}}(t_{\mathrm{f}})\delta\mathbf{x}(t_{\mathrm{f}}) \tag{9.18a}$$

while the variation of the last term in the right-hand side of Equation 9.17 gives

$$\delta\left\{\int_{t_0}^{t_{\mathrm{f}}} \{\mathcal{H}\,[\mathbf{x}(t), \mathbf{u}(t), \boldsymbol{\lambda}(t), t] + \dot{\boldsymbol{\lambda}}^{\mathrm{T}}(t)\mathbf{x}(t)\}\,\mathrm{d}t\right\} = \int_{t_0}^{t_{\mathrm{f}}} \left[\frac{\partial\mathcal{H}}{\partial\mathbf{x}}\delta\mathbf{x}(t) + \frac{\partial\mathcal{H}}{\partial\mathbf{u}}\delta\mathbf{u}(t) + \dot{\boldsymbol{\lambda}}^{\mathrm{T}}(t)\delta\mathbf{x}(t)\right]\mathrm{d}t \tag{9.18b}$$

Therefore, combining Equations 9.18a and 9.18b, we have

$$\delta\mathcal{J}_{\mathrm{A}} = \left[\left.\frac{\partial\phi}{\partial\mathbf{x}}\right|_{t=t_{\mathrm{f}}} - \boldsymbol{\lambda}^{\mathrm{T}}(t_{\mathrm{f}})\right]\delta\mathbf{x}(t_{\mathrm{f}}) + \int_{t_0}^{t_{\mathrm{f}}} \left\{\left[\frac{\partial\mathcal{H}}{\partial\mathbf{x}} + \dot{\boldsymbol{\lambda}}^{\mathrm{T}}(t)\right]\delta\mathbf{x}(t) + \frac{\partial\mathcal{H}}{\partial\mathbf{u}}\delta\mathbf{u}(t)\right\}\mathrm{d}t \tag{9.19}$$

To satisfy Equation 9.14 (that is, $\delta\mathcal{J}_{\mathrm{A}} = 0$), it is required that all the coefficients of $\delta\mathbf{x}(t_{\mathrm{f}})$, $\delta\mathbf{x}(t)$ and $\delta\mathbf{u}(t)$ be zero because of the arbitrariness of these variations; that is:

$$\left.\frac{\partial\phi}{\partial\mathbf{x}}\right|_{t=t_{\mathrm{f}}} - \boldsymbol{\lambda}^{\mathrm{T}}(t_{\mathrm{f}}) = \mathbf{0} \tag{9.20a}$$

$$\frac{\partial \mathcal{H}}{\partial \mathbf{x}} + \dot{\boldsymbol{\lambda}}^{\mathrm{T}}(t) = \mathbf{0} \tag{9.21a}$$

and

$$\frac{\partial \mathcal{H}\,[\mathbf{x}(t), \mathbf{u}(t), \boldsymbol{\lambda}(t), t]}{\partial \mathbf{u}} = \mathbf{0} \tag{9.22}$$

Equations 9.20a and 9.21a can be rewritten respectively in an alternative form

$$\boldsymbol{\lambda}(t_{\mathrm{f}}) = \left(\frac{\partial \phi[\mathbf{x}(t_{\mathrm{f}}), t_{\mathrm{f}}]}{\partial \mathbf{x}}\right)^{\mathrm{T}} \tag{9.20b}$$

$$\dot{\boldsymbol{\lambda}}(t) = -\left(\frac{\partial \mathcal{H}\,[\mathbf{x}(t), \mathbf{u}(t), \boldsymbol{\lambda}(t), t]}{\partial \mathbf{x}}\right)^{\mathrm{T}} \tag{9.21b}$$

Equations 9.20a–9.22 compose the *Euler–Lagrange equations* for optimal control, where Equation 9.20b serves as the terminal condition for Equation 9.21b. It is seen here that, to solve the optimal control problem, the differential Equations 9.21a and 9.21b with respect to the Lagrange multiplier vector $\boldsymbol{\lambda}(t)$ must be solved simultaneously with the state-control Equation 9.11. (Thus, $\boldsymbol{\lambda}(t)$ can be called an *adjoint vector* and Equations 9.21a and 9.21b *adjoint equations*.)

Combining Equations 9.14 and 9.11, we find that

$$\frac{\partial \mathcal{H}}{\partial \boldsymbol{\lambda}^{\mathrm{T}}} = \mathbf{f}[\mathbf{x}(t), \mathbf{u}(t), t] = \dot{\mathbf{x}} \tag{9.23a}$$

That is:

$$\dot{\mathbf{x}} = \frac{\partial \mathcal{H}}{\partial \boldsymbol{\lambda}^{\mathrm{T}}} \tag{9.23b}$$

On the other hand, Equations 9.21a and 9.21b give

$$\dot{\boldsymbol{\lambda}}^{\mathrm{T}}(t) = -\frac{\partial \mathcal{H}}{\partial \mathbf{x}} \tag{9.24}$$

Clearly, these two equations compose a dual equation set which is analogous to the Hamiltonian equation in analytical dynamics and thus also called the *Hamilton canonical equation*. Thus, the methodology employed here is also referred to as the *Hamiltonian system formula* (Yong and Zhou, 1999).

Solving simultaneously the Hamilton Equations 9.23b and 9.24 and the stationary value Equation 9.22, the control law and the state vector that optimize the performance index can be obtained. To be clear, the solution flow can be stated as:

(a) solve the stationary value Equation 9.22 to establish an expression of $\mathbf{u}(t)$ in terms of $\mathbf{x}(t)$ and $\boldsymbol{\lambda}(t)$; that is, $\mathbf{u}(t) = \mathcal{K}[\mathbf{x}(t), \boldsymbol{\lambda}(t), t]$;
(b) substitute this relation in Equations 9.23b and 9.24 to eliminate $\mathbf{u}(\cdot)$ in their right-hand sides;

(c) solve the simultaneous Equations 9.23b and 9.24 with only $\mathbf{x}(t)$ and $\boldsymbol{\lambda}(t)$ as unknowns under the initial and terminal conditions specified respectively by Equations 9.11 and 9.20b to obtain the state vector $\mathbf{x}(t)$ and the adjoint vector $\boldsymbol{\lambda}(t)$;
(d) substitute $\boldsymbol{\lambda}(t)$ in $\mathbf{u}(t) = \mathcal{K}[\mathbf{x}(t), \boldsymbol{\lambda}(t), t]$ to obtain the control process $\mathbf{u}(t) = \mathcal{G}[\mathbf{x}(t), t]$.

It is noted that in the step (c) a two-point boundary-value problem is encountered. In particular, the adjoint equation, Equation 9.24, should be solved backward in time with terminal conditions as specified in Equation 9.20b. As is well known in common dynamical systems governed by state equations with given initial conditions, the future information is not needed in solving the state equation physically. However, in the optimal control of dynamical systems, certain future information must be available to guide the evolution of the state-control process along the optimal trajectory. The reason why is that, in optimal control, an expected result at the terminal time is prescribed in advance.

9.2.2 Linear Quadratic Control

Optimal control of linear systems is much more tractable. Now we consider the case Equation 9.11 describes a linear system in the form

$$\dot{\mathbf{x}} = \mathbf{A}\mathbf{x}(t) + \mathbf{B}\mathbf{u}(t) + \mathbf{L}\boldsymbol{\xi}(t) \tag{9.25}$$

where $\mathbf{A} = [A_{ij}]_{n\times n}$ is the system matrix, $\mathbf{B} = [B_{ij}]_{n\times m}$ is the control influence matrix, $\mathbf{L} = [L_{ij}]_{n\times r}$ is the force influence matrix and $\boldsymbol{\xi}(t) = [\xi_1(t), \xi_2(t), \ldots, \xi_r(t)]^{\mathrm{T}}$ is the r-dimensional deterministic excitation vector. In this case, the operator $\mathbf{f}[\cdot]$ in Equation 9.11 is given as

$$\mathbf{f}[\mathbf{x}(t), \mathbf{u}(t), t] = \mathbf{A}\mathbf{x}(t) + \mathbf{B}\mathbf{u}(t) + \mathbf{L}\boldsymbol{\xi}(t) \tag{9.26}$$

where **A**, **B** and **L** may be functions of time. For notational simplicity, t will not occur explicitly, but **A**, **B** and **L** can be understood as $\mathbf{A}(t)$, $\mathbf{B}(t)$ and $\mathbf{L}(t)$ respectively. In the present stage we assume the control $\mathbf{u}(t)$ is unbounded.

We consider the performance index Equation 9.12 in a quadratic form as in Equation 9.8:

$$\mathcal{J} = \frac{1}{2}\mathbf{x}^{\mathrm{T}}(t_{\mathrm{f}})\mathbf{S}_{\mathrm{f}}\mathbf{x}(t_{\mathrm{f}}) + \frac{1}{2}\int_{t_0}^{t_{\mathrm{f}}} [\mathbf{x}^{\mathrm{T}}(t)\mathbf{Q}\mathbf{x}(t) + \mathbf{u}^{\mathrm{T}}(t)\mathbf{R}\mathbf{u}(t)]\,\mathrm{d}t \tag{9.27}$$

where $\mathbf{S}_{\mathrm{f}} = [S_{\mathrm{f},ij}]_{n\times n}$ and $\mathbf{Q} = [Q_{ij}]_{n\times n}$ are symmetric, positive semi-definite matrices and $\mathbf{R} = [R_{ij}]_{m\times m}$ is a symmetric, positive definite matrix. Again, **Q** and **R** can be understood as $\mathbf{Q}(t)$ and $\mathbf{R}(t)$ respectively if necessary. Thus, the terminal function and the Lagrangian in Equation 9.12 are given respectively by

$$\phi[\mathbf{x}(t_{\mathrm{f}}), t_{\mathrm{f}}] = \frac{1}{2}\mathbf{x}^{\mathrm{T}}(t_{\mathrm{f}})\mathbf{S}_{\mathrm{f}}\mathbf{x}(t_{\mathrm{f}}) \tag{9.28}$$

and

$$\mathcal{L}[\mathbf{x}(t), \mathbf{u}(t), t] = \frac{1}{2}\left[\mathbf{x}^{\mathrm{T}}(t)\mathbf{Q}\mathbf{x}(t) + \mathbf{u}^{\mathrm{T}}(t)\mathbf{R}\mathbf{u}(t)\right] \tag{9.29}$$

Because in the present problem the state-control equation is linear and the performance index is quadratic, the corresponding optimal control is referred to as the *linear quadratic* (LQ) control problem (Stengel, 1994; Williams and Lawrence, 2007).

Using the Euler–Lagrange equations, introducing Equations 9.29 and 9.26 into Equations 9.16 and 9.22 yields

$$\mathbf{R}\mathbf{u}(t) + \boldsymbol{\lambda}^{\mathrm{T}}(t)\mathbf{B} = \mathbf{0} \tag{9.30}$$

or

$$\mathbf{u}(t) = -\mathbf{R}^{-1}\mathbf{B}^{\mathrm{T}}\boldsymbol{\lambda}(t) \tag{9.31}$$

In addition, introducing the terminal function Equation 9.28 into Equations 9.20a and 9.20b gives the terminal condition for the adjoint vector:

$$\boldsymbol{\lambda}(t_{\mathrm{f}}) = \mathbf{S}_{\mathrm{f}}\mathbf{x}(t_{\mathrm{f}}) \tag{9.32}$$

while the adjoint equations, Equations 9.21a and 9.21b, become

$$\dot{\boldsymbol{\lambda}}(t) = -\mathbf{A}^{\mathrm{T}}\boldsymbol{\lambda}(t) - \mathbf{Q}\mathbf{x}(t) \tag{9.33}$$

The physical meaning indicated above is clearer. In fact, the adjoint vector $\boldsymbol{\lambda}(t)$ in Equation 9.31 contains the information over the future interval $[t, t_{\mathrm{f}}]$, because the terminal condition is specified. Therefore, Equation 9.31 guides the evolution of the control efforts while Equation 9.33 indicates that the adjoint vector is fed back to itself; simultaneously, the state vector is fed back.

Note that the adjoint equation (Equation 9.33) and the state-control equation (Equation 9.25) compose a linear simultaneous equation set. The linear relationship exhibited in the terminal condition in Equation 9.32 implies that the linear relationship should also hold for all t; thus, we can assume

$$\boldsymbol{\lambda}(t) = \mathbf{S}(t)\mathbf{x}(t) \tag{9.34}$$

where $\mathbf{S}(t) = [S_{ij}(t)]_{n\times n}$ is to be determined.

Combining Equations 9.34 and 9.31 gives

$$\mathbf{u}(t) = -\mathbf{R}^{-1}\mathbf{B}^{\mathrm{T}}\mathbf{S}(t)\mathbf{x}(t) = -\mathbf{G}_{\mathrm{con}}(t)\mathbf{x}(t) \tag{9.35}$$

This is the control law for LQ control, where $\mathbf{G}_{\mathrm{con}}(t) = [G_{\mathrm{con},ij}(t)]_{m\times n}$ is the control gain matrix and given by

$$\mathbf{G}_{\mathrm{con}}(t) = \mathbf{R}^{-1}\mathbf{B}^{\mathrm{T}}\mathbf{S}(t) \tag{9.36}$$

Equation 9.35 means that the LQ control employs the linear state feedback control law. To determine $\mathbf{S}(t)$, introducing Equation 9.34 into Equation 9.33, we have

$$\dot{\mathbf{S}}(t)\mathbf{x}(t) + \mathbf{S}(t)\dot{\mathbf{x}}(t) = -\mathbf{A}^{\mathrm{T}}\mathbf{S}(t)\mathbf{x}(t) - \mathbf{Q}\mathbf{x}(t) \tag{9.37}$$

Substituting Equation 9.25, the state-control equation, this equation becomes

$$\dot{\mathbf{S}}(t)\mathbf{x}(t) + \mathbf{S}(t)[\mathbf{A}\mathbf{x}(t) + \mathbf{B}\mathbf{u}(t) + \mathbf{L}\boldsymbol{\xi}(t)] = -\mathbf{A}^{\mathrm{T}}\mathbf{S}(t)\mathbf{x}(t) - \mathbf{Q}\mathbf{x}(t) \tag{9.38}$$

Introducing the control law (Equation 9.35) into it and letting the input $\boldsymbol{\xi}(t) \equiv \mathbf{0}$,[1] eliminating $\mathbf{x}(t)$ on both sides, we have

$$\dot{\mathbf{S}}(t) = -\mathbf{A}^{\mathrm{T}}\mathbf{S}(t) - \mathbf{S}(t)\mathbf{A} + \mathbf{S}(t)\mathbf{B}\mathbf{R}^{-1}\mathbf{B}^{\mathrm{T}}\mathbf{S}(t) - \mathbf{Q} \tag{9.39}$$

with the terminal condition given by combining Equation 9.34 with Equation 9.32:

$$\mathbf{S}(t_{\mathrm{f}}) = \mathbf{S}_{\mathrm{f}} \tag{9.40}$$

Equation 9.39 is the *matrix Riccati equation*. Note that $\mathbf{S}_{\mathrm{f}}$ is a symmetric matrix, and that the right-hand side of Equation 9.39 is also symmetric, $\mathbf{S}(t)$ must also be a symmetric matrix. Solution of the matrix Riccati equation is usually not easy. A variety of approaches have been investigated (Petkov *et al.*, 1991; Stengel, 1994; Adeli and Saleh, 1999). The precise integration method proposed by Zhong (2004) deserves recommendation for its high accuracy and acceptable efficiency.

9.2.3 The Minimum Principle and Hamilton–Jacobi-Bellman Equation

9.2.3.1 The Minimum Principle

Denote the optimal control and the corresponding state vector respectively by $\mathbf{u}^*(t)$ and $\mathbf{x}^*(t)$. Because $\mathbf{u}^*(\cdot)$ should minimize the performance index in the admissible domain $\Omega_{\mathbf{u}}$

$$\mathcal{J}[\mathbf{u}^*(\cdot)] = \min_{\mathbf{u}\in\Omega_{\mathbf{u}}} \mathcal{J}[\mathbf{u}(\cdot)] \tag{9.41}$$

there must be

$$\mathcal{J}[\mathbf{u}^*(\cdot) + \delta\mathbf{u}] - \mathcal{J}[\mathbf{u}^*(\cdot)] \geq 0 \tag{9.42}$$

Here, we note that the notation $\mathcal{J}[\mathbf{u}(\cdot)]$ is equivalent to $\mathcal{J}(\mathbf{u})$, where $\mathbf{u}(\cdot)$ denotes a time history over time interval $[t_0, t_{\mathrm{f}}]$; that is, $\mathbf{u}(\cdot) \triangleq \{\mathbf{u}(t) \in \mathbb{R}^m : t \in [t_0, t_{\mathrm{f}}]\}$.

It follows from Equation 9.16 that

$$\begin{aligned}\mathcal{J}(\mathbf{u}^* + \delta\mathbf{u}) - \mathcal{J}(\mathbf{u}^*) &= \mathcal{J}_{\mathrm{A}}[\mathbf{u}^*(\cdot) + \delta\mathbf{u}] - \mathcal{J}_{\mathrm{A}}[\mathbf{u}^*(\cdot)] \\ &= \int_{t_0}^{t_{\mathrm{f}}} \{\mathcal{H}[\mathbf{x}^*(t), \mathbf{u}^*(t) + \delta\mathbf{u}, \boldsymbol{\lambda}^*(t), t] - \mathcal{H}[\mathbf{x}^*(t), \mathbf{u}^*(t), \boldsymbol{\lambda}^*(t), t]\}\,\mathrm{d}t \\ &= \int_{t_0}^{t_{\mathrm{f}}} \left\{\frac{\partial \mathcal{H}^*}{\partial \mathbf{u}}\delta\mathbf{u} + \delta\mathbf{u}^{\mathrm{T}}\frac{\partial^2 \mathcal{H}^*}{\partial \mathbf{u}^2}\delta\mathbf{u} + \text{terms of higher order}\right\}\mathrm{d}t\end{aligned} \tag{9.43}$$

where the first term in the integrand will disappear according to Equation 9.22, which requires $\partial\mathcal{H}/\partial\mathbf{u} = 0$. Thus, when the terms of higher order are ignored, to satisfy Equation 9.42 it is required that

[1] In the case $\boldsymbol{\xi}(t) \neq \mathbf{0}$, the situation will be much more involved. The control law (Equation 9.35) needs to be modified with an additional term of input feedforward. The interested reader can refer to, say, Yang *et al.* (1987) and Soong (1990). We will also consider this case later in Section 9.3.3.2.

$$\frac{\partial^2 \mathcal{H}}{\partial \mathbf{u}^2} \geq 0 \tag{9.44}$$

This gives the sufficient condition for optimality, whereas Equation 9.22 is only the necessary condition.

Combining Equations 9.42 and 9.43 also gives

$$\mathcal{H}^* = \mathcal{H}[\mathbf{x}^*(t), \mathbf{u}^*(t), \boldsymbol{\lambda}^*(t), t] \leq \mathcal{H}[\mathbf{x}^*(t), \mathbf{u}(t), \mathcal{H}^*(t), t] \tag{9.45}$$

where $\mathbf{u}(t) \in \Omega_{\mathbf{u}}$ is any admissible, neighboring (nonoptimal) control history. This equation states the *Pontryagin minimum principle*, which asserts that minimizing the Hamiltonian along the optimal state trajectory is equivalent to minimizing the performance index. The minimum principle was called the maximum principle in their original context, proposed first by Pontryagin and his co-workers (Pontryagin *et al.*, 1964; Gamkrelidze, 1999; Yong and Zhou, 1999; Naidu, 2003).

The procedures developed in the preceding sections are essentially based on the minimum principle via calculus of variations. However, the minimum principle contains simultaneously the necessary and sufficient conditions, and it holds in a range wider than the Euler–Lagrange equations do because of the looser requirement on the differentiability and the boundary conditions.

9.2.3.2 Hamilton–Jacobi–Bellman Equation

Another approach to finding the optimal control history is through dynamic programming via solution of the Hamilton–Jacobi–Bellman (HJB) equation. In this approach, we need to introduce a value function, which is closely related to the performance index.

For clarity, we rewrite the performance index in a form with the initial condition as the explicit arguments

$$\mathcal{J}[\mathbf{x}(t_0), t_0; \mathbf{u}(\cdot)] = \phi[\mathbf{x}(t_f), t_f] + \int_{t_0}^{t_f} \mathcal{L}[\mathbf{x}(\tau), \mathbf{u}(\tau), \tau]\, d\tau \tag{9.46a}$$

or in an alternative form, when replacing t_0 by any $t \in [t_0, t_f]$:

$$\begin{aligned} \mathcal{J}[\mathbf{x}(t), t; \mathbf{u}(\cdot)] &= \phi[\mathbf{x}(t_f), t_f] + \textstyle\int_t^{t_f} \mathcal{L}[\mathbf{x}(\tau), \mathbf{u}(\tau), \tau]\, d\tau \\ &= \phi[\mathbf{x}(t_f), t_f] - \textstyle\int_{t_f}^{t} \mathcal{L}[\mathbf{x}(\tau), \mathbf{u}(\tau), \tau]\, d\tau \end{aligned} \tag{9.46b}$$

Note here that we also replace the dummy variable t in the integral by τ to avoid confusion.

The value function is defined as the minimum of $\mathcal{J}[\mathbf{x}^*(t), t; \mathbf{u}(\cdot)]$ when the state history $\mathbf{x}(t)$ is on the optimal trajectory $\mathbf{x}^*(\cdot)$; that is:

$$\begin{aligned} \mathcal{V}[\mathbf{x}^*(t), t] &= \min_{\mathbf{u}} \{\mathcal{J}[\mathbf{x}^*(t), t; \mathbf{u}(\cdot)]\} \\ &= \phi[\mathbf{x}^*(t_f), t_f] - \textstyle\int_{t_f}^{t} \mathcal{L}[\mathbf{x}^*(\tau), \mathbf{u}^*(\tau), \tau]\, d\tau \\ &= \min_{\mathbf{u}} \{\phi[\mathbf{x}^*(t_f), t_f] - \textstyle\int_{t_f}^{t} \mathcal{L}[\mathbf{x}^*(\tau), \mathbf{u}(\tau), \tau]\, d\tau\} \end{aligned} \tag{9.47}$$

Compared with the performance index (Equation 9.3), we find that the value function is the performance index when the lower limit t_0 is changed to an intermediate value t, $t_0 \leq t \leq t_f$, and we have the initial and terminal values given, according to Equation 9.47, by

$$\mathcal{V}[\mathbf{x}^*(t_0), t_0] = \min_{\mathbf{u}}\{\mathcal{J}[\mathbf{x}(t_0), t_0; \mathbf{u}(\cdot)]\} = \mathcal{J}[\mathbf{x}(t_0), t_0; \mathbf{u}^*(\cdot)] \tag{9.48a}$$

and

$$\mathcal{V}[\mathbf{x}^*(t_f), t_f] = \phi[\mathbf{x}^*(t_f), t_f] \tag{9.48b}$$

respectively. The *principle of optimality* asserts that a trajectory of global optimality must also be of local optimality. That is, for any t_1, $t \le t_1 \le t_f$, we have

$$\begin{aligned}\mathcal{V}[\mathbf{x}^*(t), t] &= -\int_{t_1}^{t} \mathcal{L}[\mathbf{x}^*(\tau), \mathbf{u}^*(\tau), \tau]\,d\tau + \mathcal{V}[\mathbf{x}^*(t_1; \mathbf{x}^*(t), \mathbf{u}^*(\cdot)), t_1] \\ &= \min_{\mathbf{u}}\left\{\mathcal{V}[\mathbf{x}(t_1; \mathbf{x}^*(t), \mathbf{u}(\cdot)), t_1] - \int_{t_1}^{t} \mathcal{L}[\mathbf{x}(\tau; \mathbf{x}^*(t), \mathbf{u}(\cdot)), \mathbf{u}(\tau), \tau]\,d\tau\right\}\end{aligned} \tag{9.49}$$

Here, we use $\mathbf{x}(t_1; \mathbf{x}(t), \mathbf{u}(\cdot))$ to represent that $\mathbf{x}(t_1)$, which is used as the initial condition for the time interval $[t_1, t_f]$, is the value on the trajectory of the control pair $(\mathbf{x}(\cdot), \mathbf{u}(\cdot))$ which starts with the initial value of $\mathbf{x}(t)$.

Intuitively, this can be understood as shown schematically in Figure 9.1.

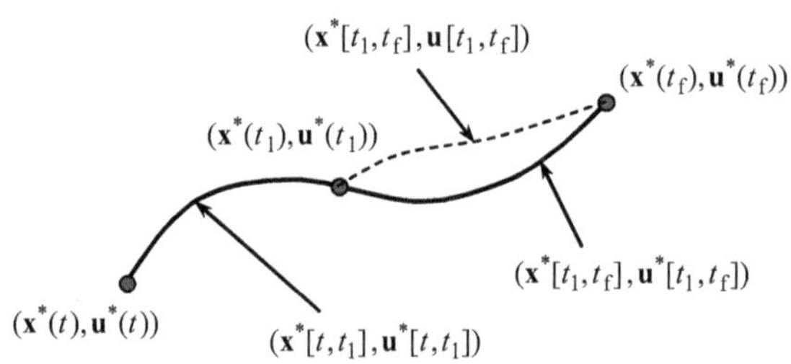

Figure 9.1 Schematical demonstration of the principle of optimality.

Suppose the solid curve is a trajectory of optimality denoted by $(\mathbf{x}^*[t, t_f], \mathbf{u}^*[t, t_f])$, where the three points denote the initial point $(\mathbf{x}^*(t), \mathbf{u}^*(t))$, an intermediate point $(\mathbf{x}^*(t_1), \mathbf{u}^*(t_1))$ and the terminal point $(\mathbf{x}^*(t_f), \mathbf{u}^*(t_f))$. Then, the principle of optimality asserts that if we consider the problem of finding the optimal trajectory with the initial condition $\mathbf{x}^*(t_1)$, then the trajectory $(\mathbf{x}^*[t_1, t_f], \mathbf{u}^*[t_1, t_f])$ is the solution. Otherwise, if we suppose a trajectory different from $(\mathbf{x}^*[t_1, t_f], \mathbf{u}^*[t_1, t_f])$ – for instance, $(\mathbf{x}^*[t_1, t_f], \mathbf{u}[t_1, t_f])$ is the optimal trajectory with the initial condition $\mathbf{x}^*(t_1)$ – then the performance index of the trajectory $(\mathbf{x}^*[t_1, t_f], \mathbf{u}[t_1, t_f])$ must be smaller than that of $(\mathbf{x}^*[t_1, t_f], \mathbf{u}^*[t_1, t_f])$; thus, the performance index of the combined trajectory $(\mathbf{x}^*[t, t_1], \mathbf{u}^*[t, t_1]) + (\mathbf{x}^*[t_1, t_f], \mathbf{u}[t_1, t_f])$ will be smaller than that of $(\mathbf{x}^*[t, t_1], \mathbf{u}^*[t, t_1]) + (\mathbf{x}^*[t_1, t_f], \mathbf{u}[t_1, t_f])$. The latter combined curve is the trajectory $(\mathbf{x}^*[t, t_f], \mathbf{u}^*[t, t_f])$. This implies that $(\mathbf{x}^*[t, t_f], \mathbf{u}^*[t, t_f])$ is not the optimal trajectory, which leads to a confliction. Rigorous proof of the principle of optimality can be found in, say, Yong and Zhou (1999). Actually, it is easy to see, from the definition in Equation 9.47, that

$$\begin{aligned}\mathcal{V}[\mathbf{x}^*(t), t] \le \mathcal{J}[\mathbf{x}^*(t), t; \mathbf{u}(\cdot)] &= \mathcal{J}[\mathbf{x}(t_1; \mathbf{x}^*(t), \mathbf{u}(\cdot)), t_1; \mathbf{u}(\cdot)] - \int_{t_1}^{t} \mathcal{L}[\mathbf{x}(\tau), \mathbf{u}(\tau), \tau]\,d\tau \\ &\le \min_{\mathbf{u}}\{\mathcal{V}[\mathbf{x}(t_1; \mathbf{x}^*(t), \mathbf{u}(\cdot)), t_1] \\ &\quad - \int_{t_1}^{t} \mathcal{L}[\mathbf{x}(\tau; \mathbf{x}^*(t), \mathbf{u}(\cdot)), \mathbf{u}(\tau), \tau]\,d\tau\}\end{aligned} \tag{9.50a}$$

On the other hand, for any $\varepsilon > 0$, there exists a control history $\mathbf{u}_\varepsilon(\cdot)$ such that

$$\begin{aligned}\mathcal{V}[\mathbf{x}^*(t),t]+\varepsilon &\geq \mathcal{J}[\mathbf{x}^*(t),t;\mathbf{u}_\varepsilon(\cdot)]\\ &= \mathcal{J}[\mathbf{x}(t_1;\mathbf{x}^*(t),\mathbf{u}_\varepsilon(\cdot)),t_1;\mathbf{u}(\cdot)] - \textstyle\int_{t_1}^{t}\mathcal{L}[\mathbf{x}(\tau),\mathbf{u}_\varepsilon(\tau),\tau]\,\mathrm{d}\tau]\\ &\geq \mathcal{V}[\mathbf{x}(t_1;\mathbf{x}^*(t),\mathbf{u}_\varepsilon(\cdot)),t_1] - \textstyle\int_{t_1}^{t}\mathcal{L}[\mathbf{x}(\tau;\mathbf{x}^*(t),\mathbf{u}_\varepsilon(\cdot)),\mathbf{u}_\varepsilon(\tau),\tau]\,\mathrm{d}\tau\\ &\geq \min_{\mathbf{u}}\left\{\mathcal{V}[\mathbf{x}(t_1;\mathbf{x}^*(t),\mathbf{u}(\cdot)),t_1] - \textstyle\int_{t_1}^{t}\mathcal{L}[\mathbf{x}(\tau;\mathbf{x}^*(t),\mathbf{u}(\cdot)),\mathbf{u}(\tau),\tau]\,\mathrm{d}\tau\right\}\end{aligned} \tag{9.50b}$$

Combining Equations 9.50a and 9.50b immediately leads to Equation 9.49.

By Equation 9.49 (or Equation 9.50a), we have

$$\frac{\mathcal{V}[\mathbf{x}^*(t),t]-\mathcal{V}[\mathbf{x}^*(t_1),t_1]}{t-t_1} \leq -\frac{1}{t-t_1}\int_{t_1}^{t}\mathcal{L}[\mathbf{x}^*(\tau),\mathbf{u}(\tau),\tau]\,\mathrm{d}\tau \tag{9.51a}$$

Let $t_1 \downarrow t$, which means t_1 tends to t from the right-hand side, we get the total derivative of the value function along the optimal control history:

$$\frac{\mathrm{d}\mathcal{V}[\mathbf{x}^*(t),t]}{\mathrm{d}t} \leq -\mathcal{L}[\mathbf{x}^*(t),\mathbf{u}^*(t),t] \tag{9.51b}$$

Being a function of $\mathbf{x}$ and t, the total derivative of $\mathcal{V}$ is given by

$$\frac{\mathrm{d}\mathcal{V}[\mathbf{x}^*(t),t]}{\mathrm{d}t} = \frac{\partial\mathcal{V}[\mathbf{x}^*(t),t]}{\partial t} + \frac{\partial\mathcal{V}[\mathbf{x}^*(t),t]}{\partial\mathbf{x}}\dot{\mathbf{x}}(t) \tag{9.52a}$$

When the state-control equation, Equation 9.11, is substituted, it becomes

$$\frac{\mathrm{d}\mathcal{V}[\mathbf{x}^*(t),t]}{\mathrm{d}t} = \frac{\partial\mathcal{V}[\mathbf{x}^*(t),t]}{\partial t} + \frac{\partial\mathcal{V}[\mathbf{x}^*(t),t]}{\partial\mathbf{x}}\mathbf{f}[\mathbf{x}^*(t),\mathbf{u}^*(t),t] \tag{9.52b}$$

Combining Equations 9.51b and 9.52b, we have

$$\frac{\partial\mathcal{V}[\mathbf{x}^*(t),t]}{\partial t} \leq -\mathcal{L}[\mathbf{x}^*(t),\mathbf{u}^*(t),t] - \frac{\partial\mathcal{V}[\mathbf{x}^*(t),t]}{\partial\mathbf{x}}\mathbf{f}[\mathbf{x}^*(t),\mathbf{u}^*(t),t] \tag{9.53}$$

If we define the Hamiltonian by

$$\mathcal{H}\left[\mathbf{x}(t),\mathbf{u}(t),\frac{\partial\mathcal{V}[\mathbf{x}(t),t]}{\partial\mathbf{x}},t\right] = \mathcal{L}[\mathbf{x}(t),\mathbf{u}(t),t] + \frac{\partial\mathcal{V}[\mathbf{x}(t),t]}{\partial\mathbf{x}}\mathbf{f}[\mathbf{x}(t),\mathbf{u}(t),t] \tag{9.54}$$

then Equation 9.53 becomes

$$\begin{aligned}\frac{\partial\mathcal{V}[\mathbf{x}^*(t),t]}{\partial t} &\leq -\mathcal{H}\left[\mathbf{x}^*(t),\mathbf{u}^*(t),\frac{\partial\mathcal{V}[\mathbf{x}^*(t),t]}{\partial\mathbf{x}},t\right]\\ &\leq -\min_{\mathbf{u}}\left\{\mathcal{H}\left[\mathbf{x}^*(t),\mathbf{u}(t),\frac{\partial\mathcal{V}[\mathbf{x}^*(t),t]}{\partial\mathbf{x}},t\right]\right\}\end{aligned} \tag{9.55}$$

On the other hand, when using Equation 9.50b and employing the likewise derivation, we have

$$\frac{\mathrm{d}\mathcal{V}[\mathbf{x}^*(t), t]}{\mathrm{d}t} + \varepsilon \geq -\mathcal{L}[\mathbf{x}^*(t), \mathbf{u}_\varepsilon(t), t] \tag{9.56}$$

Thus, when introducing Equations 9.52b and 9.54, there is

$$\begin{aligned}\frac{\partial\mathcal{V}[\mathbf{x}^*(t), t]}{\partial t} + \varepsilon &\geq -\mathcal{H}[\mathbf{x}^*(t), \mathbf{u}_\varepsilon(t), t] \\ &\geq -\min_{\mathbf{u}}\{\mathcal{H}[\mathbf{x}^*(t), \mathbf{u}(t), t]\}\end{aligned} \tag{9.57}$$

Combining Equations 9.55 and 9.57, we finally reach

$$\frac{\partial\mathcal{V}[\mathbf{x}^*(t), t]}{\partial t} = -\min_{\mathbf{u}}\{\mathcal{H}[\mathbf{x}^*(t), \mathbf{u}(t), t]\} \tag{9.58}$$

This is the celebrated *HJB equation*, which is due to Bellman's work on dynamic programming (Bellman, 1957; Naidu, 2003). Note that the HJB equation is a partial differential equation. The initial and terminal values of the value function are given respectively by Equations 9.48a and 9.48b. However, the initial value given in Equation 9.48a is unknown until the solution is obtained, whereas the terminal value condition given by Equation 9.48b, which is repeated here as Equation 9.59

$$\mathcal{V}[\mathbf{x}^*(t_\mathrm{f}), t_\mathrm{f}] = \phi[\mathbf{x}^*(t_\mathrm{f}), t_\mathrm{f}] \tag{9.59}$$

is available beforehand because $\phi[\cdot]$ is a known function. Thus, the HJB equation, Equation 9.58, should be solved backward in time with the above terminal condition.

Actually, we note that the Hamiltonian used in Equation 9.54 is essentially consistent with what is used in Equation 9.16, if we note that the partial derivative vector $\partial\mathcal{V}/\partial\mathbf{x}$ here is identical to the adjoint vector $\boldsymbol{\lambda}^\mathrm{T}(t)$ which is determined by Equations 9.21a and 9.21b.

In addition to the variational principle, dynamic programming provides an alternative approach to finding the optimal control. For instance, in terms of LQ control, with the terminal function and the Lagrangian given respectively by Equations 9.28 and 9.29, the terminal condition (Equation 9.59) becomes

$$\mathcal{V}[\mathbf{x}(t_\mathrm{f}), t_\mathrm{f}] = \frac{1}{2}\mathbf{x}^\mathrm{T}(t_\mathrm{f})\mathbf{S}_\mathrm{f}\mathbf{x}(t_\mathrm{f}) \tag{9.60}$$

Note that now the HJB equation, Equation 9.58, is linear, it is reasonable to assume that the value function takes a form similar to Equation 9.60 (Naidu, 2003):

$$\mathcal{V}[\mathbf{x}(t), t] = \frac{1}{2}\mathbf{x}^\mathrm{T}(t)\mathbf{S}(t)\mathbf{x}(t) \tag{9.61}$$

where $\mathbf{S}(t) = [S_{ij}(t)]_{n\times n}$ is a symmetric, positive definite matrix to be determined.

Thus, minimizing the Hamiltonian in Equation 9.54 requires

$$\frac{\partial\mathcal{H}}{\partial\mathbf{u}} = \mathbf{0}, \quad \text{where} \quad \mathcal{H} = \frac{1}{2}\left[\mathbf{x}^\mathrm{T}\mathbf{Q}\mathbf{x} + \mathbf{u}^\mathrm{T}\mathbf{R}\mathbf{u}\right] + \mathbf{x}^\mathrm{T}\mathbf{S}[\mathbf{A}\mathbf{x} + \mathbf{B}\mathbf{u}] \tag{9.62}$$

This gives the control law

$$\mathbf{u}(t) = \mathbf{R}^{-1}\mathbf{B}^{\mathrm{T}}\mathbf{S}(t)\mathbf{x}(t) \tag{9.63}$$

which is exactly the same as Equation 9.35.

Introducing the control law (Equation 9.63) into the HJB equation (Equation 9.58) and eliminating the factors $\mathbf{x}^{\mathrm{T}}(t)$ and $\mathbf{x}(t)$ on both sides will yield the matrix Riccati equation identical to Equation 9.39.

The HJB equation and the minimum principle are, of course, equivalent in a sense. Actually, as mentioned before, the Euler–Lagrange equations (Equations 9.23a, 9.23b and 9.24) via the variational principle based on the minimum principle are the counterparts of the Hamilton equations in mechanics. Likewise, the HJB equation via the principle of optimality is the counterpart of the Hamilton–Jacobi equation in mechanics (Lanczos, 1970). Thus, the Euler–Lagrange equations (Equations 9.23a, 9.23b and 9.24), which are ordinary differential equations, are the characteristics equations of the partial differential equation – HJB equation (on the characteristics equation, see Section 6.6.1). However, in the sense of conventional differentiation, the smoothness of the value function is required in the HJB equation, while in the sense of super- and sub-differentials and the viscocity solution, this requirement can be loosened and the connection between the HJB equation and the minimum principle can be established (Zhou, 1990; Vinter, 2000).

9.3 Stochastic Optimal Control

If randomness is involved in the system parameters and excitations, then we encounter the problem of stochastic optimal control. Almost simultaneously with modern control theory and stochastic process theory, stochastic control attracts the attention of researchers (Åström, 1970). A variety of stochastic control approaches based on different criteria have been initiated by investigators in a wide range of science and engineering disciplines, such as neighboring optimal control (Stengel, 1994), LQ Gaussian (LQG) control (Yong and Zhou, 1999), covariance control (Yang, 1975; Hotz and Skelton, 1987), PDF tracing control (Sun, 2006), optimal control based on the Hamilton formula (Zhu, 2006) and the thought for reliability-based control (Scruggs, *et al.*, 2006), and so on. In this section we will first consider the optimal control of white-noise-excited stochastic systems (Stengel, 1994; Yong and Zhou, 1999) and then come to the theoretical frame of GDEE-based control.

9.3.1 Stochastic Optimal Control of Nonlinear Systems: Classical Theory

Consider a nonlinear structural system excited by the white noise processes

$$\dot{\mathbf{X}} = \mathbf{f}[\mathbf{X}(t), \mathbf{U}(t), t] + \mathbf{L}\boldsymbol{\xi}(t) \tag{9.64}$$

where $\boldsymbol{\xi}(t) = [\xi_1(t), \xi_2(t), \ldots, \xi_r(t)]^{\mathrm{T}}$ is an r-dimensional stochastic process vector with the mean vector and the covariance matrix given by

$$\mathcal{E}[\boldsymbol{\xi}(t)] = \mathbf{0} \quad \text{and} \quad \mathcal{E}[\boldsymbol{\xi}(t)\boldsymbol{\xi}^{\mathrm{T}}(\tau)] = \mathbf{D}(t)\delta(t-\tau) \tag{9.65}$$

respectively, where $\mathbf{D}(t) = [D_{ij}(t)]_{r\times r}$ is a symmetric, positive semi-definite matrix. Equation 9.64 can be rewritten as an Itô stochastic differential equation:[2]

$$\mathrm{d}\mathbf{X} = \mathbf{f}[\mathbf{X}(t), \mathbf{U}(t), t]\,\mathrm{d}t + \mathbf{L}\,\mathrm{d}\mathbf{W}(t) \tag{9.66}$$

where $\mathbf{W}(t)$ is a Wiener process vector, of which the statistics of increments are given by

$$\mathcal{E}[\mathrm{d}\mathbf{W}(t)] = \mathbf{0} \quad \text{and} \quad \mathcal{E}[\mathrm{d}\mathbf{W}(t)\,\mathrm{d}\mathbf{W}^{\mathrm{T}}(t)] = \mathbf{D}(t)\,\mathrm{d}t \tag{9.67}$$

The performance index (Equation 9.3) is now a random variable. Therefore, the expected value can be used instead:

$$\mathcal{J} = \mathcal{E}\left\{\phi[\mathbf{X}(t_{\mathrm{f}}), t_{\mathrm{f}}] + \int_{t_0}^{t_{\mathrm{f}}} \mathcal{L}[\mathbf{X}(t), \mathbf{U}(t), t]\,\mathrm{d}t\right\} \tag{9.68}$$

In the case of stochastic control, although the variational principle can still be applied to develop a set of stochastic differential equation as the Euler–Lagrange equations (Yong and Zhou, 1999), the HJB equation is simpler and will be developed here.

Like the performance index, the value function defined in Equation 9.47 is now also a random variable. Thus, here it is reasonable to define the value function as the expected value:

$$\begin{aligned}\mathcal{V}[\mathbf{X}^*(t), t] &= \min\{\mathcal{J}[\mathbf{X}(t), t]\} = \mathcal{J}[\mathbf{X}^*(t), t] \\ &= \mathcal{E}\left\{\phi[\mathbf{X}^*(t_{\mathrm{f}}), t_{\mathrm{f}}] - \textstyle\int_{t_{\mathrm{f}}}^{t} \mathcal{L}[\mathbf{X}^*(\tau), \mathbf{U}^*(\tau), \tau]\mathrm{d}\tau\right\} \\ &= \min_{\mathbf{U}} \mathcal{E}\{\phi[\mathbf{X}^*(t_{\mathrm{f}}), t_{\mathrm{f}}] - \textstyle\int_{t}^{t_1} \mathcal{L}[\mathbf{X}^*(\tau), \mathbf{U}(\tau), \tau]\mathrm{d}\tau\}\end{aligned} \tag{9.69}$$

As a mean-square integral, the differentiation and the expectation operator are interchangeable; thus, the differentiation of the value function is given by

$$\mathrm{d}\mathcal{V}[\mathbf{X}^*(t), t] = -\mathcal{E}\{\mathcal{L}[\mathbf{X}^*(t), \mathbf{U}^*(t), t]\}\,\mathrm{d}t \tag{9.70}$$

Because the value of $\mathbf{X}^*(t)$ and $\mathbf{U}^*(t)$ at time t can be measured and become known exactly when the measured noise is ignored, from Equation 9.70 we have

$$\mathrm{d}\mathcal{V}[\mathbf{X}^*(t), t] = -\mathcal{L}[\mathbf{X}^*(t), \mathbf{U}^*(t), t]\,\mathrm{d}t \tag{9.71}$$

On the other hand, being a function of $\mathbf{X}$ and t, as the counterpart of Equations 9.52a and 9.52b, and noting that the Itô lemma (see Equation 5.196) should be employed because $\mathbf{X}(t)$ is associated with an Itô differential (Equation 9.66), the differentiation of $\mathcal{V}$ is given by

$$\mathrm{d}\mathcal{V}[\mathbf{X}^*(t), t] = \mathcal{E}\left\{\frac{\partial \mathcal{V}}{\partial t}\,\mathrm{d}t + \frac{\partial \mathcal{V}}{\partial \mathbf{X}}\,\mathrm{d}\mathbf{X} + \frac{1}{2}\,\mathrm{d}\mathbf{X}^{\mathrm{T}}\frac{\partial^2 \mathcal{V}}{\partial \mathbf{X}^2}\,\mathrm{d}\mathbf{X}\right\} \tag{9.72}$$

[2]For details of the Itô stochastic differential equation, refer to Section 5.6.1.

Replacing $\mathbf{dX}$ by the Itô differentiation (Equation 9.66), Equation 9.72 becomes

$$\begin{aligned}
\mathrm{d}\mathcal{V}[\mathbf{X}^*(t),t] &= \mathcal{E}\left\{\frac{\partial \mathcal{V}}{\partial t}\mathrm{d}t+\frac{\partial \mathcal{V}}{\partial \mathbf{X}}[\mathbf{f}\,\mathrm{d}t+\mathbf{L}\,\mathrm{d}\mathbf{W}(t)]+\frac{1}{2}[\mathbf{f}\,\mathrm{d}t+\mathbf{L}\,\mathrm{d}\mathbf{W}(t)]^{\mathrm{T}}\frac{\partial^2 \mathcal{V}}{\partial \mathbf{X}^2}[\mathbf{f}\,\mathrm{d}t+\mathbf{L}\,\mathrm{d}\mathbf{W}(t)]\right\} \\
&= \frac{\partial \mathcal{V}}{\partial t}\mathrm{d}t+\frac{\partial \mathcal{V}}{\partial \mathbf{X}}\mathbf{f}+\frac{1}{2}\mathcal{E}\left\{[\mathbf{f}\,\mathrm{d}t+\mathbf{L}\,\mathrm{d}\mathbf{W}(t)]^{\mathrm{T}}\frac{\partial^2 \mathcal{V}}{\partial \mathbf{X}^2}[\mathbf{f}\,\mathrm{d}t+\mathbf{L}\,\mathrm{d}\mathbf{W}(t)]\right\} \\
&= \frac{\partial \mathcal{V}}{\partial t}\mathrm{d}t+\frac{\partial \mathcal{V}}{\partial \mathbf{X}}\mathbf{f}+\frac{1}{2}\mathrm{Tr}\left(\frac{\partial^2 \mathcal{V}}{\partial \mathbf{X}^2}\mathcal{E}[\mathbf{f}\,\mathrm{d}t+\mathbf{L}\,\mathrm{d}\mathbf{W}(t)][\mathbf{f}\,\mathrm{d}t+\mathbf{L}\,\mathrm{d}\mathbf{W}(t)]^{\mathrm{T}}\right) \\
&= \frac{\partial \mathcal{V}}{\partial t}\mathrm{d}t+\frac{\partial \mathcal{V}}{\partial \mathbf{X}}\mathbf{f}\,\mathrm{d}t+\frac{1}{2}\mathrm{Tr}\left(\frac{\partial^2 \mathcal{V}}{\partial \mathbf{X}^2}\mathbf{L}\mathbf{D}(t)\mathbf{L}^{\mathrm{T}}\right)\mathrm{d}t
\end{aligned} \tag{9.73}$$

Here, use has been made of the matrix identity

$$\mathbf{x}^{\mathrm{T}}\mathbf{A}\mathbf{x} = \mathrm{Tr}(\mathbf{A}\mathbf{x}\mathbf{x}^{\mathrm{T}}) \tag{9.74}$$

where $\mathrm{Tr}(\cdot)$ is the trace of a matrix.

Combining Equations 9.73 and 9.71, we have

$$-\mathcal{L}[\mathbf{X}^*(t),\mathbf{U}^*(t),t]\,\mathrm{d}t = \frac{\partial \mathcal{V}}{\partial t}\mathrm{d}t+\frac{\partial \mathcal{V}}{\partial \mathbf{X}}\mathbf{f}\,\mathrm{d}t+\frac{1}{2}\mathrm{Tr}\left(\frac{\partial^2 \mathcal{V}}{\partial \mathbf{X}^2}\mathbf{L}\mathbf{D}(t)\mathbf{L}^{\mathrm{T}}\right)\mathrm{d}t \tag{9.75}$$

and thus

$$\frac{\partial \mathcal{V}}{\partial t} = -\left\{\mathcal{L}[\mathbf{X}^*(t),\mathbf{U}^*(t),t]+\frac{\partial \mathcal{V}}{\partial \mathbf{X}}\mathbf{f}+\frac{1}{2}\mathrm{Tr}\left(\frac{\partial^2 \mathcal{V}}{\partial \mathbf{X}^2}\mathbf{L}\mathbf{D}(t)\mathbf{L}^{\mathrm{T}}\right)\right\} \tag{9.76}$$

or

$$\frac{\partial \mathcal{V}}{\partial t} = -\min_{\mathbf{U}}\left\{\mathcal{L}[\mathbf{X}^*(t),\mathbf{U}(t),t]+\frac{\partial \mathcal{V}}{\partial \mathbf{X}}\mathbf{f}+\frac{1}{2}\mathrm{Tr}\left(\frac{\partial^2 \mathcal{V}}{\partial \mathbf{X}^2}\mathbf{L}\mathbf{D}(t)\mathbf{L}^{\mathrm{T}}\right)\right\} \tag{9.77}$$

Note from Equation 9.77 that the right-hand side part to be minimized in the stochastic context is always not smaller than the deterministic counterpart (Equation 9.54), because the effect of the white noise induces a nonnegative correction term; that is, the third term in the bracket.

When defining a *generalized Hamiltonian*

$$\mathcal{H}_{\mathrm{G}}\left[\mathbf{X}(t),\mathbf{U}(t),\frac{\partial \mathcal{V}}{\partial \mathbf{X}},\frac{\partial^2 \mathcal{V}}{\partial \mathbf{X}^2},t\right] = \mathcal{L}[\mathbf{X}(t),\mathbf{U}(t),t]+\frac{\partial \mathcal{V}}{\partial \mathbf{X}}\mathbf{f}+\frac{1}{2}\mathrm{Tr}\left(\frac{\partial^2 \mathcal{V}}{\partial \mathbf{X}^2}\mathbf{L}\mathbf{D}(t)\mathbf{L}^{\mathrm{T}}\right) \tag{9.78}$$

Equation 9.77 becomes

$$\frac{\partial \mathcal{V}}{\partial t} = -\min_{\mathbf{U}}\left\{\mathcal{H}_{\mathrm{G}}\left[\mathbf{X}^*(t),\mathbf{U}(t),\frac{\partial \mathcal{V}}{\partial \mathbf{X}},\frac{\partial^2 \mathcal{V}}{\partial \mathbf{X}^2},t\right]\right\} \tag{9.79}$$

This is the HJB equation in the stochastic context, which is of course the counterpart of Equation 9.58, where the Hamiltonian $\mathcal{H}(\cdot)$ is replaced by the generalized Hamiltonian $\mathcal{H}_G(\cdot)$.

To solve a stochastic optimal control problem with the approach elaborated above, the control law, which might be linear or nonlinear feedback control, is first determined by minimizing the right-hand side of Equations 9.77 and 9.79, then the control law is introduced into Equations 9.77 and 9.79 and the state-control equation (Equation 9.64) to give the control gains and simultaneously the state response.

9.3.2 Linear Quadratic Gaussian Control

Consider the case where the system in Equation 9.64 is a linear system with the state-control equation

$$\dot{\mathbf{X}} = \mathbf{AX}(t) + \mathbf{BU}(t) + \mathbf{L}\boldsymbol{\xi}(t) \tag{9.80}$$

where $\mathbf{A} = [A_{ij}(t)]_{n\times n}$, $\mathbf{B} = [B_{ij}(t)]_{n\times m}$ and $\mathbf{L} = [L_{ij}(t)]_{n\times r}$ are the same as in Equation 9.25. Now the operator $\mathbf{f}[\cdot]$ in Equation 9.64 is given by

$$\mathbf{f}[\mathbf{X}(t), \mathbf{U}(t), t] = \mathbf{AX}(t) + \mathbf{BU}(t) \tag{9.81}$$

If the terminal function $\phi[\cdot]$ and the Lagrangian $\mathcal{L}[\cdot]$ take the quadratic forms

$$\phi[\mathbf{X}(t_f), t_f] = \frac{1}{2}\mathbf{X}^T(t_f)\mathbf{S}_f\mathbf{X}(t_f) \tag{9.82}$$

and

$$\mathcal{L}[\mathbf{X}(t), \mathbf{U}(t), t] = \frac{1}{2}\left[\mathbf{X}^T(t)\mathbf{QX}(t) + \mathbf{U}^T(t)\mathbf{RU}(t)\right] \tag{9.83}$$

then Equation 9.79 becomes

$$\frac{\partial \mathcal{V}}{\partial t} = -\min_{\mathbf{U}}\left\{\frac{1}{2}\left[\mathbf{X}^{*T}\mathbf{QX}^* + \mathbf{U}^T\mathbf{RU}\right] + \frac{\partial \mathcal{V}}{\partial \mathbf{X}}(\mathbf{AX}^* + \mathbf{BU}) + \frac{1}{2}\mathrm{Tr}\left(\frac{\partial^2 \mathcal{V}}{\partial \mathbf{X}^2}\mathbf{LD}(t)\mathbf{L}^T\right)\right\} \tag{9.84}$$

with the terminal condition given by

$$\mathcal{V}[\mathbf{X}^*(t_f), t_f] = \frac{1}{2}\mathbf{X}^{*T}(t_f)\mathbf{S}_f\mathbf{X}^*(t_f) \tag{9.85}$$

Being a linear system (Equation 9.80) and a linear partial differential equation, then, compared with the deterministic counterpart (Equation 9.58), it is reasonable to assume that the value function takes the form

$$\mathcal{V}[\mathbf{X}(t), t] = \frac{1}{2}\mathbf{X}^T(t)\mathbf{S}(t)\mathbf{X}(t) + v(t) \tag{9.86}$$

where $\mathbf{S}(t) = [S_{ij}(t)]_{n\times n}$ is a symmetric, positive-definite matrix and $v(t)$ is a correction term due to the correction term in the HJB equation (Equation 9.84) compared with the deterministic counterpart. Our task is to determine $\mathbf{S}(t)$ and $v(t)$ and then obtain the control law and solution of the optimal problem.

Introducing Equation 9.86 into Equation 9.84 and noting that

$$\frac{\partial \mathcal{V}}{\partial \mathbf{X}} = \mathbf{X}^{\mathrm{T}}(t)\mathbf{S}(t) \quad \text{and} \quad \frac{\partial^2 \mathcal{V}}{\partial \mathbf{X}^2} = \mathbf{S}(t) \tag{9.87}$$

we have

$$\frac{\partial \mathcal{V}}{\partial t} = -\min_{\mathbf{U}} \frac{1}{2}\left\{ \left[\mathbf{X}^{\mathrm{T}}\mathbf{Q}\mathbf{X} + \mathbf{U}^{\mathrm{T}}\mathbf{R}\mathbf{U}\right] + 2\mathbf{X}^{\mathrm{T}}\mathbf{S}(\mathbf{A}\mathbf{X} + \mathbf{B}\mathbf{U}) + \mathrm{Tr}(\mathbf{S}(t)\mathbf{L}\mathbf{D}(t)\mathbf{L}^{\mathrm{T}}) \right\} \tag{9.88}$$

The minimization requires that $\partial \mathcal{H}_{\mathrm{G}}/\partial \mathbf{U} = 0$, which gives the control law

$$\mathbf{U}(t) = -\mathbf{R}^{-1}\mathbf{B}^{\mathrm{T}}\mathbf{S}(t)\mathbf{X}(t) \tag{9.89}$$

Thus, substituting Equations 9.89 and 9.86 in Equation 9.88 results in

$$\begin{aligned} \frac{\partial \mathcal{V}}{\partial t} &= \frac{1}{2}\mathbf{X}^{\mathrm{T}}(t)\dot{\mathbf{S}}(t)\mathbf{X}(t) + \dot{v}(t) \\ &= -\frac{1}{2}\left[\mathbf{X}^{\mathrm{T}}\mathbf{Q}\mathbf{X} + \mathbf{X}^{\mathrm{T}}\mathbf{S}\mathbf{B}\mathbf{R}^{-1}\mathbf{B}^{\mathrm{T}}\mathbf{S}\mathbf{X} + 2\mathbf{X}^{\mathrm{T}}\mathbf{S}(\mathbf{A} - \mathbf{B}\mathbf{R}^{-1}\mathbf{B}^{\mathrm{T}}\mathbf{S})\mathbf{X}\right] - \frac{1}{2}\mathrm{Tr}(\mathbf{S}(t)\mathbf{L}\mathbf{D}(t)\mathbf{L}^{\mathrm{T}}) \\ &= -\frac{1}{2}\mathbf{X}^{\mathrm{T}}(\mathbf{Q} + 2\mathbf{S}\mathbf{A} - \mathbf{S}\mathbf{B}\mathbf{R}^{-1}\mathbf{B}^{\mathrm{T}}\mathbf{S})\mathbf{X} - \frac{1}{2}\mathrm{Tr}(\mathbf{S}(t)\mathbf{L}\mathbf{D}(t)\mathbf{L}^{\mathrm{T}}) \end{aligned} \tag{9.90}$$

Further, comparing the coefficients of **X** in the first equality and the third equality and letting them be equal, we have

$$\dot{\mathbf{S}}(t) = -2\mathbf{S}\mathbf{A} + \mathbf{S}\mathbf{B}\mathbf{R}^{-1}\mathbf{B}^{\mathrm{T}}\mathbf{S} - \mathbf{Q} \tag{9.91}$$

and

$$\dot{v}(t) = -\frac{1}{2}\mathrm{Tr}(\mathbf{S}(t)\mathbf{L}\mathbf{D}(t)\mathbf{L}^{\mathrm{T}}) \tag{9.92}$$

Note that

$$2\mathbf{S}\mathbf{A} = (\mathbf{S}\mathbf{A} + \mathbf{A}^{\mathrm{T}}\mathbf{S}) + (\mathbf{S}\mathbf{A} - \mathbf{A}^{\mathrm{T}}\mathbf{S}) \tag{9.93}$$

where the first term on the right-hand side is symmetric and the second term is nonsymmetric. Because all the terms but the first one in the right-hand side of Equation 9.91 are symmetric, the first term must also be symmetric; thus, the second term in the right-hand side of Equation 9.93 must disappear. Consequently, Equation 9.91 becomes[3]

$$\dot{\mathbf{S}}(t) = -\mathbf{A}^{\mathrm{T}}\mathbf{S} - \mathbf{S}\mathbf{A} + \mathbf{S}\mathbf{B}\mathbf{R}^{-1}\mathbf{B}^{\mathrm{T}}\mathbf{S} - \mathbf{Q} \tag{9.94}$$

This is the same as the matrix Riccati equation and identical to the deterministic counterpart in Equation 9.39, of which the terminal condition is given by

[3]The analysis used here also holds for the deterministic counterpart when the HJB equation is used to treat the LQ controls at the end of Section 9.2.3.2, where the last step to obtain the matrix Riccati equation is left to the reader.

$$\mathbf{S}(t_\mathrm{f}) = \mathbf{S}_\mathrm{f} \tag{9.95}$$

On the other hand, it follows from Equation 9.92 that

$$v(t) = -\int_{t_\mathrm{f}}^{t} \frac{1}{2}\mathrm{Tr}(\mathbf{S}(t)\mathbf{LD}(t)\mathbf{L}^\mathrm{T})\,\mathrm{d}t = \frac{1}{2}\int_{t}^{t_\mathrm{f}} \mathrm{Tr}(\mathbf{S}(t)\mathbf{LD}(t)\mathbf{L}^\mathrm{T})\,\mathrm{d}t \tag{9.96}$$

which is due to the correction term in the HJB equation (Equation 9.84); therefore, the value function in Equation 9.86 is actually given by

$$\mathcal{V}[\mathbf{X}(t), t] = \frac{1}{2}\mathbf{X}^\mathrm{T}(t)\mathbf{S}(t)\mathbf{X}(t) + \frac{1}{2}\int_{t}^{t_\mathrm{f}} \mathrm{Tr}(\mathbf{S}(t)\mathbf{LD}(t)\mathbf{L}^\mathrm{T})\,\mathrm{d}t \tag{9.97}$$

Certainly, it is always no less than its deterministic counterpart (Equation 9.61).

In the preceding section, the Itô stochastic differential equation is considered and the Gaussian assumption on the excitations is used. Such a developed stochastic control is thus called *LQG* control.

Here we see that the LQG control gain matrix is identical to the deterministic LQ control. This means that, for a zero-mean Gaussian excited linear system, the control gain matrix can be computed offline and then used in practical applications or simulations.

In addition, using a perturbation technique, the optimal control of an additive zero-mean Gaussian excited nonlinear system can be achieved by combining deterministic optimal control of a nominal deterministic nonlinear system and LQG control of a quasi-linearized system perturbed along the nominal optimal control history (Stengel, 1994). However, one should be careful to appreciate that it applies only for a low degree of variation in the excitations.

In the classical theory of stochastic optimal control, the control-state equation is modeled as an Itô stochastic differential equation. Furthermore, the performance index takes an expectation of some type of index. Such treatments make it straightforward to extend deterministic optimal control theory to stochastic optimal control theory without substantial difficulties. On the other hand, it also implies that such developed stochastic optimal control theory only applies in the cases of weak disturbance by white-noise excitation, excluding the nonstationary strong excitations such as earthquakes, strong wind and large sea waves frequently encountered in civil engineering. Moreover, taking only an expectation as the performance index implies that such a control obtained is optimal essentially in the sense of variance, not in the sense of reliability. It is not adequate for precise control of performance of engineering structures.

9.3.3 Probability Density Evolution Analysis of Stochastic Optimal Control Systems

9.3.3.1 General Principle

As pointed out in Section 9.1.1, when general random fields or stochastic processes are involved in nonlinear systems, the controlled system is governed by

$$\dot{\mathbf{X}} = \mathbf{f}[\mathbf{X}(t), \mathbf{U}(t), \boldsymbol{\zeta}(\varpi), \boldsymbol{\xi}(\varpi, t), t] \tag{9.98}$$

where $\boldsymbol{\zeta}(\varpi)$ and $\boldsymbol{\xi}(\varpi, t)$ are respectively the random fields and stochastic processes involved and ϖ represents the embedded random event. By introducing orthogonal decomposition of random fields and stochastic processes (see Chapter 3), Equation 9.98 becomes

$$\dot{\mathbf{X}} = \mathbf{f}[\mathbf{X}(t), \mathbf{U}(t), t, \mathbf{\Theta}(\varpi)] \tag{9.99}$$

where $\mathbf{\Theta}(\varpi) = (\Theta_1(\varpi), \Theta_2(\varpi), \ldots, \Theta_s(\varpi))$ is a random vector with known probability density $p_{\mathbf{\Theta}}(\boldsymbol{\theta}) = p_{\mathbf{\Theta}}(\theta_1, \theta_2, \ldots, \theta_s)$. Note that, in contrast to classical stochastic control theory, when we consider the random state-control equation (Equation 9.99), it is no longer the Itô type stochastic differential equation.

Because of the randomness involved in the system, the control history $\mathbf{U}(\cdot)$ and the corresponding state $\mathbf{X}(\cdot)$ are both stochastic processes dependent on $\mathbf{\Theta}(\varpi)$. Thus, the performance index

$$\mathcal{J}[\mathbf{X}(t_0), \mathbf{U}(\mathbf{\Theta}, \cdot)] = \phi[\mathbf{X}(t_f), t_f] + \int_{t_0}^{t_f} \mathcal{L}[\mathbf{X}(\mathbf{\Theta}, t), \mathbf{U}(\mathbf{\Theta}, t), t]\, dt \tag{9.100}$$

is a random variable with dependence on $\mathbf{\Theta}(\varpi)$.

In classical stochastic control theory of nonlinear systems, the performance index is defined as the expected value of Equation 9.100, as shown in the preceding sections (see Equation 9.68). However, this is not the only option. In contrast, a direct, sample-based treatment, if possible, might be more reasonable. This is the case when the GDEE elaborated in Chapters 6 and 7 is incorporated.

The variational principle in the context of the stochastic controlled system in Equation 9.99 will be employed here. The problem to be tackled here is to minimize the stochastic performance index $\mathcal{J}[\mathbf{X}(t_0), \mathbf{U}(\mathbf{\Theta}, \cdot)]$ in Equation 9.100 in the sense of a sample under the dynamic constraint imposed by Equation 9.99. In this case, a Lagrange multiplier vector $\boldsymbol{\lambda}(\mathbf{\Theta}, t) = [\lambda_1(\mathbf{\Theta}, t), \lambda_2(\mathbf{\Theta}, t), \ldots, \lambda_n(\mathbf{\Theta}, t)]^{\mathrm{T}}$ can be introduced. Note here that the adjoint vector $\boldsymbol{\lambda}(\mathbf{\Theta}, t)$ has a dependence on $\mathbf{\Theta}$.

Denote the random Hamiltonian by

$$\mathcal{H}\,[\mathbf{X}(t), \mathbf{U}(t), \boldsymbol{\lambda}(t), t, \mathbf{\Theta}] = \mathcal{L}[\mathbf{X}(\mathbf{\Theta}, t), \mathbf{U}(\mathbf{\Theta}, t), t] + \boldsymbol{\lambda}^{\mathrm{T}}(\mathbf{\Theta}, t)\mathbf{f}[\mathbf{X}(t), \mathbf{U}(t), t, \mathbf{\Theta}] \tag{9.101}$$

Using the variational principle in a way similar to Section 9.2.1, we obtain

$$\dot{\boldsymbol{\lambda}}(t) = -\left[\frac{\partial\,\mathcal{H}\,[\mathbf{X}(t), \mathbf{U}(t), \boldsymbol{\lambda}(t), t, \mathbf{\Theta}]}{\partial \mathbf{X}}\right]^{\mathrm{T}} \tag{9.102}$$

$$\boldsymbol{\lambda}(t_f) = \left[\left.\frac{\partial \phi}{\partial \mathbf{X}}\right|_{t=t_f}\right]^{\mathrm{T}} \tag{9.103}$$

and

$$\frac{\partial\,\mathcal{H}\,[\mathbf{X}(t), \mathbf{U}(t), \boldsymbol{\lambda}(t), t, \mathbf{\Theta}]}{\partial \mathbf{U}} = \mathbf{0} \tag{9.104}$$

Equations 9.102–9.104 are the stochastic Euler–Lagrange equations of the controlled system. Combined with the state-control equation (Equation 9.99), the stochastic optimal control problem can be resolved.

Without loss of generality, the solution obtained of the state vector and the control process and their derivative processes can be expressed in the following form:

$$\mathbf{X} = \mathbf{H}_{\mathbf{X}}(\boldsymbol{\Theta}, t) \tag{9.105a}$$

$$\dot{\mathbf{X}} = \mathbf{h}_{\mathbf{X}}(\boldsymbol{\Theta}, t) \tag{9.105b}$$

$$\mathbf{U} = \mathbf{H}_{\mathbf{U}}(\boldsymbol{\Theta}, t) \tag{9.106a}$$

$$\dot{\mathbf{U}} = \mathbf{h}_{\mathbf{U}}(\boldsymbol{\Theta}, t) \tag{9.106b}$$

With Equations 9.105a–9.106b understood as the Lagrangian description of the controlled system, according to the probability density evolution theory elaborated in Section 6.5.2, the GDEE with respect to any component of the state vector $\mathbf{X}(t)$ and the control history $\mathbf{U}(t)$ can be obtained. For instance, if we denote the joint probability density of $(X_\ell(t), \boldsymbol{\Theta})$ by $p_{X_\ell\boldsymbol{\Theta}}(x_\ell, \boldsymbol{\theta}, t)$, where $X_\ell(t)$ is the ℓth component of $\mathbf{X}(t)$, $1 \le \ell \le n$, then we have the GDEE

$$\frac{\partial p_{X_\ell\boldsymbol{\Theta}}(x_\ell, \boldsymbol{\theta}, t)}{\partial t} + h_{\mathbf{X},\ell}(\boldsymbol{\theta}, t)\frac{\partial p_{X_\ell\boldsymbol{\Theta}}(x_\ell, \boldsymbol{\theta}, t)}{\partial x_\ell} = 0 \tag{9.107a}$$

where $h_{\mathbf{X},\ell}(\cdot)$ is the ℓth component of $\mathbf{h}_{\mathbf{X}}(\cdot)$.

Likewise, the joint density of the control history component $U_\ell(t)$ (the ℓth component of $\mathbf{U}(t)$) and $\boldsymbol{\Theta}$ satisfies

$$\frac{\partial p_{U_\ell\boldsymbol{\Theta}}(u_\ell, \boldsymbol{\theta}, t)}{\partial t} + h_{\mathbf{U},\ell}(\boldsymbol{\theta}, t)\frac{\partial p_{U_\ell\boldsymbol{\Theta}}(u_\ell, \boldsymbol{\theta}, t)}{\partial u_\ell} = 0 \tag{9.108a}$$

Solving these equations, we can get the density functions of $X_\ell(t)$ and $U_\ell(t)$ respectively by

$$p_{X_\ell}(x_\ell, t) = \int_{\Omega_{\boldsymbol{\Theta}}} p_{X_\ell\boldsymbol{\Theta}}(x_\ell, \boldsymbol{\theta}, t)\,\mathrm{d}\boldsymbol{\theta} \tag{9.108b}$$

$$p_{U_\ell}(u_\ell, t) = \int_{\Omega_{\boldsymbol{\Theta}}} p_{U_\ell\boldsymbol{\Theta}}(u_\ell, \boldsymbol{\theta}, t)\,\mathrm{d}\boldsymbol{\theta} \tag{9.108b}$$

Similar to Section 6.6.2, the probability density evolution analysis of a stochastic optimal control system includes the following steps:

Step 1: Specify the representative point set $\mathcal{P}_{\text{sel}} = \{\boldsymbol{\theta}_q = (\theta_{1,q}, \theta_{2,q}, \ldots, \theta_{s,q});\ q = 1, 2, \ldots, n_{\text{sel}}\}$ in the space $\Omega_{\boldsymbol{\Theta}}$ and the corresponding assigned probabilities P_q, $q = 1, 2, \ldots, n_{\text{sel}}$, as elaborated in Sections 7.2–7.4.

Step 2: Solve the simultaneous Equations 9.99 and 9.102–9.104 at the representative points to obtain the quantities of interest; for instance, the state $\mathbf{X}(\boldsymbol{\theta}_q, t)$ and its derivative process $\dot{\mathbf{X}}(\boldsymbol{\theta}_q, t)$, the control history $\mathbf{U}(\boldsymbol{\theta}_q, t)$ and its derivative process $\dot{\mathbf{U}}(\boldsymbol{\theta}_q, t)$.

Step 3: Substitute the quantities obtained, solve the GDEE, say Equations 9.107a and 9.108a, by the numerical methods elaborated in Section 7.1 and obtain the joint densities $p_{X_\ell\boldsymbol{\Theta}}(x_\ell, \boldsymbol{\theta}_q, t)$ and $p_{U_\ell\boldsymbol{\Theta}}(u_\ell, \boldsymbol{\theta}_q, t)$.

Step 4: Repeat steps 2 and 3 running over all q and sum the results to obtain the desired probability densities by

$$p_{X_\ell}(x_\ell, t) = \sum_{q=1}^{n_{\text{sel}}} p_{X_\ell \mathbf{\Theta}}(x_\ell, \boldsymbol{\theta}_q, t) \quad \text{and} \quad p_{U_\ell}(u_\ell, t) = \sum_{q=1}^{n_{\text{sel}}} p_{U_\ell \mathbf{\Theta}}(u_\ell, \boldsymbol{\theta}_q, t) \tag{9.109}$$

9.3.3.2 Probability Density Evolution Analysis of Linear Quadratic Control of Stochastic Systems

For a linear system with a quadratic performance index, the control problem will be much more tractable. Consider the random state-control equation of a linear system:

$$\dot{\mathbf{X}} = \mathbf{A}\mathbf{X}(t) + \mathbf{B}\mathbf{U}(t) + \mathbf{L}\boldsymbol{\xi}(\mathbf{\Theta}, t) \tag{9.110}$$

which is the stochastic counterpart of Equation 9.25. Here, $\mathbf{A} = [A_{ij}]_{n\times n}$ is the system matrix, $\mathbf{B} = [B_{ij}]_{n\times m}$ is the control influence matrix, $\mathbf{L} = [L_{ij}]_{n\times r}$ is the force influence matrix and $\boldsymbol{\xi}(\mathbf{\Theta}, t) = [\xi_1(\mathbf{\Theta}, t), \xi_2(\mathbf{\Theta}, t), \ldots, \xi_r(\mathbf{\Theta}, t)]^{\mathrm{T}}$ is the r-dimensional excitation vector, which can be represented by random functions of $\mathbf{\Theta}$ through the methodology in Chapter 3. For simplicity, in the present stage we do not consider the randomness involved in the structural parameters.

Consider a quadratic performance index:

$$\mathcal{J}[\mathbf{X}(t_0), \mathbf{U}(\mathbf{\Theta}, \cdot)] = \frac{1}{2}\mathbf{X}^{\mathrm{T}}(t_{\mathrm{f}})\mathbf{S}_{\mathrm{f}}\mathbf{X}(t_{\mathrm{f}}) + \frac{1}{2}\int_{t_0}^{t_{\mathrm{f}}} [\mathbf{X}^{\mathrm{T}}(\mathbf{\Theta}, t)\mathbf{Q}\mathbf{X}(\mathbf{\Theta}, t) + \mathbf{U}^{\mathrm{T}}(\mathbf{\Theta}, t)\mathbf{R}\mathbf{U}(\mathbf{\Theta}, t)]\,\mathrm{d}t \tag{9.111}$$

where $\mathbf{S}_{\mathrm{f}} = [S_{\mathrm{f},ij}]_{n\times n}$ and $\mathbf{Q} = [Q_{ij}]_{n\times n}$ are symmetric, positive semi-definite matrices, $\mathbf{R} = [R_{ij}]_{m\times m}$ is a symmetric, positive definite matrix and $\mathbf{U}(\mathbf{\Theta}, \cdot)$ represents a time history dependent on $\mathbf{\Theta}$.

In this context, the Hamiltonian defined in Equation 9.101 becomes

$$\begin{aligned}\mathcal{H}\,[\mathbf{X}(t), \mathbf{U}(t), \boldsymbol{\lambda}(t), t, \mathbf{\Theta}] = \frac{1}{2}&[\mathbf{X}^{\mathrm{T}}(\mathbf{\Theta}, t)\mathbf{Q}\mathbf{X}(\mathbf{\Theta}, t) + \mathbf{U}^{\mathrm{T}}(\mathbf{\Theta}, t)\mathbf{R}\mathbf{U}(\mathbf{\Theta}, t)]\\ &+\boldsymbol{\lambda}^{\mathrm{T}}(\mathbf{\Theta}, t)[\mathbf{A}\mathbf{X}(\mathbf{\Theta}, t) + \mathbf{B}\mathbf{U}(\mathbf{\Theta}, t) + \mathbf{L}\boldsymbol{\xi}(\mathbf{\Theta}, t)]\end{aligned} \tag{9.112}$$

Thus, the Euler–Lagrange equations in Equations 9.102–9.104 become

$$\dot{\boldsymbol{\lambda}}(t) = -\mathbf{A}^{\mathrm{T}}\boldsymbol{\lambda}(t) - \mathbf{Q}\mathbf{X}(t) \tag{9.113}$$

$$\boldsymbol{\lambda}(t_{\mathrm{f}}) = \mathbf{S}_{\mathrm{f}}\mathbf{X}(t_{\mathrm{f}}) \tag{9.114}$$

and

$$\mathbf{U}^{\mathrm{T}}\mathbf{R} + \boldsymbol{\lambda}^{\mathrm{T}}\mathbf{B} = \mathbf{0} \tag{9.115}$$

respectively. From Equation 9.115, we get the control law

$$\mathbf{U}(t) = -\mathbf{R}^{-1}\mathbf{B}^{\mathrm{T}}\boldsymbol{\lambda}(t) \tag{9.116}$$

Substituting this in the state-control equations (Equation 9.110) yields

$$\dot{\mathbf{X}} = \mathbf{A}\mathbf{X}(t) - \mathbf{B}\mathbf{R}^{-1}\mathbf{B}^{\mathrm{T}}\boldsymbol{\lambda}(t) + \mathbf{L}\boldsymbol{\xi}(\mathbf{\Theta}, t) \tag{9.117}$$

whose initial condition $\mathbf{X}(t_0) = \mathbf{X}_0$ is specified.

We note that the two-point boundary value problem consisting of Equations 9.113, 9.114 and 9.117 is a linear equation set, and the terminal condition (Equation 9.114) is also in a linear relationship; therefore, it is reasonable to suppose that the relationship between the solutions of $\boldsymbol{\lambda}(t)$ and $\mathbf{X}(t)$ takes the linear form

$$\boldsymbol{\lambda}(t) = \mathbf{S}(t)\mathbf{X}(t) + \boldsymbol{\psi}(t) \tag{9.118}$$

where $\mathbf{S}(t) = [S_{ij}(t)]_{n \times n}$ and $\boldsymbol{\psi}(t) = [\psi_1(t), \psi_2(t), \ldots, \psi_n(t)]^{\mathrm{T}}$ are to be determined.

Differentiating Equation 9.118 with respect to time t yields

$$\dot{\boldsymbol{\lambda}} = \dot{\mathbf{S}}\mathbf{X} + \mathbf{S}\dot{\mathbf{X}} + \dot{\boldsymbol{\psi}}(t) \tag{9.119}$$

Substituting the state-control equation (Equation 9.117) in this gives

$$\dot{\boldsymbol{\lambda}} = \dot{\mathbf{S}}\mathbf{X} + \mathbf{S}[\mathbf{A}\mathbf{X} - \mathbf{B}\mathbf{R}^{-1}\mathbf{B}^{\mathrm{T}}\boldsymbol{\lambda} + \mathbf{L}\boldsymbol{\xi}(\boldsymbol{\Theta}, t)] + \dot{\boldsymbol{\psi}}(t) \tag{9.120}$$

Replacing the left-hand side by the right-hand side of Equation 9.113 and introducing Equation 9.118 to eliminate $\boldsymbol{\lambda}$, we reach

$$-\mathbf{A}^{\mathrm{T}}[\mathbf{S}\mathbf{X} + \boldsymbol{\psi}(t)] - \mathbf{Q}\mathbf{X} = \dot{\mathbf{S}}\mathbf{X} + \mathbf{S}\{\mathbf{A}\mathbf{X} - \mathbf{B}\mathbf{R}^{-1}\mathbf{B}^{\mathrm{T}}[\mathbf{S}\mathbf{X} + \boldsymbol{\psi}(t)] + \mathbf{L}\boldsymbol{\xi}(\Theta, t)\} + \dot{\boldsymbol{\psi}}(t) \tag{9.121}$$

Making the coefficient matrices of $\mathbf{X}$ be zero, we obtain

$$\dot{\mathbf{S}} = -\mathbf{A}^{\mathrm{T}}\mathbf{S} - \mathbf{S}\mathbf{A} + \mathbf{S}\mathbf{B}\mathbf{R}^{-1}\mathbf{B}^{\mathrm{T}}\mathbf{S} - \mathbf{Q} \tag{9.122}$$

and

$$\dot{\boldsymbol{\psi}}(t) = [-\mathbf{A}^{\mathrm{T}} + \mathbf{S}\mathbf{B}\mathbf{R}^{-1}\mathbf{B}^{\mathrm{T}}]\boldsymbol{\psi}(t) - \mathbf{S}\mathbf{L}\boldsymbol{\xi}(\boldsymbol{\Theta}, t) \tag{9.123}$$

Equation 9.122 is a matrix Riccati equation, of which the terminal condition is specified according to Equation 9.114 by

$$\mathbf{S}(t_{\mathrm{f}}) = \mathbf{S}_{\mathrm{f}} \tag{9.124}$$

Once $\mathbf{S}(t)$ is determined by solving Equation 9.122, we can introduce it into Equation 9.123 and then solve it backward in time with the terminal condition, according to Equation 9.114:

$$\boldsymbol{\psi}(t_{\mathrm{f}}) = \mathbf{0} \tag{9.125}$$

The control law may be given by combining Equations 9.116 and 9.118:

$$\begin{aligned} \mathbf{U}(t) &= -\mathbf{R}^{-1}\mathbf{B}^{\mathrm{T}}\mathbf{S}(t)\mathbf{X}(t) - \mathbf{R}^{-1}\mathbf{B}^{\mathrm{T}}\boldsymbol{\psi}(\boldsymbol{\Theta}, t) \\ &= -\mathbf{C}(t)\mathbf{X}(t) - \mathbf{C}_{\mathrm{Input}}(t)\boldsymbol{\psi}(\boldsymbol{\Theta}, t) \end{aligned} \tag{9.126}$$

Here, the gains

$$\mathbf{C}(t) = \mathbf{R}^{-1}\mathbf{B}^{\mathrm{T}}\mathbf{S}(t) \quad \text{and} \quad \mathbf{C}_{\mathrm{Input}}(t) = \mathbf{R}^{-1}\mathbf{B}^{\mathrm{T}}(t) \tag{9.127a,b}$$

are respectively the state feedback gain matrix and the input feedforward gain matrix. The latter term is due to the fact that $\boldsymbol{\psi}(t)$ results from the existence of the input load $\boldsymbol{\xi}(\boldsymbol{\Theta}, t)$. Incidentally, we note that if the excitation disappears (that is, $\boldsymbol{\xi}(\boldsymbol{\Theta}, t) \equiv \mathbf{0}$), then from Equations 9.123 and 9.125 it is seen that the solution process $\boldsymbol{\psi}(t) \equiv \mathbf{0}$. Thus, the input feedforward term disappears if the input disappears. The solution now becomes the conventional 'pure' state feedback control.

In Equation 9.126, the first part of the control force comes from the linear state feedback, where the control gain is available in advance offline and the information of the state comes from the online measured data or optimal estimate. The second part of the control force, however, cannot be obtained through the online measured information because a backward differential equation needs to be solved and this differs from sample to sample. Some techniques could be employed to treat this part reasonably. A possible option is to use the mean effect; that is, the control law is given by

$$\mathbf{U}(t) = -\mathbf{C}(t)\mathbf{X}(t) - \mathbf{C}_{\text{Input}}(t)\bar{\boldsymbol{\psi}}(t) \tag{9.128}$$

where $\bar{\boldsymbol{\psi}}(t) = \mathcal{E}[\boldsymbol{\psi}(\boldsymbol{\Theta}, t)]$. Note that for an earthquake, usually we have $\mathcal{E}[\boldsymbol{\xi}(\boldsymbol{\Theta}, t)] = \mathbf{0}$; thus, from Equations 9.123 and 9.125, we have $\mathcal{E}[\boldsymbol{\psi}(\boldsymbol{\Theta}, t)] = \mathbf{0}$. In this case, the control law becomes

$$\mathbf{U}(t) = -\mathbf{C}(t)\mathbf{X}(t) \tag{9.129}$$

By substituting the control law (Equation 9.129) in the state-control equation (Equation 9.110), the state vector and the optimal control force process can be obtained. The GDEEs Equations 9.107a–9.108b can thus be solved to obtain the PDF of the state vector and the control history.

It is worth pointing out that, in the design stage of the controller, all the above computations can be carried out offline. In practical implementations, the only online computations needed are in Equation 9.129, for which the computational effort is small.

Selection of the Weighting Matrices

In the the Riccati equation, the weighting matrices **Q** and **R** are involved in the coefficients. Different weighting matrices will of course lead to different control laws. Although it is very important to select appropriate weighting matrices, between which the relative relationship accounts for the trade-off between the controlled response and the input control demand, there is no available rational criterion. In most cases the selection of the weighting matrices is based on engineering experience and determined through a trial-and-error strategy (Stengel, 1994; Zhang and Xu, 2001; Agranovich *et al.*, 2004). This problem will be more significant in optimal control of stochastic systems because of the variation of the system parameters and the fluctuation of the excitations.

In the context of civil engineering, the weighting matrices **Q** usually take the block form

$$\mathbf{Q} = \begin{bmatrix} \mathbf{Q}_{\text{d}} & 0 & 0 \\ 0 & \mathbf{Q}_{\text{v}} & 0 \\ 0 & 0 & \mathbf{Q}_{\text{a}} \end{bmatrix} \tag{9.130}$$

where $\mathbf{Q}_{\text{d}}$, $\mathbf{Q}_{\text{v}}$ and $\mathbf{Q}_{\text{a}}$ are diagonal matrices of appropriate dimension, corresponding to the weights on the displacements, velocities and accelerations. For simplicity, we can use times of unit matrices of appropriate dimension; thus:

$$\mathbf{Q} = \begin{bmatrix} \gamma_d \mathbf{I} & 0 & 0 \\ 0 & \gamma_v \mathbf{I} & 0 \\ 0 & 0 & \gamma_a \mathbf{I} \end{bmatrix} \tag{9.131}$$

where γ_d, γ_v and γ_a are the corresponding weighting coefficients. If we suppose the weighting matrix $\mathbf{R}$ also takes the form of multiples of unit matrices of appropriate dimension, namely

$$\mathbf{R} = \gamma_{\mathbf{R}} \mathbf{I} \tag{9.132}$$

then the four parameters γ_d, γ_v, γ_a and $\gamma_{\mathbf{R}}$ determine the weighting matrices $\mathbf{Q}$ and $\mathbf{R}$. Therefore, if we establish a criterion for the selection of the weighting matrices, we come to a parametric optimality problem.

To consider the optimal control problem where the accelerations are weighted, a more involved filtering process should be taken into account. Consistent with the preceding sections, we now set $\gamma_a = 0$ for simplicity. Further, we let $\gamma_{\mathbf{Q}} \triangleq \gamma_d = \gamma_v$. The weighting matrix $\mathbf{Q}$ is now

$$\mathbf{Q} = \gamma_{\mathbf{Q}} \mathbf{I} \tag{9.133}$$

where $\mathbf{I}$ is the unit matrix of appropriate dimension.

Because only the relative relationship between $\mathbf{Q}$ and $\mathbf{R}$ matters, now only one parameter, the weighting ratio

$$g_W = \frac{\gamma_{\mathbf{Q}}}{\gamma_{\mathbf{R}}} \tag{9.134}$$

is involved.

Example 9.1. Probability Density Evolution Analysis of a Linear SDOF System Consider an SDOF linear system subjected to random ground-motion excitations. Using the random Fourier function model for seismic ground motion (see Equation 3.41), 221 representative ground-motion acceleration histories can be generated (Li and Ai, 2006). Figure 9.2 shows the relationships of the maximum acceleration and the maximum control force versus the weighting ratio of the 221 representative excitations. Figure 9.3 shows the mean maximum quantities (the maximum acceleration, relative displacement and the control force) versus the weighting ratio g_W. It is seen that the tendency of the mean maximum relative displacement versus the weighting

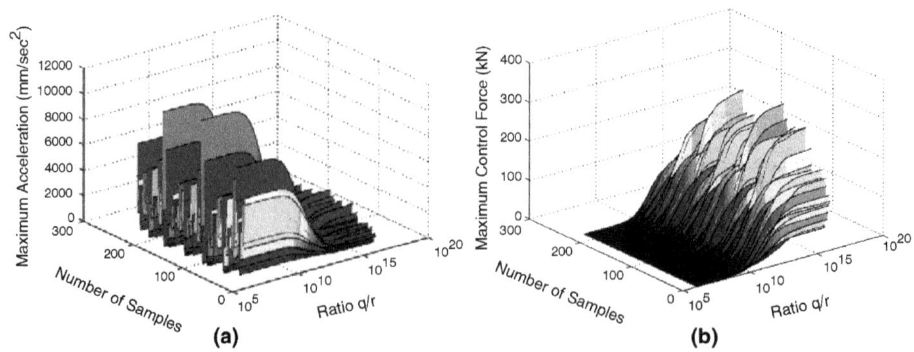

Figure 9.2 Maximum quantities versus weighting ratio ($\gamma_{\mathbf{Q}} = 100$).

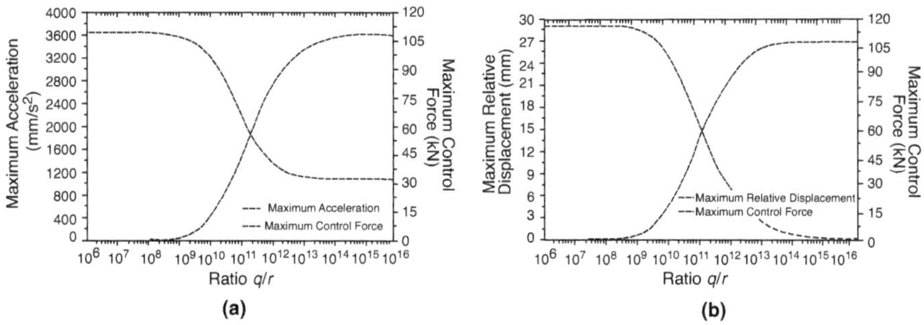

Figure 9.3 Mean maximum quantities versus weighting ratio ($\gamma_Q = 100$).

ratio (Figure 9.3b) is somewhat opposite to that of the mean maximum acceleration (and control force) versus the weighting ratio. From the figures, a trade-off among the relative displacement, acceleration and control force means that it is reasonable to choose the weighting ratio $g_W = 1.0 \times 10^{11}$.

Figure 9.4 shows the standard deviations of the relative displacement and the acceleration responses of the uncontrolled and controlled systems. Obviously, the control greatly reduces

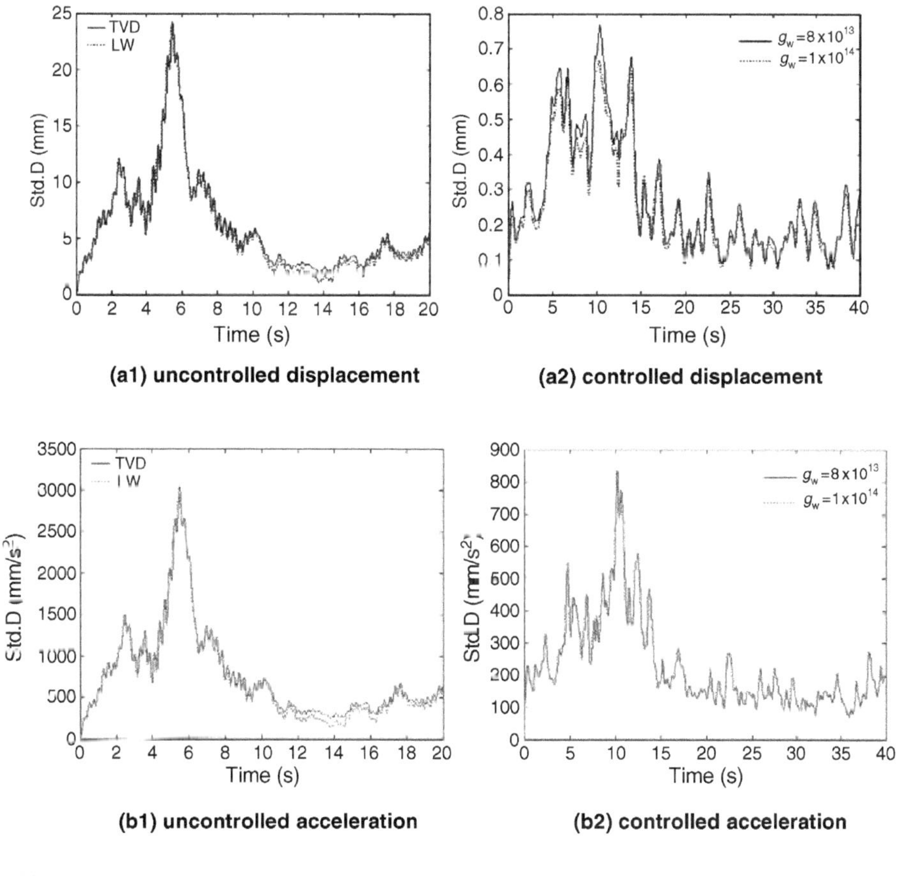

Figure 9.4 Standard deviations of the responses of a controlled and uncontrolled system.

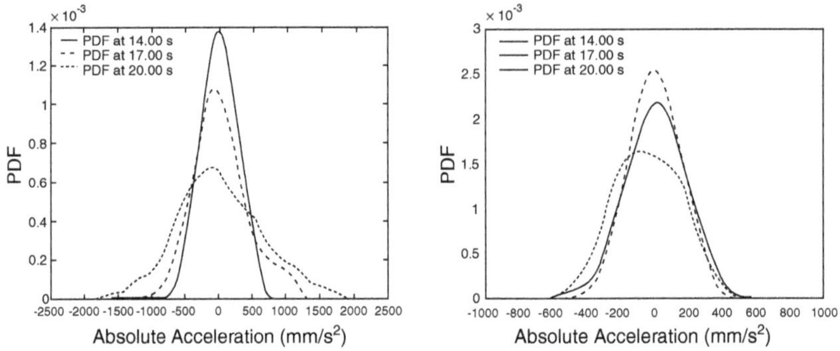

(a) uncontrolled acceleration **(b) controlled acceleration**

Figure 9.5 Probability density at different time instants.

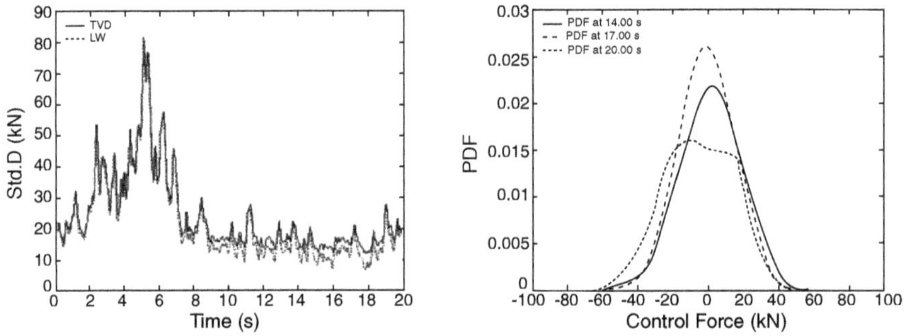

(a) standard diviation **(b) probability density**

Figure 9.6 Probabilistic information of the control force.

the standard deviation of the displacement by about 12 times. Similarly, the control also reduces the standard deviation of the acceleration response by about four times.

Figure 9.5a and b shows the probability densities of the acceleration of the uncontrolled and controlled systems at some instants of time. Clearly, the distribution range of the acceleration is greatly narrowed.

As previously pointed out, the control force process is also a stochastic process, even if the control gains are deterministic. Figure 9.6 shows the standard deviation and the probability density of the control force at some instants of time. From the figure, it is seen that during the stage of strong ground motion, larger control forces are needed to suppress the response of the controlled structure. In addition, at different instants of time, the PDFs are quite distinct. □

9.4 Reliability-Based Control of Structural Systems

9.4.1 Reliability of Controlled Structural Systems

Reliability of the control is one of the critical issues in stochastic control. In the past decade or so, some investigators have proposed a few frameworks to incorporate reliability theory in stochastic control (Spencer *et al.*, 1994; Field and Bergman, 1998; Scott May and Beck, 1998; Battaini *et al.*, 2000; Yuen and Beck, 2003; Scruggs *et al.*, 2006; Zhu, 2006; Li *et al.*, 2008). Depending on the purpose, the problem involves two aspects: the first is to minimize the performance index with the reliability of the synthesized system not lower than a prescribed level (P1, for brevity); the second is to maximize the reliability of the synthesized system with some indices of the system bounded in a prescribed range (P2, for brevity). Zhu (2006) has done some work in P2, while most other studies have focused on P1. In this section, we will only deal with P1.

Without loss of generality, we consider the controlled stochastic system in Equation 9.99, repeated here as Equation 9.135 for convenience:

$$\dot{\mathbf{X}} = \mathbf{f}[\mathbf{X}(t), \mathbf{U}(t), t, \boldsymbol{\Theta}(\varpi)] \tag{9.135}$$

Once we determine a control law, say in the form of

$$\mathbf{U}(t) = \mathcal{G}[\mathbf{X}(t), t, \boldsymbol{\Theta}, \boldsymbol{\kappa}] \tag{9.136}$$

where $\boldsymbol{\kappa}$ is an undetermined parametric vector related to the feedback gains, then substituting Equation 9.136 in the dynamic Equation 9.135 yields

$$\begin{aligned}\dot{\mathbf{X}} &= \mathbf{f}\{\mathbf{X}(t), \mathcal{G}[\mathbf{X}(t), t, \boldsymbol{\Theta}, \boldsymbol{\kappa}], t, \boldsymbol{\Theta}(\varpi)\} \\ &= \tilde{\mathbf{f}}[\mathbf{X}(t), t, \mathcal{G}, \boldsymbol{\kappa}, \boldsymbol{\Theta}(\varpi)]\end{aligned} \tag{9.137}$$

Solving these equations we get the Lagrangian description (formal solution) of the state vector

$$\mathbf{X} = \mathbf{H}_{\mathrm{X}}(\boldsymbol{\Theta}, t, \mathcal{G}, \boldsymbol{\kappa}) \tag{9.138}$$

where $\mathbf{H}_{\mathbf{X}}(\cdot) = [H_{X,1}(\cdot), H_{X,2}(\cdot), \ldots H_{X,n}(\cdot)]^{\mathrm{T}}$.

Thus, we can further obtain the reliability (or probability of failure) of the system by employing the approach based on the absorbing boundary condition or extreme-value distribution elaborated in Chapter 8. Here, we exemplify the approach based on the extreme-value distribution. Considering the ℓth component of $\mathbf{X}$, the extreme value is given by

$$W(\boldsymbol{\Theta}, \boldsymbol{\kappa}) = \underset{0 \le t \le T}{\mathrm{ext}}\{X_\ell(t)\} = \underset{0 \le t \le T}{\mathrm{ext}}\{\mathbf{H}_{\mathrm{X}}(\boldsymbol{\Theta}, t, \boldsymbol{\kappa})\} \tag{9.139}$$

Therefore, we can then construct a virtual stochastic process

$$Z(\tau) = \tilde{\varphi}[W(\boldsymbol{\Theta}, \boldsymbol{\kappa}), \tau] = \varphi(\boldsymbol{\Theta}, \tau, \boldsymbol{\kappa}) \tag{9.140}$$

which satisfies the conditions

$$Z(\tau)|_{\tau=0} = 0 \quad \text{and} \quad Z(\tau)|_{\tau=\tau_c} = W(\boldsymbol{\Theta}, \boldsymbol{\kappa}) \tag{9.141}$$

Thus, the joint density of $(Z(\tau), \boldsymbol{\Theta})$, denoted by $p_{Z\boldsymbol{\Theta}}(z, \boldsymbol{\theta}, \tau)$, satisfies the generalized density evolution equation (see Equations 6.123a and 6.123b)

$$\frac{\partial p_{Z\boldsymbol{\Theta}}(z, \boldsymbol{\theta}, \tau)}{\partial \tau} + \dot{\varphi}(\boldsymbol{\theta}, \tau, \boldsymbol{\kappa})\frac{\partial p_{Z\boldsymbol{\Theta}}(z, \boldsymbol{\theta}, \tau)}{\partial z} = 0 \tag{9.142}$$

Solving it under the initial condition

$$p_{Z\mathbf{\Theta}}(z,\boldsymbol{\theta},\tau)|_{\tau=0} = \delta(z)p_{\mathbf{\Theta}}(\boldsymbol{\theta}) \tag{9.143}$$

and then using the integration we have

$$p_Z(z,\tau) = \int_{\Omega_{\mathbf{\Theta}}} p_{Z\mathbf{\Theta}}(z,\boldsymbol{\theta},\tau)\,\mathrm{d}\boldsymbol{\theta} \tag{9.144}$$

Equations 9.142–9.144 can be solved via the numerical algorithms elaborated in Chapter 7. On doing the above, we get the reliability

$$R(\mathcal{G},\boldsymbol{\kappa}) = \int_{\Omega_s} p_W(w,\mathcal{G},\boldsymbol{\kappa})\,\mathrm{d}w = \int_{\Omega_s} p_Z(z,\tau_c,\mathcal{G},\boldsymbol{\kappa})\,\mathrm{d}z \tag{9.145}$$

and simultaneously the probability of failure

$$P_f(\mathcal{G},\boldsymbol{\kappa}) = 1 - R(\mathcal{G},\boldsymbol{\kappa}) \tag{9.146}$$

where Ω_s is the safe domain.

Denote the desired reliability by R_D. Thus, if

$$P_f(\mathcal{G},\boldsymbol{\kappa}) = 1 - R(\mathcal{G},\boldsymbol{\kappa}) \le 1 - R_D = P_{f,D} \tag{9.147}$$

where $P_{f,D} = 1 - R_D$ is the highest acceptable probability of failure, then the designed control with the form $\mathcal{G}$ and parameter vector $\boldsymbol{\kappa}$ satisfies the desired purpose. If $\mathcal{G}$ is first determined by some method, then the optimal parameter $\boldsymbol{\kappa}$ should be searched for. For instance, let $\boldsymbol{\kappa}^{(j)}, j=0, 1, 2, \ldots$, denote the parameter vector used in the jth iteration. We might choose the $\boldsymbol{\kappa}^{(j+1)}$ by

$$\frac{\|\boldsymbol{\kappa}^{(j+1)} - \boldsymbol{\kappa}^{(j)}\|}{\|\boldsymbol{\kappa}^{(j)} - \boldsymbol{\kappa}^{(j-1)}\|} = \frac{P_f(\boldsymbol{\kappa}^{(j)}) - P_{f,D}}{P_f(\boldsymbol{\kappa}^{(j)}) - P_f(\boldsymbol{\kappa}^{(j-1)})} \tag{9.148}$$

where $\|\cdot\|$ is the norm of the vector. For the single-parametric problem, Equation 9.148 is easy to realize. However, for general multiparametric problems, it may be much more complicated. Engineering experiences are very important in selecting the new parameters.

9.4.2 Determination of Control Criterion

Actually, in stochastic optimal control, what is really important is how to determine the control laws. In other words, how to design the controller, including the mechanism $\mathcal{G}$ and the control parameters $\boldsymbol{\kappa}$. In the preceding sections, we determine the control laws in fact based on some modifications on the sample counterpart. However, this might not ensure optimality. A more reasonable way might be based on the concept of reliability or exceeding probability.

For instance, if $Z(t)$ is a physical quantity to be considered in the controlled system, then denote the extreme value

$$Z_{\text{ext}} = \underset{t\in[t_0,t_f]}{\text{ext}}\, Z(\mathbf{\Theta},t) \tag{9.149}$$

The control law might be determined by minimizing the performance index

$$\mathcal{J}_1[\mathcal{G}, \boldsymbol{\kappa}] = \mathcal{E}[Z_{\text{ext}}] \tag{9.150}$$

or minimizing the performance index

$$\mathcal{J}_2[\mathcal{G}, \boldsymbol{\kappa}] = \mathcal{E}[Z_{\text{ext}}] + \alpha\sigma[Z_{\text{ext}}] \tag{9.151}$$

where $\alpha \geq 0$ is a coefficient. Alternatively, the control law can also be determined by minimizing the performance index

$$\mathcal{J}_3[\mathcal{G}, \boldsymbol{\kappa}] = \Pr\{Z_{\text{ext}} > Z_{\text{b}}\} \tag{9.152}$$

where Z_{b} is a prescribed threshold.

Clearly, the above performance indices will yield different control laws, which is in need of further investigation.

Appendix A: Dirac Delta Function

A.1 Definition

The Dirac delta function has various physical backgrounds in many different disciplines. If a function $f(x)$ satisfies the two conditions

$$f(x) = \begin{cases} \infty & \text{for } x = x_0 \\ 0 & \text{otherwise} \end{cases} \tag{A.1}$$

$$\int_{-\infty}^{\infty} f(x)\,\mathrm{d}x = 1 \tag{A.2}$$

then $f(x)$ is called the *Dirac delta function* and usually denoted by

$$f(x) = \delta(x - x_0) \tag{A.3}$$

It is also usually called the *Dirac function* for simplicity, and can be illustrated by Figure A.1.

The condition in Equation A.1 says that the Dirac function is zero except at the point x_0, where its value is infinite. The condition in Equation A.2 means that the total area below the curve of the function is unity. Clearly, it is a function with some interesting features. Mathematically, rigorously speaking, the Dirac function belongs to the family of distribution functions (also called generalized functions) (Zemanian, 1965; Zayed, 1996).

To give a more visual image, we consider the PDF of a uniformly distributed random variable

$$p_{\text{Uniform}}(x; a) = \begin{cases} \dfrac{1}{2a} & \text{for } \mu - a \le x \le \mu + a \\ 0 & \text{otherwise} \end{cases} \tag{A.4}$$

It is easy to verify that

$$\lim_{a \to 0} p_{\text{Uniform}}(x; a) = \delta(x - \mu) \tag{A.5}$$

which shows that, as the distribution width of a uniformly distributed random variable narrows and tends to zero, the uniform distribution tends to a Dirac delta function. Of course, the

Stochastic Dynamics of Structures Jie Li and Jianbing Chen

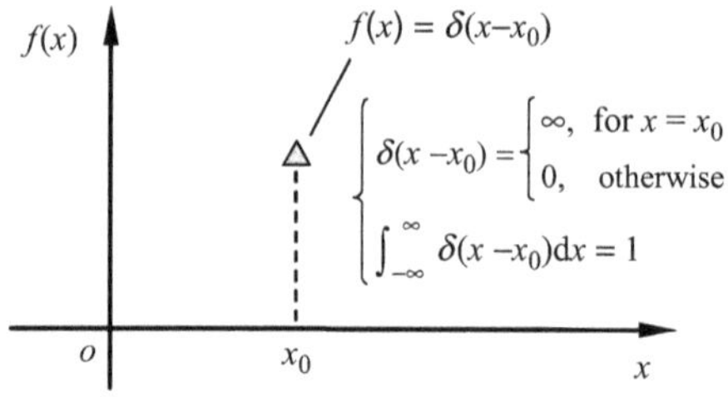

Figure A.1 The Dirac delta function.

uniformly distributed random variable reduces simultaneously to a deterministic variable (Figure A.2a).

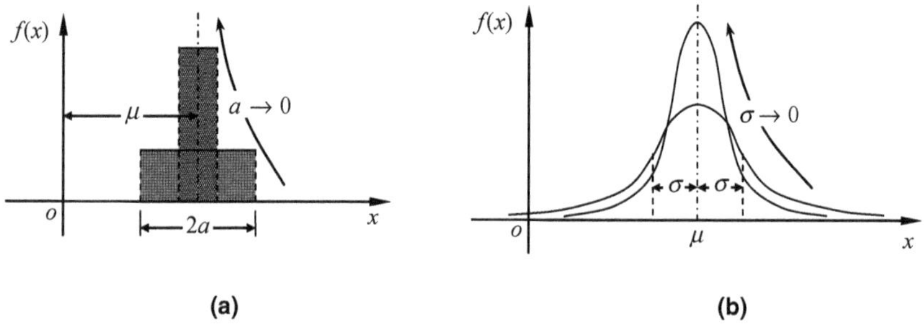

Figure A.2 The Dirac delta function as a limit of other functions: (a) uniform distribution; (b) normal distribution.

Likewise, if we consider the probability density of a normally distributed random variable

$$p_{\text{Normal}}(x;\sigma) = \frac{1}{\sqrt{2\pi}\sigma} \mathrm{e}^{-(x-\mu)^2/2\sigma^2} \tag{A.6}$$

then we can verify that

$$\lim_{\sigma \to 0} p_{\text{Normal}}(x;\sigma) = \delta(x-\mu) \tag{A.7}$$

Thus, as the standard deviation of a normally distributed random variable narrows and tends to zero, the normal distribution also tends to a Dirac delta function. Simultaneously, the normally distributed random variable reduces to a deterministic variable (Figure A.2b).

A.2 Integration and Differentiation

According to the conditions in Equations A.1 and A.2 in the definition, it is easy to verify that for a general function $g(x)$ the integral involving a Dirac delta function is given by

$$\int_{-\infty}^{\infty} g(x)\delta(x-x_0)\,\mathrm{d}x = g(x_0) \tag{A.8}$$

This indicates that a Dirac delta function identifies the value of the adjoint function in the integrand.

Based on this, if we consider the Fourier transform of $\delta(t-t_0)$, then we have

$$\mathcal{F}[\delta(t-t_0)] = \int_{-\infty}^{\infty} \delta(t-t_0)\mathrm{e}^{-i\omega t}\,\mathrm{d}t = \mathrm{e}^{-i\omega t_0} \tag{A.9}$$

Conversely, we know that the inverse Fourier transform on Equation A.9 gives

$$\mathcal{F}^{-1}\left[\mathrm{e}^{-i\omega t_0}\right] = \frac{1}{2\pi}\int_{-\infty}^{\infty} \mathrm{e}^{-i\omega t_0}\mathrm{e}^{i\omega t}\,\mathrm{d}\omega = \frac{1}{2\pi}\int_{-\infty}^{\infty} \mathrm{e}^{i\omega(t-t_0)}\,\mathrm{d}\omega = \delta(t-t_0) \tag{A.10}$$

If we let $t_0 = 0$, then Equations A.9 and A.10 become

$$\mathcal{F}[\delta(t)] = 1 \tag{A.11a}$$

and

$$\mathcal{F}^{-1}[1] = \delta(t) \tag{A.11b}$$

respectively. According to Equation A.8, we know that for a time history $x(t)$ there is

$$x(t) = \int_{-\infty}^{\infty} x(\tau)\delta(\tau - t)\,\mathrm{d}\tau \tag{A.12}$$

That is, the Dirac delta function can be regarded as a basis unit of a general time history.

The Dirac delta function can also be regarded as the derivative of a discontinuous function – the unit step function (Heaviside's function)

$$u(t-t_0) = \begin{cases} 1 & \text{for } t \geq t_0 \\ 0 & \text{otherwise} \end{cases} \tag{A.13}$$

If we denote $\dot{u}(t-t_0) = \mathrm{d}u(t-t_0)/\mathrm{d}t$, then it is easy to verify that

$$\int_{-\infty}^{\infty} \dot{u}(t-t_0)\,\mathrm{d}t = u(\infty) - u(-\infty) = 1 \tag{A.14a}$$

$$\dot{u}(t-t_0) = \begin{cases} \infty & \text{for } t = t_0 \\ 0 & \text{otherwise} \end{cases} \tag{A.14b}$$

Comparing these with Equations A.1 and A.2, it is clear that

$$\dot{u}(t-t_0) = \delta(t-t_0) \tag{A.15}$$

Thus, the Dirac function can be regarded as the derivative of the unit step function.

We now consider the integral of a Dirac function involving a compound function; for instance:

$$\mathcal{I}[g(x)] = \int_{-\infty}^{\infty} g(x)\delta[\varphi(x) - x_0]\,\mathrm{d}x \tag{A.16a}$$

If we change the integral variable x by $\varphi^{-1}(y)$, then Equation A.16a becomes

$$\mathcal{I}[g(x)] = \int_{-\infty}^{\infty} |J| g[\varphi^{-1}(y)]\delta[y - x_0]\,\mathrm{d}y = |J| g[\varphi^{-1}(x_0)] \tag{A.16b}$$

where $|J| = |\mathrm{d}\varphi^{-1}/\mathrm{d}y|$ is the Jacobian. Here, it should be noted that the Jacobian cannot be omitted.

We now consider the integral of Dirac's function when the x_0 take only integer values over a neighboring small interval near the integers:

$$\mathcal{I}_{ij} = \lim_{\varepsilon \to 0} \int_{j-\varepsilon}^{j+\varepsilon} \delta(x - i)\,\mathrm{d}x = \lim_{\varepsilon \to 0} \int_{i-\varepsilon}^{i+\varepsilon} \delta(x - j)\,\mathrm{d}x = \begin{cases} 1 & \text{for } i = j \\ 0 & \text{otherwise} \end{cases} \tag{A.17}$$

when i and j are integers; the above integral $\mathcal{I}_{ij}$ is called the Kronecker delta and is denoted by

$$\delta_{ij} = \begin{cases} 1 & \text{for } i = j \\ 0 & \text{otherwise} \end{cases} \tag{A.18}$$

Thus, it is seen that the Dirac delta function can be regarded as the continuous version of the Kronecker delta and, in turn, the latter can be regarded as the discretized version of the former.

A.3 Common Physical Backgrounds

A.3.1 Probability Distribution of Discrete Random Variables

If X is a discrete random variable with the distribution that

$$\Pr\{X = x_j\} = P_j \qquad j = 1, 2, \cdots, n \tag{A.19}$$

and $\sum_{j=1}^{n} P_j = 1$, then the PDF of X can be written as

$$p_X(x) = \sum_{j=1}^{n} P_j \delta(x - x_j) \tag{A.20}$$

Actually, according to the definition in Section 2.1.1, the CDF of the random variable with the distribution in Equation A.19 is given by

$$F_X(x) = \sum_{j=1}^{n} P_j u(x - x_j) \tag{A.21}$$

where $u(\cdot)$ is the unit step function defined in Equation A.14. Noting $p_X(x) = \mathrm{d}F_X(x)/\mathrm{d}x$ (see Equation 2.1) and Equation A.15, this immediately yields Equation A.20.

Thus, introducing the Dirac delta function, the discrete and the continuous random variable can be treated in a unified theory frame. In this context, a discrete random variable can be viewed as the limit of a sequence of continuous random variables; this can be understood very clear particularly when we note Equations A.4–A.7.

A.3.2 Concentrated and Distributed Loads

Consider a simply supported beam AB in Figure A.3. A distributed vertical load $w(x)$ and n concentrated forces $F_1, F_2, \cdots, F_n$ are applied.

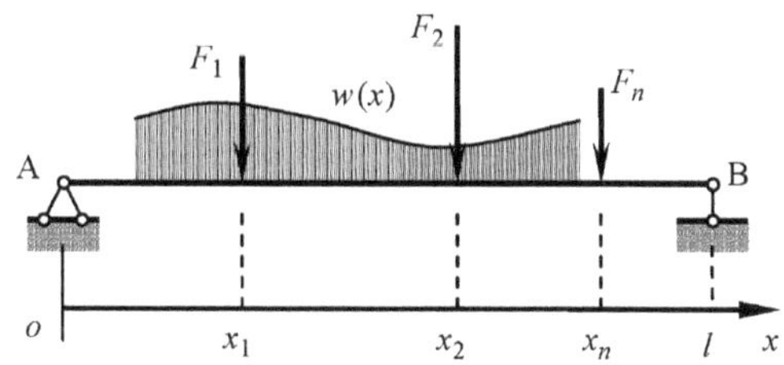

Figure A.3 Loads on a simply supported beam.

The load applied on the beam can be written in a unified way as

$$q(x) = -w(x) - \sum_{j=1}^{n} F_j \delta(x - x_j) \tag{A.22}$$

Actually, it is easy to know that if the reaction force of support A is R_A, then the shear force of the beam at x is given by

$$Q(x) = R_A - \int_0^x w(x)\, dx - \sum_{j=1}^{n} F_j u(x - x_j) \tag{A.23}$$

Noting $q(x) = dQ(x)/dx$ and Equation A.15, this immediately leads to Equation A.22. This is actually the mathematical expression of the physical sense that a concentrated force is an idealized situation when the applied area of the distributed force is very small.

A.3.3 Unit Impulse Function

If $f(t)$ is a time history of a force applied on an initially rest mass particle m, then the impulse of the force is given by

$$\kappa = \int_0^t f(t)\, dt = mv \tag{A.24}$$

where v is the velocity of the mass particle. If the time duration of the force is shortened, then to achieve the same velocity v the force should be enlarged. In the limit, as $t \rightarrow 0$, for a specified v we have

$$mv = \lim_{t \rightarrow 0} \int_0^t f(t)\, dt = \text{constant} \tag{A.25}$$

Therefore, we find that

$$f(t) = mv\delta(t) \tag{A.26}$$

This indicates that $\delta(t)$ is a unit impulse function. The applied $\delta(t)$ produces a sudden increment of the momentum on the mass point. We have elaborated this fact in Section 5.2.1.

A.3.4 Unit Harmonic Function

Consider the inverse Fourier transform of a function $g(\omega) = 2\pi\delta(\omega - \omega_0)$, we have

$$\mathcal{F}^{-1}[2\pi\delta(\omega - \omega_0)] = \frac{1}{2\pi}\int_{-\infty}^{\infty} 2\pi\delta(\omega - \omega_0)e^{i\omega t}\,d\omega = e^{i\omega_0 t} \tag{A.27}$$

Thus, the Fourier transform of a unit harmonic function gives

$$\mathcal{F}[e^{i\omega_0 t}] = \int_{-\infty}^{\infty} e^{i\omega_0 t}e^{-i\omega t}\,d\omega = \int_{-\infty}^{\infty} e^{-i(\omega-\omega_0)t}\,d\omega = 2\pi\delta(\omega - \omega_0) \tag{A.28}$$

The physical sense is clear: in a unit harmonic function the frequency content is so simple that only one single frequency is involved; this is just what is implied in Equation A.28.

Further, in the case $g(\omega) = \pi[\delta(\omega - \omega_0) + \delta(\omega + \omega_0)]$, Equations A.27 and A.28 become

$$\mathcal{F}^{-1}\left\{\pi[\delta(\omega - \omega_0) + \delta(\omega + \omega_0)]\right\} = \frac{1}{2\pi}\int_{-\infty}^{\infty} \pi[\delta(\omega - \omega_0) + \delta(\omega + \omega_0)]e^{i\omega t}\,d\omega = \cos\omega_0 t \tag{A.29}$$

and

$$\begin{aligned}\mathcal{F}[\cos\omega_0 t] &= \textstyle\int_{-\infty}^{\infty} \cos\omega_0 t e^{-i\omega t}\,dt \\ &= \pi[\delta(\omega - \omega_0) + \delta(\omega + \omega_0)]\end{aligned} \tag{A.30}$$

respectively.

Appendix B: Orthogonal Polynomials

B.1 Basic Concepts

Let $[a, b]$ denote a finite or infinite interval and a function $w(x)$ is defined over it. If $w(x)$ satisfies the following properties:

(a) $w(x) \geq 0,\ x \in [a, b]$
(b) $\int_a^b w(x)\,\mathrm{d}x > 0$ and
(c) $\int_a^b x^n w(x)\,\mathrm{d}x,\ n = 1, 2, \ldots$ exists

then $w(x)$ is called a weight function over $[a, b]$.

For the n-order polynomials, whose first-term coefficient $a_n \neq 0$:

$$f_n(x) = a_n x^n + \ldots + a_1 x + a_0 \qquad n = 0, 1, 2, \ldots \tag{B.1}$$

if they satisfy

$$\int_a^b w(x) f_n(x) f_m(x)\,\mathrm{d}x = 0 \qquad n \neq m;\ n, m = 0, 1, 2, \ldots \tag{B.2}$$

then the sequence of polynomials $f_n(x), f_m(x), \ldots$ is orthogonal over $[a, b]$ with respect to $w(x)$, and $f_n(x)$ are orthogonal polynomials over $[a, b]$ with weight function $w(x)$.

In the case $n = m$, there is

$$\int_a^b w(x) f_n^2(x)\,\mathrm{d}x = h_n \tag{B.3}$$

where h_n is a positive number.

Let

$$\varphi_n(x) = \frac{f_n(x)}{\sqrt{h_n}} \tag{B.4}$$

Stochastic Dynamics of Structures Jie Li and Jianbing Chen

Then from Equations B.2 and B.3, there is

$$\int_a^b w(x)\varphi_n(x)\varphi_m(x)\,\mathrm{d}x = \begin{cases} 1 & \text{if } n = m \\ 0 & \text{otherwise} \end{cases} \quad n, m = 0, 1, 2, \ldots \tag{B.5}$$

Here, $\varphi_n(x)$ are called standard orthogonal functions with weight $w(x)$.

Using

$$\tilde{\varphi}_n(x) = \frac{\varphi_n(x)}{\sqrt{w(x)}} \quad n, m = 0, 1, 2, \ldots \tag{B.6}$$

then $\varphi_n(x)$ is transformed into standard orthogonal functions in common cases.

For a space of functions with weight $w(x)$, the inner product is defined by

$$\langle f, g \rangle = \int_a^b w(x) f(x) g(x)\,\mathrm{d}x \tag{B.7}$$

where $f(x)$ and $g(x)$ are points in the space.

It can be proved that a space with the inner product defined as above is a Hilbert space. Thus, any function $f(x)$ in it can be expanded into a generalized Fourier series; that is:

$$f(x) = \sum_{i=0}^{\infty} a_i \varphi_i(x) \tag{B.8}$$

where the coefficients

$$a_i = \langle f, \varphi_i \rangle = \int_a^b w(x) f(x) \varphi_i(x)\,\mathrm{d}x \tag{B.9}$$

are considered as the projections of $f(x)$ on the basis functions $\varphi_i(x)$.

Obviously, the weighted orthogonal decomposition mentioned above is an extension of the orthogonal decomposition in the Hilbert spaces. As the weight function $w(x) = 1$, it is reduced to the orthogonal decomposition in the Hilbert spaces. On the other hand, if we first make the transformation

$$\tilde{f}(x) = \frac{f(x)}{\sqrt{w(x)}} \tag{B.10}$$

then the orthogonal decomposition of $\tilde{f}(x)$ also belongs to the orthogonal decomposition in the Hilbert spaces. Note that $\tilde{f}(x)$ form a Hilbert space. It follows that the weighted Hilbert spaces are equivalent to the transformed Hilbert spaces.

There are a variety of orthogonal polynomials, each of which comes from a specific generating function (Andrews *et al.*, 2000). Denoting these generating functions by $G(x, t)$, in general there is

$$G(x, t) = \sum_{n=0}^{\infty} f_n(x) t^n \tag{B.11}$$

This implies that the orthogonal polynomials are essentially the coefficients of the series expansions of the generating functions with respect to the parameter t.

Among the properties of orthogonal polynomials, the property of recursion is most useful. For three orthogonal polynomials $f_{n-1}(x)$, $f_n(x)$ and $f_{n+1}(x)$ defined in Equation B.1, they satisfy the recursion relation

$$f_{n+1}(x) = \frac{a_{n+1}}{a_n}(x - A_n)f_n(x) - \frac{a_{n+1}a_{n-1}}{a_n^2}B_n f_{n-1}(x) \tag{B.12}$$

where

$$A_n = \frac{1}{h_n}\int_a^b xw(x)f_n^2(x)\,\mathrm{d}x \tag{B.13}$$

$$B_n = \frac{h_n}{h_{n-1}} \tag{B.14}$$

Rearranging Equation B.12 and expressing it in terms of the weighted standard orthogonal functions yields

$$x\varphi_n(x) = \alpha_n\varphi_{n-1}(x) + \beta_n\varphi_n(x) + \gamma_n\varphi_{n+1}(x) \tag{B.15}$$

where

$$\alpha_n = \frac{a_{n-1}h_n^{1/2}}{a_n h_n^{1/2}} \tag{B.16}$$

$$\beta_n = \frac{1}{h_n}\int_a^b xw(x)f_n^2(x)\,\mathrm{d}x \tag{B.17}$$

$$\gamma_n = \frac{a_n h_{n+1}^{1/2}}{a_{n+1}h_n^{1/2}} \tag{B.18}$$

B.2 Common Orthogonal Polynomials

The so-called 'common' here means that these orthogonal polynomials are closely related to the contents in this book. They are Hermite, Legendre and Gegenbauer polynomials respectively (Andrews *et al.*, 2000; Dunkl and Xu, 2001).

B.2.1 Hermite Polynomials $H_{e_n}(x)$

The Hermite polynomials are orthogonal in $(-\infty, \infty)$ with respect to the weight function $\mathrm{e}^{-x^2/2}$. Since the weight function $\mathrm{e}^{-x^2/2}$ is just the density function of the standard normal distribution, $H_{e_n}(x)$ play an important role in the expansion of functions on probability spaces associated with the normal distributions. Hermite polynomials are generated by

$$H_{e_n}(x) = (-1)^n e^{x^2/2} \frac{d^n}{dx^n}(e^{-x^2/2}) \qquad n = 0, 1, 2, \ldots \tag{B.19}$$

It can be verified that

$$h_n = \int_{-\infty}^{\infty} e^{-x^2/2} H_{e_n}^2(x)\,dx = \sqrt{2\pi} n! \tag{B.20}$$

and the recursion coefficients, corresponding to Equation B.15, are given by

$$\alpha_n = \sqrt{n} \qquad \beta_n = 0 \qquad \gamma_n = \sqrt{n+1} \tag{B.21}$$

Using Equation B.19, the first few polynomials are as follows:

$$\begin{cases} H_{e_0} = 1 \\ H_{e_1} = x \\ H_{e_2} = x^2 - 1 \\ H_{e_3} = x^3 - 3x \\ H_{e_4} = x^4 - 6x^2 + 3 \\ \ldots \end{cases} \tag{B.22}$$

In fact, the recursion formula relative to $H_{e_n}(x)$ can be rewritten as

$$\begin{cases} H_{e_{n+1}}(x) = xH_{e_n}(x) - nH_{e_{n-1}}(x) \\ H_{e_0} = 1,\ H_{e_1} = x \end{cases} \tag{B.23}$$

Graphs of the first six Hermite polynomials are shown in Figure B.1.

B.2.2 Legendre Polynomials $P_n(x)$

The Legendre polynomials are orthogonal with respect to weight function $w(x) = 1$ over the interval $[-1, 1]$. As noted, their weight function and the defining interval are analogous to the density function of the uniform distribution and its interval respectively. Thus, for the probability space associated with the uniform distributions, the Legendre polynomials are of great significance.

The Legendre polynomials are generated by

$$P_n(x) = \frac{1}{2^n n!} \frac{d^n}{dx^n}(x^2 - 1)^n \qquad n = 0, 1, 2, \ldots \tag{B.24}$$

We can verify that

$$h_n = \int_{-1}^{1} P_n^2(x)\,dx = \frac{2}{2n+1} \tag{B.25}$$

and give the recursion coefficients in Equation B.15:

$$\alpha_n = \frac{n}{\sqrt{4n^2 - 1}} \qquad \beta_n = 0 \qquad \gamma_n = \frac{n+1}{\sqrt{4n^2 + 8n + 3}} \tag{B.26}$$

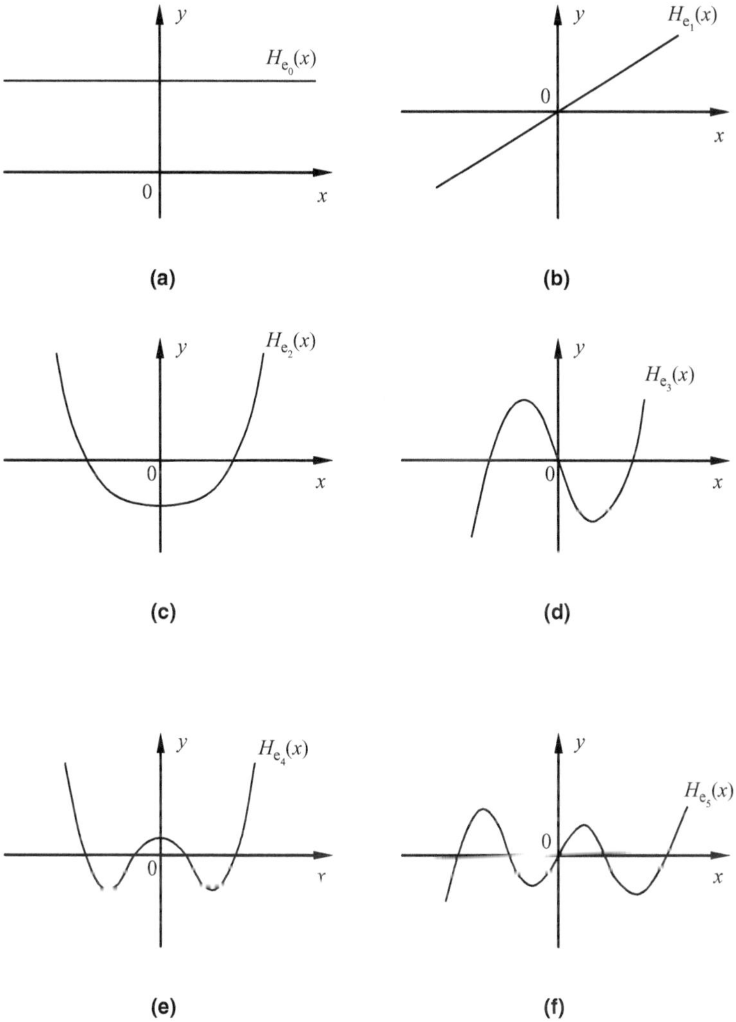

Figure B.1 Orthogonal polynomials.

Using Equation B.24, the first few polynomials are as follows

$$\begin{cases} P_0(x) = 1 \\ P_1(x) = x \\ P_2(x) = \frac{1}{2}(3x^2 - 1) \\ P_3(x) = \frac{1}{2}(5x^3 - 3x) \\ P_4(x) = \frac{1}{8}(35x^4 - 30x^2 + 3) \\ \cdots \end{cases} \tag{B.27}$$

A Legendre polynomial of any order can be obtained from the following recursive formula:

$$P_{n+1}(x) = \frac{2n+1}{n+1} x P_n(x) - \frac{n}{n+1} P_{n-1}(x) \tag{B.28}$$

Graphs of the first six Legendre polynomials are similar to those shown in Figure B.1.

B.2.3 Gegenbauer Polynomials $C_n^{(\alpha)}(x)$

The Gegenbauer polynomials are orthogonal with respect to weight function $(1-x^2)^{\alpha-(1/2)}$ over the interval $[-1, 1]$, and in general are expressed by

$$C_n^{(\alpha)}(x) = \frac{(-1)^n (2\alpha)^n}{2^n n! (\alpha+1/2)^n (1-x^2)^{\alpha-(1/2)}} \frac{\mathrm{d}^n}{\mathrm{d}x^n} (1-x^2)^{n+\alpha-(1/2)} \qquad \alpha > \frac{1}{2},\ n = 0, 1, 2, \ldots \tag{B.29}$$

where α is a given parameter.

It can be proved that in this case there exists

$$h_n = \int_{-1}^{1} (1-x^2)^{\alpha-(1/2)} [C_n^{\alpha}(x)]^2 \,\mathrm{d}x = \frac{\pi 2^{1-2\alpha} \Gamma(n+2\alpha)}{n!(n+\alpha)[\Gamma(\alpha)]^2} \tag{B.30}$$

where $\Gamma(\cdot)$ denotes the Gamma function.

In like manner, for Equation B.15, there are

$$\begin{cases} \alpha_n = \dfrac{n+2\alpha-1}{2(n+\alpha)} \sqrt{\dfrac{n(n+\alpha)\Gamma(n-1+2\alpha)}{(n+1+\alpha)\Gamma(n+2\alpha)}} \\ \beta_n = 0 \\ \gamma_n = \dfrac{n+1}{2(n+\alpha)} \sqrt{\dfrac{(n+\alpha)\Gamma(n+1+2\alpha)}{(n+1)(n+1+\alpha)\Gamma(n+2\alpha)}} \end{cases} \tag{B.31}$$

The first few polynomials are as follows:

$$\begin{cases} C_0^{(\alpha)}(x) = 1 \\ C_1^{(\alpha)}(x) = 2\alpha x \\ C_2^{(\alpha)}(x) = 2\alpha(1+\alpha)x^2 - \alpha \\ C_3^{(\alpha)}(x) = \dfrac{4}{3}\alpha(1+\alpha)(2+\alpha)x^3 - 2\alpha(1-\alpha)x \\ \cdots \end{cases} \tag{B.32}$$

and the others can be obtained by the recursive formula

$$C_{n+1}^{(\alpha)}(x) = \frac{2(n+\alpha)}{n+1} x C_n^{(\alpha)}(x) - \frac{n+2\alpha-1}{n+1} C_{n-1}^{(\alpha)}(x) \tag{B.33}$$

In some other books, Gegenbauer polynomials are also called hyper-spherical polynomials. Graphs of the first six Gegenbauer polynomials are also similar to those shown in Figure B.1.

Appendix C: Relationship between Power Spectral Density and Random Fourier Spectrum

C.1 Spectra via Sample Fourier Transform

Consider two real-valued stationary stochastic processes $X(\varpi, t)$ and $Y(\varpi, t)$, where ϖ denotes the embedded random event. We define the finite Fourier transforms of a sample of $X(t)$ and $Y(t)$ over the time interval $[-T, T]$ as

$$X_{\pm T}(\varpi, \omega) = \int_{-T}^{T} X(\varpi, t)\mathrm{e}^{-\mathrm{i}\omega t}\,\mathrm{d}t \tag{C.1}$$

and

$$Y_{\pm T}(\varpi, \omega) = \int_{-T}^{T} Y(\varpi, t)\mathrm{e}^{-\mathrm{i}\omega t}\,\mathrm{d}t \tag{C.2}$$

respectively; then, the cross-PSD can be obtained by

$$S_{XY}(\omega) = \lim_{T\to\infty} \frac{1}{2T}\mathcal{E}\left[X_{\pm T}(\varpi, \omega)Y^*_{\pm T}(\varpi, \omega)\right] \tag{C.3}$$

Here, $\mathcal{E}[\cdot]$ is the ensemble average with respect to ϖ; the superscript asterisk denotes complex conjugate.

Equation C.3 can be proved as follows.

Stochastic Dynamics of Structures Jie Li and Jianbing Chen

Proof: Using Equations C.1 and C.2, we have

$$\begin{aligned}
\lim_{T\to\infty}\frac{1}{2T}\mathcal{E}\left[X_{\pm T}(\varpi,\omega)Y^*_{\pm T}(\varpi,\omega)\right] &= \lim_{T\to\infty}\frac{1}{2T}\mathcal{E}\left\{\int_{-T}^{T}X(\varpi,t)\mathrm{e}^{-\mathrm{i}\omega t}\,\mathrm{d}t\left[\int_{-T}^{T}Y(\varpi,t)\mathrm{e}^{-\mathrm{i}\omega t}\,\mathrm{d}t\right]^*\right\} \\
&= \lim_{T\to\infty}\frac{1}{2T}\mathcal{E}\left[\int_{-T}^{T}X(\varpi,t_1)\mathrm{e}^{-\mathrm{i}\omega t_1}\,\mathrm{d}t_1\int_{-T}^{T}Y(\varpi,t_2)\mathrm{e}^{\mathrm{i}\omega t_2}\,\mathrm{d}t_2\right] \\
&= \lim_{T\to\infty}\frac{1}{2T}\mathcal{E}\left[\int_{-T}^{T}\int_{-T}^{T}X(\varpi,t_1)Y(\varpi,t_2)\mathrm{e}^{\mathrm{i}\omega(t_2-t_1)}\,\mathrm{d}t_1\,\mathrm{d}t_2\right] \\
&= \lim_{T\to\infty}\frac{1}{2T}\left\{\int_{-T}^{T}\int_{-T}^{T}\mathcal{E}[X(\varpi,t_1)Y(\varpi,t_2)]\mathrm{e}^{\mathrm{i}\omega(t_2-t_1)}\,\mathrm{d}t_1\,\mathrm{d}t_2\right\} \\
&= \lim_{T\to\infty}\frac{1}{2T}\left[\int_{-T}^{T}\int_{-T}^{T}R_{XY}(t_1,t_2)\mathrm{e}^{\mathrm{i}\omega(t_2-t_1)}\,\mathrm{d}t_1\,\mathrm{d}t_2\right]
\end{aligned} \tag{C.4}$$

Now, change the region of integration from (t_1, t_2) to (t_1, τ), where $\tau = t_2 - t_1$, $d\tau = dt_2$. This changes the limits of integration as shown in Figure C.1. Thus, Equation C.4 is changed to

$$\begin{aligned}
&\lim_{T\to\infty}\frac{1}{2T}\left[\int_{-T}^{T}\int_{-T}^{T}R_{XY}(t_1,t_2)\mathrm{e}^{\mathrm{i}\omega(t_2-t_1)}\,\mathrm{d}t_1\,\mathrm{d}t_2\right] \\
&= \lim_{T\to\infty}\frac{1}{2T}\left\{\int_{-2T}^{0}\left[\int_{-\tau-T}^{T}R_{XY}(\tau)\mathrm{e}^{\mathrm{i}\omega\tau}\,\mathrm{d}t_1\right]\mathrm{d}\tau+\int_{0}^{2T}\left[\int_{-T}^{T-\tau}R_{XY}(\tau)\mathrm{e}^{\mathrm{i}\omega\tau}\,\mathrm{d}t_1\right]\mathrm{d}\tau\right\} \\
&= \lim_{T\to\infty}\frac{1}{2T}\left[\int_{0}^{2T}(2T-\tau)R_{XY}(\tau)\mathrm{e}^{-\mathrm{i}\omega\tau}\,\mathrm{d}\tau+\int_{-2T}^{0}(2T+\tau)R_{XY}(\tau)\mathrm{e}^{-\mathrm{i}\omega\tau}\,\mathrm{d}\tau\right] \\
&= \lim_{T\to\infty}\left[\int_{-2T}^{0}\left(1+\frac{\tau}{2T}\right)R_{XY}(\tau)\mathrm{e}^{-\mathrm{i}\omega\tau}\,\mathrm{d}\tau+\int_{0}^{2T}\left(1-\frac{\tau}{2T}\right)R_{XY}(\tau)\mathrm{e}^{-\mathrm{i}\omega\tau}\,\mathrm{d}\tau\right] \\
&= \int_{-\infty}^{\infty}R_{XY}(\tau)\mathrm{e}^{-\mathrm{i}\omega\tau}\,\mathrm{d}\tau \\
&= S_{XY}(\omega)
\end{aligned} \tag{C.5}$$

□

This proves Equation C.3.

It follows from Equation C.3 immediately that the auto power spectral density is given by

$$\begin{aligned}
S_X(\omega) &= \lim_{T\to\infty}\frac{1}{2T}\mathcal{E}\left[X_{\pm T}(\varpi,\omega)X^*_{\pm T}(\varpi,\omega)\right] \\
&= \lim_{T\to\infty}\frac{1}{2T}\mathcal{E}\left[\left|X_{\pm T}(\varpi,\omega)\right|^2\right]
\end{aligned} \tag{C.6}$$

Because Equations C.1 and C.2 can be understood as random Fourier spectra, Equations C.3 and C.6 establish the relationship between the PSD and the random Fourier spectra. In other words, they bridge the gap between the PSD and the sample properties. From this point of view, it is convenient to derive the frequency properties of random vibration systems, as has been elaborated in Chapter 5.

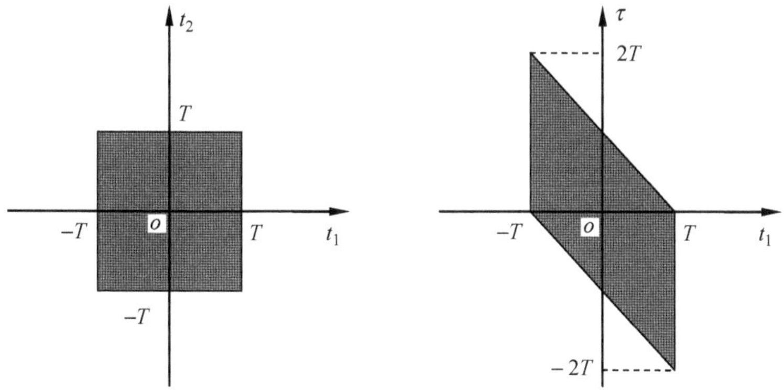

Figure C.1 Change of integral region.

C.2 Spectra via One-sided Finite Fourier Transform

In practice, the measured data are usually over the time interval $[0, T]$. We define the sample one-sided finite Fourier spectra of the real-valued $X(t)$ and $Y(t)$ over the time interval $[0, T]$ as

$$X_T(\varpi, \omega) = \int_0^T X(\varpi, t)\mathrm{e}^{-\mathrm{i}\omega t}\,\mathrm{d}t \tag{C.7}$$

and

$$Y_T(\varpi, \omega) = \int_0^T Y(\varpi, t)\mathrm{e}^{-\mathrm{i}\omega t}\,\mathrm{d}t \tag{C.8}$$

respectively; then, the cross-PSD can be obtained by

$$S_{XY}(\omega) = \lim_{T\to\infty}\frac{1}{T}\mathcal{E}\big[X_T(\varpi, \omega)Y_T^*(\varpi, \omega)\big] \tag{C.9}$$

This can be proved as follows.

Proof: Using Equations C.7 and C.8, we have

$$\begin{aligned}
\lim_{T\to\infty}\mathcal{E}\left[\frac{1}{T}X_T(\varpi, \omega)Y_T^*(\varpi, \omega)\right] &= \lim_{T\to\infty}\frac{1}{T}\mathcal{E}\left\{\left[\int_0^T X(\varpi, t_1)\mathrm{e}^{-\mathrm{i}\omega t_1}\,\mathrm{d}t_1\right]\left[\int_0^T Y(\varpi, t_2)\mathrm{e}^{\mathrm{i}\omega t_2}\,\mathrm{d}t_2\right]\right\} \\
&= \lim_{T\to\infty}\frac{1}{T}\mathcal{E}\left[\int_0^T\int_0^T X(\varpi, t_1)Y(\varpi, t_2)\mathrm{e}^{\mathrm{i}\omega(t_2-t_1)}\,\mathrm{d}t_1\,\mathrm{d}t_2\right] \\
&= \lim_{T\to\infty}\frac{1}{T}\left\{\int_0^T\int_0^T \mathcal{E}[X(\varpi, t_1)Y(\varpi, t_2)]\mathrm{e}^{\mathrm{i}\omega(t_2-t_1)}\,\mathrm{d}t_1\,\mathrm{d}t_2\right\} \\
&= \lim_{T\to\infty}\frac{1}{T}\left[\int_0^T\int_0^T R_{XY}(t_1, t_2)\mathrm{e}^{\mathrm{i}\omega(t_2-t_1)}\,\mathrm{d}t_1\,\mathrm{d}t_2\right]
\end{aligned} \tag{C.10}$$

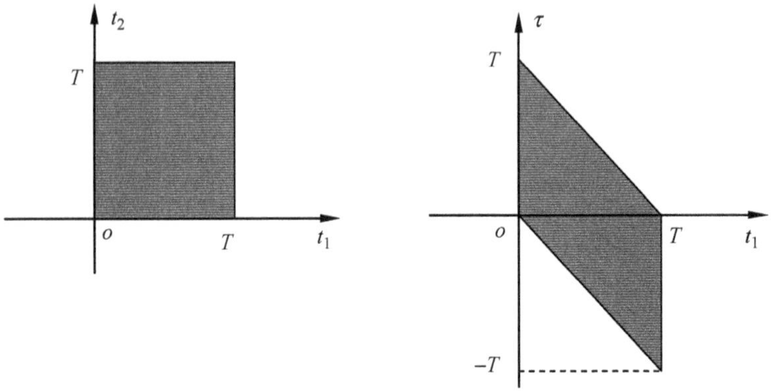

Figure C.2 Change of integral region.

Now, change the region of integration from (t_1,t_2) to (t_1,τ), where $\tau = t_2 - t_1$, $d\tau = dt_2$. This changes the limits of integration as shown in Figure C.2. Thus, Equation C.10 can be rearranged into

$$\begin{aligned}
&\lim_{T\to\infty}\frac{1}{T}\left[\int_0^T\int_0^T R_{XY}(t_1,t_2)\mathrm{e}^{\mathrm{i}\omega(t_2-t_1)}\,\mathrm{d}t_1\,\mathrm{d}t_2\right]\\
&=\lim_{T\to\infty}\frac{1}{T}\left\{\int_0^T\left[\int_0^{T-\tau}R_{XY}(\tau)\mathrm{e}^{\mathrm{i}\omega\tau}\mathrm{d}t_1\right]\mathrm{d}\tau+\int_{-T}^0\left[\int_{-\tau}^T R_{XY}(\tau)\mathrm{e}^{\mathrm{i}\omega\tau}\,\mathrm{d}t_1\right]\mathrm{d}\tau\right\}\\
&=\lim_{T\to\infty}\frac{1}{T}\left[\int_0^T(T-\tau)R_{XY}(\tau)\mathrm{e}^{\mathrm{i}\omega\tau}\,\mathrm{d}\tau+\int_{-T}^0(T+\tau)R_{XY}(\tau)\mathrm{e}^{\mathrm{i}\omega\tau}\,\mathrm{d}\tau\right]\\
&=\lim_{T\to\infty}\left[\int_{-T}^0\left(1+\frac{\tau}{T}\right)R_{XY}(\tau)\mathrm{e}^{-\mathrm{i}\omega\tau}\,\mathrm{d}\tau+\int_0^T\left(1-\frac{\tau}{T}\right)R_{XY}(\tau)\mathrm{e}^{-\mathrm{i}\omega\tau}\,\mathrm{d}\tau\right]\\
&=\int_{-\infty}^{\infty}R_{XY}(\tau)\mathrm{e}^{-\mathrm{i}\omega\tau}\,\mathrm{d}\tau\\
&=S_{XY}(\omega)
\end{aligned}\tag{C.11}$$

This proves Equation C.9. □

Of course, the physical sense of Equations C.7–C.9 is similar to Equations C.1–C.3. Further, if we define standardized spectra by

$$\widehat{X}_T(\varpi,\omega)=\frac{X_T(\varpi,\omega)}{\sqrt{T}}=\frac{1}{\sqrt{T}}\int_0^T X(\varpi,t)\mathrm{e}^{-\mathrm{i}\omega t}\,\mathrm{d}t \tag{C.12}$$

and

$$\widehat{Y}_T(\varpi,\omega)=\frac{Y_T(\varpi,\omega)}{\sqrt{T}}=\frac{1}{\sqrt{T}}\int_0^T Y(\varpi,t)\mathrm{e}^{-\mathrm{i}\omega t}\,\mathrm{d}t \tag{C.13}$$

respectively, then according to Equation C.9 we have

$$S_{XY}(\omega) = \lim_{T \to \infty} \mathcal{E}[\widehat{X_T}(\varpi, \omega) \widehat{Y_T}^*(\varpi, \omega)] \tag{C.14}$$

In particular, these relations establish the theoretical basis for the random Fourier spectrum methodology in Chapter 3 (for example, Equation 3.12). In addition, for practical applications they are particularly useful when we want to obtain the PSD of stochastic processes if a set of sample time histories is available (Bendat and Piersol, 2000).

Appendix D: Orthonormal Base Vectors

To satisfy Equations 7.134 and 7.135, the orthonormal base vectors for the coordinate transformation in dimensions of $s = 4, 5, \ldots, 23$ could be chosen as follows:

$$\begin{cases} \mathbf{e}_1 = \dfrac{1}{\sqrt{2}}(1, -1, 0, 0, 0) \\ \mathbf{e}_2 = \dfrac{1}{\sqrt{2}}(0, 0, 1, -1, 0) \\ \mathbf{e}_3 = \dfrac{1}{2}(1, 1, -1, -1, 0) \\ \mathbf{e}_4 = \dfrac{1}{\sqrt{20}}(1, 1, 1, 1, -4) \end{cases} \quad \text{for} \quad s = 4 \tag{D.1}$$

$$\begin{cases} \mathbf{e}_1 = \dfrac{1}{\sqrt{2}}(1, -1, 0, 0, 0, 0) \\ \mathbf{e}_2 = \dfrac{1}{\sqrt{2}}(0, 0, 1, -1, 0, 0) \\ \mathbf{e}_3 = \dfrac{1}{\sqrt{2}}(0, 0, 0, 0, 1, -1) \\ \mathbf{e}_4 = \dfrac{1}{2}(1, 1, -1, -1, 0, 0) \\ \mathbf{e}_5 = \dfrac{1}{\sqrt{12}}(1, 1, 1, 1, -2, -2) \end{cases} \quad \text{for} \quad s = 5 \tag{D.2}$$

Stochastic Dynamics of Structures Jie Li and Jianbing Chen

$$\begin{cases} \mathbf{e}_1 = \frac{1}{\sqrt{2}}(1, -1, 0, 0, 0, 0, 0) \\ \mathbf{e}_2 = \frac{1}{\sqrt{2}}(0, 0, 1, -1, 0, 0, 0) \\ \mathbf{e}_3 = \frac{1}{\sqrt{2}}(0, 0, 0, 0, 1, -1, 0) \\ \mathbf{e}_4 = \frac{1}{2}(1, 1, -1, -1, 0, 0, 0) \\ \mathbf{e}_5 = \frac{1}{\sqrt{12}}(1, 1, 1, 1, -2, -2, 0) \\ \mathbf{e}_6 = \frac{1}{\sqrt{42}}(1, 1, 1, 1, 1, 1, -6) \end{cases} \quad \text{for} \quad s = 6 \tag{D.3}$$

$$\begin{cases} \mathbf{e}_1 = \frac{1}{\sqrt{2}}(1, -1, 0, 0, 0, 0, 0, 0) \\ \mathbf{e}_2 = \frac{1}{\sqrt{2}}(0, 0, 1, -1, 0, 0, 0, 0) \\ \mathbf{e}_3 = \frac{1}{\sqrt{2}}(0, 0, 0, 0, 1, -1, 0, 0) \\ \mathbf{e}_4 = \frac{1}{\sqrt{2}}(0, 0, 0, 0, 0, 0, 1, -1) \\ \mathbf{e}_5 = \frac{1}{2}(1, 1, -1, -1, 0, 0, 0, 0) \\ \mathbf{e}_6 = \frac{1}{2}(0, 0, 0, 0, 1, 1, -1, -1) \\ \mathbf{e}_7 = \frac{1}{\sqrt{8}}(1, 1, 1, 1, -1, -1, -1, -1) \end{cases} \quad \text{for} \quad s = 7 \tag{D.4}$$

$$\begin{cases} \mathbf{e}_1 = \frac{1}{\sqrt{2}}(1, -1, 0, 0, 0, 0, 0, 0, 0) \\ \mathbf{e}_2 = \frac{1}{\sqrt{2}}(0, 0, 1, -1, 0, 0, 0, 0, 0) \\ \mathbf{e}_3 = \frac{1}{\sqrt{2}}(0, 0, 0, 0, 1, -1, 0, 0, 0) \\ \mathbf{e}_4 = \frac{1}{\sqrt{2}}(0, 0, 0, 0, 0, 0, 1, -1, 0) \\ \mathbf{e}_5 = \frac{1}{2}(1, 1, -1, -1, 0, 0, 0, 0, 0) \\ \mathbf{e}_6 = \frac{1}{2}(0, 0, 0, 0, 1, 1, -1, -1, 0) \\ \mathbf{e}_7 = \frac{1}{\sqrt{8}}(1, 1, 1, 1, -1, -1, -1, -1, 0) \\ \mathbf{e}_8 = \frac{1}{\sqrt{72}}(1, 1, 1, 1, 1, 1, 1, 1, -8) \end{cases} \quad \text{for} \quad s = 8 \tag{D.5}$$

$$\begin{cases} \mathbf{e}_1 = \frac{1}{\sqrt{2}}(1, -1, 0, 0, 0, 0, 0, 0, 0, 0) \\ \mathbf{e}_2 = \frac{1}{\sqrt{2}}(0, 0, 1, -1, 0, 0, 0, 0, 0, 0) \\ \mathbf{e}_3 = \frac{1}{\sqrt{2}}(0, 0, 0, 0, 1, -1, 0, 0, 0, 0) \\ \mathbf{e}_4 = \frac{1}{\sqrt{2}}(0, 0, 0, 0, 0, 0, 1, -1, 0, 0) \\ \mathbf{e}_5 = \frac{1}{\sqrt{2}}(0, 0, 0, 0, 0, 0, 0, 0, 1, -1) \\ \mathbf{e}_6 = \frac{1}{2}(1, 1, -1, -1, 0, 0, 0, 0, 0, 0) \\ \mathbf{e}_7 = \frac{1}{2}(0, 0, 0, 0, 1, 1, -1, -1, 0, 0) \\ \mathbf{e}_8 = \frac{1}{\sqrt{8}}(1, 1, 1, 1, -1, -1, -1, -1, 0, 0) \\ \mathbf{e}_9 = \frac{1}{\sqrt{40}}(1, 1, 1, 1, 1, 1, 1, 1, -4, -4) \end{cases} \quad \text{for} \quad s = 9 \tag{D.6}$$

$$\begin{cases} \mathbf{e}_1 = \frac{1}{\sqrt{2}}(1, -1, 0, 0, 0, 0, 0, 0, 0, 0, 0) \\ \mathbf{e}_2 = \frac{1}{\sqrt{2}}(0, 0, 1, -1, 0, 0, 0, 0, 0, 0, 0) \\ \mathbf{e}_3 = \frac{1}{\sqrt{2}}(0, 0, 0, 0, 1, -1, 0, 0, 0, 0, 0) \\ \mathbf{e}_4 = \frac{1}{\sqrt{2}}(0, 0, 0, 0, 0, 0, 1, -1, 0, 0, 0) \\ \mathbf{e}_5 = \frac{1}{\sqrt{2}}(0, 0, 0, 0, 0, 0, 0, 0, 1, -1, 0) \\ \mathbf{e}_6 = \frac{1}{2}(1, 1, -1, -1, 0, 0, 0, 0, 0, 0, 0) \\ \mathbf{e}_7 = \frac{1}{2}(0, 0, 0, 0, 1, 1, -1, -1, 0, 0, 0) \\ \mathbf{e}_8 = \frac{1}{\sqrt{8}}(1, 1, 1, 1, -1, -1, -1, -1, 0, 0, 0) \\ \mathbf{e}_9 = \frac{1}{\sqrt{40}}(1, 1, 1, 1, 1, 1, 1, 1, -4, -4, 0) \\ \mathbf{e}_{10} = \frac{1}{\sqrt{110}}(1, 1, 1, 1, 1, 1, 1, 1, 1, 1, -10) \end{cases} \quad \text{for} \quad s = 10 \tag{D.7}$$

$$\begin{cases} \mathbf{e}_1 = \frac{1}{\sqrt{2}}(1, -1, 0, 0, 0, 0, 0, 0, 0, 0, 0, 0) \\ \mathbf{e}_2 = \frac{1}{\sqrt{2}}(0, 0, 1, -1, 0, 0, 0, 0, 0, 0, 0, 0) \\ \mathbf{e}_3 = \frac{1}{\sqrt{2}}(0, 0, 0, 0, 1, -1, 0, 0, 0, 0, 0, 0) \\ \mathbf{e}_4 = \frac{1}{\sqrt{2}}(0, 0, 0, 0, 0, 0, 1, -1, 0, 0, 0, 0) \\ \mathbf{e}_5 = \frac{1}{\sqrt{2}}(0, 0, 0, 0, 0, 0, 0, 0, 1, -1, 0, 0) \\ \mathbf{e}_6 = \frac{1}{\sqrt{2}}(0, 0, 0, 0, 0, 0, 0, 0, 0, 0, 1, -1) \\ \mathbf{e}_7 = \frac{1}{2}(1, 1, -1, -1, 0, 0, 0, 0, 0, 0, 0, 0) \\ \mathbf{e}_8 = \frac{1}{2}(0, 0, 0, 0, 1, 1, -1, -1, 0, 0, 0, 0) \\ \mathbf{e}_9 = \frac{1}{2}(0, 0, 0, 0, 0, 0, 0, 0, 1, 1, -1, -1) \\ \mathbf{e}_{10} = \frac{1}{\sqrt{8}}(1, 1, 1, 1, -1, -1, -1, -1, 0, 0, 0, 0) \\ \mathbf{e}_{11} = \frac{1}{\sqrt{24}}(1, 1, 1, 1, 1, 1, 1, 1, -2, -2, -2, -2) \end{cases} \quad \text{for} \quad s = 11 \tag{D.8}$$

$$
\left\{
\begin{aligned}
\mathbf{e}_1 &= \frac{1}{\sqrt{2}}(1,\,-1,0,0,0,0,0,0,0,0,0,0,0) \\
\mathbf{e}_2 &= \frac{1}{\sqrt{2}}(0,0,1,\,-1,0,0,0,0,0,0,0,0,0) \\
\mathbf{e}_3 &= \frac{1}{\sqrt{2}}(0,0,0,0,1,\,-1,0,0,0,0,0,0,0) \\
\mathbf{e}_4 &= \frac{1}{\sqrt{2}}(0,0,0,0,0,0,1,\,-1,0,0,0,0,0) \\
\mathbf{e}_5 &= \frac{1}{\sqrt{2}}(0,0,0,0,0,0,0,0,1,\,-1,0,0,0) \\
\mathbf{e}_6 &= \frac{1}{\sqrt{2}}(0,0,0,0,0,0,0,0,0,0,1,\,-1,0) \\
\mathbf{e}_7 &= \frac{1}{2}(1,1,\,-1,\,-1,0,0,0,0,0,0,0,0,0) \\
\mathbf{e}_8 &= \frac{1}{2}(0,0,0,0,1,1,\,-1,\,-1,0,0,0,0,0) \\
\mathbf{e}_9 &= \frac{1}{2}(0,0,0,0,0,0,0,0,1,1,\,-1,\quad 1,0) \\
\mathbf{e}_{10} &= \frac{1}{\sqrt{8}}(1,1,1,1,\,-1,\,-1,\,-1,\,-1,0,0,0,0,0) \\
\mathbf{e}_{11} &= \frac{1}{\sqrt{24}}(1,1,1,1,1,1,1,1,\,-2,\,-2,\,-2,\,-2,0) \\
\mathbf{e}_{12} &= \frac{1}{\sqrt{156}}(1,1,1,1,1,1,1,1,1,1,1,1,\,-12)
\end{aligned}
\right.
\qquad \text{for} \quad s = 12 \tag{D.9}
$$

$$
\begin{cases}
\mathbf{e}_1 = \dfrac{1}{\sqrt{2}}(1,\ -1,0,0,0,0,0,0,0,0,0,0,0,0,0) \\
\mathbf{e}_2 = \dfrac{1}{\sqrt{2}}(0,0,1,\ -1,0,0,0,0,0,0,0,0,0,0,0) \\
\mathbf{e}_3 = \dfrac{1}{\sqrt{2}}(0,0,0,0,1,\ -1,0,0,0,0,0,0,0,0,0) \\
\mathbf{e}_4 = \dfrac{1}{\sqrt{2}}(0,0,0,0,0,0,1,\ -1,0,0,0,0,0,0,0) \\
\mathbf{e}_5 = \dfrac{1}{\sqrt{2}}(0,0,0,0,0,0,0,0,1,\ -1,0,0,0,0,0) \\
\mathbf{e}_6 = \dfrac{1}{\sqrt{2}}(0,0,0,0,0,0,0,0,0,0,1,\ -1,0,0) \\
\mathbf{e}_7 = \dfrac{1}{\sqrt{2}}(0,0,0,0,0,0,0,0,0,0,0,0,1,\ -1) \\
\mathbf{e}_8 = \dfrac{1}{2}(1,1,\ -1,\ -1,0,0,0,0,0,0,0,0,0,0) \\
\mathbf{e}_9 = \dfrac{1}{2}(0,0,0,0,1,1,\ -1,\ -1,0,0,0,0,0,0) \\
\mathbf{e}_{10} = \dfrac{1}{2}(0,0,0,0,0,0,0,0,1,1,\ -1,\ -1,0,0) \\
\mathbf{e}_{11} = \dfrac{1}{\sqrt{8}}(1,1,1,1,\ -1,\ -1,\ -1,\ -1,0,0,0,0,0,0) \\
\mathbf{e}_{12} = \dfrac{1}{\sqrt{24}}(1,1,1,1,1,1,1,1,\ -2,\ -2,\ -2,\ -2,0,0) \\
\mathbf{e}_{13} = \dfrac{1}{\sqrt{84}}(1,1,1,1,1,1,1,1,1,1,1,1,\ -6,\ -6)
\end{cases}
\quad \text{for} \quad s = 13 \qquad \text{(D.10)}
$$

$$\left\{\begin{aligned}
\mathbf{e}_1 &= \frac{1}{\sqrt{2}}(1,\,-1,0,0,0,0,0,0,0,0,0,0,0,0,0)\\
\mathbf{e}_2 &= \frac{1}{\sqrt{2}}(0,0,1,\,-1,0,0,0,0,0,0,0,0,0,0,0)\\
\mathbf{e}_3 &= \frac{1}{\sqrt{2}}(0,0,0,0,1,\,-1,0,0,0,0,0,0,0,0,0)\\
\mathbf{e}_4 &= \frac{1}{\sqrt{2}}(0,0,0,0,0,0,1,\,-1,0,0,0,0,0,0,0)\\
\mathbf{e}_5 &= \frac{1}{\sqrt{2}}(0,0,0,0,0,0,0,0,1,\,-1,0,0,0,0,0)\\
\mathbf{e}_6 &= \frac{1}{\sqrt{2}}(0,0,0,0,0,0,0,0,0,0,1,\,-1,0,0,0)\\
\mathbf{e}_7 &= \frac{1}{\sqrt{2}}(0,0,0,0,0,0,0,0,0,0,0,0,1,\,-1,0)\\
\mathbf{e}_8 &= \frac{1}{2}(1,1,\,-1,\,-1,0,0,0,0,0,0,0,0,0,0,0)\\
\mathbf{e}_9 &= \frac{1}{2}(0,0,0,0,1,1,\,-1,\,-1,0,0,0,0,0,0,0)\\
\mathbf{e}_{10} &= \frac{1}{2}(0,0,0,0,0,0,0,0,1,1,\,-1,\,-1,0,0,0)\\
\mathbf{e}_{11} &= \frac{1}{\sqrt{8}}(1,1,1,1,\,-1,\,-1,\,-1,\,-1,0,0,0,0,0,0,0)\\
\mathbf{e}_{12} &= \frac{1}{\sqrt{24}}(1,1,1,1,1,1,1,1,\,-2,\,-2,\,-2,\,-2,0,0,0)\\
\mathbf{e}_{13} &= \frac{1}{\sqrt{84}}(1,1,1,1,1,1,1,1,1,1,1,1,\,-6,\,-6,0)\\
\mathbf{e}_{14} &= \frac{1}{\sqrt{210}}(1,1,1,1,1,1,1,1,1,1,1,1,1,1,\,-14)
\end{aligned}\right. \qquad \text{for} \quad s = 14 \qquad \text{(D.11)}$$

$$
\begin{cases}
\mathbf{e}_1 = \dfrac{1}{\sqrt{2}}(1, -1, 0, 0, 0, 0, 0, 0, 0, 0, 0, 0, 0, 0, 0, 0) \\
\mathbf{e}_2 = \dfrac{1}{\sqrt{2}}(0, 0, 1, -1, 0, 0, 0, 0, 0, 0, 0, 0, 0, 0, 0, 0) \\
\mathbf{e}_3 = \dfrac{1}{\sqrt{2}}(0, 0, 0, 0, 1, -1, 0, 0, 0, 0, 0, 0, 0, 0, 0, 0) \\
\mathbf{e}_4 = \dfrac{1}{\sqrt{2}}(0, 0, 0, 0, 0, 0, 1, -1, 0, 0, 0, 0, 0, 0, 0, 0) \\
\mathbf{e}_5 = \dfrac{1}{\sqrt{2}}(0, 0, 0, 0, 0, 0, 0, 0, 1, -1, 0, 0, 0, 0, 0, 0) \\
\mathbf{e}_6 = \dfrac{1}{\sqrt{2}}(0, 0, 0, 0, 0, 0, 0, 0, 0, 0, 1, -1, 0, 0, 0, 0) \\
\mathbf{e}_7 = \dfrac{1}{\sqrt{2}}(0, 0, 0, 0, 0, 0, 0, 0, 0, 0, 0, 0, 1, -1, 0, 0) \\
\mathbf{e}_8 = \dfrac{1}{\sqrt{2}}(0, 0, 0, 0, 0, 0, 0, 0, 0, 0, 0, 0, 0, 0, 1, -1) \\
\mathbf{e}_9 = \dfrac{1}{2}(1, 1, -1, -1, 0, 0, 0, 0, 0, 0, 0, 0, 0, 0, 0, 0) \\
\mathbf{e}_{10} = \dfrac{1}{2}(0, 0, 0, 0, 1, 1, -1, -1, 0, 0, 0, 0, 0, 0, 0, 0) \\
\mathbf{e}_{11} = \dfrac{1}{2}(0, 0, 0, 0, 0, 0, 0, 0, 1, 1, -1, -1, 0, 0, 0, 0) \\
\mathbf{e}_{12} = \dfrac{1}{2}(0, 0, 0, 0, 0, 0, 0, 0, 0, 0, 0, 0, 1, 1, -1, -1) \\
\mathbf{e}_{13} = \dfrac{1}{\sqrt{8}}(1, 1, 1, 1, -1, -1, -1, -1, 0, 0, 0, 0, 0, 0, 0, 0) \\
\mathbf{e}_{14} = \dfrac{1}{\sqrt{8}}(0, 0, 0, 0, 0, 0, 0, 0, 1, 1, 1, 1, -1, -1, -1, -1) \\
\mathbf{e}_{15} = \dfrac{1}{4}(-1, -1, -1, -1, 1, 1, 1, 1, -1, -1, -1, -1, 1, 1, 1, 1)
\end{cases}
\quad \text{for} \quad s = 15
\tag{D.12}
$$

$$
\left\{
\begin{aligned}
\mathbf{e}_1 &= \frac{1}{\sqrt{2}}(1,-1,0,0,0,0,0,0,0,0,0,0,0,0,0,0,0)\\
\mathbf{e}_2 &= \frac{1}{\sqrt{2}}(0,0,1,-1,0,0,0,0,0,0,0,0,0,0,0,0,0)\\
\mathbf{e}_3 &= \frac{1}{\sqrt{2}}(0,0,0,0,1,-1,0,0,0,0,0,0,0,0,0,0,0)\\
\mathbf{e}_4 &= \frac{1}{\sqrt{2}}(0,0,0,0,0,0,1,-1,0,0,0,0,0,0,0,0,0)\\
\mathbf{e}_5 &= \frac{1}{\sqrt{2}}(0,0,0,0,0,0,0,0,1,-1,0,0,0,0,0,0,0)\\
\mathbf{e}_6 &= \frac{1}{\sqrt{2}}(0,0,0,0,0,0,0,0,0,0,1,-1,0,0,0,0,0)\\
\mathbf{e}_7 &= \frac{1}{\sqrt{2}}(0,0,0,0,0,0,0,0,0,0,0,0,1,-1,0,0,0)\\
\mathbf{e}_8 &= \frac{1}{\sqrt{2}}(0,0,0,0,0,0,0,0,0,0,0,0,0,0,1,-1,0)\\
\mathbf{e}_9 &= \frac{1}{2}(1,1,-1,-1,0,0,0,0,0,0,0,0,0,0,0,0,0)\\
\mathbf{e}_{10} &= \frac{1}{2}(0,0,0,0,1,1,-1,-1,0,0,0,0,0,0,0,0,0)\\
\mathbf{e}_{11} &= \frac{1}{2}(0,0,0,0,0,0,0,0,1,1,-1,-1,0,0,0,0,0)\\
\mathbf{e}_{12} &= \frac{1}{2}(0,0,0,0,0,0,0,0,0,0,0,0,1,1,-1,-1,0)\\
\mathbf{e}_{13} &= \frac{1}{\sqrt{8}}(1,1,1,1,-1,-1,-1,-1,0,0,0,0,0,0,0,0,0)\\
\mathbf{e}_{14} &= \frac{1}{\sqrt{8}}(0,0,0,0,0,0,0,0,1,1,1,1,-1,-1,-1,-1,0)\\
\mathbf{e}_{15} &= \frac{1}{4}(-1,-1,-1,-1,1,1,1,1,-1,-1,-1,-1,1,1,1,1,0)\\
\mathbf{e}_{16} &= \frac{1}{\sqrt{272}}(1,1,1,1,1,1,1,1,1,1,1,1,1,1,1,1,-16)
\end{aligned}
\right. \quad \text{for} \quad s=16 \tag{D.13}
$$

$$
\left\{
\begin{aligned}
\mathbf{e}_1 &= \frac{1}{\sqrt{2}}(1,\ -1,0,0,0,0,0,0,0,0,0,0,0,0,0,0,0,0) \\
\mathbf{e}_2 &= \frac{1}{\sqrt{2}}(0,0,1,\ -1,0,0,0,0,0,0,0,0,0,0,0,0,0,0) \\
\mathbf{e}_3 &= \frac{1}{\sqrt{2}}(0,0,0,0,1,\ -1,0,0,0,0,0,0,0,0,0,0,0,0) \\
\mathbf{e}_4 &= \frac{1}{\sqrt{2}}(0,0,0,0,0,0,1,\ -1,0,0,0,0,0,0,0,0,0,0) \\
\mathbf{e}_5 &= \frac{1}{\sqrt{2}}(0,0,0,0,0,0,0,0,1,\ -1,0,0,0,0,0,0,0,0) \\
\mathbf{e}_6 &= \frac{1}{\sqrt{2}}(0,0,0,0,0,0,0,0,0,0,1,\ -1,0,0,0,0,0,0) \\
\mathbf{e}_7 &= \frac{1}{\sqrt{2}}(0,0,0,0,0,0,0,0,0,0,0,0,1,\ -1,0,0,0,0) \\
\mathbf{e}_8 &= \frac{1}{\sqrt{2}}(0,0,0,0,0,0,0,0,0,0,0,0,0,0,1,\ -1,0,0) \\
\mathbf{e}_9 &= \frac{1}{\sqrt{2}}(0,0,0,0,0,0,0,0,0,0,0,0,0,0,0,0,1,\ -1) \\
\mathbf{e}_{10} &= \frac{1}{2}(1,1,\ -1,\ -1,0,0,0,0,0,0,0,0,0,0,0,0,0,0) \\
\mathbf{e}_{11} &= \frac{1}{2}(0,0,0,0,1,1,\ -1,\ -1,0,0,0,0,0,0,0,0,0,0) \\
\mathbf{e}_{12} &= \frac{1}{2}(0,0,0,0,0,0,0,0,1,1,\ -1,\ -1,0,0,0,0,0,0) \\
\mathbf{e}_{13} &= \frac{1}{2}(0,0,0,0,0,0,0,0,0,0,0,0,1,1,\ -1,\ -1,0,0) \\
\mathbf{e}_{14} &= \frac{1}{\sqrt{8}}(1,1,1,1,\ -1,\ -1,\ -1,\ -1,0,0,0,0,0,0,0,0,0,0) \\
\mathbf{e}_{15} &= \frac{1}{\sqrt{8}}(0,0,0,0,0,0,0,0,1,1,1,1,\ -1,\ -1,\ -1,\ -1,0,0) \\
\mathbf{e}_{16} &= \frac{1}{4}(-1,\ -1,\ -1,\ -1,1,1,1,1,\ -1,\ -1,\ -1,\ -1,1,1,1,1,0,0) \\
\mathbf{e}_{17} &= \frac{1}{12}(1,1,1,1,1,1,1,1,1,1,1,1,1,1,1,1,\ -8,\ -8)
\end{aligned}
\right. \qquad \text{for} \quad s = 17
\tag{D.14}
$$

$$
\left\{
\begin{aligned}
\mathbf{e}_1 &= \frac{1}{\sqrt{2}}(1,-1,0,0,0,0,0,0,0,0,0,0,0,0,0,0,0,0,0)\\
\mathbf{e}_2 &= \frac{1}{\sqrt{2}}(0,0,1,-1,0,0,0,0,0,0,0,0,0,0,0,0,0,0,0)\\
\mathbf{e}_3 &= \frac{1}{\sqrt{2}}(0,0,0,0,1,-1,0,0,0,0,0,0,0,0,0,0,0,0,0)\\
\mathbf{e}_4 &= \frac{1}{\sqrt{2}}(0,0,0,0,0,0,1,-1,0,0,0,0,0,0,0,0,0,0,0)\\
\mathbf{e}_5 &= \frac{1}{\sqrt{2}}(0,0,0,0,0,0,0,0,1,-1,0,0,0,0,0,0,0,0,0)\\
\mathbf{e}_6 &= \frac{1}{\sqrt{2}}(0,0,0,0,0,0,0,0,0,0,1,-1,0,0,0,0,0,0,0)\\
\mathbf{e}_7 &= \frac{1}{\sqrt{2}}(0,0,0,0,0,0,0,0,0,0,0,0,1,-1,0,0,0,0,0)\\
\mathbf{e}_8 &= \frac{1}{\sqrt{2}}(0,0,0,0,0,0,0,0,0,0,0,0,0,0,1,-1,0,0,0)\\
\mathbf{e}_9 &= \frac{1}{\sqrt{2}}(0,0,0,0,0,0,0,0,0,0,0,0,0,0,0,0,1,-1,0)\\
\mathbf{e}_{10} &= \frac{1}{2}(1,1,-1,-1,0,0,0,0,0,0,0,0,0,0,0,0,0,0,0)\\
\mathbf{e}_{11} &= \frac{1}{2}(0,0,0,0,1,1,-1,-1,0,0,0,0,0,0,0,0,0,0,0)\\
\mathbf{e}_{12} &= \frac{1}{2}(0,0,0,0,0,0,0,0,1,1,-1,-1,0,0,0,0,0,0,0)\\
\mathbf{e}_{13} &= \frac{1}{2}(0,0,0,0,0,0,0,0,0,0,0,0,1,1,-1,-1,0,0,0)\\
\mathbf{e}_{14} &= \frac{1}{\sqrt{8}}(1,1,1,1,-1,-1,-1,-1,0,0,0,0,0,0,0,0,0,0,0)\\
\mathbf{e}_{15} &= \frac{1}{\sqrt{8}}(0,0,0,0,0,0,0,0,1,1,1,1,-1,-1,-1,-1,0,0,0)\\
\mathbf{e}_{16} &= \frac{1}{4}(-1,-1,-1,-1,1,1,1,1,-1,-1,-1,-1,1,1,1,1,0,0,0)\\
\mathbf{e}_{17} &= \frac{1}{4}(1,1,1,1,1,1,1,1,-1,-1,-1,-1,-1,-1,-1,-1,0,0,0)\\
\mathbf{e}_{18} &= \frac{1}{\sqrt{342}}(1,1,1,1,1,1,1,1,1,1,1,1,1,1,1,1,1,1,-18)
\end{aligned}
\right. \qquad \text{for} \quad s=18
\tag{D.15}
$$

$$
\left\{
\begin{aligned}
\mathbf{e}_1 &= \frac{1}{\sqrt{2}}(1,-1,0,0,0,0,0,0,0,0,0,0,0,0,0,0,0,0,0,0)\\
\mathbf{e}_2 &= \frac{1}{\sqrt{2}}(0,0,1,-1,0,0,0,0,0,0,0,0,0,0,0,0,0,0,0,0)\\
\mathbf{e}_3 &= \frac{1}{\sqrt{2}}(0,0,0,0,1,-1,0,0,0,0,0,0,0,0,0,0,0,0,0,0)\\
\mathbf{e}_4 &= \frac{1}{\sqrt{2}}(0,0,0,0,0,0,1,-1,0,0,0,0,0,0,0,0,0,0,0,0)\\
\mathbf{e}_5 &= \frac{1}{\sqrt{2}}(0,0,0,0,0,0,0,0,1,-1,0,0,0,0,0,0,0,0,0,0)\\
\mathbf{e}_6 &= \frac{1}{\sqrt{2}}(0,0,0,0,0,0,0,0,0,0,1,-1,0,0,0,0,0,0,0,0)\\
\mathbf{e}_7 &= \frac{1}{\sqrt{2}}(0,0,0,0,0,0,0,0,0,0,0,0,1,-1,0,0,0,0,0,0)\\
\mathbf{e}_8 &= \frac{1}{\sqrt{2}}(0,0,0,0,0,0,0,0,0,0,0,0,0,0,1,-1,0,0,0,0)\\
\mathbf{e}_9 &= \frac{1}{\sqrt{2}}(0,0,0,0,0,0,0,0,0,0,0,0,0,0,0,0,1,-1,0,0)\\
\mathbf{e}_{10} &= \frac{1}{\sqrt{2}}(0,0,0,0,0,0,0,0,0,0,0,0,0,0,0,0,0,0,1,-1)\\
\mathbf{e}_{11} &= \frac{1}{2}(1,1,-1,-1,0,0,0,0,0,0,0,0,0,0,0,0,0,0,0,0)\\
\mathbf{e}_{12} &= \frac{1}{2}(0,0,0,0,1,1,-1,-1,0,0,0,0,0,0,0,0,0,0,0,0)\\
\mathbf{e}_{13} &= \frac{1}{2}(0,0,0,0,0,0,0,0,1,1,-1,-1,0,0,0,0,0,0,0,0)\\
\mathbf{e}_{14} &= \frac{1}{2}(0,0,0,0,0,0,0,0,0,0,0,0,1,1,-1,-1,0,0,0,0)\\
\mathbf{e}_{15} &= \frac{1}{2}(0,0,0,0,0,0,0,0,0,0,0,0,0,0,0,0,1,1,-1,-1)\\
\mathbf{e}_{16} &= \frac{1}{\sqrt{8}}(1,1,1,1,-1,-1,-1,-1,0,0,0,0,0,0,0,0,0,0,0,0)\\
\mathbf{e}_{17} &= \frac{1}{\sqrt{8}}(0,0,0,0,0,0,0,0,1,1,1,1,-1,-1,-1,-1,0,0,0,0)\\
\mathbf{e}_{18} &= \frac{1}{4}(1,1,1,1,1,1,1,1,-1,-1,-1,-1,-1,-1,-1,-1,0,0,0,0)\\
\mathbf{e}_{19} &= \frac{1}{\sqrt{80}}(1,1,1,1,1,1,1,1,1,1,1,1,1,1,1,1,-4,-4,-4,-4)
\end{aligned}
\right.
\qquad \text{for} \quad s = 19
\tag{D.16}
$$

$$
\left\{
\begin{aligned}
\mathbf{e}_1 &= \frac{1}{\sqrt{2}}(1,-1,0,0,0,0,0,0,0,0,0,0,0,0,0,0,0,0,0,0,0)\\
\mathbf{e}_2 &= \frac{1}{\sqrt{2}}(0,0,1,-1,0,0,0,0,0,0,0,0,0,0,0,0,0,0,0,0,0)\\
\mathbf{e}_3 &= \frac{1}{\sqrt{2}}(0,0,0,0,1,-1,0,0,0,0,0,0,0,0,0,0,0,0,0,0,0)\\
\mathbf{e}_4 &= \frac{1}{\sqrt{2}}(0,0,0,0,0,0,1,-1,0,0,0,0,0,0,0,0,0,0,0,0,0)\\
\mathbf{e}_5 &= \frac{1}{\sqrt{2}}(0,0,0,0,0,0,0,0,1,-1,0,0,0,0,0,0,0,0,0,0,0)\\
\mathbf{e}_6 &= \frac{1}{\sqrt{2}}(0,0,0,0,0,0,0,0,0,0,1,-1,0,0,0,0,0,0,0,0,0)\\
\mathbf{e}_7 &= \frac{1}{\sqrt{2}}(0,0,0,0,0,0,0,0,0,0,0,0,1,-1,0,0,0,0,0,0,0)\\
\mathbf{e}_8 &= \frac{1}{\sqrt{2}}(0,0,0,0,0,0,0,0,0,0,0,0,0,0,1,-1,0,0,0,0,0)\\
\mathbf{e}_9 &= \frac{1}{\sqrt{2}}(0,0,0,0,0,0,0,0,0,0,0,0,0,0,0,0,1,-1,0,0,0)\\
\mathbf{e}_{10} &= \frac{1}{\sqrt{2}}(0,0,0,0,0,0,0,0,0,0,0,0,0,0,0,0,0,0,1,-1,0)\\
\mathbf{e}_{11} &= \frac{1}{2}(1,1,-1,-1,0,0,0,0,0,0,0,0,0,0,0,0,0,0,0,0,0)\\
\mathbf{e}_{12} &= \frac{1}{2}(0,0,0,0,1,1,-1,-1,0,0,0,0,0,0,0,0,0,0,0,0,0)\\
\mathbf{e}_{13} &= \frac{1}{2}(0,0,0,0,0,0,0,0,1,1,-1,-1,0,0,0,0,0,0,0,0,0)\\
\mathbf{e}_{14} &= \frac{1}{2}(0,0,0,0,0,0,0,0,0,0,0,0,1,1,-1,-1,0,0,0,0,0)\\
\mathbf{e}_{15} &= \frac{1}{2}(0,0,0,0,0,0,0,0,0,0,0,0,0,0,0,0,1,1,-1,-1,0)\\
\mathbf{e}_{16} &= \frac{1}{\sqrt{8}}(1,1,1,1,-1,-1,-1,-1,0,0,0,0,0,0,0,0,0,0,0,0,0)\\
\mathbf{e}_{17} &= \frac{1}{\sqrt{8}}(0,0,0,0,0,0,0,0,1,1,1,1,-1,-1,-1,-1,0,0,0,0,0)\\
\mathbf{e}_{18} &= \frac{1}{4}(1,1,1,1,1,1,1,1,-1,-1,-1,-1,-1,-1,-1,-1,0,0,0,0,0)\\
\mathbf{e}_{19} &= \frac{1}{\sqrt{80}}(1,1,1,1,1,1,1,1,1,1,1,1,1,1,1,1,-4,-4,-4,-4,0)\\
\mathbf{e}_{20} &= \frac{1}{\sqrt{420}}(1,1,1,1,1,1,1,1,1,1,1,1,1,1,1,1,1,1,1,1,-20)
\end{aligned}
\right. \qquad \text{for} \quad s = 20 \tag{D.17}
$$

$$
\begin{cases}
\mathbf{e}_1 = \dfrac{1}{\sqrt{2}}(1,-1,0,0,0,0,0,0,0,0,0,0,0,0,0,0,0,0,0,0,0,0) \\
\mathbf{e}_2 = \dfrac{1}{\sqrt{2}}(0,0,1,-1,0,0,0,0,0,0,0,0,0,0,0,0,0,0,0,0,0,0) \\
\mathbf{e}_3 = \dfrac{1}{\sqrt{2}}(0,0,0,0,1,-1,0,0,0,0,0,0,0,0,0,0,0,0,0,0,0,0) \\
\mathbf{e}_4 = \dfrac{1}{\sqrt{2}}(0,0,0,0,0,0,1,-1,0,0,0,0,0,0,0,0,0,0,0,0,0,0) \\
\mathbf{e}_5 = \dfrac{1}{\sqrt{2}}(0,0,0,0,0,0,0,0,1,-1,0,0,0,0,0,0,0,0,0,0,0,0) \\
\mathbf{e}_6 = \dfrac{1}{\sqrt{2}}(0,0,0,0,0,0,0,0,0,0,1,-1,0,0,0,0,0,0,0,0,0,0) \\
\mathbf{e}_7 = \dfrac{1}{\sqrt{2}}(0,0,0,0,0,0,0,0,0,0,0,0,1,-1,0,0,0,0,0,0,0,0) \\
\mathbf{e}_8 = \dfrac{1}{\sqrt{2}}(0,0,0,0,0,0,0,0,0,0,0,0,0,0,1,-1,0,0,0,0,0,0) \\
\mathbf{e}_9 = \dfrac{1}{\sqrt{2}}(0,0,0,0,0,0,0,0,0,0,0,0,0,0,0,0,1,-1,0,0,0,0) \\
\mathbf{e}_{10} = \dfrac{1}{\sqrt{2}}(0,0,0,0,0,0,0,0,0,0,0,0,0,0,0,0,0,0,1,-1,0,0) \\
\mathbf{e}_{11} = \dfrac{1}{\sqrt{2}}(0,0,0,0,0,0,0,0,0,0,0,0,0,0,0,0,0,0,0,0,1,-1) \\
\mathbf{e}_{12} = \dfrac{1}{2}(1,1,-1,-1,0,0,0,0,0,0,0,0,0,0,0,0,0,0,0,0,0,0) \\
\mathbf{e}_{13} = \dfrac{1}{2}(0,0,0,0,1,1,-1,-1,0,0,0,0,0,0,0,0,0,0,0,0,0,0) \\
\mathbf{e}_{14} = \dfrac{1}{2}(0,0,0,0,0,0,0,0,1,1,-1,-1,0,0,0,0,0,0,0,0,0,0) \\
\mathbf{e}_{15} = \dfrac{1}{2}(0,0,0,0,0,0,0,0,0,0,0,0,1,1,-1,-1,0,0,0,0,0,0) \\
\mathbf{e}_{16} = \dfrac{1}{2}(0,0,0,0,0,0,0,0,0,0,0,0,0,0,0,0,1,1,-1,-1,0,0) \\
\mathbf{e}_{17} = \dfrac{1}{\sqrt{8}}(1,1,1,1,-1,-1,-1,-1,0,0,0,0,0,0,0,0,0,0,0,0,0,0) \\
\mathbf{e}_{18} = \dfrac{1}{\sqrt{8}}(0,0,0,0,0,0,0,0,1,1,1,1,-1,-1,-1,-1,0,0,0,0,0,0) \\
\mathbf{e}_{19} = \dfrac{1}{4}(1,1,1,1,1,1,1,1,-1,-1,-1,-1,-1,-1,-1,-1,0,0,0,0,0,0) \\
\mathbf{e}_{20} = \dfrac{1}{\sqrt{80}}(1,1,1,1,1,1,1,1,1,1,1,1,1,1,1,1,-4,-4,-4,-4,0,0) \\
\mathbf{e}_{21} = \dfrac{1}{\sqrt{220}}(1,1,1,1,1,1,1,1,1,1,1,1,1,1,1,1,1,1,1,1,-10,-10)
\end{cases}
\quad \text{for } s=21
\tag{D.18}
$$

$$
\begin{cases}
\mathbf{e}_1 = \frac{1}{\sqrt{2}}(1,-1,0,0,0,0,0,0,0,0,0,0,0,0,0,0,0,0,0,0,0,0,0) \\
\mathbf{e}_2 = \frac{1}{\sqrt{2}}(0,0,1,-1,0,0,0,0,0,0,0,0,0,0,0,0,0,0,0,0,0,0,0) \\
\mathbf{e}_3 = \frac{1}{\sqrt{2}}(0,0,0,0,1,-1,0,0,0,0,0,0,0,0,0,0,0,0,0,0,0,0,0) \\
\mathbf{e}_4 = \frac{1}{\sqrt{2}}(0,0,0,0,0,0,1,-1,0,0,0,0,0,0,0,0,0,0,0,0,0,0,0) \\
\mathbf{e}_5 = \frac{1}{\sqrt{2}}(0,0,0,0,0,0,0,0,1,-1,0,0,0,0,0,0,0,0,0,0,0,0,0) \\
\mathbf{e}_6 = \frac{1}{\sqrt{2}}(0,0,0,0,0,0,0,0,0,0,1,-1,0,0,0,0,0,0,0,0,0,0,0) \\
\mathbf{e}_7 = \frac{1}{\sqrt{2}}(0,0,0,0,0,0,0,0,0,0,0,0,1,-1,0,0,0,0,0,0,0,0,0) \\
\mathbf{e}_8 = \frac{1}{\sqrt{2}}(0,0,0,0,0,0,0,0,0,0,0,0,0,0,1,-1,0,0,0,0,0,0,0) \\
\mathbf{e}_9 = \frac{1}{\sqrt{2}}(0,0,0,0,0,0,0,0,0,0,0,0,0,0,0,0,1,-1,0,0,0,0,0) \\
\mathbf{e}_{10} = \frac{1}{\sqrt{2}}(0,0,0,0,0,0,0,0,0,0,0,0,0,0,0,0,0,0,1,-1,0,0,0) \\
\mathbf{e}_{11} = \frac{1}{\sqrt{2}}(0,0,0,0,0,0,0,0,0,0,0,0,0,0,0,0,0,0,0,0,1,-1,0) \\
\mathbf{e}_{12} = \frac{1}{2}(1,1,-1,-1,0,0,0,0,0,0,0,0,0,0,0,0,0,0,0,0,0,0,0) \\
\mathbf{e}_{13} = \frac{1}{2}(0,0,0,0,1,1,-1,-1,0,0,0,0,0,0,0,0,0,0,0,0,0,0,0) \\
\mathbf{e}_{14} = \frac{1}{2}(0,0,0,0,0,0,0,0,1,1,-1,-1,0,0,0,0,0,0,0,0,0,0,0) \\
\mathbf{e}_{15} = \frac{1}{2}(0,0,0,0,0,0,0,0,0,0,0,0,1,1,-1,-1,0,0,0,0,0,0,0) \\
\mathbf{e}_{16} = \frac{1}{2}(0,0,0,0,0,0,0,0,0,0,0,0,0,0,0,0,1,1,-1,-1,0,0,0) \\
\mathbf{e}_{17} = \frac{1}{\sqrt{8}}(1,1,1,1,-1,-1,-1,-1,0,0,0,0,0,0,0,0,0,0,0,0,0,0,0) \\
\mathbf{e}_{18} = \frac{1}{\sqrt{8}}(0,0,0,0,0,0,0,0,1,1,1,1,-1,-1,-1,-1,0,0,0,0,0,0,0) \\
\mathbf{e}_{19} = \frac{1}{4}(1,1,1,1,1,1,1,1,-1,-1,-1,-1,-1,-1,-1,-1,0,0,0,0,0,0,0) \\
\mathbf{e}_{20} = \frac{1}{\sqrt{80}}(1,1,1,1,1,1,1,1,1,1,1,1,1,1,1,1,-4,-4,-4,-4,0,0,0) \\
\mathbf{e}_{21} = \frac{1}{\sqrt{220}}(1,1,1,1,1,1,1,1,1,1,1,1,1,1,1,1,1,1,1,1,-10,-10,0) \\
\mathbf{e}_{22} = \frac{1}{\sqrt{506}}(1,1,1,1,1,1,1,1,1,1,1,1,1,1,1,1,1,1,1,1,1,1,-22)
\end{cases}
\quad \text{for } s=22
\tag{D.19}
$$

$$
\begin{cases}
\mathbf{e}_1 = \dfrac{1}{\sqrt{2}}(1,-1,0,0,0,0,0,0,0,0,0,0,0,0,0,0,0,0,0,0,0,0,0,0) \\
\mathbf{e}_2 = \dfrac{1}{\sqrt{2}}(0,0,1,-1,0,0,0,0,0,0,0,0,0,0,0,0,0,0,0,0,0,0,0,0) \\
\mathbf{e}_3 = \dfrac{1}{\sqrt{2}}(0,0,0,0,1,-1,0,0,0,0,0,0,0,0,0,0,0,0,0,0,0,0,0,0) \\
\mathbf{e}_4 = \dfrac{1}{\sqrt{2}}(0,0,0,0,0,0,1,-1,0,0,0,0,0,0,0,0,0,0,0,0,0,0,0,0) \\
\mathbf{e}_5 = \dfrac{1}{\sqrt{2}}(0,0,0,0,0,0,0,0,1,-1,0,0,0,0,0,0,0,0,0,0,0,0,0,0) \\
\mathbf{e}_6 = \dfrac{1}{\sqrt{2}}(0,0,0,0,0,0,0,0,0,0,1,-1,0,0,0,0,0,0,0,0,0,0,0,0) \\
\mathbf{e}_7 = \dfrac{1}{\sqrt{2}}(0,0,0,0,0,0,0,0,0,0,0,0,1,-1,0,0,0,0,0,0,0,0,0,0) \\
\mathbf{e}_8 = \dfrac{1}{\sqrt{2}}(0,0,0,0,0,0,0,0,0,0,0,0,0,0,1,-1,0,0,0,0,0,0,0,0) \\
\mathbf{e}_9 = \dfrac{1}{\sqrt{2}}(0,0,0,0,0,0,0,0,0,0,0,0,0,0,0,0,1,-1,0,0,0,0,0,0) \\
\mathbf{e}_{10} = \dfrac{1}{\sqrt{2}}(0,0,0,0,0,0,0,0,0,0,0,0,0,0,0,0,0,0,1,-1,0,0,0,0) \\
\mathbf{e}_{11} = \dfrac{1}{\sqrt{2}}(0,0,0,0,0,0,0,0,0,0,0,0,0,0,0,0,0,0,0,0,1,-1,0,0) \\
\mathbf{e}_{12} = \dfrac{1}{\sqrt{2}}(0,0,0,0,0,0,0,0,0,0,0,0,0,0,0,0,0,0,0,0,0,0,1,-1) \\
\mathbf{e}_{13} = \dfrac{1}{2}(1,1,-1,-1,0,0,0,0,0,0,0,0,0,0,0,0,0,0,0,0,0,0,0,0) \\
\mathbf{e}_{14} = \dfrac{1}{2}(0,0,0,0,1,1,-1,-1,0,0,0,0,0,0,0,0,0,0,0,0,0,0,0,0) \\
\mathbf{e}_{15} = \dfrac{1}{2}(0,0,0,0,0,0,0,0,1,1,-1,-1,0,0,0,0,0,0,0,0,0,0,0,0) \\
\mathbf{e}_{16} = \dfrac{1}{2}(0,0,0,0,0,0,0,0,0,0,0,0,1,1,-1,-1,0,0,0,0,0,0,0,0) \\
\mathbf{e}_{17} = \dfrac{1}{2}(0,0,0,0,0,0,0,0,0,0,0,0,0,0,0,0,1,1,-1,-1,0,0,0,0) \\
\mathbf{e}_{18} = \dfrac{1}{\sqrt{8}}(1,1,1,1,-1,-1,-1,-1,0,0,0,0,0,0,0,0,0,0,0,0,0,0,0,0) \\
\mathbf{e}_{19} = \dfrac{1}{\sqrt{8}}(0,0,0,0,0,0,0,0,1,1,1,1,-1,-1,-1,-1,0,0,0,0,0,0,0,0) \\
\mathbf{e}_{20} = \dfrac{1}{4}(1,1,1,1,1,1,1,1,-1,-1,-1,-1,-1,-1,-1,-1,0,0,0,0,0,0,0,0) \\
\mathbf{e}_{21} = \dfrac{1}{\sqrt{80}}(1,1,1,1,1,1,1,1,1,1,1,1,1,1,1,1,-4,-4,-4,-4,0,0,0,0) \\
\mathbf{e}_{22} = \dfrac{1}{\sqrt{220}}(1,1,1,1,1,1,1,1,1,1,1,1,1,1,1,1,1,1,1,1,-10,-10,0,0) \\
\mathbf{e}_{23} = \dfrac{1}{\sqrt{264}}(1,1,1,1,1,1,1,1,1,1,1,1,1,1,1,1,1,1,1,1,1,1,-11,-11)
\end{cases}
\quad \text{for } s=23
\tag{D.20}
$$

Appendix E: Probability in a Hyperball

The probability in a hyperball of radius r in dimension s is

$$F(r,s)=\int_{||\mathbf{x}||\le r}p(\mathbf{x})\,\mathrm{d}\mathbf{x}=\int_{x_1^2+x_2^2+\ldots+x_s^2\le r}p(x_1,x_2,\ldots,x_s)\,\mathrm{d}x_1\,\mathrm{d}x_2\,\ldots\,\mathrm{d}x_s \tag{E.1}$$

where $p(\mathbf{x})=p(x_1,x_2,\ldots,x_s)$ is the joint PDF of a set of standardized random variables. Here, the independent normal distribution is considered; that is:

$$p(\mathbf{x})=\frac{1}{(2\pi)^{s/2}}\exp\left(-\frac{x_1^2+x_2^2+\ldots+x_s^2}{2}\right) \tag{E.2}$$

Using a multiple integral equality:

$$\int_{x_1^2+x_2^2+\ldots+x_s^2\le r^2}f\left(\sqrt{x_1^2+x_2^2+\ldots+x_s^2}\right)\mathrm{d}x_1\,\mathrm{d}x_2\,\ldots\,\mathrm{d}x_s=r^s\frac{\pi^{s/2}}{\Gamma(s/2)}\int_0^1u^{(s/2)-1}f(r\sqrt{u})\,\mathrm{d}u \tag{E.3}$$

where $f(\cdot)$ is any arbitrary integrable function, we can get

$$\begin{aligned}F(r,s)&=\int_{||\mathbf{x}||\le r}p(\mathbf{x})\,\mathrm{d}\mathbf{x}=\int_{x_1^2+x_2^2+\ldots+x_s^2\le r^2}f\left(\sqrt{x_1^2+x_2^2+\ldots+x_s^2}\right)\mathrm{d}x_1\,\mathrm{d}x_2\,\ldots\,\mathrm{d}x_s\\&=r^s\frac{\pi^{s/2}}{\Gamma(s/2)}\int_0^1u^{(s/2)-1}\frac{1}{(2\pi)^{s/2}}\exp\left(-\frac{(r\sqrt{u})^2}{2}\right)\mathrm{d}u\\&=\frac{r^s}{2^{s/2}\Gamma(s/2)}\int_0^1u^{(s/2)-1}\exp\left(-\frac{r^2u}{2}\right)\mathrm{d}u\\&=\frac{1}{\Gamma(s/2)}\int_0^{r^2/2}x^{(s/2)-1}\mathrm{e}^{-x}\,\mathrm{d}x\end{aligned} \tag{E.4}$$

Stochastic Dynamics of Structures Jie Li and Jianbing Chen

where the Gamma function is (Zayed, 1996; Andrews *et al.*, 2000)

$$\Gamma(t) = \int_0^{\infty} x^{t-1} \mathrm{e}^{-x} \, \mathrm{d}x \tag{E.5}$$

If we define

$$\tilde{\Gamma}(t, \gamma) = \int_0^{\gamma} x^{t-1} \mathrm{e}^{-x} \, \mathrm{d}x \tag{E.6}$$

then Equation E.4 becomes

$$F(r, s) = \frac{\tilde{\Gamma}(s/2, r^2/2)}{\Gamma(s/2)} = \frac{\tilde{\Gamma}(s/2, r^2/2)}{\tilde{\Gamma}(s/2, \infty)} \tag{E.7}$$

where use has been made of $\Gamma(t) = \tilde{\Gamma}(t, \infty)$ according to Equations E.5 and E.6.

For convenience, in the following sections, Equation E.4 is integrated by considering the cases s as even and as odd numbers.

E.1 The Case s is Even

In the case $s = 2m \, (m \geq 1)$, Equation E.7 becomes

$$F(r, 2m) = \frac{\tilde{\Gamma}(m, r^2/2)}{\Gamma(m)} = \frac{\tilde{\Gamma}(m, r^2/2)}{\tilde{\Gamma}(m, \infty)} \tag{E.8}$$

The integral in Equation E.6 gives

$$\tilde{\Gamma}(m, \gamma) = G_1(m, 0) - G_1(m, \gamma) \tag{E.9}$$

where

$$G_1(m, \gamma) = -\int_0^{\gamma} x^{m-1} \mathrm{e}^{-x} \, \mathrm{d}x = \mathrm{e}^{-x} \left\{ \sum_{j=1}^{m} \left[x^{m-j} \prod_{k=1}^{j-1} (m-k) \right] \right\} \tag{E.10}$$

in which the convention is used that $\prod_{k=1}^{0} (m-k) = 1$.

Therefore:

$$F(r, 2m) = \frac{\tilde{\Gamma}(m, r^2/2)}{\tilde{\Gamma}(m, \infty)} = \frac{1}{\Gamma(m)} \left[G_1(m, 0) - G_1(m, r^2/2) \right] = 1 - \frac{G_1(m, r^2/2)}{\Gamma(m)} \tag{E.11}$$

E.2 The Case s is Odd

In the case $s = 2m + 1 (m \geq 0)$, Equation (E.7) becomes

$$F(r, 2m+1) = \frac{\tilde{\Gamma}(m + \frac{1}{2}, r^2/2)}{\Gamma(m + \frac{1}{2})} = \frac{\tilde{\Gamma}(m + \frac{1}{2}, r^2/2)}{\tilde{\Gamma}(m + \frac{1}{2}, \infty)} \tag{E.12}$$

According to Equation E.6, we can find that

$$\tilde{\Gamma}(\tfrac{1}{2}, r^2/2) = \int_0^{r^2/2} x^{-1/2} \mathrm{e}^{-x}\, \mathrm{d}x = 2\int_0^{r/\sqrt{2}} \mathrm{e}^{-z^2}\, \mathrm{d}z = 2\sqrt{\pi}(\Phi(r) - \tfrac{1}{2}) \tag{E.13}$$

where $\Phi(r) = \int_{-\infty}^{r} (1/\sqrt{2\pi})\mathrm{e}^{-z^2/2}\, \mathrm{d}z$ is the CDF of the standardized normal distribution.

Likewise, we get

$$\begin{aligned} \tilde{\Gamma}(1+\tfrac{1}{2}, r^2/2) &= \int_0^{r^2/2} x^{1/2} \mathrm{e}^{-x}\, \mathrm{d}x = -x^{1/2}\mathrm{e}^{-x}\Big|_{x=0}^{x=r^2/2} + \int_0^{r^2/2} \mathrm{e}^{-x}\, \mathrm{d}x^{1/2} \\ &= -\frac{r}{\sqrt{2}}\mathrm{e}^{-r^2/2} + \int_0^{r/\sqrt{2}} \mathrm{e}^{-z^2}\, \mathrm{d}z \\ &= \sqrt{\pi}\left[\Phi(r) - \tfrac{1}{2}\right] - \frac{r}{\sqrt{2}}\mathrm{e}^{-r^2/2} \end{aligned} \tag{E.14}$$

and then for $m \geq 2$:

$$\tilde{\Gamma}(m+\tfrac{1}{2}, r^2/2) = \frac{\tilde{\Gamma}(1+\tfrac{1}{2}, r^2/2)}{2^{m-1}} \prod_{j=2}^{m} (2j-1) - G_2(m, r^2/2) \tag{E.15}$$

where

$$G_2(m,x) = \mathrm{e}^{-x}\left\{\sum_{j=2}^{m} x^{m-(1/2)-(j-2)}\left[\prod_{k=2}^{j-1} m - \tfrac{1}{2} - (k-2)\right]\right\} \tag{E.16}$$

If we use the convention that $\sum_{j=2}^{m} f_j(\cdot) = 0$ for $m=0$ and $m=1$, and $\prod_{k=2}^{1} f_k(\cdot) = 1$, it is seen that $G_2(m,x) = 0$ for $m=0$ and $m=1$.

Substituting Equation E.15 in Equation E.12 will then yield

$$\begin{aligned} F(r, 2m+1) &= \frac{\tilde{\Gamma}(m+\tfrac{1}{2}, r^2/2)}{\Gamma(m+\tfrac{1}{2})} = \frac{\tilde{\Gamma}(m+\tfrac{1}{2}, r^2/2)}{\tilde{\Gamma}(m+\tfrac{1}{2}, \infty)} \\ &= 1 - \frac{G_2(m, r^2/2)}{\Gamma(m+\tfrac{1}{2})} + 2\left(\Phi(r) - 1 - \frac{r}{\sqrt{2\pi}}\mathrm{e}^{-r^2/2}\right) \end{aligned} \tag{E.17}$$

where $\Gamma(m+\tfrac{1}{2}) = (\sqrt{\pi}/2^m)\prod_{j=1}^{m} (2j-1)$.

Combining Equations E.11 and E.17 finally gives

$$\begin{aligned} F(r,s) &= \frac{\tilde{\Gamma}(s/2, r^2/2)}{\tilde{\Gamma}(s/2, \infty)} \\ &= \begin{cases} 1 - \dfrac{G_1(m, r^2/2)}{\Gamma(m)} & \text{for } \quad s = 2m;\ m \geq 1 \\ 1 - \dfrac{G_2(m, r^2/2)}{\Gamma(m+\frac{1}{2})} + 2\left[\Phi(r) - 1 - (1-\delta_{0m})\dfrac{r}{\sqrt{2\pi}}\mathrm{e}^{-r^2/2}\right] & \text{for } \quad s = 2m+1;\ m \geq 0 \end{cases} \end{aligned} \tag{E.18}$$

where $G_1(\cdot)$ and $G_2(\cdot)$ are given by Equations E.10 and E.16 respectively and δ is the Kronecker delta.

Clearly, it follows from Equation E.7 that

$$F(r,s)|_{r=0} = 0 \qquad F(r,s)|_{r\to\infty} = 1 \tag{E.19}$$

The probabilities $F(r,s)$ for different s given by Equation E.18 are shown in Figure 7.2b.

E.3 Monotonic Features of $F(r, s)$

E.3.1 Monotonic Feature of F(r, s) with Respect to the Radius r

According to Equation E.18, we can get

$$\frac{\partial F(r,2m)}{\partial r} = \frac{1}{2^{m-1}\Gamma(m)} r^{2m-1} \mathrm{e}^{-r^2/2} \tag{E.20}$$

and further:

$$\frac{\partial^2 F(r,2m)}{\partial r^2} = \frac{r^{2m-2}\mathrm{e}^{-r^2/2}}{2^{m-1}\Gamma(m)}\left[(2m-1)-r^2\right] \tag{E.21}$$

Obviously, it is seen that

$$\begin{aligned} \frac{\partial F(r,2m)}{\partial r} &= 0 \qquad \text{for} \quad r=0 \\ \frac{\partial F(r,2m)}{\partial r} &> 0 \qquad \text{for} \quad r>0 \end{aligned} \tag{E.22}$$

$$\frac{\partial^2 F(r,2m)}{\partial r^2} = 0 \qquad \text{for} \quad r=0 \text{ and } r=\sqrt{2m-1} \tag{E.23}$$

The above two formulae mean that, in the case s is even, $F(r, s)$ is increasing monotonically against the radius while has an inflexion at the position $r = \sqrt{2m-1}$.

Likewise, from Equation E.18 we can get that

$$\frac{\partial F(r,2m+1)}{\partial r} = \sqrt{\frac{2}{\pi}} r^{2m} \mathrm{e}^{-r^2/2} \tag{E.24}$$

$$\frac{\partial^2 F(r,2m+1)}{\partial r^2} = \sqrt{\frac{2}{\pi}}(2m-r^2) r^{2m-1} \mathrm{e}^{-r^2/2} \tag{E.25}$$

Therefore, we have

$$\begin{aligned} \frac{\partial F(r,2m+1)}{\partial r} &= 0 \qquad \text{for} \quad r=0 \\ \frac{\partial F(r,2m+1)}{\partial r} &> 0 \qquad \text{for} \quad r>0 \end{aligned} \tag{E.26}$$

$$\frac{\partial^2 F(r, 2m+1)}{\partial r^2} = 0 \qquad \text{for} \quad r = 0 \text{ and } r = \sqrt{2m} \tag{E.27}$$

These two formulae indicate that in the case s is odd $F(r,s)$ is increasing monotonically against the radius while has an inflexion at the position $r = \sqrt{2m}$.

Equations E.22, E.23 and E.26 and E.27 could be unified to

$$\begin{aligned} \frac{\partial F(r,s)}{\partial r} &= 0 \qquad \text{for} \quad r = 0 \\ \frac{\partial F(r,s)}{\partial r} &> 0 \qquad \text{for} \quad r > 0 \end{aligned} \tag{E.28}$$

$$\frac{\partial^2 F(r,s)}{\partial r^2} = 0 \qquad \text{for} \quad r = 0 \text{ and } r = \sqrt{s-1} \tag{E.29}$$

respectively. Therefore, in any case $F(r, s)$ increases monotonically against the radius and has an inflection at the position $r = \sqrt{s-1}$.

E.3.2 Monotonic Feature of F(r,s) with Respect to the Dimensions

From Equation E.18, the recursive relations follow

$$F(r, 2(m+1)) = F(r, 2m) - \frac{1}{\Gamma(m+1)}\left(\frac{r^2}{2}\right)^m e^{-r^2/2} \qquad m \geq 1 \tag{E.30}$$

$$F(r, 2m+3) = F(r, 2m+1) - \frac{1}{\Gamma[(2m+3)/2]}\left(\frac{r^2}{2}\right)^{m+(1/2)} e^{-r^2/2} \qquad m \geq 0 \tag{E.31}$$

The above two formulae could be written in a unified form as

$$F(r, s+2) = F(r, s) - \frac{1}{\Gamma[(s/2)+1]}\left(\frac{r^2}{2}\right)^{s/2} e^{-r^2/2} \qquad s \geq 1 \tag{E.32}$$

Obviously, there is

$$F(r, s+2) < F(r, s) \qquad \text{for } r > 0 \tag{E.33}$$

This means that the probability contained in a hyperball of the same radius decreases, at least in a jumping way, as the dimension increases.

If we denote

$$F_{s_1 s_2}(r) = F(r, s_1) - F(r, s_2) \tag{E.34}$$

then it can be proved that

$$F_{43} < 0 \qquad F_{32} < 0 \qquad F_{21} < 0 \qquad \text{for } r > 0 \tag{E.35}$$

and the minimum values of $F_{43}(r)$, $F_{32}(r)$ and $F_{21}(r)$ occur at $r = \sqrt{2/\pi}$, $\sqrt{\pi/2}$ and $2\sqrt{2/\pi}$ respectively.

According to the above observations, we guess that

$$F_{s+1,s}(r) < 0 \qquad \text{for } r > 0,\ s \geq 1 \tag{E.36}$$

holds and the minimum values of $F_{s+1,s}(r)$ occur at increasing value of r against s.

The monotonic features of the probability contained in a hyperball, $F(r, s)$, can be observed clearly in Figure 7.2b.

Appendix F: Spectral Moments

In the first-passage probability analysis based on the excursion assumption, computation of spectral parameters and passage rate is useful. They will be introduced herein.

According to Equations 2.81a and 2.81b in Chapter 2, the auto-PSD $S_X(\omega)$ of a stationary stochastic process is an even function; consequently, the nth-order spectral moment of the stationary stochastic process can be defined as

$$\alpha_n = \int_0^{\infty} \omega^n G_X(\omega)\,\mathrm{d}\omega \qquad n = 0, 1, 2, \ldots \tag{F.1}$$

in which $G_X(\omega)$ is the single-sided PSD.

By the equation, the following spectral parameters can be further defined:

$$\gamma_1 = \frac{\alpha_1}{\alpha_0} \tag{F.2}$$

$$\gamma_2 = \sqrt{\frac{\alpha_2}{\alpha_0}} \tag{F.3}$$

$$q = \left(1 - \frac{\alpha_1^2}{\alpha_0 \alpha_2}\right)^{1/2} = \left(\frac{\gamma_2^2 - \gamma_1^2}{\gamma_2^2}\right)^{1/2} \tag{F.4}$$

where γ_1 is the frequency at the area centric of $G_X(\omega)$, generally indicating where the spectral density concentrates, γ_2 is the gyration radius of $G_X(\omega)$ with respect to the coordinate origin and q is the gyration radius of $G_X(\omega)$ with respect to the frequency γ_1. The value of q varies over 0–1. The smaller the value of q, the narrower the figure of $G_X(\omega)$ is; in contrast, the greater the value of q, the wider the figure of $G_X(\omega)$ is. Usually, the stochastic process with $0 \le q \le 0.35$ is called a narrow-band stochastic process, whereas it is a white noise if $q = 1$.

The spectral parameters are illustrated in Figure F.1.

For nonstationary stochastic processes, when the concept of evolutionary power spectral density is introduced (see Section 5.3.2), the nth-order spectral moment can be defined as

$$\alpha_n(t) = \int_0^{\infty} \omega^n G_X(\omega, t)\,\mathrm{d}\omega \qquad n = 0, 1, 2 \ldots \tag{F.5}$$

Stochastic Dynamics of Structures Jie Li and Jianbing Chen

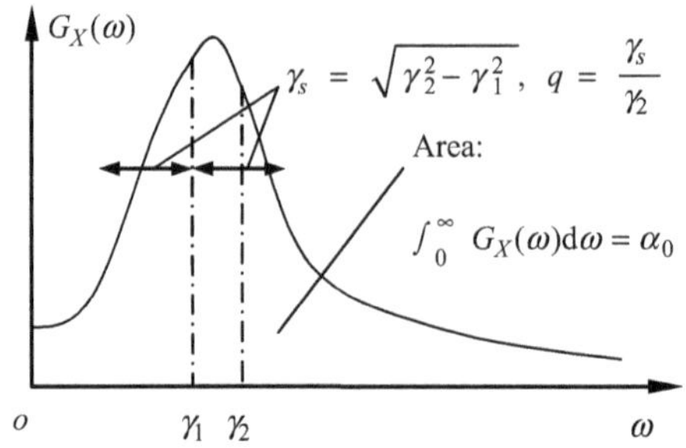

Figure F.1 Spectral parameters.

Evidently, the spectral moments of nonstationary stochastic processes are functions of time. Consequently, the spectral parameters defined by the spectral moments are also related to time:

$$\gamma_1(t) = \frac{\alpha_1(t)}{\alpha_0(t)} \tag{F.6}$$

$$\gamma_2(t) = \sqrt{\frac{\alpha_2(t)}{\alpha_0(t)}} \tag{F.7}$$

$$q(t) = \left(1 - \frac{\alpha_1^2(t)}{\alpha_0(t)\alpha_2(t)}\right)^{1/2} = \left(\frac{\gamma_2^2(t) - \gamma_1^2(t)}{\gamma_2^2(t)}\right)^{1/2} \tag{F.8}$$

If $G_X(\omega, t)$ is understood as an instant PSD, then at a specified time point the above spectral parameters have the geometric interpretation shown in Figure F.1.

Appendix G: Generator Vectors in the Number Theoretical Method

As discussed in Section 7.3.3, in the number theoretical method a uniform point set $\mathcal{P}_{\text{NTM}} = \{\mathbf{x}_k = (x_{1,k}, x_{2,k}, \ldots, x_{s,k}) : k = 1, 2, \ldots, n\}$ over the hypercube $C^s = [0,1]^s$ can be generated by Equation 7.140 (repeated here for convenience):

$$\begin{aligned} \hat{x}_{j,k} &= (2kQ_j - 1)\text{mod}(2n) \qquad j = 1, 2, \ldots, s;\ k = 1, 2, \ldots, n \\ x_{j,k} &= \frac{\hat{x}_{j,k}}{2n} \end{aligned} \tag{G.1}$$

or equivalently:

$$x_{j,k} = \frac{2kQ_j - 1}{2n} - \text{int}\left(\frac{2kQ_j - 1}{2n}\right) \qquad j = 1, 2, \ldots, s;\ k = 1, 2, \ldots, n \tag{G.2}$$

Here, the integer vector $(n, Q_1, Q_2, \ldots, Q_s)$ is called the generator vector.

Tables G.1–G.12 are the generators which can be used to generate good uniform point sets (Hua and Wang, 1978).

Table G.1 $s = 2$ ($n = F_m$, $Q_1 = 1$, $Q_2 = F_{m-1}$). (Reproduced with permission from Hua Luo-Keng and Wang Yuan. *Applications of Number Theory to Approximate Analysis*. Science Press, Beijing. © 1978 Yuan Wang)

n	8	13	21	34	55	89	144	233	377	610
Q_2	5	8	13	21	34	55	89	144	233	377
n	987	1 597	2 584	4 181	6 765	10 946	17 711	28 657	46 368	75 025
Q_2	610	987	1 597	2 584	4 181	6 765	10 946	17 711	28 657	46 368

Table G.2 $s=3$, $Q_1=1$. (Reproduced with permission from Hua Luo-Keng and Wang Yuan. *Applications of Number Theory to Approximate Analysis.* Science Press, Beijing. © 1978 Yuan Wang)

n	35	101	135	185	266	418	597	828	1 010
Q_2	11	40	29	26	27	90	63	285	140
Q_3	16	85	42	64	69	130	169	358	237
n	1 220	1 459	1 626	1 958	2 440	3 237	4 044	5 037	6 066
Q_2	319	256	572	202	638	456	400	580	600
Q_3	510	373	712	696	1 002	1 107	1 054	1 997	1 581
n	8 191	10 007	20 039	28 117	39 029	57 091	82 001	140 052	314 694
Q_2	739	544	5 704	19 449	10 607	48 188	21 252	34 590	77 723
Q_3	5 515	5 733	12 319	5 600	26 871	21 101	67 997	112 313	252 365

Table G.3 $s=4$, $Q_1=1$. (Reproduced with permission from Hua Luo-Keng and Wang Yuan. *Applications of Number Theory to Approximate Analysis.* Science Press, Beijing. © 1978 Yuan Wang)

n	307	562	701	1 019	2 129	3 001	4 001	5 003	6 007
Q_2	42	53	82	71	766	174	113	792	1 351
Q_3	229	89	415	765	1 281	266	766	1 889	5 080
Q_4	101	221	382	865	1 906	1 269	2 537	191	3 086
n	8 191	10 007	20 039	28 117	39 029	57 091	82 001	100 063	147 312
Q_2	2 448	1 206	19 668	17 549	30 699	52 590	57 270	92 313	136 641
Q_3	5 939	3 421	17 407	1 900	34 367	48 787	58 903	24 700	116 072
Q_4	7 859	2 842	14 600	24 455	605	38 790	17 672	95 582	76 424

Table G.4 $s=5$, $Q_1=1$. (Reproduced with permission from Hua Luo-Keng and Wang Yuan. *Applications of Number Theory to Approximate Analysis.* Science Press, Beijing. © 1978 Yuan Wang)

n	1 069	1 543	2 129	3 001	4 001	5 003	6 007	8 191
Q_2	63	58	618	408	1 534	840	509	1 386
Q_3	762	278	833	1 409	568	117	780	4 302
Q_4	970	694	1 705	1 681	3 095	3 593	558	7 715
Q_5	177	134	1 964	1 620	2 544	1 311	1 693	3 735
n	10 007	15 019	20 039	33 139	51 097	71 053	100 063	374 181
Q_2	198	10 641	11 327	32 133	44 672	33 755	90 036	343 867
Q_3	9 183	2 640	11 251	17 866	45 346	65 170	77 477	255 381
Q_4	6 967	6 710	12 076	21 281	7 044	12 740	27 253	310 881
Q_5	8 507	784	18 677	32 247	14 242	6 878	6 222	115 892

Table G.5 $s=6$, $Q_1=1$. (Reproduced with permission from Hua Luo-Keng and Wang Yuan. *Applications of Number Theory to Approximate Analysis.* Science Press, Beijing. © 1978 Yuan Wang)

n	2 129	3 001	4 001	5 003	6 007	8 191	10 007	15 019
Q_2	41	233	1 751	2 037	312	1 632	2 240	8 743
Q_3	1 681	271	1 235	1 882	1 232	1 349	4 093	8 358
Q_4	793	122	1 945	1 336	5 943	6 380	1 908	6 559
Q_5	578	1 417	844	4 803	4 060	1 399	931	2 795
Q_6	279	51	1 475	2 846	5 250	6 070	3 984	772
n	20 039	33 139	51 097	71 053	100 063	114 174	302 686	
Q_2	5 557	18 236	9 931	18 010	43 307	107 538	285 095	
Q_3	150	1 831	7 551	3 155	15 440	88 018	233 344	
Q_4	11 951	19 143	29 683	50 203	39 114	15 543	41 204	
Q_5	2 461	5 522	44 446	6 065	43 534	80 974	214 668	
Q_6	9 179	22 910	17 340	13 328	29 955	56 747	150 441	

Table G.6 $s=7$, $Q_1=1$. (Reproduced with permission from Hua Luo-Keng and Wang Yuan. *Applications of Number Theory to Approximate Analysis.* Science Press, Beijing. © 1978 Yuan Wang)

n	3 997	11 215	15 019	24 041	33 139	46 213	57 091	71 053
Q_2	3 888	10 909	12 439	1 833	7 642	37 900	35 571	31 874
Q_3	3 564	10 000	2 983	18 190	9 246	17 534	45 299	36 082
Q_4	3 034	8 512	8 607	21 444	5 584	41 873	51 436	13 810
Q_5	2 311	6 485	7 041	23 858	23 035	32 280	34 679	6 605
Q_6	1 417	3 976	7 210	1 135	32 241	15 251	1 472	68 784
Q_7	375	1 053	6 741	12 929	30 396	26 909	8 065	9 848
n	84 523	100 063	172 155	234 646	462 891	769 518	957 838	
Q_2	82 217	39 040	167 459	228 245	450 265	748 528	931 711	
Q_3	75 364	62 047	153 499	209 218	412 730	686 129	854 041	
Q_4	64 149	89 839	130 657	178 084	351 310	584 024	726 949	
Q_5	48 878	6 347	99 554	135 691	267 681	444 998	553 900	
Q_6	29 969	30 892	61 040	83 197	164 124	272 843	339 614	
Q_7	7 936	64 404	18 165	22 032	43 464	72 255	89 937	

Table G.7 $s=8$, $Q_1=1$. (Reproduced with permission from Hua Luo-Keng and Wang Yuan. *Applications of Number Theory to Approximate Analysis.* Science Press, Beijing. © 1978 Yuan Wang)

n	3 997	11 215	24 041	28 832	33 139	46 213	57 091	71 053
Q_2	3 888	10 909	17 441	27 850	3 520	5 347	17 411	60 759
Q_3	3 564	10 000	21 749	24 938	29 553	30 775	46 802	26 413
Q_4	3 034	8 512	5 411	20 195	3 239	35 645	9 779	24 409
Q_5	2 311	6 485	12 326	13 782	1 464	11 403	16 807	48 215
Q_6	1 417	3 976	3 144	5 918	16 735	16 894	35 302	51 048
Q_7	375	1 053	21 024	25 703	19 197	32 016	1 416	19 876
Q_8	3 211	9 010	6 252	15 781	3 019	16 600	47 755	29 096
n	84 523	100 063	172 155	234 646	462 891	769 518	957 838	
Q_2	82 217	4 344	167 459	228 245	450 265	748 528	931 711	
Q_3	75 364	58 492	153 499	209 218	412 730	686 129	854 041	
Q_4	64 149	29 291	130 657	178 084	351 310	584 024	726 949	
Q_5	48 878	60 031	99 554	135 691	267 681	444 998	553 900	
Q_6	29 969	10 486	61 040	83 197	164 124	272 843	339 614	
Q_7	7 936	22 519	18 165	22 032	43 464	72 255	89 937	
Q_8	67 905	60 985	138 308	188 512	371 882	618 224	769 518	

Table G.8 $s=9$, $Q_1=1$. (Reproduced with permission from Hua Luo-Keng and Wang Yuan. *Applications of Number Theory to Approximate Analysis*. Science Press, Beijing. © 1978 Yuan Wang)

n	3 997	11 215	33 139	42 570	46 213	57 091	71 053
Q_2	3 888	10 909	68	41 409	8 871	20 176	26 454
Q_3	3 564	10 000	4 624	37 957	40 115	12 146	13 119
Q_4	3 034	8 512	16 181	32 308	20 065	23 124	27 174
Q_5	2 311	6 485	6 721	24 617	30 352	2 172	17 795
Q_6	1 417	3 976	26 221	15 094	15 654	33 475	22 805
Q_7	375	1 053	26 661	3 997	42 782	5 070	43 500
Q_8	3 211	9 010	23 442	34 200	17 966	42 339	45 665
Q_9	1 962	5 506	3 384	20 901	33 962	36 122	49 857
n	100 063	159 053	172 155	234 646	462 891	769 518	957 838
Q_2	70 893	60 128	167 459	228 245	450 265	748 528	931 711
Q_3	53 211	101 694	153 499	209 218	412 730	686 129	854 041
Q_4	12 386	23 300	130 657	178 084	351 310	584 024	726 949
Q_5	27 873	43 576	99 554	135 691	267 681	444 998	553 900
Q_6	56 528	57 659	61 040	83 197	164 124	272 843	339 614
Q_7	16 417	42 111	18 165	22 032	43 464	72 255	89 937
Q_8	17 628	85 501	138 308	188 512	371 882	618 224	769 518
Q_9	14 997	93 062	84 523	115 204	227 266	377 811	470 271

Table G.9 $s=10$, $Q_1=1$. (Reproduced with permission from Hua Luo-Keng and Wang Yuan. *Applications of Number Theory to Approximate Analysis*. Science Press, Beijing. © 1978 Yuan Wang)

n	4 661	13 587	24 076	58 358	85 633
Q_2	4 574	13 334	23 628	57 271	37 677
Q_3	4 315	12 579	22 290	54 030	35 345
Q_4	3 889	11 337	20 090	48 695	3 864
Q_5	3 304	9 631	17 066	41 366	54 821
Q_6	2 570	7 492	13 276	32 180	74 078
Q_7	1 702	4 961	8 790	21 307	30 354
Q_8	715	2 084	3 692	8 950	57 935
Q_9	4 289	12 502	22 153	53 697	51 906
Q_{10}	3 122	9 100	16 125	39 086	56 279
n	103 661	115 069	130 703	155 093	805 098
Q_2	45 681	65 470	64 709	90 485	790 101
Q_3	57 831	650	53 373	20 662	745 388
Q_4	80 987	95 039	17 385	110 048	671 792
Q_5	9 718	77 293	5 244	102 308	570 685
Q_6	51 556	98 366	29 008	148 396	443 949
Q_7	55 377	70 366	52 889	125 399	293 946
Q_8	37 354	74 605	66 949	124 635	123 470
Q_9	4 353	55 507	51 906	10 480	740 795
Q_{10}	27 595	49 201	110 363	44 198	539 222

Table G.10 $s = 11$, $Q_1 = 1$. (Reproduced with permission from Hua Luo-Keng and Wang Yuan. *Applications of Number Theory to Approximate Analysis*. Science Press, Beijing. © 1978 Yuan Wang)

n	4 661	13 587	24 076	58 358	297 974
Q_2	4 574	13 334	23 628	57 271	294 481
Q_3	4 315	12 579	22 290	54 030	284 041
Q_4	3 889	11 337	20 090	48 695	266 778
Q_5	3 304	9 631	17 066	41 366	242 894
Q_6	2 570	7 492	13 276	32 180	212 668
Q_7	1 702	4 961	8 790	21 307	176 456
Q_8	715	2 084	3 692	8 950	134 682
Q_9	4 289	12 502	22 153	53 697	87 835
Q_{10}	3 122	9 100	16 125	39 086	36 464
Q_{11}	1 897	5 529	9 797	23 747	279 147
n	689 047	1 243 423	2 226 963	7 494 007	
Q_2	685 041	1 228 845	2 200 854	7 354 408	
Q_3	646 274	1 185 282	2 122 833	6 838 211	
Q_4	582 461	1 113 244	1 993 814	6 253 169	
Q_5	494 796	1 013 577	1 815 311	5 312 043	
Q_6	384 914	887 449	1 589 415	4 132 365	
Q_7	254 860	736 338	1 318 777	2 736 109	
Q_8	107 051	562 016	1 006 567	1 149 286	
Q_9	642 292	366 527	656 448	6 895 461	
Q_{10}	467 527	152 163	272 523	5 019 180	
Q_{11}	284 044	1 164 860	2 086 257	3 049 402	

Table G.11 $s = 12, 13, 14$, $Q_1 = 1$. (Reproduced with permission from Hua Luo-Keng and Wang Yuan. *Applications of Number Theory to Approximate Analysis*. Science Press, Beijing. © 1978 Yuan Wang)

n	18 984	53 328	77 431	297 974	1 243 423
Q_2	18 761	52 703	76 523	294 481	1 228 845
Q_3	18 096	50 834	73 810	284 041	1 185 282
Q_4	16 996	47 745	69 324	266 778	1 113 244
Q_5	15 475	43 470	63 118	242 894	1 013 577
Q_6	13 549	38 061	55 264	212 668	887 449
Q_7	11 242	31 580	45 854	176 456	736 338
Q_8	8 581	24 104	34 998	134 682	562 016
Q_9	5 596	15 720	22 825	87 835	366 527
Q_{10}	2 323	6 526	9 476	36 464	152 163
Q_{11}	17 785	49 959	72 539	279 147	1 164 860
Q_{12}	14 053	39 477	57 320	220 583	920 477
Q_{13}	10 158	28 534	41 430	159 433	665 302
Q_{14}	6 143	17 255	25 054	96 414	402 327
n	2 428 705	14 753 436	19 984 698	34 248 063	
Q_2	2 400 231	14 580 465	19 750 396	33 846 536	
Q_3	2 315 141	14 063 582	19 050 236	32 646 662	
Q_4	2 174 435	13 208 845	17 892 427	30 662 508	
Q_5	1 979 761	12 026 276	16 290 543	27 917 337	

(continued)

Table G.11 *(Continued)*

Q_6	1 733 402	10 529 739	14 263 366	24 443 334
Q_7	1 438 245	8 736 780	11 834 661	20 281 228
Q_8	1 097 753	6 668 420	9 032 903	15 479 816
Q_9	715 916	4 348 908	5 890 941	10 095 390
Q_{10}	297 211	1 805 439	2 445 610	4 191 077
Q_{11}	2 275 252	13 821 268	18 722 002	32 084 164
Q_{12}	1 797 913	10 921 619	14 794 199	25 353 030
Q_{13}	1 299 495	7 893 924	10 692 946	18 324 655
Q_{14}	785 841	4 773 681	6 466 329	11 081 440

Table G.12 $s = 15, 16, 17, 18, Q_1 = 1$. (Reproduced with permission from Hua Luo-Keng and Wang Yuan. *Applications of Number Theory to Approximate Analysis.* Science Press, Beijing. © 1978 Yuan Wang)

n	70 864	139 489	1 139 691	2 422 957
Q_2	70 353	138 484	1 131 480	2 398 094
Q_3	68 825	135 476	1 106 904	2 323 761
Q_4	66 291	130 487	1 066 142	2 200 720
Q_5	62 768	123 553	1 009 487	2 030 234
Q_6	58 283	114 724	937 347	1 814 052
Q_7	52 867	104 063	850 242	1 554 392
Q_8	46 559	91 647	748 799	1 253 920
Q_9	39 405	77 566	633 750	915 717
Q_{10}	31 457	61 921	505 923	543 256
Q_{11}	22 772	44 825	366 239	140 357
Q_{12}	13 412	26 401	215 705	2 134 112
Q_{13}	3 445	6 781	55 406	1 683 011
Q_{14}	63 806	125 597	1 026 186	1 214 641
Q_{15}	52 844	104 019	849 882	733 806
Q_{16}	41 501	81 691	667 455	
Q_{17}	29 859	58 775	480 219	
Q_{18}	18 002	35 435	289 522	
n	4 395 774	14 271 038	55 879 244	
Q_2	4 364 102	14 168 215	55 476 633	
Q_3	4 269 316	13 860 486	54 271 700	
Q_4	4 112 097	13 350 069	52 273 127	
Q_5	3 893 578	12 640 642	49 495 314	
Q_6	3 615 335	11 737 315	45 958 274	
Q_7	3 279 371	10 646 597	41 687 493	
Q_8	2 888 108	9 376 347	36 713 742	
Q_9	2 444 365	7 935 718	31 072 856	
Q_{10}	1 951 338	6 335 088	24 805 477	
Q_{11}	1 412 580	4 585 990	17 956 764	
Q_{12}	831 972	2 701 027	10 576 061	
Q_{13}	213 699	693 780	50 314 090	
Q_{14}	3 957 988	12 849 750	41 669 876	
Q_{15}	3 277 986	10 642 098	32 725 430	
Q_{16}	2 574 365	8 357 770	23 545 197	
Q_{17}	1 852 197	6 013 224	14 195 319	
Q_{18}	1 116 683	3 625 352	2 716 545	

References and Bibliography

[1] Abramowitz, M. and Stegun, I.A. (1964) *Handbook of Mathematical Functions*, Dover Publications.

[2] Adeli, H. and Saleh, A. (1999) *Control, Optimization, and Smart Structures*, John Wiley & Sons, Inc., New York.

[3] Agranovich, G., Ribakov, Y. and Blostotsky, B. (2004) A numerical method for choice of weighting matrices in active controlled structures. *The Structural Design of Tall and Special Buildings*, **13**, 55–72.

[4] Aleksandrov, A.D., Kolmogorov, A.N. and Lavrent'ev, M.A. (1999) *Mathematics: Its Content, Methods and Meaning*, Dover Publications.

[5] Amin, M. and Ang, A.H.-S. (1968) Nonstationary stochastic model of earthquake motions. *Journal of Engineering Mechanics Division, ASCE*, **94** (EM2), 559–583.

[6] Anderson, J.D. Jr (1995) *Computational Fluid Dynamics*, McGraw-Hill.

[7] Andrews, G.E., Askey, R. and Roy, R. (2000) *Special Functions*, Cambridge University Press.

[8] Ang, A.H.-S. and Tang, W.H. (1984) *Probability Concepts in Engineering Planning and Design (Vol. II: Decision, Risk, and Reliability)*, John Wiley & Sons, Inc., New York.

[9] Arnold, V.I. (1978) *Mathematical Methods of Classical Mechanics*, 2nd edn, Springer-Verlag.

[10] Åström, K.J. (1970) *Introduction to Stochastic Control Theory*, Academic Press, New York.

[11] Augusti, G., Baratta, A. and Casciati, F. (1984) *Probabilistic Methods in Structural Engineering*, Chapman and Hall.

[12] Baber, T.T. and Wen, Y.K. (1981) Random vibration of hysteretic, degrading systems. *Journal of the Engineering Mechanics Division*, **107** (6), 1069–1087.

[13] Baber, T.T. and Noori, M.N. (1985) Random vibration of degrading, pinching systems. *Journal of Engineering Mechanics*, **111** (8), 1010–1026.

[14] Baber, T.T. and Noori, M.N. (1986) Modeling general hysteresis behavior and random vibration application. *Journal of Vibration, Acoustics, Stress, and Reliability in Design*, **108**, 411–420.

[15] Baecher, G.B. and Ingra, T.S. (1981) Stochastic FEM in settlement predictions. *Journal of Geotechnical Engineering Division (ASCE)*, **107** (GT4), 449–463.

[16] Barndorff-Nielsen, O.E., Kendall, W.S. and van Lieshout, M.N.M. (1999) *Stochastic Geometry*, CRC Press.

[17] Bathe, K.J. and Wilson, E.L. (1985) *Numerical Methods in Finite Element Analysis*, Science Press.

[18] Battaini, M., Casciati, F. and Faravelli, L. (2000) Some reliability aspects in structural control. *Probabilistic Engineering Mechanics*, **15**, 101–107.

[19] Bellman, R. (1957) *Dynamic Programming*, Princeton University Press.

[20] Belytschko, T., Liu, W.K. and Moran, B. (2000) *Nonlinear Finite Elements for Continua and Structures*, John Wiley & Sons, New York.

[21] Benaroya, H. and Rehak, M. (1988) Finite element methods in probabilistic structural analysis: a selective review. *Applied Mechanic Reviews*, **41** (5), 201–213.

[22] Bendat, J.S. and Piersol, A.G. (2000) *Random Data: Analysis and Measurement Procedures*, 3rd edn, John Wiley & Sons, Inc., New York.

[23] Bontempi, F. and Faravelli, L. (1998) Lagrangian/Eulerian description of dynamic system. *Journal of Engineering Mechanics*, **124** (8), 901–911.

[24] Booton, R.C. (1954) Nonlinear control systems with random inputs. *IRE Transactions on Circuit Theory*, **CT-1**, 9–18.

[25] Bouc, R. (1967) Forced vibration of mechanical systems with hysteresis. Presented at *Proceedings of the 4th International Conference on Nonlinear Oscillations, Prague, Czechoslovakia.*

[26] Boyce, E.W. and Goodwin, B.E. (1964) Random transverse vibration of elastic beams. *SIAM Journal*, **12** (3), 613–629.

[27] Bracewell, R.N. (1986) *The Hartley Transform*, Oxford University Press, New York.

[28] Burg, J.P. (1978) Maximum entropy spectral analysis, in *Proceedings of the 37th Meeting of the Society of Exploration Geophysicists, 1967, Modern Spectrum Analysis* (ed. D.G. Chiders), IEEE Press, New York, pp. 34–41.

[29] Casciati, F., Magonette, G. and Marazzi, F. (2006) *Technology of Semiactive Devices and Applications in Vibration Mitigation*, John Wiley & Sons, Ltd, Chichester.

[30] Caughey, T.K. (1960) Random excitation of a system with bilinear hysteresis. *Journal of Applied Mechanics ASME*, **27**, 649–652.

[31] Caughey, T.K. (1963) Equivalent linearization techniques. *Journal of the Acoustical Society of America*, **35**, 1706–1711 (reference is made to presentations of the procedure in lectures delivered in 1953 at the California Institute of Technology).

[32] Caughey, T. K. (1971) Nonlinear theory of random vibrations. *Advances in Applied Mechanics*, **11**, 209–253.

[33] Cercignani, C. (1998) *Ludwig Boltzmann: the Man Who Trusted Atoms*, Oxford University Press.

[34] Chen, J.B., Ghanem, R. and Li, J. (2009) Partition of the probability-assigned space in probability density evolution analysis of nonlinear stochastic structures. *Probabilistic Engineering Mechanics*, **24** (1), 27–42.

[35] Chen, J.B. and Li, J. (2004) Difference method for probability density evolution equation of stochastic structural response. *Chinese Quarterly of Mechanics*, **25** (1), 21–28 (in Chinese).

[36] Chen, J.B. and Li, J. (2004) Extreme value distribution and dynamic reliability of stochastic structures. *The 21st International Congress of Theoretical and Applied Mechanics (ICTAM04), ICTAM04 Abstracts and CD-ROM Proceedings, Warsaw, Poland, August 15–21*, p. 366.

[37] Chen, J.B. and Li, J. (2005a) Dynamic response and reliability analysis of nonlinear stochastic structures. *Probabilistic Engineering Mechanics*, **20** (1), 33–44.

[38] Chen, J.B. and Li, J. (2005b) The extreme value distribution and reliability of nonlinear stochastic structures. *Earthquake Engineering and Engineering Vibration*, **4** (2), 275–286.

[39] Chen, J.B. and Li, J. (2007a) The extreme value distribution and dynamic reliability analysis of nonlinear structures with uncertain parameters. *Structural Safety*, **29**, 77–93.

[40] Chen, J.B. and Li, J. (2007b) Development-process-of-nonlinearity-based reliability evaluation of structures. *Probabilistic Engineering Mechanics*, **22** (3), 267–275.

[41] Chen, J.B. and Li, J. (2007c) Joint probability density function of the stochastic responses of nonlinear structures. *Earthquake Engineering & Engineering Vibration*, **6** (1), 35–48.

[42] Chen, J.B. and Li, J. (2008) Strategy of selecting representative points via tangent spheres in the probability density evolution method. *International Journal for Numerical Methods in Engineering*, **74** (13), 1988–2014.

[43] Chen, J.B. and Li, J. (2009) A note on the principle of preservation of probability and probability density evolution equation. *Probabilistic Engineering Mechanics*, **24** (1), 51–59.

[44] Cheng, F.Y. (2001) *Matrix Analysis of Structural Dynamics: Applications and Earthquake Engineering*, Marcel Dekker Inc., New York.

[45] Cheng, Q.X. *et al.* (1983) *Real Function and Functional Analysis Fundamentals*, Higher Education Press (in Chinese).

[46] Chu, S.Y., Soong, T.T. and Reinhorn, A.M. (2005) *Active, Hybrid and Semi-active Structural Control,* John Wiley & Sons, Ltd.

[47] Chung, K.L. (1974) *A Course in Probability Theory,* 2nd edn, Academic Press.

[48] Clough, R.W. and Penzien, J. (1993) *Dynamics of Structures,* 2nd edn, McGraw-Hill, Inc.

[49] Coleman, J.J. (1959) Reliability of aircraft structures in resisting chance failure. *Operations Research,* **7** (5): 639–645.

[50] Collins, J.D. and Thompson, W.T. (1969) The eigenvalue problem for structural systems with uncertain parameters. *AIAA Journal,* **7** (4), 642–648.

[51] Conway, J.H. and Sloane, N.J.A. (1999) *Sphere Packings, Lattices and Groups,* 3rd edn, Springer-Verlag.

[52] Cornell, C.A. (1971) First order uncertainty analysis in soils deformation and stability, in *Statistics and Probability in Civil Engineering,* Proceedings of the First International Conference on Applications of Statistics and Probability to Soil and Structural Engineering, Hong Kong, 1971 (ed. P. Lumb), Hong Kong University Press, London (distributed by Oxford University Press).

[53] Courant, R., Friedrichs, K. and Lewy, H. (1928) On the partial difference equations of mathematical physics. *Mathematische Annalen,* **100**, 32–74 (in German). (Reprinted (1967), English translation of the 1928 German original, *IBM Journal of Research and Development,* **11**, 215–234.)

[54] Cramer, H. (1966) On the intersections between the trajectories of a normal stationary stochastic process and a high level. *Arkiv für Mathematik,* **6**, 337–349.

[55] Crandall, S.H. (1958) *Random Vibration,* MIT Press, Cambridge, MA.

[56] Crandall, S.H. (2006) A half-century of stochastic equivalent linearization. *Structural Control and Health Monitoring,* **13**, 27–40.

[57] Crandall, S.H., Chandiramani, K.L. and Cook, R.G. (1966) Some first passage problems in random vibrations. *Journal of Applied Mechanics, Transactions of ASME,* **33** (3), 532–538.

[58] Dafermos, C.M. (2000) *Hyperbolic Conservation Laws in Continuum Physics,* Springer-Verlag, Berlin.

[59] Davenport, A.G. (1961) The spectrum of horizontal gustiness near the ground in high winds. *Quarterly Journal of the Royal Meteorological Society,* **87**, 194–211.

[60] Davenport, A.G. (1968) The dependence of wind load on meteorological parameters, in *International Research Seminar Wind Effects on Buildings and Structures, Ottawa, Canada, 11–15, September 1967, Proceedings, Vol. I,* University of Toronto Press, Toronto.

[61] Dendrou, B. and Houstis, E. (1978) An inference finite element model for field problems. *Applied Mathematical Modelling,* **1**, 109–114

[62] Deodatis, G. (1990) Bounds on response variability of stochastic finite element systems. *Journal of Engineering Mechanics ASCE,* **116** (3), 565–586.

[63] Der Kiureghian, A. (1980) Structural response to stationary excitation. *Journal of the Engineering Mechanics Division,* **106** (EM6), 1195–1213.

[64] Der Kiureghian, A., Li, C.C. and Zhang, Y. (1991) Recent developments in stochastic finite elements. *Reliability and Optimization of Structural Systems, Proceedings of 4th IFIP WG 7.5 Conference, Munich, Germany, September* (eds R. Rackwitz and P. Thoft-Christensen), Lecture Notes in Engineering 76, Springer-Verlag, Berlin, pp. 19–38.

[65] Der Kiureghian, A. and Ke, J.B. (1988) The stochastic finite element method in structural reliability. *Probabilistic Engineering Mechanics,* **3** (2), 83–91.

[66] Donelan, M.A., Hamilton, J. and Hui, W.H. (1985) Directional spectra of wind-generated waves. *Philosophical Transactions of the Royal Society of London, Series A,* **315**, 509–562.

[67] Dostupov, B.G. and Pugachev, V.S. (1957) The equation for the integral of a system of ordinary differential equations containing random parameters. *Automatika i Telemekhanika,* **18**, 620–630 (in Russian).

[68] Dunkl, C.F. and Xu, Y. (2001) *Orthogonal Polynomials of Several Variables,* Cambridge University Press.

[69] Dyke, S.J., Spencer, B.F., Jr, Sain, M.K. and Carlson, J.D. (1996) Modeling and control of magnetorheological dampers for seismic response reduction. *Smart Materials and Structures,* **5**, 565–575.

[70] Einstein, A. (1905) Über die von der molecular-kinetischen Theorie der Wärme geforderte Bewegung von in ruhenden Flüssigkeiten suspendierten Teilchen. *Annalen der Physik (Leipzig),* **17**, 549 (in German).

[71] Elishakoff, I. and Colajanni, P. (1997) Stochastic linearization critically re-examined. *Chaos, Solitons and Fractals*, **8** (12), 1957–1972.

[72] Engels, H. (1980) *Numerical Quadrature and Cubature*, Academic Press, London.

[73] Farlow, S.J. (1993) *Partial Differential Equations for Scientists and Engineers*, Dover Publications Inc.

[74] Fan, F.G. and Ahmadi, G. (1990) Nonstationary Kanai–Tajimi models for El Centro 1940 and Mexico City 1985 earthquakes. *Probabilistic Engineering Mechanics*, **5**, 171–181.

[75] Fang, K.T. and Wang, Y. (1994) *Number-Theoretic Methods in Statistics*, Chapman & Hall, London.

[76] Fang, T. and Wang, Z.N. (1986) Complex modal analysis of random vibrations. *AIAA Journal*, **24** (2), 342–344.

[77] Fang, T., Zhang, T.S. and Wang, Z.N. (1991) Complex modal analysis of nonstationary random response. *Proceedings of the 9th International Modal Analysis Conference (IMAC), Florence, Italy Apr. 15–18, Vol. 1 (A93-29227 10-39)*, pp. 12–18.

[78] Fang, T. and Sun, M. (1997) A unified approach to two types of evolutionary random response problems in engineering. *Archive of Applied Mechanics*, **67**, 496–506.

[79] Fang, T., Li, J.Q. and Sun, M.N. (2002) A universal solution for evolutionary random response problems. *Journal of Sound and Vibration*, **253** (4), 909–916.

[80] Feng, Q.M. and Hu, Y.X. (1981) Mathematical modeling of ground motions with spatial correlation. *Journal of Earthquake Engineering and Engineering Vibration*, **1** (2), 1–8 (in Chinese).

[81] Field, R.V. Jr and Bergman, L.A. (1998) Reliability-based approach to linear convariance control design. *Journal of Engineering Mechanics*, **124** (2), 193–199.

[82] Finkenstädt, B. and Rootzén, H. (2004) *Extreme Values in Finance, Telecommunications, and the Environment*, Chapman and Hall/CRC, Boca Raton, FL.

[83] Fokker, A.D. (1914) Die mittlere Energie rotierender elektrischer Dipole im Strahlungsfeld. *Annalen der Physik (Leipzig)*, **43**, 810–820.

[84] Foliente, G.C. (1995) Hysteresis modeling of wood joints and structural systems. *Journal of Structural Engineering*, **121** (6), 1013–1022.

[85] Friedman, J.H. and Wright, M.H. (1981) A nested partitioning procedure for numerical multiple integration. *ACM Transactions on Mathematical Software*, **7** (1), 76–92.

[86] Fudan University, (1979) *Probability Theory (1st volume: Probability Theory Fundamentals)*, People's Education Press (in Chinese).

[87] Fudan University, (1981) *Probability Theory (3rd volume: Random Processes)*, People's Education Press (in Chinese).

[88] Fung, Y.C. (1994) *A First Course in Continuum Mechanics*, Prentice-Hall, Inc.

[89] Fung, Y.C. and Tong, P. (2001) *Classical and Computational Solid Mechanics*, World Scientific, Singapore.

[90] Gamkrelidze, R.V. (1999) Discovery of the maximum principle. *Journal of Dynamical and Control Systems*, **5** (4), 437–451.

[91] Gardiner, C.W. (1983) *Handbook of Stochastic Methods for Physics, Chemistry and the Natural Sciences*, 2nd edn, Springer, Berlin.

[92] Genz, A. (1986) Fully symmetric interpolatory rules for multiple integrals. *SIAM Journal on Numerical Analysis*, **23** (6), 1273–1283.

[93] Ghanem, R.G. and Spanos, P.D. (1990) Polynomial chaos in stochastic finite elements. *Journal of Applied Mechanics*, **57**, 197–202.

[94] Ghanem, R. G. and Spanos, P.D. (1991) *Stochastic Finite Elements: A Spectral Approach*, Springer-Verlag, Berlin.

[95] Ghanem, R.G. and Spanos, P.D. (1991) Spectral stochastic finite-element formulation for reliability analysis. *Journal of Engineering Mechanics ASCE*, **117** (10), 2351–2372.

[96] Gibbs, J.W. (1981) *Elementary Principles in Statistical Mechanics*, 1902. (Reprinted by) Ox Bow Press.

[97] Gihman, I.I. and Skorohod, A.V. (1979) *The Theory of Stochastic Processes III*, Springer-Verlag, Berlin (Russian original edition published by Nauka, Moscow, 1975).

[98] Golub, G.H. and van Loan, C.F. (1996) *Matrix Computations*, 3rd edn, Johns Hopkins University Press.

[99] Godlewski, E. and Raviart, P.-A. (1996) *Numerical Approximation of Hyperbolic Systems of Conservation Laws*, Springer-Verlag, New York.

[100] Graff, K.F. (1975) *Wave Motion in Elastic Solids*, Oxford University Press, London.

[101] Haber, S. (1970) Numerical evaluation of multiple integrals. *SIAM Review*, **12** (4), 481–526.

[102] Haken, H. (2004) *Synergetics: Introduction and Advanced Topics*, Springer.

[103] Hales, T.C. (2006) Historic overview of the Kepler conjecture. *Discrete and Computational Geometry*, **36**, 5–20.

[104] Hales, T.C. and Ferguson, S.P. (2006) A formulation of the Kepler conjecture. *Discrete and Computational Geometry*, **36**, 21–69.

[105] Hart, G.C. and Collins, J.D. (1970) The treatment of randomness in finite element modeling. *SAE Shock and Vibrations Symposium, Society of Automotive Engineers, Los Angeles, CA*, pp. 2509–2519.

[106] Harten, A. (1983) High resolution schemes for hyperbolic conservation laws. *Journal of Computational Physics*, **49** (3), 357–393.

[107] Hasselmann, K., Barnett, T.P., Bouws, E. *et al.* (1973) Measurements of wind-wave growth and swell decay during the Joint North Sea Wave Project (JONSWAP). *Ergänzungsheft zurDeutsche Hydrographische Zeitschrift, Reihe A*, **8** (12), 1–95.

[108] Hasselman, T.K. and Hart, G.E. (1972) Model analysis of random structural systems. *Journal of the Engineering Mechanics Division, ASCE*, **98** (2), 561–586.

[109] He, T.X. (2001) *Dimensionality Reducing Expansion of Multivariate Integration*, Birkhäuser, Boston, MA.

[110] Hedstrom, G.W. (1975) Models of difference schemes for $u_t + u_x = 0$ by partial differential equations. *Mathematics of Computation*, **29** (132), 969–977.

[111] Hien, T.D. and Kleiber, M. (1990) Finite element analysis based on stochastic Hamilton variational principle. *Computers & Structures*, **37** (6), 893–902.

[112] Hisada, T. and Nakagiri, S. (1981) Stochastic finite element method developed for structural safety and reliability, in *Structural Safety and Reliability: Proceedings of ICOSSAR '81, the 3rd International Conference on Structural Safety and Reliability, Trondheim* (eds T. Moan and M. Shinozuka), Elsevier, Amsterdam, pp. 395–408.

[113] Hisada, T. and Nakagiri, S. (1982) Stochastic finite element analysis of uncertain structural systems. *Proceedings of the Fourth International Conference in Australia on Finite Element Methods, University of Melbourne, Australia, August 1982* (eds P.J. Hoadley and L.K. Stevens), p. 133.

[114] Hisada, T., Nakagiri, S. and Nagasaki, T. (1983) Stochastic finite element analysis of uncertain intrinsic stresses caused by structural misfits. *7th International Conference on Structural Mechanics in Reactor Technology, Chicago.*

[115] Hisada, T. and Nakagiri, S. (1985) Role of the stochastic finite element method in structural safety and reliability. *Structural safety and reliability: Proceedings of ICOSSAR 85, the 4th International Conference on Structural Safety and Reliability, Kobe, Japan* (eds I. Konishi, A.H-S. Ang and M. Shinozuka), International Association for Structural Safety and Reliability.

[116] Hodge, P.G. Jr (1950) On the method of characteristics. *The American Mathematical Monthly*, **57** (9), 621–623.

[117] Hotz, A. and Skelton, R.E. (1987) Covariance control theory. *International Journal of Control*, **46** (1), 13–32.

[118] Housner, G.W., Bergman, L.A., Caughey, T.K. *et al.* (1997) Structural control: past, present, and future. *Journal of Engineering Mechanics*, **123** (9), 897–971.

[119] Hsiang, W.-Y. (2002) *Least Action Principle of Crystal of Dense Packing Type and the Proof of Kepler's Conjecture*, World Scientific, River Edge, NJ.

[120] Hu, Y.X. and Zhou, X.Y. (1962) Seismic response of elastic system under stationary and stationarized ground motion, *Research Reports on Earthquake Engineering, Vol. I*, Science Press, Beijing (in Chinese)

[121] Hua, L.K. and Wang, Y. (1981) *Applications of Number Theory to Numerical Analysis*, Springer, Berlin.

[122] Hua, L.K. and Wang, Y. (1978) *Applications of Number Theory to Approximate Analysis*, Science Press, Beijing (In Chinese).

[123] Huang, N.E., Shen, Z., Long, S.R. *et al.* (1998) The empirical mode decomposition and the Hilbert spectrum for nonlinear and non-stationary time series analysis. *Proceedings of the Royal Society of London, Series A: Mathematical and Physical Sciences*, **454**, 903–995.

[124] Itô, K. (1942) Differential equations determining a Markoff process. *Zenkoku Sizyo Sugaku Danwaka-si*, No. 1077.

[125] Itô, K. (1944) Stochastic integral. *Proceedings of the Imperial Academy of Tokyo*, **20**, 519–524.

[126] Itô, K. (1957) *Essentials of Stochastic Processes*, Iwanami Shoten, Tokyo (reprinted in English by the American Mathematical Society, 2006).

[127] Itô, K. and McKean, H.P. Jr. (1965) *Diffusion Processes and Their Sample Paths*, Springer, Berlin.
[128] Iwan, W.D. and Jensen, H. (1993) On the dynamic response of continuous systems including model uncertainty. *Journal of Applied Mechanics*, **60**, 484–490.
[129] Jansen, L.M. and Dyke, S.J. (2000) Semiactive control strategies for MR dampers: comparative study. *Journal of Engineering Mechanics*, **126** (8), 795–803.
[130] Jennings, P.C., Housner, G.W. and Tsai, N.C. (1968) *Simulated earthquake motions. Report of Earthquake Engineering Research Laboratory*, California Institute of Technology.
[131] Jensen, H. and Iwan, W.D. (1991) Response variability in structural dynamics. *Earthquake Engineering and Structural Dynamics*, **20**, 949–959.
[132] Jensen, H. and Iwan, W.D. (1992) Response of systems with uncertain parameters to stochastic excitation. *Journal of Engineering Mechanics ASCE*, **118** (5), 1012–1025.
[133] Kaimal, J.C., Wyngaard, J.C., Izumi, Y. and Coté, O.R. (1972) Spectral characteristics of surface-layer turbulence. *Quarterly Journal of the Royal Meteorological Society*, **98**, 563–589.
[134] Kallenberg, O. (2002) *Foundations of Modern Probability*, 2nd edn, Springer-Verlag.
[135] Kanai, K. (1957) Semi-empirical formula for the seismic characteristics of the ground. *Bulletin of the Earthquake Research Institute, University of Tokyo*, **35**, 309–325.
[136] Kazakov, I.E. (1954) An approximate method for the statistical investigation of nonlinear systems. *Trudi Voenna-Vozdushnoi Inzheneroi Akademii imeni Professora N.E., Zhukovskogo*, **394**, 1–52 (in Russian).
[137] Kleiber, M. and Hien, T.D. (1992) *The Stochastic Finite Element Method: Basic Perturbation Technique and Computer Implementation*, John Wiley & Sons, Ltd.
[138] Klein, M. (1990) *Mathematical Thought from Ancient to Modern Times (New Edition)*, Oxford University Press.
[139] Kolmogorov, A.N. (1931) Über die analytischen Methoden in der Wahrscheinlichkeitsrechnung. *Mathematische Annalen*, **104**, 415–458.
[140] Kolmogorov, A.N. (1956) *Grundbegriffe der Wahrscheinlichkeitsrechnung. 1933. (Foundations of the Theory of Probability*, 2nd English edn, Chelsea Publishing Company, New York.
[141] Kolmogorov, A.N. and Fomin, S.V. (1999) *Elements of the Theory of Functions and Functional Analysis*, Dover Publications, Inc., New York.
[142] Korn, G.A. and Korn, T.M. (1968) *Mathematical Handbook for Scientists and Engineers*, McGraw-Hill, New York.
[143] Kozin, F. (1961) On the probability densities of the output of some random systems. *Journal of Applied Mechanics*, **28** (2), 161–164.
[144] Kubo, T. and Penzien, J. (1979) Analysis of three-dimensional strong ground motions along principal axes, San Fernando earthquake. *Earthquake Engineering and Structural Dynamics*, **7** (3), 265–278.
[145] Kubo, T. and Penzien, J. (1979) Simulation of three-dimensional strong ground motions along principal axes, San Fernando earthquake. *Earthquake Engineering and Structural Dynamics*, **7** (3), 279–294.
[146] Kulikovskii, A.G., Pogorelov, N.V. and Semenov, A.Y. (2001) *Mathematical Aspects of Numerical Solution of Hyperbolic Systems*, Chapman & Hall/CRC, Boca Raton, FL.
[147] Lanczos, C. (1970) *The Variational Principles of Mechanics*, 4th edn, University of Toronto Press.
[148] Langevin, P. (1908) Sur la theorie du mouvement brownien. *Comptes Rendus de l'Academie des Sciences Paris C*, **146**, 530–532.
[149] Lasota, A. and Mackey, M.C. (1994) *Chaos, Fractals, and Noise: Stochastic Aspects of Dynamics*, 2nd edn, Springer-Verlag, New York.
[150] Lawrence, M.A. (1987) Basic random variables in finite element analysis. *International Journal for Numerical Methods in Engineering*, **24**, 1849–1863.
[151] LeFloch, P.G. (2002) *Hyperbolic Systems of Conservation Laws*, Birkhäuser Verlag, Berlin.
[152] LeVeque, R.J. (1992) *Numerical Methods for Conservation Laws*, 2nd edn, Birkhäuser Verlag, Berlin.
[153] Li, C.-C. and Der Kiureghian, A. (1993) Optimal discretization of random fields. *Journal of Engineering Mechanics*, **119** (6), 1136–1154.
[154] Li, J. (1995a) Expanded system method of stochastic structural analysis: (I) expanded system equations. *Earthquake Engineering and Engineering Vibration*, **15** (3), 111–118 (in Chinese).
[155] Li, J. (1995b) Expanded system method of stochastic structural analysis: (II) structural dynamic analysis. *Earthquake Engineering and Engineering Vibration*, **15** (4), 27–35 (in Chinese).
[156] Li, J. (1995c) Linear representation and linear cut-off of stochastic dynamic matrix. *World Earthquake Engineering*, **10** (2), 8–12 (in Chinese).

[157] Li, J. (1996) *Stochastic Structural Systems: Analysis and Modeling*, Science Press, Beijing (in Chinese).
[158] Li, J. (1996) Expanded system method for compound random vibration. *Acta Mechanica Sinica*, **28** (1), 66–75 (in Chinese).
[159] Li, J. (2006) A physical approach to stochastic dynamical systems. *Science Paper Online*, **1** (2), 95–104 (in Chinese).
[160] Li, J. and Ai, X.Q. (2006) Study on random model of earthquake ground motion based on physical process. *Earthquake Engineering and Engineering Vibration*, **26** (5), 21–26 (in Chinese).
[161] Li, J. and Ai, X.Q. (2007) A new random model of earthquake motion considering stochastic physical process. *Proceedings of the International Symposium on Innovation & Sustainability of Structures in Civil Engineering, Shanghai, China, Nov. 28–30*, pp. 1198–1204.
[162] Li, J., Chen, H.M. and Chen, J.B. (2007) Studies on seismic performances of the prestressed egg-shaped digester with shaking table test. *Engineering Structures*, **29** (4), 552–566.
[163] Li, J. and Chen, J.B. (2003) Probability density evolution method for dynamic response analysis of stochastic structures. *Proceeding of the Fifth International Conference on Stochastic Structural Dynamics, Hangzhou, China*, pp. 309–316.
[164] Li, J. and Chen, J.B. (2004a) Probability density evolution method for dynamic response analysis of structures with uncertain parameters. *Computational Mechanics*, **34**, 400–409.
[165] Li, J. and Chen, J.B. (2004b) Seismic dynamics response analysis of stochastic structures exhibiting nonlinearity. *Proceedings of the Third International Conference on Earthquake Engineering (3ICEE): New Frontier and Research Transformation, Nanjing, China, 19–20 October*, pp. 270–274.
[166] Li, J. and Chen, J.B. (2005a) Dynamic response and reliability analysis of structures with uncertain parameters. *International Journal for Numerical Methods in Engineering*, **62** (2), 289–315.
[167] Li, J. and Chen, J.B. (2005b) Stochastic seismic response analysis and reliability evaluation of hysteretic structures. Invited paper. *Proceedings of the International Symposium on Innovation & Sustainability of Structures in Civil Engineering, Nanjing, China, Nov. 20–22*, pp. 218–230.
[168] Li, J. and Chen, J.B. (2006a) The probability density evolution method for dynamic response analysis of non-linear stochastic structures. *International Journal for Numerical Methods in Engineering*, **65**, 882–903.
[169] Li, J. and Chen, J.B. (2006b) The dimension-reduction strategy via mapping for the probability density evolution analysis of nonlinear stochastic systems. *Probabilistic Engineering Mechanics*, **21** (4), 442–453.
[170] Li, J. and Chen, J.B. (2006c) Generalized density evolution equation in stochastic systems. *Progress in Natural Science*, **16** (6), 712–719 (in Chinese).
[171] Li, J. and Chen, J.B. (2007a) The number theoretical method in response analysis of nonlinear stochastic structures. *Computational Mechanics*, **39** (6), 693–708.
[172] Li, J. and Chen, J.B. (2007b) Probability density evolution equations: a historical investigation. *Proceedings of the International Symposium on Advances in Urban Safety, Nanjing, China.*
[173] Li, J. and Chen, J.B. (2008) The principle of preservation of probability and the generalized density evolution equation. *Structural Safety*, **30**, 65–77.
[174] Li, J., Chen, J.B. and Fan, W.L. (2007) The equivalent extreme-value event and evaluation of the structural system reliability. *Structural Safety*, **29**, 112–131.
[175] Li, J., Chen, J.B. and Peng, Y.B. (2008) PDEM-based semi-active control of stochastic systems. *Proceedings of Earth and Space 2008 (CD-ROM), Long Beach, CA, USA, March 3–5.*
[176] Li, J., Chen, J.B. and Zhang, L.L. (2007) PDEM-based modeling of random variable and stochastic process, in *Proceedings of the Second International Conference on Structural Condition Assessment, Monitoring and Improvement, Changsha, China*, vol. **II**, Science Press, Beijing, pp. 1463–1468.
[177] Li, J. and Li, G.Q. (1992) *Introduction to Earthquake Engineering*, Earthquake Press, Beijing (in Chinese).
[178] Li, J. and Liao, S.T. (2001) Response analysis of stochastic parameter structures under non-stationary random excitation. *Computational Mechanics*, **27** (1), 61–68.
[179] Li, J. and Liu, Z.J. (2006) Decomposition of stochastic processes based on standard orthogonal basis. *Journal of Tongji University*, **34** (10), 1279–1283 (in Chinese).
[180] Li, J. and Liu, Z.J. (2008) Orthogonal expansion method of random fields of wind velocity fluctuations. *China Civil Engineering Journal*, **41** (2), 49–53 (in Chinese).
[181] Li, J. and Wei, X. (1995) Dynamic reliability study of stochastic structures based on the expanded order system method. *Second China–Japan Symposium on Structural Optimization and Mechanical Systems, Beijing*, pp. 116–121.

[182] Li, J. and Wei, X. (1996) Recursive condensation method for dynamic analysis of stochastic structures. *Acta Mechanica Solida Sinica*, **17** (3), 263–267 (in Chinese).
[183] Li, J. and Zhang, L.L. (2007) Random Fourier spectrum of the measured wind field. *Journal of Vibration Engineering*, **20** (1), 66–72 (in Chinese).
[184] Li, Z. and Katubura, H. (1990) Markovian hysteretic characteristic of structures. *Journal of Engineering Mechanics ASCE*, **116** (8), 1798–1811.
[185] Lin, J.H. (1985) Deterministic algorithm for analysis of stochastic response. *Earthquake Engineering and Engineering Vibration*, **5** (1), 89–93 (in Chinese).
[186] Lin, J.H., Zhang, W.S. and Williams, F.W. (1994a) Pseudo-excitation algorithm for nonstationary random seismic responses. *Engineering Structures*, **16**, 270–276.
[187] Lin, J.H., Zhang, W.S. and Li, J.J. (1994b) Structural responses to arbitrarily coherent stationary random excitations. *Computers & Structures*, **50**, 629–633.
[188] Lin, J.H., Li, J.J., Zhang, W.S. and Williams, F.W. (1997) Non-stationary random seismic responses of multi-support structures in evolutionary inhomogeneous random fields. *Earthquake Engineering and Structural Dynamics*, **26**, 135–145.
[189] Lin, Y.K. (1967) *Probabilistic Theory of Structural Dynamics*, McGraw-Hill Book Company, New York.
[190] Lin, Y.K. and Cai, G.Q. (1995) *Probabilistic Structural Dynamics: Advanced Theory and Applications*, McGraw-Hill, New York.
[191] Liu, J.S. (2001) *Monte Carlo Strategies in Scientific Computing*, Springer-Verlag.
[192] Liu, P.L. and Liu, K.G. (1993) Selection of random field mesh in finite element reliability analysis. *Journal of Engineering Mechanics ASCE*, **119** (4), 667–680.
[193] Liu, S.C. (1970) Evolutionary power spectral density of strong-motion earthquakes. *Bulletin of the Seismological Society of America*, **60** (3), 891–900.
[194] Liu, W.K., Belytschko, T. and Mani, A. (1985) A computational method for the determination of the probabilistic distribution of the dynamic response of structures. *Computer-Aided Engineering ASME*, **98** (5), 243–248.
[195] Liu, W.K., Belytschko, T. and Mani, A. (1986a) Probabilistic finite element methods for nonlinear structural dynamics. *Computer Methods in Applied Mechanics and Engineering*, **56**, 61–81.
[196] Liu, W.K., Belytschko, T. and Mani, A. (1986b) Random fields finite elements. *International Journal for Numerical Methods in Engineering*, **23**, 1831–1845.
[197] Liu, W.K., Belytschko, T. and Mani, A. (1987) Applications of probabilistic finite element methods in elastic/plastic dynamics. *Journal of Engineering for Industry–Transactions of the ASME*, **109** (1), 2–8.
[198] Liu, W.K., Bestefield, G. and Belytschko, T. (1988a) Variational approach to probabilistic finite elements. *Journal of Engineering Mechanics ASCE*, **114** (12), 2115–2133.
[199] Liu, W.K., Bestefield, G. and Belytschko, T. (1988b) Transient probabilistic systems. *Computer Methods in Applied Mechanics and Engineering*, **67**, 27–54.
[200] Liu, W.K., Belytschko, T. and Chen, J.S. (1988) Nonlinear versions of flexurally superconvergent elements. *Computer Methods in Applied Mechanics and Engineering*, **68** (3), 259–310.
[201] Liu, W.K., Ong, J.S. and Uras, R.A. (1985) Finite element stabilization matrices: a unification approach. *Computer Methods in Applied Mechanics and Engineering*, **53**, 13–46.
[202] Liu, Z.J. and Li, J. (2006) The orthogonal expansion of stochastic processes for earthquake ground motion based on the Clough–Penzien power spectrum. *Proceedings of the Ninth International Symposium on Structural Engineering for Young Experts, Fuzhou, China*, pp. 2158–2164.
[203] Liu, Z.J. and Li, J. (2008) Orthogonal expansion of stochastic processes for wind velocity. *Journal of Vibration Engineering*, **21** (1), 96–101 (in Chinese).
[204] Loève, M. (1948) *Fonctions Aleatoires du Second Ordre. Supplement au livre P. Levy. Processus Stochastic et Mouvement Brownien*, Gauthier Villars, Paris.
[205] Loève, M. (1977) *Probability Theory*, Springer-Verlag, Berlin.
[206] Loh, C.H. (1991) Spatial variability of seismic waves and its engineering application. *Structural Safety*, **10** (1–3), 95–111.
[207] Loh, C.H. and Yeh, Y.T. (1988) Spatial variation and stochastic modeling of seismic differential ground movement. *Earthquake Engineering & Structural Dynamics*, **16** (5), 583–596.
[208] Longuet-Higgins, M.S., Cartwright, D.E. and Smith, N.D. (1963) Observation of the directional spectrum of sea waves using the motion of a floating buoy. *Proceedings of the Conference on Ocean Wave Spectra*, Prentice Hall, Englewood Cliffs, NJ, pp. 111–136.

[209] Lumley, J.L. and Panofsky, H.A. (1964) *The Structure of Atmospheric Turbulence*, Interscience Publishers, New York.

[210] Lutes, L.D. and Sarkani, S. (2004) *Random Vibrations: Analysis of Structural and Mechanical Systems*, Elsevier, Amsterdam.

[211] Lützen, J. (1990) *Joseph Liouville 1809–1882: Master of Pure and Applied Mathematics*, Springer-Verlag, New York.

[212] Ma, F., Zhang, H., Bockstedte, A. *et al.* (2004) Parameter analysis of the differential model of hysteresis. *Journal of Applied Mechanics*, **71**, 342–349.

[213] Madsen, H.O., Krenk, S. and Lind, N.C. (1986) *Methods of Structural Safety*, Prentice-Hall.

[214] Madsen, H.O. and Tvedt, L. (1990) Methods for time-dependent reliability and sensitivity analysis. *Journal of Engineering Mechanics ASCE*, **116** (10), 2118–2135.

[215] Manin, Y.I. and Panchishkin, (2005) *Introduction to Modern Number Theory*, 2nd edn, Springer-Verlag, Berlin.

[216] Martinet, J. (2003) *Perfect Lattices in Euclidean Spaces*, Springer-Verlag, Berlin.

[217] McKay, M.D., Beckman, R.J. and Conover, W.J. (1979) A comparison of three methods for selecting values of input variables in the analysis of output from a computer code. *Technometrics*, **21** (2), 239–245.

[218] McShane, E.J. (1989) The calculus of variations from the beginning through optimal control theory. *SIAM Journal of Control and Optimization*, **27** (5), 916–939.

[219] Melchers, R.E. (1999) *Structural Reliability Analysis and Prediction*, 2nd edn, John Wiley & Sons, Ltd, Chichester.

[220] Mitchell, A.R. and Griffiths, D.F. (1980) *The Finite Difference Method in Partial Differential Equations*, John Wiley & Sons, Ltd, Chichester.

[221] Mitsuyasu, H., Tasai, F., Suhara, T. *et al.* (1975) Observations of the directional spectrum of ocean waves using a cloverleaf buoy. *Journal of Physical Oceanography*, **5**, 750–760.

[222] Morison, J.R., O'Brien, M.P., Johnson, J.W. and Schaef, S.A. (1950) The force exerted by surface waves on piles. *Transactions of the American Institute of Mining and Metallurgical, Engineering*, **189**, 149–154.

[223] Moyal, J.E. (1949) Stochastic processes and statistical physics. *Journal of the Royal Statistical Society, Series B (Methodological)*, **11** (2), 150–210.

[224] Naess, A. and Johnsen, J.M. (1993) Response statistics of nonlinear, compliant offshore structures by the path integral solution method. *Probabilistic Engineering Mechanics*, **8** (2), 91–106.

[225] Naess, A. and Moe, V. (2000) Efficient path integral methods for nonlinear dynamics systems. *Probabilistic Engineering Mechanics*, **15** (2), 221–231.

[226] Naidu, D.S. (2003) *Optimal Control Systems*, CRC Press, Boca Raton, FL.

[227] Nakagiri, S. and Hisada, T. (1981) Finite element stress analysis extended to stochastic treatment in problems of structural safety and reliability, in *Transactions of the 6th International Conference on Structural Mechanics in Reactor Technology, Palais des Congrès, Paris, 17–21 August*, North-Holland, Amsterdam.

[228] Nakagiri, S. and Hisada, T. (1982) Stochastic finite element method applied to structural analysis with uncertain parameters. *Proceedings of the Fourth International Conference in Australia on Finite Element Methods, University of Melbourne, Australia, August* (eds P.J. Hoadley and L.K. Stevens), pp. 206–211.

[229] Nakagiri, S., Hisada, T. and Nagasaki, T. (1984) Stochastic stress analysis of assembled structure. *ASEM, 5th International Conference on Pressure Vessel Technology, Vol. 1, Design and Analysis*, p. 197.

[230] Nakagiri, S., Hisada, T. and Toshimitsu, K. (1984) Stochastic time history analysis of structural vibration with uncertain damping. *ASME, Probabilistic Structural Analysis*, **PVP93**, 109.

[231] Nayfeh, A.H. (2000) *Perturbation Methods*, Wiley-Interscience.

[232] Nayfeh, A.H. and Mook, D.T. (1995) *Nonlinear Oscillations*, Wiley-Interscience.

[233] Neumann, G. (1952) *On Wind Generated Ocean Waves with Special Reference to the Problem of Wave Forecasting*. New York University, College of Engineering, Department of Meteorology, 136 pp.

[234] Newland, D.E. (1993) *An Introduction to Random Vibration, Spectral Analysis & Wavelet Analysis*, 3rd edn, Longman, London.

[235] Nie, J.S. and Ellingwood, B.R. (2004) A new directional simulation method for system reliability. Part I: application of deterministic point sets. *Probabilistic Engineering Mechanics*, **19**, 425–436.

[236] Niederreiter, H. (1992) *Random Number Generation and Quasi-Monte Carlo Methods*, Capital City Press, Montpelier.

[237] Nise, N.S. (2000) *Control Systems Engineering*, 3rd edn, John Wiley & Sons, Inc., New York.

[238] Øksendal, B. (2005) *Stochastic Differential Equations: An Introduction with Applications*, 6th edn, Springer-Verlag, Berlin.

[239] Oliveira, C.S., Hao, H. and Penzien, J. (1991) Ground motion modeling for multiple-input structural analysis. *Structural Safety*, **10**, 79–93.
[240] Peng, Y.B. and Li, J. (2007) Probability density evolution method and pseudo excitation method for random seismic response analysis. *Proceedings of the International Symposium on Innovation & Sustainability of Structures in Civil Engineering, Shanghai, China, Nov. 28–30*, pp. 462–471.
[241] Penzien, J. and Watabe, M. (1975) Characteristics of 3-dimensional earthquake ground motions. *Earthquake Engineering and Structural Dynamics*, **3** (4), 365–373.
[242] Pesch, H.J. and Bulirsch, R. (1994) The maximum principle, Bellman's equation, and Carathéodory's work. *Journal of Optimization Theory and Applications*, **80** (2), 199–225.
[243] Petkov, PHr. Christov, N.D. and Konstantinov, M.M. (1991) *Computational Methods for Linear Control Systems*, Prentice Hall, New York.
[244] Petrovsky, I.G. (1954) *Lectures on Partial Differential Equations*, Interscience Publishers, Inc., New York.
[245] Pierson, W.J. and Moskowitz, L. (1964) A proposed spectral form for fully developed wind seas based on the similarity theory of S.A. Kitaigorodskii. *Journal of Geophysical Research*, **69**, 5181–5190.
[246] Planck, M. (1917) Uber einen Satz der statistichen Dynamik und eine Erweiterung in der Quantumtheorie. *Sitzungberichte der Preussischen Akadademie der Wissenschaften*, 324–341.
[247] Pontryagin, L.S., Boltyanskii, V.G., Gamkrelidze, R.V. and Mishenko, E.F. (1964) *The Mathematical Theory of Optimal Processes* (translated by D.E. Brown), Macmillan, New York.
[248] Priestley, M.B. (1965) Evolutionary spectra and non-stationary processes. *Journal of the Royal Statistical Society, Series B (Methodological)*, **27** (2), 204–237.
[249] Priestley, M.B. (1967) Power spectral analysis of non-stationary random processes. *Journal of Sound and Vibration*, **6** (1), 86–97.
[250] Proppe, C., Pradlwarte, H.J. and Schuëller, G.I. (2003) Equivalent linearization and Monte Carlo simulation in stochastic dynamics. *Probabilistic Engineering Mechanics*, **18**, 1–15.
[251] Qu, T.J., Wang, J.J. and Wang, Q.X. (1996) An applied model of seismic power spectrum considering spatial variation. *Acta Seismologica Sinica*, **18** (1), 55–62 (in Chinese).
[252] Rice, S.O. (1944) Mathematical analysis of random noise. *Bell System Technical Journal*, **23**, 282–332.
[253] Risken, H. (1984) *The Fokker–Planck Equation: Methods of Solution and Applications*, Springer-Verlag, Berlin.
[254] Roberts, J.B. and Spanos, P.D. (1990) *Random Vibration and Statistical Linearization*, John Wiley & Sons, Ltd, Chichester.
[255] Robinstein, R.Y. (1981) *Simulation and the Monte Carlo Method*, John Wiley & Sons, Inc., New York.
[256] Roe, P.L. (1986) Characteristic-based schemes for the Euler equations. *Annual Reviews on Fluid Mechanics*, **18**, 337–365.
[257] Ruiz, P. and Penzien, J. (1969) Probabilistic study of the behavior of structures during earthquakes, Report No. EERC 69-03, *Earthquake Engineering Research Center, University of California, Berkeley, CA*.
[258] Rye, H. (1974) Wave group formation among storm waves. *Proceedings of the 14th Coastal Engineering Conference, Copenhagen*, pp. 164–183.
[259] Sarra, S.A. (2003) The method of characteristics & conservation laws. *Journal of Online Mathematics and its Applications*, http://mathdl.maa.org/mathDL/4/?pa=content&sa=viewDocument&nodeId=389 (accessed 3 February 2008).
[260] Schenk, C.A. and Schuëller, G.I. (2005) *Uncertainty Assessment of Large Finite Element Systems*, Springer, Berlin.
[261] Schuëller, G.I. (ed.) (1997) A state-of-the-art report on computational stochastic mechanics. *Probabilistic Engineering Mechanics*, **12** (4), 197–321.
[262] Schuëller, G.I. (2001) Computational stochastic mechanics – recent advances. *Computers and Structures*, **79**, 2225–2234.
[263] Scott May, B. and Beck, J.L. (1998) Probabilistic control for the active mass driver benchmark structural model. *Earthquake Engineering and Structural Dynamics*, **27**, 1331–1346.
[264] Scruggs, J.T., Taflanidis, A.A. and Beck, J.L. (2006) Reliability-based control optimization for active isolation systems. *Structural Control and Health Monitoring*, **13**, 705–723.
[265] Shinozuka, M. and Jan, C.M. (1972) Digital simulation of random processes and its applications. *Journal of Sound and Vibration*, **25**, 111–128.
[266] Shinozuka, M. (1972) Monte Carlo solution of structural dynamics. *International Journal of Computers and Structures*, **2**, 855–874.
[267] Shinozuka, M. and Wen, Y.K. (1972) Monte Carlo solution of nonlinear vibrations. *AIAA Journal*, **10** (1), 37–40.

[268] Shinozuka, M. (1974) Digital simulation of random processes in engineering mechanics with the aid of the FFT technique, in *Stochastic Problems in Mechanics*, University of Waterloo Press, Waterloo, Canada, pp. 277–286.
[269] Shinozuka, M. and Lenoe, E. (1976) A probabilistic model for spatial distribution of material properties. *Engineering Fracture Mechanics*, **8** (1), 217–227.
[270] Shinozuka, M. and Deodatis, G. (1988) Response variability of stochastic finite element systems. *Journal of Engineering Mechanics, ASCE*, **116** (3), 499–519.
[271] Shinozuka, M. and Deodatis, G. (1991) Simulation of stochastic processes by spectral representation. *Applied Mechanics Reviews*, **44** (4), 191–203.
[272] Shinozuka, M. and Deodatis, G. (1996) Simulation of multi-dimensional Gaussian stochastic fields by spectral representation. *Applied Mechanics Reviews*, **49** (1), 29–53.
[273] Shu, C.W. (1988) Total-variation-diminishing time discretization. *SIAM Journal on Scientific and Statistical Computating*, **9** (6), 1073–1084.
[274] Simiu, E. (1974) Wind spectra and dynamic along wind response. *Journal of the Structural Division–ASCE*, **100** (ST9), 1897–1910.
[275] Simiu, E. and Scanlan, R.H. (1996) *Wind Effects on Structures: Fundamentals and Applications to Design*, John Wiley & Sons, Inc.
[276] Sims, N.D., Stanway, R. and Johnson, A.R. (1999) Vibration control using smart fluids: a state-of-the-art review. *Shock and Vibration Digest*, **31** (3), 195–205.
[277] Skorokhod, A.V., Hoppensteadt, F.C. and Salehi, H. (2002) *Random Perturbation Methods with Applications in Science and Engineering*, Springer-Verlag, New York.
[278] Sloan, I.H. (1985) Lattice methods for multiple integration. *Journal of Computational and Applied Mathematics*, **12–13**, 131–143.
[279] Sloan, I.H. and Kachoyan, P.J. (1987) Lattice methods for multiple integration: theory, error analysis and examples. *SIAM Journal of Numerical Analysis*, **24** (1), 116–128.
[280] Smith, G.D. (1985) *Numerical Solution of Partial Differential Equations: Finite Difference Methods*, 3rd edn, Clarendon Press, Oxford.
[281] Sobol', I.M. (1998) On quasi-Monte Carlo integrations. *Mathematics and Computers in Simulation*, **47**, 103–112.
[282] Sobolev, S.L. (1964) *Partial Differential Equations of Mathematical Physics*, Pergamon Press.
[283] Sobolev, S.L. and Vaskevich, (1997) *The Theory of Cubature Formulus*, Kluwer Acadamic Publishers, Dordrecht, Netherlands.
[284] Soong, T.T. (1973) *Random Differential Equations in Science and Engineering*, Academic Press, New York.
[285] Soong, T.T. (1990) *Active Structural Control: Theory and Practice*, John Wiley & Sons, Inc., New York.
[286] Soong, T.T. and Bogdanoff, J. (1963) On the natural frequencies of a disordered linear chain of n degree of freedom. *International Journal of Mechanical Sciences*, **5**, 237–265.
[287] Spanos, P.D. and Ghanem, R.G. (1989) Stochastic finite element expansion for random media. *Journal of Engineering Mechanics, ASCE*, **115** (5), 1035–1053.
[288] Spencer, B.F. Jr and Bergman, L.A. (1993) On the numerical solution on the Fokker-Planck equation for nonlinear stochastic systems. *Nonlinear Dynamics*, **4**, 357–372.
[289] Spencer, B.F., Jr., Dyke, S.J., Sain, M.K. and Carlson, J.D. (1997) Phenomenological model for magnetorheological dampers. *Journal of Engineering Mechanics*, **123** (3), 230–238.
[290] Spencer, B.F., Jr., Kaspari, D.C. Jr and Sain, M.K. (1994) Structural control design: a reliability-based approach. *Proceedings of the American Control Conference, Baltimore, Maryland*, pp. 1062–1066
[291] Stengel, R.F. (1994) *Optimal Control and Estimation*, Dover Publications, Inc., New York.
[292] Stratonovich, R.L. (1963) *Introduction to the Theory of Random Noise*, Gordon and Breach, New York.
[293] Stricwerda, J.C. (1989) *Finite Difference Schemes and Partial Differential Equations*, Wadsworth & Brooks, Belmont, CA.
[294] Strogatz, S.H. (1994) *Nonlinear Dynamics and Chaos with Applications to Physics, Biology, Chemistry and Engineering*, Westview Press.
[295] Su, Y.C. and Wu, Q.G. (1992) *Introduction to Numerical Methods of Singular Perturbation Theory*, Chongqing Press (in Chinese).
[296] Sweby, P.K. (1984) High resolution schemes using flux limiters for hyperbolic conservation laws. *SIAM Journal of Numerical Analysis*, **21** (5), 995–1011.
[297] Syski, R. (1967) Stochastic differential equations, in *Modern Nonlinear Equations* (ed. T.L. Saaty), McGraw-Hill, New York, Chapter 8.

[298] Sun, J.Q. (2006) *Stochastic Dynamics and Control,* Elsevier, Amsterdam.

[299] Sun, T.C. (1979) A finite element method for random differential equations with random coefficients. *SIAM Journal on Numerical Analysis,* **16** (6), 1019–1035.

[300] Tajimi, H. (1960) A statistical method of determining the maximum response of a building structure during an earthquake. *Proceedings of the Second World Conference on Earthquake Engineering, Tokyo, Japan, July,* Vol. **II**.

[301] Takada, T. (1989) Weighted integral method in multi-dimensional stochastic finite element analysis. *Probabilistic Engineering Mechanics,* **5** (4), 158–166.

[302] Takada, T. (1990) Weighted integral method in stochastic finite element analysis. *Probabilistic Engineering Mechanics,* **5** (3), 146–156.

[303] Trefethen, L.N. (1982) Group velocity in finite difference schemes. *SIAM Review,* **24** (2), 113–136.

[304] Uhlenbeck, G.E. and Ornstein, L.S. (1930) On the theory of Brownion motion. *Physical Review,* **36** (5), 823–841.

[305] Van Leer, B. (1974) Towards the ultimate conservative difference scheme. II. Monotonicity and conservation combined in second-order scheme. *Journal of Computational Physics,* **14** (4), 361–370.

[306] Vanmarcke, E.H. (1972) Properties of spectral moments with applications to random vibration. *Journal of the Engineering Mechanics Division ASCE,* **98**, 425–446.

[307] Vanmarcke, E.H. (1975) On the distribution of the first-passage time for normal stationary random process. *Journal of Applied Mechanics, ASME,* **42**, 215–220.

[308] Vanmarcke, E.H. (1983) *Random Fields: Analysis and Synthesis,* MIT Press, Cambridge, MA.

[309] Vanmarcke, E.H. *et al.* (1986) Random fields and stochastic finite elements. *Structural Safety,* **3**, 143–166.

[310] Vinter, R.B. (2000) *Optimal Control,* Birkhäuser, Boston, MA.

[311] Wang, G.Y. (1981) *Applied Analytical Dynamics,* People's Education Press (in Chinese).

[312] Wang, Z.K. (1976) *Fundamentals of Probability Theory and Applications,* Science Press (in Chinese).

[313] Warming, R.F. and Hyett, B.J. (1974) The modified equation approach to the stability and accuracy analysis of finite-difference methods. *Journal of Computational Physics,* **14**, 159–179.

[314] Washizu, K. (1975) *Variational Method in Elasticity and Plasticity,* 2nd edn, Pergamon Press.

[315] Wehner, W.F. and Wolf, W.G. (1983) Numerical evaluation of path-integral solutions to the Fokker–Planck equations. *Physical Review A,* **27** (5), 2663–2670.

[316] Wen, S.C. and Yu, Z.W. (1985) *Ocean Wave Theory and Computation Principle,* Science Press, Beijing (in Chinese).

[317] Wen, S., Zhang, D., Sun, S. *et al.* (1994a) Form of deep-water wind-wave frequency spectrum: I. Derivation of spectrum. *Progress in Natural Science,* **4** (4), 407–427.

[318] Wen, S., Zhang, D., Sun, S. *et al.* (1994b) Form of deep-water wind-wave frequency spectrum: II. Comparison with existing spectra and observations. *Progress in Natural Science,* **4** (5), 586–596.

[319] Wen, S.C., Wu, K.J., Guan, C.L. *et al.* (1995) A proposed directional function and wind-wave directional spectrum. *Acta Oceanologica Sinica,* **14** (2), 155–166.

[320] Wen, Y.K. (1976) Method for random vibration of hysterestic systems. *Journal of the Engineering Mechanics Division,* **102** (2), 249–263.

[321] Wesseling, P. (2001) *Principles of Computational Fluid Dynamics,* Springer-Verlag, Berlin.

[322] Wiener, N. (1923) Differential space. *Journal of Mathematical Physics,* **58**, 131–174.

[323] Williams, R.L. II and Lawrence, D.A. (2007) *Linear State-Space Control Systems,* John Wiley & Sons, Inc., New Jersey.

[324] Wong, E. and Zakai, M. (1965) On the relation between ordinary and stochastic differential equations. *International Journal of Engineering Science,* **3** (2), 213–229.

[325] Xu, Y. (1998) Orthogonal polynomials and cubature formulae on spheres and on balls. *SIAM Journal of Numerical Analysis,* **29** (3), 779–793.

[326] Yamazaki, F., Shinozuka, M. and Dasgupta, G. (1985) *Neumann expansion for stochastic finite element analysis – technical report,* Department of Civil Engineering and EngineeringMechanics, Columbia University, New York.

[327] Yamazaki, F., Shinozuka, M. and Dasgupta, G. (1988) Neumann expansion for stochastic finite element analysis. *Journal of Engineering Mechanics-ASCE,* **114** (8), 1335–1345.

[328] Yang, J.N. (1975) Application of optimal control theory to civil engineering structures. *Journal of the Engineering Mechanics Division,* **101** (EM6), 819–838.

[329] Yang, J.N., Akbarpour, A. and Ghaemmaghami, P. (1987) New optimal control algorithms for structural control. *Journal of Engineering Mechanics,* **113** (9), 1369–1386.

[330] Yao, J.T.P. (1972) Concept of structural control. *Journal of the Structural Division, ASCE*, **98** (ST7), 1567–1574.

[331] Yong, J.M. and Zhou, X.Y. (1999) *Stochastic Controls: Hamiltonian Systems and HJB Equations*, Springer, New York.

[332] Yu, Y.X. and Liu, S.X. (1994) Observations and estimations of directional spectra of SFA waves. *The Ocean Engineering*, **12** (2), 1–11 (in Chinese).

[333] Yuen, K.V. and Beck, J.L. (2003) Reliability-based robust control for uncertain dynamical systems using feedback of incomplete noisy response measurements. *Earthquake Engineering and Structural Dynamics*, **32**, 751–770.

[334] Zayed, A.I. (1996) *Handbook of Function and Generalized Function Transformations*, CRC Press.

[335] Zemanian, A.H. (1965) *Distribution Theory and Transform Analysis*, McGraw-Hill Book Company, New York.

[336] Zhang, L.L. and Li, J. (2006) Cross stochastic Fourier spectrum function of turbulence wind velocity. *Journal of Architecture and Civil Engineering*, **23** (2), 57–61 (in Chinese).

[337] Zhang, L.L., Li, J. and Peng, Y.B. (2008) Dynamic response and reliability analysis of tall buildings subject to wind-loading. *Journal of Wind Engineering and Industrial Aerodynamics*, **96**, 25–40.

[338] Zhang, W.S. and Xu, Y.L. (2001) Closed form solution for along wind response of actively controlled tall buildings with LQG controllers. *Journal of Wind Engineering and Industrial Aerodynamics*, **89**, 785–807.

[339] Zhang, X.T. (1985) *Structural Analysis of Wind Pressure and Vibration*, Tongji University Press, Shanghai (in Chinese).

[340] Zhao, Y.G. and Ang, A.H.-S. (2003) System reliability assessment by method of moments. *Journal of Structural Engineering*, **129** (10), 1341–1349.

[341] Zhong, W.X. (2004) *Duality System in Applied Mechanics and Optimal Control*, Kluwer Academic Publishers, Boston.

[342] Zhou, X.Y. (1990) Maximum principle, dynamic programming, and their connection in deterministic control. *Journal of Optimization and Applications*, **65** (2), 363–373.

[343] Zhu, W.Q. (1992) *Random Vibration*, Science Press, Beijing (in Chinese).

[344] Zhu, W.Q. (2003) *Nonlinear Stochastic Dynamics and Control in Hamiltonian Formulation*, Science Press, Beijing (in Chinese).

[345] Zhu, W.Q. (2006) Nonlinear stochastic dynamics and control in Hamiltonian formulation. *Applied Mechanics Reviews*, **59**, 230–248.

[346] Zhu, W.Q. and Wu, W.Q. (1990) On the local overage of random field in stochastic finite elements analysis. *Acta Mechanica Solida China*, **3** (1), 27–42.

[347] Zhu, W.Q., Ren, Y.J. and Wu, W.Q. (1992) Stochastic FEM based on local averages of random fields. *Journal of Engineering Mechanics ASCE*, **118** (3), 496–511.

[348] Zienkiewicz, O.C. and Taylor, R.L. (2004) *The Finite Element Method*, 5th edn, Elsevier (Singapore) Pte Ltd.

[349] Zong, C.M. (1999) *Sphere Packings*, Springer-Verlag, New York.

Index

σ–algebra, 7
absorbing boundary condition, 295, 296
acceptance-rejection sampling, 94–95
adjoint equation, 317–319
adjoint vector, 317–319, 324, 331
affine transformation, 272, 273
amplification factor, 243–245
apparent wave velocity, 51
assigned probability, 226, 256, 260
asymptotic sequence, 100, 106
attracting domain, 202
auto correlation function, 20–22, 25, 26, 30, 138, 139, 142
auto-covariance function, 21, 30
average wind speed, 57, 68

base vector, 268, 270, 361
Bouc–Wen model, 170, 278, 279
boundary layer, 44, 56–64
Brownian motion, 2–4, 173, 183, 191, 205

cell-mapping method, 190
central moment, 10, 11, 20
Chapman–Kolmogorov equation, 183, 184, 199, 207, 209
characteristic curves, 201, 221–223, 235
characteristic function, 11, 19
coherency function, 51–53, 64
complex conjugate, 22, 38, 148, 150, 152, 159, 161–163, 355
conditional PDF, 12, 13
continuous random variable, 8, 10, 11, 93, 192, 346
contour, 2, 203, 275, 281, 283, 296
control criterion, 340
correlation coefficient, matrix, 15, 20, 34, 97
Courant–Friedrichs–Lewy condition, 236
covariance, 14–17, 20, 21, 23, 24, 27, 28, 30, 31, 34, 35, 36, 45, 73, 95, 109, 110
  control, 325
covering, 251, 253–257, 262, 266, 267, 269, 270
  radius, 254–257, 269
cross-correlation function, 20, 21, 23
cross-covariance function, 21
cross-power spectral density function, 23, 24, 151
cumulative distribution function (CDF), 8–10, 12, 93, 94, 192, 259

Davenport spectrum, 58, 59, 78
decomposition of correlation matrix, 16–17
density-related transformation, 262, 272–274
derivate moment, 183, 205, 242
derived stochastic process, 150
diffusion coefficient, 183
diffusion equation, 3, 4, 191
Dirac delta function, 8, 134, 202, 215, 217, 224, 343–346
direction spectrum, 70–72
discrepancy, 257, 258, 261, 262, 265, 270, 271
discrete random variable, 8, 34, 346
dispersion, 240, 242, 243, 245, 247, 248, 250
dissipation, 240, 242, 245, 247, 248
divergence theorem, 198
Dostupov–Pugachev equation, 6, 192, 209–213, 215, 216, 218
drift coefficient, 183
dual point set, 269
Duhamel's integral, 134–138
dynamic excitation, 5, 43–44, 46

Eigenvalue, 16, 27, 36, 73, 74, 108, 143, 145, 209
Eigenvector, 16, 36, 72, 73, 74, 109, 115, 142, 143, 145
element stiffness matrix, 81–84, 86
ensemble average, 4, 138, 355
equivalent extreme-value event, 300, 303–305, 307, 308
Euler–Lagrange equations, 317, 319, 321, 325, 326, 331, 333
evolutionary PSD, 157, 158, 159, 160, 163
evolutionary spectral analysis, 156–160
evolutionary stochastic process, 156–158, 159, 163
excursion, 288–291
expansion-contraction transformation, 274–278
expectation, 10, 11, 14, 15, 20, 21, 25, 30, 33
extreme value distribution, 296, 299, 339

F-discrepancy, 257, 259–262, 276, 278
failure probability, 286, 295, 300, 306, 310
Fibonacci sequence, 271
finite difference method, 231, 260
finite-dimensional distributions, 18, 21
finite element, 79, 80, 81, 88, 103, 106
finite Fourier transform, 24, 355, 357
first order discrepancy, 261
first-order partial differential equation, 231
first-passage, 286–288, 290, 292, 293, 295, 297, 308
fluctuating wind speed, 56–64, 77, 78
Fokker–Planck–Kolmogorov equation, 4, 5, 174–190, 197, 203–208
Fourier transform, 11, 19, 23, 24, 26, 345, 348
frequency response function, 146–156, 158, 160

Gegenbauer polynomials, 351, 354
generalized density evolution equation, 5, 6, 192, 213–229
generating function, 350, 351
generator vector, 385
global coordinate, 84, 85, 86, 89, 104, 106
global reliability, 287, 288
group velocity, 246

Hamilton–Jacobi–Bellman equation, 320, 321
Hamiltonian, 316, 317, 321, 323, 324, 327, 328, 331, 333
  canonical equation, 317
  formulation, 189
  principle, 90
Hartley function, 74–75, 77
Hartley transform, 74, 75
Hermite polynomials, 41, 351, 352
Hilbert space, 37, 38, 39, 41, 350
homogeneous random field, 31, 35
hyper-ball, 276, 277, 377, 381, 382
hyperbolic partial differential equation, 231, 232
hyper-cube, 257, 262, 271, 273, 276, 277
hysteretic structure, 167–170

impulse response function, 134–138, 141, 146, 147, 154, 157
independence, 15
initial condition, 191, 198, 201, 202, 219, 221–223, 225, 227, 229
inverse Fourier transform, 23, 24, 345, 348
isotropic random field, 31
Itô calculus, 177
Itô integral, 4, 174–179, 186
Itô stochastic differential equation, 174–179, 181

Jacobian, 193, 195, 201, 202, 210, 212
joint probability density function, 12, 14, 20
JONSWAP spectrum, 68, 69, 70

Kaimail spectrum, 59
Kanai–Tajimi spectrum, 48
Karhunen–Loève decomposition, 27–29, 35, 72, 74
Kepler's conjecture, 251, 252
kissing number, 251, 252, 262, 264, 269

Lagrange equation, 89, 90
Lagrangian, 314, 315, 318, 324, 328, 332, 339
Langevin equation, 4
lattice, 252, 254, 262, 266, 267, 269, 270, 272, 284
law of large numbers, 91
Lax–Richtmyer equivalent theorem, 233
Lax–Wendroff scheme, 238–240, 242–250, 284
Legendre polynomials, 352, 354
level-crossing, 288, 300
limit state function, 285–287, 300, 308, 309
linear congruential generator, 91, 93
linear quadratic Gaussian control, 328
Liouville equation, 191, 192, 198–204, 211, 212, 220, 221, 227, 228
local average, 32, 33, 34
local coordinate, 82–86, 88
local truncation error, 242

map, 192–196, 201, 204, 211
marginal distribution, 12
Markov process, 183, 184, 191, 198, 199, 208, 292
Markovian crossing assumption, 290, 292
Markovian system, 198–209
material derivative, 196
mean-square calculus, 5, 25
measurable function, 8, 181
method of characteristics, 185, 201, 221–223
metric space, 36, 37
mixed shuffle method, 92–93
modal superposition, 108, 142–145, 153, 162, 163
modified partial differential equation, 241
modulated envelope function, 49
moment function, 10, 11, 19, 20, 30, 33, 133, 138, 141, 142, 144
monotonicity preserving, 247, 248
Monte Carlo simulation, 79, 300
Morison formula, 66, 67

nonanticipating function, 177, 182, 186, 204, 207
normed linear space, 36, 37, 38
number theoretical
  method, 262, 265, 270, 385
  net, 270, 280
numerical flux, 236, 238, 242, 248

one-sided difference scheme, 232–237, 240, 242, 247
one-step linear scheme, 236, 243
one-way wave, 224
optimal control, 313–315, 317–321, 323–325, 328, 330–332, 335, 336, 340
order-expanded system method, 116–124
Ornstein–Uhlenbeck process, 188
orthogonal decomposition, 36–41, 72–78
orthogonal expansion theory, 113–128
orthogonal polynomial, expansion, 79, 114, 117, 118, 126, 349–351, 353

packing radius, 254, 255, 257, 268, 269
parallel system, 288
partition of space, 251, 254
path-integral method, 189
performance index, 314–322, 326, 330, 331, 333, 339, 341
phase velocity, 246
phenomenological tradition, 4, 5
physical approach, 5
physical stochastic model, 53–55
physical tradition, 5
Pierson–Moscowitz (P–M) spectrum, 68
Poisson crossing assumption, 290
Pontryagin minimum principle, 321
power spectral density, 23, 58, 68, 146–160, 355, 356
principle of energy equivalence, 76, 78
principle of optimality, 322, 325
principle of preservation of probability, 6, 189, 192–198, 206, 212, 213, 220, 223
probability density function (PDF), 8–13, 15, 17, 20, 40, 171–172
  evolution surface, 281
probability space, 8, 12, 35, 40
projection ratio, 265, 266, 270
pseudo-excitation method, 6, 146, 160–164
pseudo-random number, 92, 93, 95

random event, 7, 8
  description of the principle of preservation of probability, 195–196, 212, 223
random field, 2, 5, 29–36, 43, 44, 50, 52, 67, 95–98, 102, 213, 315, 330
random Fourier spectrum, 46–48, 61, 355, 359
random function, 15, 36, 40, 41
random perturbation, 101–102, 107
random simulation, 90–99
random variable, 2, 5, 7–18, 20, 27, 28, 29, 33–35, 40
random vector, 12, 14, 16, 17, 18, 35
random vibration, 5, 6, 133, 145, 160, 162, 163, 166–170
random walk method, 190
recursive condensation algorithm, 129–132
reliability, 285–311
  based control, 325, 338
remaining probability density, 295, 296
representative point(s) 226, 227
  set, 251, 256, 257, 261, 262, 273
representative region, 251, 255
restoring force, 165, 167, 168, 169
Riccati equation, 320, 325, 329, 334, 335
Rice's formula, 290

sample-based modeling, 47
second order discrepancy, 261
secular terms, 101, 110–113
seismic ground motion, 43, 44, 45, 48–55, 75–76

sequential orthogonal decomposition, 113, 114, 116, 117, 122, 125, 126
series system, 287
Simiu spectrum, 58, 59
spectral moment, 383, 384
sphere packings, 251–253, 262
spherical sieving, 274–276
stability, 232, 233, 235, 240
state feedback, 319, 335
State space description of the principle of preservation of probability, 192, 196–198
stationary process, 21–24, 26, 27
statistical linearization, 164–173
stochastic differential equation, 133, 174–181, 183, 184, 186, 188
stochastic process, 17–29, 30
stochastic structure, 79, 80, 95, 107, 116, 120, 128, 129, 130
  analysis, 2
Stratonovich's stochastic differential equation, 179–181
system reliability, 288, 300, 308–311

tangent sphere method, 262, 265
terminal condition, 317–320, 324, 328, 329, 334
total probability formula, 199, 202
total variation diminishing (TVD), 240, 247–250
transition probability density function, 183, 187–189, 199, 200, 202, 204, 206, 208
turbulence, 1, 5, 43, 44, 58, 63, 64
two-sided scheme, 238, 247

uncontrollability, 1
unit impulse function, 135, 347, 348

value function, 321, 323–326, 328, 330
variance, 10, 14, 15, 20, 21, 22, 25, 30, 34, 139, 140, 149, 150, 157, 158, 173, 176, 178, 185
variational approach, 315
variational principle, 81, 90, 324–326, 331
virtual structure, 104, 105, 106, 107
von Neumann analysis, 243
Voronoi cell, 226, 255, 256, 270

wave number, 52, 61, 65
weakest link, 305–307
weight function, 349–352, 354
weighting matrices, 335, 336
Wen's spectrum, 70
white noise, 5, 140, 149, 156, 174, 179, 181
Wiener process, 4, 175, 176, 183, 185
wind speed, 1, 2, 5, 56–64, 68, 69, 77, 78
wind wave, 65–72
Wong–Zakai correction, 180

www.ingramcontent.com/pod-product-compliance
Ingram Content Group UK Ltd.
Pitfield, Milton Keynes, MK11 3LW, UK
UKHW050943290926
15470UKWH00005B/1603

9 780470 824245